Ecological Studies

Analysis and Synthesis

Volume 223

Ecological Studies is Springer's premier book series treating all aspects of ecology. These volumes, either authored or edited collections, appear several times each year. They are intended to analyze and synthesize our understanding of natural and managed ecosystems and their constituent organisms and resources at different scales from the biosphere to communities, populations, individual organisms and molecular interactions. Many volumes constitute case studies illustrating and synthesizing ecological principles for an intended audience of scientists, students, environmental managers and policy experts. Recent volumes address biodiversity, global change, landscape ecology, air pollution, ecosystem analysis, microbial ecology, ecophysiology and molecular ecology.

More information about this series at http://www.springer.com/series/86

Richard K. Sutton
Editor

Green Roof Ecosystems

Springer

Editor
Richard K. Sutton
Department of Agronomy and Horticulture
University of Nebraska - Lincoln
Lincoln
Nebraska
USA

ISSN 0070-8356 ISSN 2196-971X (electronic)
Ecological Studies
ISBN 978-3-319-14982-0 ISBN 978-3-319-14983-7 (eBook)
DOI 10.1007/978-3-319-14983-7

Library of Congress Control Number: 2015940751

Springer Cham Heidelberg New York Dordrecht London

Printed on acid-free paper

Springer is part of Springer Science+Business Media (www.springer.com)

Foreword

Ralph Waldo Emerson once wrote: "The creation of a thousand forests is in one acorn." Let the knowledge, concepts and theories contained in this book be the acorn that inspires thousands of professionals to advance the technical performance of green roofs. For far too long, green roofs have been misunderstood and over simplified in terms of ecological performance. The challenges to creating diverse and resilient systems in our anthropogenic urban environments are well recognized. Due to weight, cost and loading restrictions, green roofs attempt to compress biological and ecological function into the narrowest of profiles, limiting natural processes and nutrient cycles. In response to these constraints the industry has evolved to simplistic low diversity solutions which provide less ecological services than what is possible in the urban fabric of our cities where these benefits are in greatest need.

Today's urban footprint is composed of more than twenty percent roof cover. This vast urban land cover provides an immense opportunity to solve many of our environmental concerns, especially if we convert these spaces to integrated and highly functioning living architecture. E. O. Wilson the noted American biologist and theorist stated: "We should preserve every scrap of biodiversity as priceless while we learn to use it and come to understand what it means to humanity". Furthermore, we should endeavor to create biodiversity on every surface of our cities, as it helps to fulfill the basic needs of humanity.

Despite the efforts of many within the green roof industry, roofs for the most part remain under-utilized, forgotten places with exceptional opportunities to be reclaimed and repurposed as vibrant, functional centers of nature and human enjoyment. As a green infrastructure tool, green roofs provide some of the highest quality eco-services benefits available for solving a multitude of social and environmental ills, despite the fact they are too quickly dismissed early in the design process because of a lack of understanding of their potential. Greater knowledge about ***Green Roof Ecosystems*** will only increase implementation of this vital and natural solution.

Recently a renewed interest in landscape planning seeks to link ecological services and community needs. And increasingly, public policy recognizes that creating livable and healthy communities requires connected landscapes in order to provide for clean air, clean water, public fitness, wildlife diversity and ecological benefits. The natural capital in our cities and efforts to restore it need not be considered at a single site or scale. Rather, natural ecology needs to be assessed and restored across scales. Widespread implementation of green roof technologies can set a foundation for mitigating and reversing environmental deterioration of the Anthropocene, as well as, dramatically broadening our response by providing new ways of thinking about ecological restoration. This process will be greatly enhanced by an interdisciplinary team approach to validate the robustness of the approaches underlying the restoration of ecosystem processes.

Green Roofs for Healthy Cities established the *Journal of Living Architecture* in order to identify the state of the art in green roof and green wall research, to identify the best in class, and share these findings with as many professionals as possible. This book represents a seminal compilation of research and technical knowledge about green roof ecology and how functional attributes can be enhanced. Written by over twenty leading experts and researchers in the field of green roofs, the narration covers in detail a number of important topics rarely discussed. While documenting current research, trends and theory, this book delves further to explore the next wave of evolution in green technology, defining potential paths for technological advancement and research.

This effort represents an informed and progressive way of approaching our environmental response to urban design. It makes a compelling case that the long-term health and viability of our communities depend upon highly functioning green roof ecologies that connect green spaces to create a resilient tapestry of natural diversity spanning the urban landscape. ***Green Roof Ecosystems*** will be an invaluable reference for individuals who have the desire to implement ecologically conscious green roofs, such as planners, policy makers, agencies, and professionals who have substantial interest in designing them. (i.e.; landscape architects, ecologists, engineers, architects, biologists, and other holders of environmental knowledge). Ecological intelligence expands the context of life as it enlarges who we are as a person, and this book provides a wealth of intelligence for those interested in the topic of green roofs.

Kansas City, MO

Jeffrey L. Bruce, FASLA,
LEED, ASIC, GRP

Contents

Contributors

Dr. Richard K. Sutton PLA, FASLA, GRP teaches landscape architecture and landscape design at the University of Nebraska-Lincoln and conducts research on green roof plants for low input, low cost green roofs.

Lee R. Skabelund, ASLA is an associate professor of Landscape Architecture at Kansas State University and has overseen the installation and monitoring of the Seaton Hall green roof and a campus bio-retention treatment train.

Dr. Kimberly DiGiovanni is a post-doctoral research fellow and adjunct assistant professor in the Department of Civil and Environmental Engineering at Villanova University. She is a graduate of Drexel University's doctoral program in Environmental Engineering and is an alumna of the Sustainable Water Resource Engineering Laboratory at Drexel University. Dr. DiGiovanni's research interests and focus include the eco-hydrology of natural vegetated spaces and engineered urban green spaces, including green roofs.

Dr. Olyssa Starry is Assistant Professor at Portland State University's Honors College. Her research relates various aspects of green roof structure to function through both experimentation and modeling. Her doctoral work at the University of Maryland, College Park assessed the role of different Sedum species in the green roof water cycle utilizing controlled chambers for monitoring gas exchange as well as outdoor experimental green roof platforms.

Dr. Mark T. Simmons is Director of Research at the Lady Bird Johnson Wildflower Center at the University of Texas. Restoration and design projects focus on the creation of urban prairies and forests. Scientific research projects include: green roofs, green walls, native turfgrass, urban prairies and prescribed fire.

Dr. Ishi Buffam, University of Cincinnati is an ecosystem ecologist and aquatic biogeochemist focusing on carbon and nitrogen transformations and hydrologic transport within watersheds and streams. His recent research explores the influence of vegetated roofs on urban surface water quality.

Mark Mitchell, University of Cincinnati is an ecologist studying nitrogen and phosphorus cycling dynamics in vegetated roof ecosystems. He received his M.S. from Loyola University Chicago and is working towards his Ph.D. in biology.

Dr. Brooke Byerley Best serves as botanist and editor at the Botanical Research Institute of Texas (BRIT), a 501(c)3 non-profit organization dedicated to botanical conservation and science education. Her main research focus is performance monitoring of the building's biomimicry-based living roof.

Rebecca Swadek is a plant ecologist with the New York City Department of Parks & Recreation's Natural Resources Group and a research associate with the Botanical Research Institute of Texas (BRIT). Rebecca works primarily on wetland restoration projects for the City of New York while continuing research on green roofs and prairies in her spare time.

Dr. Tony Burgess is a research associate with the Botanical Research Institute of Texas (BRIT) and a retired Professor of Professional Practice in the Department of Environmental Science at Texas Christian University (TCU). Dr. Burgess has been studying biomimicry-based design since the early 1980s when he designed and helped construct the Caravan of Dreams Desert Dome in Fort Worth, Texas, and the Desert Biome habitat for the Biosphere II Project in Tucson, Arizona.

Dr. John Lambrinos evaluates the design and function of green roofs in the Pacific Northwest. He has published regionally focused studies on appropriate green roof plant species and irrigation regimes.

Dr. Krista McGuire is an Assistant Professor of Biology at Barnard College, Columbia University and in the department of Ecology, Evolution and Environmental Biology at Columbia University. Her research focuses on feedbacks between soil microbes, plant communities, and global change.

Sara Payne received a B. A. in Biological Sciences from Barnard College. She is currently conducting blood stem cell research at Albert Einstein, and will later pursue a PhD.

Dr. Matthew Palmer is a senior lecturer in the department of Ecology, Evolution, and Environmental Biology at Columbia University. His research includes work on community and ecosystem ecology in cities.

Giulia Orazi received a B. A. in Biology from Barnard College. She will pursue a PhD in Molecular and Cellular Biology at Dartmouth, with a focus in Microbiology and Immunology.

Heema Sharma is a graduate of Barnard College of Columbia University, and received her Bachelors degree in biology in May 2013. She now does breast cancer research at the College of Staten Island.

Dr. Susan Cook-Patton studies biodiversity-ecosystem function relationships. She is interested in using ecological knowledge to foster healthy forest and urban environments. Formerly a Smithsonian Fellow, she now conducts policy analysis for the US Forest Service.

Dr. Jeremy Lundholm is a professor in the Biology and Environmental Science Departments at Saint Mary's University in Halifax, Canada. He studies plant community ecology and its application to urban ecosystems.

Dr. Nicholas (Nick) Williams is a senior lecturer at the University of Melbourne, Australia. Research interests include urban ecology and developing and evaluating green infrastructure for climate change mitigation and adaptation.

Dr. Nigel Dunnett is Reader in Urban Horticulture at the University of Sheffield. His interdisciplinary research focuses on the interface between ecology, design and horticulture in urban green space and built development.

Peter MacDonagh, FASLA, RLA is an Adjunct Faculty, University of Minnesota College of Design, registered landscape architect, and Director of Design + Science, Kestrel Design Group. He specializes in landscape restoration, urban forestry, stormwater planning, and green roof technologies.

Nathalie Shanstrom RLA is a landscape architect with Kestrel Design Group, in Minneapolis, MN, since 1999, specializing in sustainable landscape architecture.

Stephanie Carlisle is an environmental researcher at KieranTimberlake, where she investigates interactions between the constructed and natural environment, bringing data-driven analysis and complex systems theory to architectural problems at the building, landscape and urban scale.

Max Piana is an urban ecologist and environmental researcher at KieranTimberlake, where he employs a range of quantitative and qualitative assessment methods to inform high performance building and landscape design, emphasizing an understanding of site ecology.

Dr. Bradley Rowe has conducted green roof research at MSU since 2000, was the founding chair of the Green Roofs for Healthy Cities research committee, and received the GRHC Excellence in Research Award in 2008.

J. Scott MacIvor is an ecologist interested in the role and relevance of biodiversity in landscape design and city planning. He is completing a PhD in the Biology Department at York University and is an instructor in the faculty of Architecture, Landscape, and Design at the University of Toronto.

Kelly Ksiazek is an urban ecologist with Plant Biology and Conservation Department, Northwestern University and the Chicago Botanic Garden, Chicago, IL, USA. Her research interests focus on native plant and pollinator conservation in cities with a focus on green roofs habitats

Dr. Katherine Dunster, RLA, RPB leads the new Bachelor of Horticulture Science degree for the Urban Ecosystems major at School of Horticulture, Kwantlen Polytechnic University, B.C. She is also an Adjunct Professor in the School of Architecture & Landscape Architecture at the University of British Columbia.

Dr. Reid Coffman, heads the Novel Urban Ecological Systems (NUES) Cleveland Urban Design Collaborative at Kent State University and is a research associate professor and urban ecological designer. He specializes in the conceptualization and application of living architecture that optimizes ecological services.

Bruce Dvorak ASLA, RLA is an associate professor in the Department of Landscape Architecture and Urban Planning at Texas A&M University where in 2009, he established the Interdisciplinary Green Roof Research Group.

Chapter 1
Introduction to Green Roof Ecosystems

Richard K. Sutton

Abstract Green roofs have been heralded as a "sustainable building practice" in cities throughout the world as one response to mounting environmental stresses. A range of stressors plus erosion of aesthetics and human well being in urban areas have initiated policies and practices often with incentives to develop green infrastructure such as green roofs. They provide a suite of public and private benefits most of which map onto services generally provided by the ecosystem. Green roof development imbeds in environmental design processes and is constrained by both human and environmental factors.

As relatively small, simple, anthropogenic ecosystems, green roofs relate to several existing conceptual and applied ecological ideas. Understanding and applying from ecology and ecosystem studies, ecological engineering, managed ecosystems, construction ecology, urban ecology, landscape ecology, restoration ecology, reconciliation ecology, soil ecology and community ecology show green roof ecosystems can be created to cycle energy and nutrients. Furthermore, green roofs can be constructed to model an ecosystem and may provide a setting for testing ecological concepts. This book takes an ecosystems approach to describing a large number of interactions on green roofs placing them in the total human ecosystem.

Keywords Novel ecosystems · Ecosystem benefits · Ecosystem services · Design

1.1 Structure and Purview of this Book and Chapter

It has been nearly a decade since the seminal article, Green Roofs as Urban Ecosystems: Ecological Structures, Functions, and Services (Oberndorfer et al. 2007), reviewed green roofs' impacts on ecosystem services (benefits) and suggested a modest applied research agenda. That agenda focused on an ecosystem approach to

R. K. Sutton (✉)
Department of Agronomy & Horticulture, UN-Lincoln,
279 Plant Sciences, Lincoln, NE 68583-0915, USA
e-mail: rsutton1@unl.edu

R. Sutton (ed.), *Green Roof Ecosystems,* Ecological Studies 223,
DOI 10.1007/978-3-319-14983-7_1

diversifying plant assemblages, experimental studies of belowground and aboveground communities (both plant and animal) quantification and qualification of local stormwater outputs (especially roof leachate), energy, and air quality, reductions and social benefit models linked to economics and governmental policy. Meanwhile, there have been many more acres of green roofs created and research has continued apace. The following pages of this volume survey some of what has been accomplished and suggest more that should be done.

This book was assembled and written with three, somewhat overlapping yet distinct groups in mind: policy makers interested in urban sustainability and livability, designers who specify and layout green roofs to meet a wide range of stakeholder needs, and environmental professionals, researchers, and students wanting a primer on the ecological foundations and interactions occurring on and between green roofs and other living systems. While this target audience is broad, we have assumed that an interest in sustainable, ecologically prudent design connects them. Each chapter will review, examine, and analyze current knowledge about a specific area, propose unanswered questions, and suggest future research directions and applications from the perspective of the author(s).

This introductory chapter will place green roofs in the realm of policy and urban sustainability (especially its ecological underpinnings), requisite support services and resultant public and private benefits. Next, it will briefly describe green roof technology and components; then it will tie green roofs to a wide variety of ecosystem approaches studies, concepts and applications; finally it will give a brief preview of each chapter.

1.2 What is a Green Roof?

Modern green roofs, also known as vegetated green roofs (Enright 2013) or ecoroofs, are nascent, somewhat isolated, novel, anthropogenic patches consisting of membranes, engineered substrate (the growing medium), and assemblages of plants placed atop buildings or other structures. Their shallow profiles and usual detachment from the earth's surface produce strong wind exposure creating an unusual niche with few potential natural analogues (Lundholm and Richardson 2010; Sutton et al. 2012). They receive intense solar input and varied precipitation and may or may not be irrigated. Green roofs have appeared because of advanced building materials, evolving design techniques, and emerging ideas about how to make our built environment more sustainable and humane (Getter and Rowe 2006; Weiler and Scholz-Barth 2009). The modern green roof movement began in Europe in the 1980's (Köhler and Keeley 2005), and spread to North America and the rest of the world after the new Millennium. Thousands of green roofs now lay atop buildings in most urban metropolises worldwide.

1.3 Green Roof Policy as a Sustainable Practice

Normative policies describe, explain, and advocate how humans should act in organizing ourselves. Policies promote features or actions that ought to occur and thus are future-oriented. An often-quoted example of sustainable development policy comes from the World Commission on Economic Development (Brundtland Commission) (WCED 1987): "development that meets the needs of current generations without compromising the ability of future generations to meet their own needs."

Policies often become the method by which discussion about the allocation of public resources are focused, formulated and ratified. In the United States at the Federal level the Clean Water Act provides an impetus to local subdivisions for improving storm water discharges (Carter and Fowler 2008). Ratified locally, policies are crafted into laws, ordinances, and finally reflected in building codes or other types of standards (GRHC 2006b). Local policies and codes can hinder or facilitate green roofs as a sustainable building practice (Dvorak 2011).

Places where green roofs have been promoted include cities with pressing environmental problems and/or compelling visions about creating more resilient and beautiful infrastructure. Sustainable urban environments vary in the suite of issues (Tables 1.1 and 1.2) and their importance underlying policies and the ways those policies are implemented. Urban stormwater controls or ordinances mandating green roof coverage on new development often involve fees or trade offs for impermeable surfaces. Revised building codes that simplify structure weight-loading requirements, tax incentives, rebates on fees, fast-tracking the development process, density bonuses, and outright grants have all been used to encourage green development and in some cases have been specifially directed at promoting green roof implementation (Carter and Fowler 2008; Simons et al. 2009).

1.4 Benefits

Looking at the various reasons posited to promote public policies that include green roofs as a part of sustainable building development (Getter and Rowe 2006), we see many overlapping benefits. These benefits can further be subdivided into those for private and/or public green roofs (Table 1.2) (Green Roofs for Healthy Cities 2006a; Berardi et al. 2013):

Green roofs can be considered a category of a stormwater best management practice (BMP). In comparison to conventional impervious rooftops, green roofs retain greater amounts of precipitation (that eventually return to the atmosphere through evapotranspiration) and also detain precipitation allowing it to drain more slowly (Bates et al. 2009; Berndtsson 2010; Morgan et al. 2012). Retarding and holding runoff water depends the type of roof vegetation, the total volume of the substrate, its composition and the nature of the storm event (Schroll et al. 2011; Gregoire and Clausen 2011).

Table 1.1 Key issues specified for green roof adoption in twenty-five world metropolises

City	Stormwater quality	Stormwater quantity	Heat Island	Green space	Energy savings	Air quality	Bio-diversity	Urban Agr
Toronto	√	√	√	√		√		
Chicago	√	√	√		√			
New York	√	√	√		√			
Baltimore	√	√	√		√			
Berlin	√	√	√				√	
Atlanta	√	√	√			√		
Singapore			√	√	√	√		
Washington		√	√			√		
Tokyo			√		√	√		
Austin		√	√		√			
Cologne	√	√						
Seattle	√	√						
Philadelphia	√	√						
S Francisco	√			√				
Waterloo		√				√		
Munster		√		√				
Stuttgart				√		√		
London				√			√	
Montreal					√			√
Pittsburgh	√							
Seattle	√							
Minneapolis	√							
Vancouver	√							
Basel							√	

Because green roofs intercept and detain rainwater, they can initiate a train of stormwater treatments and be designed to direct the slowed runoff into cisterns, rainwater gardens, bio-swales or detention ponds. Green roofs filter out many atmospheric pollutants and nutrients borne in precipitation before they reach streams or lakes (Berndtsson et al. 2009).

Green roofs by themselves and in aggregate affect a building's and a city's energy budget. In summer, a city with enough green roofs will have its overall ambient temperature reduced (Smith and Roebber 2011; Solecki and Leichenko 2006; Gaffin et al. 2008). An individual building similarly can reduce its need for summer cooling and winter heating since green roofs act as an insulator (He and Jim 2010; Jim and He 2010; Teemusk and Mander 2010; Feng et al. 2010). Additionally, some of the retained stormwater will be transpired during the growing season to further cool a building. In the winter, dormant green roof vegetation captures additional

Table 1.2 Green roof benefits derive from their existence as functional, living ecosystems and map onto a suite of ecosystem services described by the Millennium Ecosystem Assessment (MEA 2005)

	Green roof benefits		Ecosystem services			
	Public	Private	Provision	Regulate	Cultural	Support
Stormwater quantity	√	√		Water		
Stormwater quality	√		Water	Purification		Nutr. cycling
Heat island	√			Climate		
Membrane life		√				Resilience
Building energy		√		Climate		
Noise reduction		√		Sound	Aesthetic	
Air quality	√			Air		Nutr. cycling
Biodiversity	√			Pollination	Knowledge	Nutr. cycling
Retard fire		√			Well-being	
Views; marketing	√	√			Aesthetic	
Rooftop agriculture		√	Food		Educational	Soil formation
Education opportunity		√			Educational	
Local employment		√			Well-being	
Carbon sequestration	√			Air		

precipitation and enables snow to stay on the roof, thus adding an additional layer of insulation.

Typically, most roof membranes have a lifespan of about 20 years largely because of ultraviolet light degradation and micro-tears caused by diurnal heating and cooling cycles. Green roofs protect a membrane from those deleterious effects and may double membrane life thus reducing life cycle costs and delaying worn out membranes from entering the landfill (Carter and Keeler 2008; Bianchini and Hewage 2012).

Because green roofs have a porous mass, they serve as noise attenuators (Connelly and Hodgson 2008). Depending on depth and composition they can lower the noise impact from an overhead source such as an airliner up to 10 decibels.

Buildings and their urban conglomerations reduce space for other living things such as plants (Cook-Patton and Bauerle 2012; Madre et al. 2014) insects (MacIvor and Lundholm 2011), and birds (Coffman and Waite 2011). Green roofs allow some reestablishment of habitat for a few of those organisms. Flowering plants on vegetated roofs allow the introduction of bees and support other pollinators. While green roofs can never completely replace the biodiversity and complexity of intact ecosystems, they mitigate some of those changes and may supply living corridors for insect and bird movement in cities. They vastly improve the lack of biodiversity found on white and black roofs dominating a city's impervious surfaces. Hotels with green roofs easily charge more for rooms that open on to garden terraces. Chefs seek the herbs and vegetables grown nearby to their restaurants for the reduced cost and high quality freshness. Green roofs offer such a venue. Office workers gazing onto a green roof fatigue less easily and produce more under stress-reduced workloads. Green roof aesthetics, however, go beyond the mere pleasure than might be experienced in view the surface feature of any garden (Sutton 2014). All such connections help people value the natural world, become calmer, more alert and involved as humans (Kaplan and Kaplan 1989; Kahn 1999; Louv 2012).

1.5 Green Roof Design and Technology

Most often green roofs sit atop nearly flat roofs of commercial or public buildings. Occasionally they can be found on sloping or residential roofs though they are most likely to be part of new building project when extra weight loading can be considered and accounted for in structural design. Retrofitting the structure of existing building is a difficult and expensive proposition. Weight limitations become paramount because green roofs capture and hold a portion of the precipitation that falls on them. Based on the substrate depth, green roofs are classified as extensive (< 15 cm) (< 6 in.), semi-intensive (> 10 and < 20 cm) (> 4 and < 8 in.), or intensive (> 15 cm) (6 in.) (GRHC 2006a). This general nomenclature applied above to depth classifications actually refers to the amount of maintenance expected for shallow, moderate, and deep substrate. Deeper substrate means that a wider array of plantings that include herbaceous perennials, shrubs and trees could be grown creating more of a rooftop garden, whereas, shallow substrate depths support fewer and lower-growing plant types. Most roof decks allow only minimal weight loads and so limit adoption of even extensive green roofs with shallower substrate depths. Where more weight loading can be supported, a semi-intensive or intensive roof can be used. The term extensive comes from a German to English translation of these concepts in the 2002 English translation of the FLL Guidelines for Green Roofing. It is a green roof system that "involves cultivation of vegetation in forms which create a 'Virtual Nature' landscape and requires little if any external input for either maintenance or development" (FLL 2002, p. 12). Its intention is to be extensive or wide spread in its application because of low cost, low-maintenance and ease in population with local flora (FLL 2002).

A typical green roof cross-section begins at the bottom with the building's structural system, moving up through its decking, insulation, waterproof membrane, root barrier, drainage layer, drain filter, growing substrate, and finally a living layer of plants (Fig. 1.1).

Each layer plays a role in protecting the membrane, buffering and filtering rainfall and, with plant coverage, guarding against wind and rain-caused erosion of the growing substrate. Because substrate ballasts the building's membrane and insulation, it must possess some weight, yet it must be well drained with large pore spaces to quickly allow percolation of excessive rain and lessen weight loading. Plants must be selected to withstand drought, wind, heat, and cold. If plantings fail then the substrate, becomes exposed to loss due to wind scour. Three key factors must

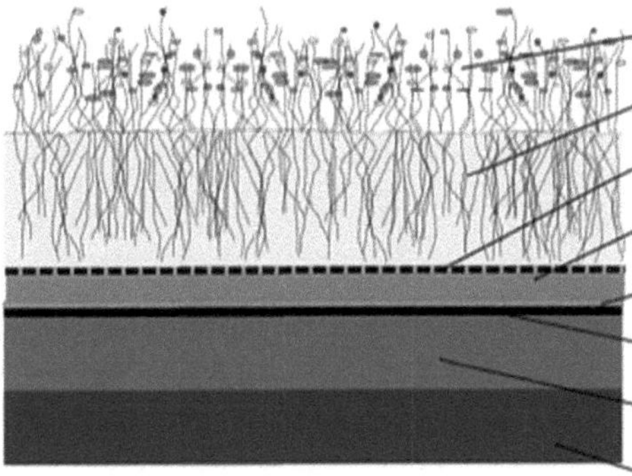

Fig. 1.1 A typical green roof cross-section shown above with its multiple layers

always be kept in mind when designing and maintaining green roofs: (1) stay with in structural loading limit, (2) protect the integrity of the waterproofing membrane, and (3) keep plants alive to protect and hold substrate in place.

A green roof design must account for horizontal as well as vertical forces. Daily wind pressure and especially wind action during storm events can cause scouring of substrate and dislodging of plantings. As the height of a green roof from the ground increases, so do ambient and storm winds, particularly if the green roof is unprotected by other building mass. Placement of membrane ballasting and scour protection may be dictated by local building codes (SPRI 2010). The height and location of parapet and building walls can create turbulent, chaotic, unpredictable, and increased speeds for wind flows (Suaris and Irwin 2010) but also reduce wind speeds. Wall and parapet height and location can also affect sun and shade patterns that should be acknowledged in layout of any designed planting.

Substrate can be layered and embedded in several different ways. The simplest is monolithic placement in a bed at the specified depth. Placement could also be built-up with layers of two or three substrates with differing drainage characteristics. The next method consists of modular tray systems either with pre-grown plants or filled with substrate and planted after placement. Trays can be made of plastic or a degradable material. One advantage of plastic material is that the tray can be picked up and moved for roof repair. A third method involves a thin, integrated, pre-planted, flexible, rubberized or plastic rug-like structure embedded with substrate and plants. It can be laid as a mat or rolled for transport and unrolled upon installation.

The plants may or may not be irrigated and supplemental water beyond rainfall may be applied by hand as needed, or by automatic spray or drip systems on the surface or embedded in the substrate. Excessive use of water for irrigation runs counter to the intent of a sustainable building.

Where green roofs have been designed for physical access other landscape amenities can be added such as paving, decking, seating, water features, arbors, and trellises. Green roof landscape design per se is beyond the scope of this book. It is suggested that readers wishing to know more about the design and construction process consult books by Osmundson (1999), Weiler and Scholz-Barth (2009), Luckett (2009), Snodgrass and MacIntyre (2010), or Daykin et al. (2013).

1.6 Design and Use

To gain public and private benefits from green roofs means adding another layer of complexity to design of buildings. It requires a design team that includes an owner, engineer, architect and landscape architect to establish parameters and oversee specification and installation of green roof materials to meet desired outcomes. Green Roof Professional (GRP) is a special certification by the Green Roofs for Healthy Cities group in North America given after completion of an exam and requiring yearly continuing education. These design professionals also rely upon a group of craftsmen and suppliers to help create a green roof. Designers must understand local policies and codes, the building's needs, location, and whether a roof will be accessible and by whom. Weight loading and roof slopes must be acknowledged and understood very early in the process. Building codes for green roofs apply not only to building envelope integrity and public health, safety and welfare during and after construction including fall safety, emergency egress, and wind and fire impacts. The development of code and performance specification requirements for green roof construction is most advanced in Europe. North America is beginning to make advances in development of codes, guidelines and other legal documents for public and private green roof construction, but many green roof elements lack guidance (Dvorak 2011).

Integrity and lifespan of the waterproofing membrane represents the second most important condition of a green roof. Everyone accessing a roof from designers to installers, maintainers, and visitors must do so in a way that protects the membrane. Improper access and use impact a membrane and can, at best, void any warranty and at worst cause roof leakage.

On green roofs, the substrate composition for physically supporting plants and supplying water and nutrients varies widely. Some designers recommend highly organic admixtures with up to twenty percent compost or peat moss, while others opt for lower amounts of organic matter in the five percent range (Friedrich 2008; Buist 2008). Many of the large-scale commercial providers of growing media in Europe and North America use an engineered media based upon the German FLL Guidelines (FLL 2002) for Green Roofs. The FLL-based guide suggests a range of materials and performance characteristics for media assuming use of a xeric plant palette. The organic fraction holds water, microbial populations and supplies nutrients and structure that while the inorganic fraction brings needed internal structure and adds overall ballasting weight. Importantly, the inorganic substrate fraction provides structure that allows rapid permeability and resists freeze-thaw cycles, and compaction. Inorganic material must be near neutral in pH, size-graded to allow rapid percolation and have very little substrate in the clay particle size range. Types of inorganic substrate material typically include heat expanded slate, shale and clay;

crushed brick or tile; volcanic ash; pumice; lava rock; perlite; sand and admixtures of these. Compost or worm castings supply the initial organic fraction with its critical complement of nutrients.

Plant selection for green roofs must consider its microclimate (Metselaar 2012), the well-drained growing substrate, plants, ecological relationships amongst themselves and fauna (Brenneisen 2006; MacIvor and Lundholm 2011), as well as aesthetic intent and use. Plants must be able to withstand wind, heat, cold, and drought (Sutton et al. 2012). For low profile, extensive green roofs, plants also need to be able to self-sow and/or fill in gaps by creeping rootstocks or stems. Effort should be taken to use locally sourced materials with minimal emergy (embodied energy).

All human-occupied landscapes require maintenance, especially carefully designed green roofs. Roof top environments can be harsh and cause stress on plants. Detailed inspections of plants for insects and diseases must occur frequently during the first 3–5 years, and continue beyond in subsequent years. During and after heat or drought spells and at the beginning and end of the growing season, plant health must be assessed and water added (as needed) and repairs made to scoured substrate, faulty irrigation and the pavements, drains, and other life-safety features. If flower displays become critical to the objectives of the green roof's design, then yearly, spring soil tests are required to reveal need for supplemental nutrients. When applying nutrients, small amounts in a slow release form should be used. Nutrients easily leach from the substrate so it is important to guard against over-fertilizing (Morgan et al. 2012).

1.7 Green Roof Ecosystems Concepts and Applications

While some readers may question juxtaposing the terms, green roof with ecosystem as we have done in this volume's title, the expansion of ecological study into a plethora of sub-disciplines suggests a wider view of what constitutes the study of ecology and displays a broad suite of compatible ecological concepts underlying a green or living roof. Below we review some of the varied ecosystem-oriented, ecological approaches that may have facility and add understanding to the practice of green roof design and assembly.

Ecology as a subject has matured to the point where it is no longer ensconced wholly in biology (Odum 1992). Broadly defined, ecology is the science of relationships between, living things and their environment. The ecosystem model explicates system inputs and outputs powered by the flow of energy (Odum 1971). Even the term, energy, has evolved into emergy (Odum and Nilsson 1996; Odum 2002) (embodied energy) and exergy (Jorgensen et al. 2004; Kibert 2002) (the useful applicable part of energy driving ecosystem processes). Feedback occurs in systems that tend to be self-organizing and hierarchical (Berryman 1989; Allen 2002; Kay 2002).

Ecosystem diagrams (Fig. 1.2) use symbols to describe flows of energy, materials and information to, from and within an ecosystem. Ecosystems are very open to

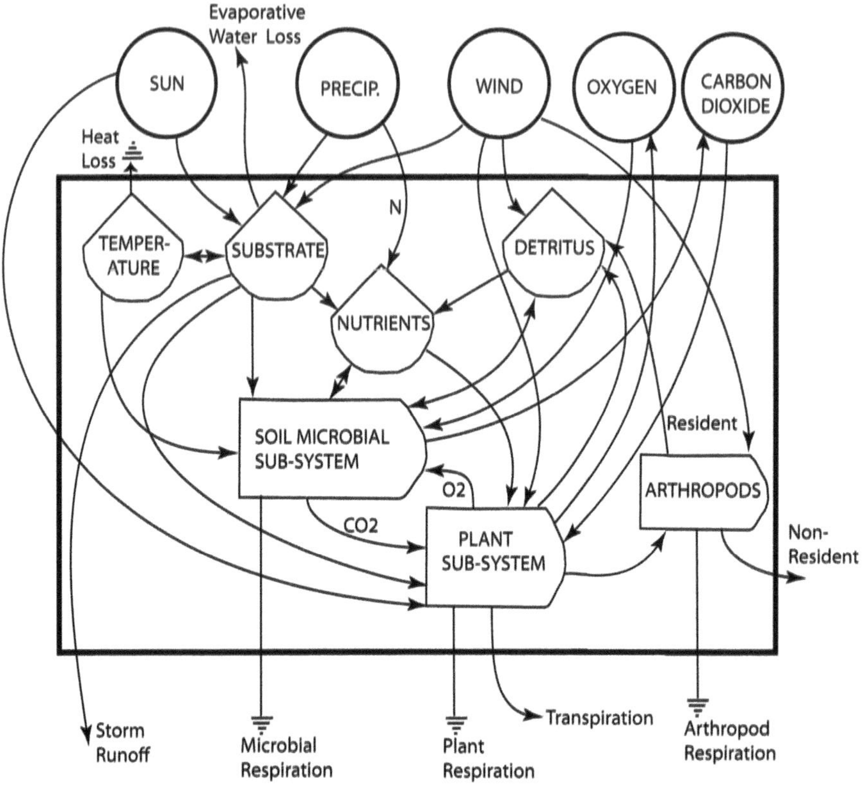

Fig. 1.2 A green roof ecosystem showing flows of energy, water, nutrients, and organisms

inputs and outputs, though a green roof ecosystem has spatial boundaries that are relatively easy to define. Within it, physical materials, like substrate or plants, become objects to, through, and from which flows occur. The usefulness of ecosystem diagrams using Odum's emergese language shows and brings invisible interactions to our attention.

These activities are rather like a transitive verb: they display action on and between physical objects. Allen (2002, p. 118) describes the design of systems processes thusly: "A critical distinction between design, embodied in mechanisms, and system dynamics, is the notion of rate dependence as opposed to rate independence. Dynamics is described as a series of rates of processes that are interrelated. Dynamics depends on rates, and a description of dynamics has to rely on rates for adequate description." A sedum plant on a green roof is not a sedum plant at a rate. It is simply a locus of activity that absorbs energy, stores it in complex biochemical compounds for later to use with water and carbon dioxide in Crassulacean acid metabolism (CAM). It does so to reduce the loss of internal water caused by opening stomata, though some *Sedum* species may switch between C3 and CAM (Sayed 1994; Cushman and Boland 2002). In the relatively cool, higher humidity found at night less water is lost in CAM. Because rates can only be described as some activity or flow measureable over time. Rates of action and flow, like energy, get stored, amplified, attenuated, dissipated, transformed, and extinguished moving through an ecosystem. These pathways or circuits remain largely invisible, and thus hard to imagine for most viewers. From an ecological perspective, when a green roof designer sees a *Sedum* plant the thought is about reduced transpiration rates and lower water use inherent on extensive green roofs. But shifting water sensitive processes to nighttime does not allow stonecrop (*Sedum* spp.) plants to escape their vulnerability to high temperatures in tropical or sub-tropical longitudes. Heat in these locations can be high enough to cause collapse of the entire *Sedum* plant colony and should eliminate it from the designer's green roof plant palette in those locations. While *Sedum* spp. are popular in Europe where they inhabit rocky soils in mountainous environments, other succulent species native to tropical or sub-tropical climates have been found to perform well in shallow substrates (Dvorak and Volder 2010; Dvorak and Volder 2012).

Designers work with, arrange, and relate physical material such as plants and substrate to create green roof ecosystems. While they do so it is imperative they pay attention to the invisible flows of energy, materials, and information. When one tugs at one portion of the ecosystem there will be a response, because it is interconnected with other parts of the system. Every change in a part of the system has an often-unknown and delayed impact on another part of the system. For example, top-dress a green roof substrate with fresh compost to improve plant growth and there may be more robust plants along with an increase in nutrients leaching from the entire system. (Eco)systems thinking means considering the ramifications of actions (so far as they can be understood) to an entire system. That is the study of ecology: imagining, understanding, describing, measuring, and linking the relationships between the entities in a system however such a system is defined or delimited.

Ecological Engineering seeks the design of sustainable ecosystems that integrate human society with its natural environment for the benefit of both (Mitsch and Jorgensen 2004). Using ecological theory and quantifiable engineering technology this approach focuses both on restoration of existing natural environments and the creation of new, self-organizing ecosystems. Mitsch and Jorgensen (2004) list five general concepts of ecological engineering that have importance for green roofs:

- Self-design [organizing] capacity of ecosystems
- Testing of ecological theory
- Systems approach
- Conserves non-renewable energy
- Supports ecosystem conservation

Self-organization relies heavily on careful introduction of assemblies into a supportive space. Not a simple input-output model, this approach relies on the capacity and proper mix of biotic and abiotic materials to encourage the emergence of self-organizing, hierarchical systems (Allen 2002; Kay 2002). Theories can be put to use, modified, or discounted after being tested in the construction of ecosystems. While ecosystems (and green roofs) can be examined and broken into detailed parts, it is their existence and operation as wholes that make them powerful entities. The hierarchical nature of systems, especially ecosystems, allow for the emergence of complexity (Allen and Starr 1982). Ecosystems science centers around the acquisition, use, change, transfer and degradation of energy, so it is understandable that constructed, anthropogenic ecosystems should operate primarily on solar energy like most natural ones.

Sustainability is a key concept driving green roof policy. Ecological engineering theorist, James Kay (2002), lists four basic design concepts for creation of more sustainable built environmental ecosystems, which can be thought of as extensions of those noted above by Mitsch and Jorgensen (2004):

- Interfacing
- Bionics
- Biotechnology
- Conservation

These basic concepts map on to green roofs: interfaces exchange energy and waste between a man-made structure and the wider environment. On a green roof, for example, the excess rainwater runoff outflows to stormwater systems; bionic design attempts to as closely as possible imitate a natural system, green roof examples being a dry, windswept talus slope, rocky seashore, or semi-arid prairie; biotechnology completes a function utilizing natural systems, for example, roof cooling through transpiration by living plants; conservation of non-renewable resources occurs when using them only to upgrade anthropogenic features–for example, using petro-chemicals to manufacture highly efficient membranes that underlie and support green roof systems extending the membrane's useful life.

Unfortunately, sustainable environmental systems displaying Kay's four design concepts are often ignored because of desire for immediate economic return, lack of knowledge (Kay 2002), and invisibility to the public (Thayer 1989).

Managed ecosystems are defined as, "one[s] where [ecosystem] processes are influenced by purposeful human decision-making" (Antle et al. 2001, p. 724). These certainly include green roofs where human decisions direct the flows and impacts of needed resources such as labor and water. These decisions appear to be strongly connected to economic circumstances and often expressed in a hierarchy of governmental policies.

Construction ecology, a sub-field of industrial ecology, focuses on the design, installation, management, and decommissioning of buildings. The building design and construction process strongly affects green roofs. Construction ecology builds the human environment '(1) [with] materials systems function[ing] in a closed loop integrated with eco-industrial and natural systems; (2) that depends solely on renewable energy resources; and (3) that fosters the preservation of natural system functions' (Kibert 2002, p. 292). These features are strongly influenced by economic decisions of the individual, firm, and government about current and future costs of energy, material, labor and the level of acceptable environmental impact.

Urban ecology may also include green roofs and be associated with, the study of the amounts and locations of organisms, their relationships with each other and with their environment, and material and energy flows within urban areas (Gaston 2010; Madre et al. 2014), or alternately and more succinctly from (Alberti 2008, p. 252), "the study of the coevolution of human ecological systems." Humans become the keystone organism for driving the urban ecosystem. So, human actions affect economic, cultural, social, and psychological aspects of the urban ecosystem.

Urban ecologists have used Long-term Ecological Research (LTER) sites based in metropolitan areas and offer three alternative and overlapping conceptual models of urban ecosystems: Baltimore (University of Maryland) uses a patch dynamic approach, Phoenix (Arizona State University) uses hierarchical modeling, and Seattle (University of Washington) uses adaptive cycles. The approach at the Seattle LTER incorporates both dynamic patches and hierarchy, implicit in linking of pattern and process with effects, changes, and scale. It is easy to see within levels how green roof patches might interact with and feedback among model compartments. Patches like anthropogenic, green roof features, represent a system for flows of information, knowledge transfers and system learning (Alberti 2008).

The small (but growing) number and dispersed nature of green roofs in the urban milieu means green roofs currently play a minor role in urban ecosystem functioning. Snep and Opdam (2010, p. 270) downplay the visual impact of green roofs claiming, "…[T]he concept of living roofs and living walls is an important part of [ecosystem functioning] but here functionality is much more important than visual quality." Where green roofs expand offerings of visually recognizable nature or provide the backdrop for human activities they do, however, become important. Visual impacts tied to green roofs link culture, ethics, aesthetics, and biodiversity (Sutton 2014), thus visibility and aesthetic relief may become critical feedbacks shaping green roof acceptance and broaden urban policy.

Landscape ecology studies the structure, function, and change of patchy ecosystems across a range of scales in time and space (Forman and Godron, 1986). It explicitly includes humans as part of the system (Naveh and Leiberman 1994; Naveh 2000). Structures include patches, corridors, matrices, and networks, all with boundaries and gradients; functions include movement of energy, materials, genes, and information (Turner 1989; Wiens 2005); change recognizes the temporal dimension of dynamic ecological systems (Wu 2013). Though Forman and Godron (1986) indicate that landscape ecological study operates over meters to kilometers-wide areas and beyond, landscape ecologists have largely ignored the smaller, meters-wide scale. This makes it difficult for a meters-wide green roof microsite to pique the interest of landscape ecology researchers. Nevertheless, the links between small sites and larger landscapes offer research opportunities.

While the approach in this book follows the ecosystem concept, there are other emerging paradigms such as those that look at landscape hierarchies as a more realistic model. Because green roof are relatively small and sparse, they may have not reached a critical threshold to be influenced by other that abiotic factors (Blandin 2013)

Restoration ecology does examine microsites because of those sites' impact on restoration theory, focus, and practice (Coulson et al. 2001; Zobel et al. 2000). Yet, restoration ecology may seem out of place when discussing green roof ecosystems, since green roofs synthesize new substitutes for pre-existing natural structures and functions occurring on the Earth's surface that predated urbanization and building. Restoration ecology expends resources like land, labor and capital to change degraded or denuded land and landscapes into functioning ecosystems. However, restoration ecology contains an important human decision making component and an ethical understanding of restoration that may help focus promotion and use of green roofs through policy, design, installation, and management (Higgs 2012).

Scale is also an important issue in restoration ecology (White and Jentsch 2004). Green roof ecosystems, often only covering a few hundred square meters, yet share important structure, process, and change characteristics with small, isolated restoration plots. For example, their small size precludes all possible viable species, offers limited potential succession paths, and may oscillate unpredictably from disturbance. The time scales applicable to green roofs are also very short. Very few green roofs older than 50 years exist and even fewer can be found older than that.

Perception and manipulation of green roofs largely occurs at a human scale—one that embroils human values. Restoration ecology, where it deals with small sites, must also confront human values (Hobbs 2007) and negotiate stakeholder needs (Gobster and Barro 2000; Hobbs 2004).

Some restoration ecologists have thought about what occurs when faced with barren sites where ecosystems must be created from scratch. These situations have been described as designer ecosystems (MacMahon 1998; Nuttle et al. 2004) or novel ecosystems (Higgs 2012; Hobbs et al. 2013). Clewell and Aronson (2013, p. 211) are less sanguine in their view of such novel systems in relation to ecological restoration and ecosystem assembly declaring, "'from-scratch' ecosystems are constructed to fulfill narrowly conceived or short-term societal needs, such as green roofs, roadside revegetation, or wastewater treatment." While green roof ecosystems certainly fall under such rubrics, they cover more natural and anthropogenic function than a bare rooftop and their existing and future potential ecological structure and function should not be so easily dismissed.

Views such as Clewell and Aronson's (2013) above also preclude the opportunity for controllable designed experiments (Felson and Pickett 2005; Sutton 2013a) for exploring ecosystem functions (e.g. biomass production) and restoration structure (e.g. biodiversity) on green roofs. For example, Rosenzweig (2003) and Loreau et al. (2001) review the importance and impact of biodiversity (Hooper et al. 2005) and its interaction with ecosystem processes, functionally important species (Ranalli and Lundholm 2008; Lundholm et al. 2010) and basic causal mechanisms. Loreau et al. (2001, p. 804) state: "A major future challenge is to determine how biodiversity dynamics, ecosystem processes, and abiotic factors interact." Properly designed green roofs could serve as a baseline in gaging those challenges and thus inform both ecological restoration and restoration ecology.

Reconciliation ecology (Rosenzweig 2003) links urban ecology and restoration ecology by calling for an additional type of nature and species diversity protection beyond traditional preservation and restoration. He proposes a win-win approach that discovers how to modify and diversify anthropogenic habitats so that they harbor a wide variety of wild species; it seeks techniques to give many species back their geographical ranges without taking away ours. Francis and Lorimer (2011) specifically describe green roofs and green walls as an application of reconciliation ecology and suggest that citizen scientists be recruited to learn about green roofs and to monitor species diversity and movement.

Soil ecology studies the pedon, a cohesive unit that covers 1–10 m^2 (10.75–107.5 ft^2) (Coleman et al. 2004) and approaches a spatial scale suitable for understanding green roof functioning. Unfortunately green roof design most often calls for low weight engineered substrate lacking many critical qualities of a living soil. The concept of the soil health (Doran and Zeiss 2000, p. 3) calls for "a living soil in which soil organism and biotic parameters (e.g. abundance, diversity, food web structure, or community stability)… [provide] useful indicators of soil quality." Acknowledgement of these concepts is largely missing from selection use and management most green roof substrates.

Movement and saturation of water and its connection with soil texture and structure form an important abiotic context for microbial communities. Very porous soils such as found in green roof substrates show wide and rapid changes in free water, relative humidity, and temperature; natural soils vary much more in space and time than green roof substrates.

Fine roots <2 mm (0.08 in.) makeup a large portion of most plant's annual, below ground biomass production and the root's rate of turnover is measured in weeks and months, not seasons (Coleman et al. 2004). In natural environments most roots remain in the upper portions of the soils and shallow extensive green roof root zones will likely permeate it completely. In either case, these fine roots provide an important source of organic matter for microbes to recycle into useable plant nutrients. Nutrients held in porous green roof substrate tend to have severe leaching (Emilsson et al. 2007) until an adequate root network has been established.

The numbers and kinds of microfauna, mesofauna, and macrofauna found in natural soils are large and diverse, but variable due to turnover of microbes the patchiness of suitable biologically useable materials. These range from soil organic matter to root exudates to fecal pellets and earthworm slime. Spotty location of those materials contributes to the heterogeneous nature of soil organic matter and its accompanying, variable suite of microbes. In the case of symbiotic mycorrhizal fungi, hyphae actually go beyond tapping mere organic matter and enter individual root cells. This behavior allows the very efficient uptake of water and nutrients with the mycorrhizal fungi extending root hair function while the host plant supplies carbohydrates (Jeffries and Barea 2001). Thus mycorrhizal fungi help plants function in low nutrient, low moisture environments (Cripps and Eddington 2005) and are most likely critical for green roof substrates and plants (John 2013).

Microbes function importantly in nutrient turnover and release, yet may be limited by available nitrogen (Coleman et al. 2004). Highly dynamic interactions in soils include soluble nutrient exchange, solubility of organic matter, movement of soluble components, and growth and turnover of microbes (Coleman et al. 2004, p. 77). Ecosystem functions interact with the numbers of soil organisms to effect the quality of soil humus (Coleman 2004). These functional and species diversity factors were severely limited on many young German green roofs (Schrader and

Table 1.3 Important community ecology concepts pertinent to green roofs

Concept
Abiotic influences
Biotic influences
Disturbance regimes
Functional groups
Filters
Competition
Invasive species/Ruderals
Symbioses and Mutualisms
Autecology
Pathogens

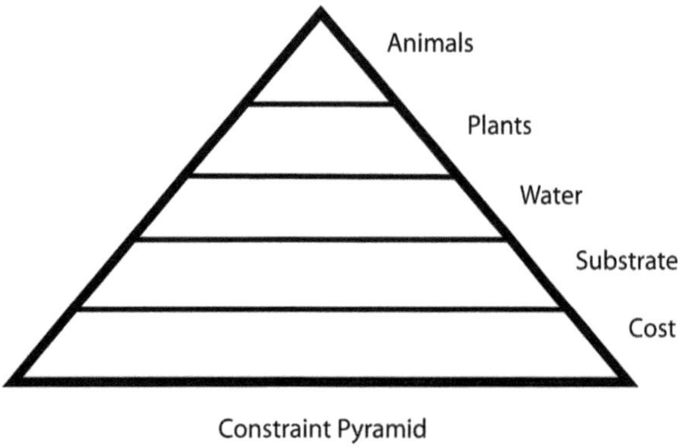

Fig. 1.3 Initially humans constrain green roof costs by specifying and controlling materials like substrate, water, and layout. These primary constraints set the stage for future changes

Boning 2006). Green roofs thus present excellent study sites to control, manipulate and examine the formation of soil microbial and micro-faunal communities and their interaction with plant assemblages.

Community Ecology serves as a basis for applied and conceptual ecological restoration. Assembly rules for identification (Temperton and Hobbs 2004), sourcing, placement, monitoring and management for species embody key human decisions in the process of community restoration that parallel those needed for creating green roof ecosystems. Assembly rules posited in restoration ecology also draw heavily on theoretical and applied concepts (Temperton et al. 2004) found in community ecology (synecology) (Table 1.3).

Green roofs may be largely governed, at least initially by abiotic factors such as precipitation, temperature, wind, insolation, and substrate (Butler and Orians 2011; Simmons et al. 2008; Molineaux et al. 2009; Dunnett and Nolan 2004; Dunnett et al. 2008a, b; Ampim et al. 2010; Rowe et al. 2012; Rowe 2011; Getter et al. 2009). Biotic influences also occur there simply due to interaction of living organisms including humans. Perhaps one of the basic human impacts comes from the original assembly of plant species to be placed on green roofs (Fig. 1.3). Careful, thoughtful, and experienced green roof designers should account for all of the conceptual aspects of the preceding community ecology list and serve as human filters in applying their plant knowledge to the design process. Most often this selection filter concentrates merely on the plants and not the environment or other organisms. Too often it ends with a narrowly conceived assemblage when often cost, immediate coverage and floriferous patterns holding sway (Sutton 2013b). Any reduction in cost may be reallocated to deepening substrate, adding irrigation, or widening plant palettes. On the other hand, water, plants, and animals are strongly constrained by the biotic and abiotic factors. After planting, disturbance regimes (called for or not), arrival of invasive species, competition for water and nutrients, and the health of individual species or stands of species become more important and are addressed under the rubric of maintenance. Other concepts from community ecology may or may not be considered during the process of species assembly and placement. Nevertheless, a tacit understanding of gradients of light, heat, moisture, wind and substrate depth should be considered fundamental in creating green roof niches. Functional traits can be used to both define the niche or for utilizing plant characteristics in supplying the green roof ecosystem feedback via facilitation (Butler and Orians 2011) or capturing and holding more precipitation (Lundholm et al. 2010). Here, at a community level, lies one more example where green roofs could be adapted as controllable, yet extensive enough areas to observe and test community ecology theories and concepts.

Concluding this brief review of salient ecological study approaches and ecosystems are three aspects of green roofs that connect them to the wider scope of ecological studies:

- Small to large size
- Simple to complex organization
- Anthropogenic to natural history

The conceptual spaces for scale, complexity and natural history (Fig. 1.4) indicate green roofs occupy a small scale, simple and largely anthropogenic realm.

1.8 Chapter Topics

The following 16 individual chapters start with monitoring and then next cover the critical abiotic factors of water, substrate, climate and microclimate. Next come chapters that examine plants, microbes, animals, and their interactions. The final two chapters provide summaries. Chapter 16 studies actual green roofs. Chapter 17 relates and synthesizes common themes and makes appropriate conclusions about green roof ecosystems, especially future design management, and research.

Chapter 2 Intensively managed ecosystems generally follow a high input, high output model and require relatively large subsidies of time, energy, and materials such as labor, water and nutrients. Since many green roofs follow this model what is known of their inputs and outputs of energy and materials? How might those be measured?

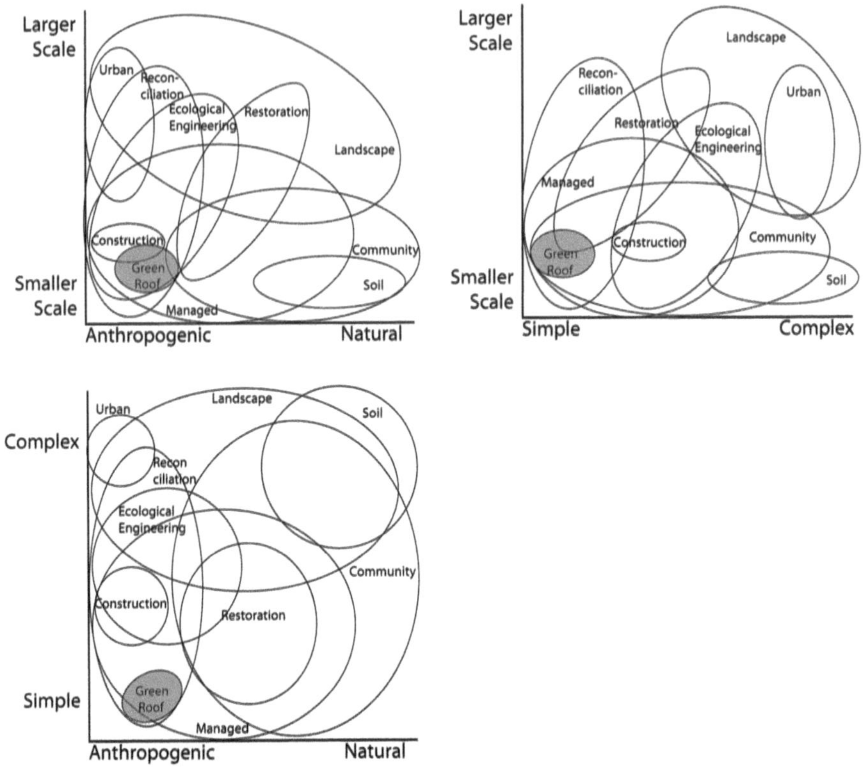

Fig. 1.4 Green roof ecosystem and ecological studies occupy a restricted portion of the conceptual and application space associated with the broader study of ecology and eco-systems

Chapter 3 Green roofs represent a synthetic ecosystem subject to unusual stresses. They not only arise from thin substrate but also, depending on the elevation, receive varying impacts from local climatic extremes of wind, heat, air humidity and substrate water content. Microclimates create abiotic gradients in which plants must grow.

Similarly, substrate composition and structural design have a direct role in ameliorating soil microclimate to accommodate appropriate plants. In warmer, non-temperate systems with greater climatic extremes (e.g., high daytime and night time temperatures, frequent flash flood events), green roofs may offer relatively larger intrinsic (e.g. cooling building, extension of roof membrane lifetime) and extrinsic (e.g., flash flood mitigation, reduction of heat island effect) benefits. But the design (including the plant palette, substrate composition and profile design) can be modified to accommodate different conditions.

Chapter 4 How green roofs use and process water is a critical component of their function and effective management. As green roof technology has spread from northern Europe's relatively cool and humid climate, successful green roof designs have had to adapt to regional variations in the timing and availability of water.

The development of regionally appropriate designs requires a mechanistic understanding of green roof water relations and plant eco-physiology. Water efficient designs effectively match environmental conditions, substrate characteristics, plant physiological needs, plant community interactions, and expert systems for applying supplemental water.

Chapter 5 In support of plant communities, green roofs have typically been constructed with compost or other nutrient-rich organic matter blended into the growing substrate. As a consequence, leaching of nitrogen (N) and especially phosphorous (P) can be high in green roof runoff, which is a disservice for downstream ecosystems. Cycling of N and P has been studied in natural ecosystems, revealing fundamental characteristics about plant-soil-air interactions, ecological stoichiometry, nutrient limitation and saturation. However, it is not known whether the principles generated from these systems translate readily or directly to constructed ecosystems like green roofs.

Chapter 6 Although typically eschewed in favor of highly engineered substrate, natural soils can provide an obvious benefit for roof systems by jump-starting a viable, self-organizing habitat. Such soils can act as microbial inoculants and serve as an additional source for plants and insects via seed banks, eggs, and larvae. However green roofs utilizing mostly natural soil with finer particles and slower internal drainage can lead to clogged drains or mass movement.

Chapter 7 Green roofs provide a number of ecosystem services such as the provision of habitats for organisms residing in and migrating throughout the city that have only recently been studied and documented. Microorganisms such as fungi and bacteria have been found to be diverse and abundant components to green roof substrate and may contribute to some of the other benefits green roofs provide such as the removal of organic pollutants from precipitation, recycle organic detritus, and help create soil structure.

Chapter 8 Many green roof designs employ a limited palette of drought-resistant *Sedum* species are assumed to be static. Comparatively few utilize diverse species assemblages or consider assemblage dynamics. However, diverse green roof plantings not only help to restore biodiversity to species-poor urban environments, but may also improve the quality of services provided by green roofs while recruiting new species and allowing existing ones to move about.

Chapter 9 The ecosystem services green roofs provide are influenced by both the engineered and biotic components of green roof systems. How might plant species and the synthetic vegetation communities created for and by them control the functioning of green roofs? Studies show that plant species can differ greatly in their ability to provide services such as roof cooling and stormwater retention. Newer work, emphasizing less-well characterized benefits such as reduction of heat loss in winter, air pollution mitigation, and carbon sequestration also shows significant effects from plant species and functional groups of plants into communities.

Chapter 10 In the United Kingdom, promotion of urban biodiversity has become a leading driver for green roof installation. A special category of vegetated roofs, known as 'Biodiverse Living Roofs' (formerly know as Brown Roofs) has become well established as the primary means by which this is achieved. These roofs attempt to create the ecological conditions of urban brownfield or post-industrial sites on the ground, which are often biodiversity hot spots in cities. Largely consisting of ruderal species, these bio-diverse roofs offer a testing grounding for applying Competitor-Stress Tolerator-Ruderal theory and questions about plant dynamics and succession.

Chapter 11 Green roofs can be currently seen as an ecologically sustainable practice, but in fact many are both unstable and vulnerable. Low-diversity systems break; they are not resilient. Within prairies many specialized plant community templates arise in hot, dry, windy places with thin, poorly developed soils. These communities with their suite of adapted plants are closely analogous to green roof conditions and provide a designer with a potential palette from which to select. Examples of an plant assembly process are applied to two such projects in Minneapolis.

Chapter 12 Despite an emerging understanding of green roofs as dynamic ecosystems, most green roof vegetative studies treat plant communities as static assemblages. An ecological perspective of green roof composition and dynamics allows for deeper examination of green roof design and maintenance practices rooted in performance, while potentially changing the ways in which designers, engineers and managers conceive of a green roof. Novel, anthropogenic ecosystems, such as green roofs, display complex growth dynamics rooted in a combination of initial site conditions (shading, thermal exposure, wind, moisture), roof design (vegetation, growing media, roof substrate, drainage), and disturbances (extreme climate conditions, weeding, disease, emergent species, fertilization).

Chapter 13 If the ecological benefits of green roofs are to be realized then plant selection and long-term plant and media performance are extremely important. Long Term Ecological Research (LTER) sites have expanded ecologists' understandings of both ecosystem and community concepts. Some sites represent human ecosystems and as such relate to long-term study of green roof ecosystems.

Chapter 14 Insects and other arthropods are essential for several ecosystem services on green roofs. Although it is assumed that arthropods are mostly desirable on green roofs, it is not clear whether green roofs adequately provide habitat.

Chapter 15 Increased biodiversity is one of the commonly stated benefits of installing extensive green roofs. Because biodiversity and its conservation is multiscalar, there are multi-scalar opportunities for ecological green roof design and management that link biodiversity conservation efforts on the ground plane while contributing to the supply of food, water, energy and other ecosystem services for the benefit of human populations.

Chapter 16 Biodiversity and general ecological criteria have been consciously used in designing some green roofs. Looking worldwide, a group of one dozen green roofs were selected to reiterate and highlight the key concepts covered in the previous chapters. This chapter provides ideas for how green roofs have been thought of and designed as ecosystems.

Chapter 17 The final chapter draws on preceding ones to identify, reiterate, highlight, and most importantly explain key ecological concepts in the context of green roofs. It identifies concepts with strong connections and application to design and management of green roof ecosystems and notes where knowledge is limited and how ecosystems conscious designers might investigate questions as green roofs are created.

Acknowledgements A UN-Lincoln Fleming Horticulture Research Grant to R. Sutton supported his editing and writing for this book.

References

Alberti M (ed) (2008) Advances in urban ecology: integrating humans and ecological processes in urban ecosystems. Springer, New York

Allen THF (2002) Applying the principles of ecological emergence to building design and construction. In: Kibert C, Sendzimir J, Guy G (eds) Construction ecology: nature as the basis for green buildings, Spon, London, pp 108–126

Allen THF, Starr T (1982) Hierarchy: perspectives for ecological complexity. University of Chicago Press, Chicago

Ampim P, Sloan J, Cabrerra R, Harp D, Jaber F (2010) Green roof growing substrates: types, ingredients, composition and properties. J Environ Hortic 28(4): 244–252

Antle J, Capalbo J, Elliott E, Hunt H, Mooney S, Paustian K (2001) Research needs for understanding and predicting the behavior of managed ecosystems: lessons form the study of agroecosystems. Ecosystems 4:723–735

Bates A, Mackay R, Greswell R, Sadler J (2009) SWITCH in Birmingham, UK: experimental investigation of the ecological and hydrological performance of extensive green roofs. Rev of Env Sci Biotech 8:295–300

Berardi U, Ghaffarian-Hoseini A, Ghaffarian-Hoseini A (2013) State-of-the-art analysis of the environmental benefits of green roofs. Appl Energy 115:411–428

Berndtsson J (2010) Green roof performance towards management of runoff water quantity and quality: a review. Ecol Eng 36(4):351–360

Berndtsson J, Bengtsson L, Jinno K (2009) Runoff water quality from intensive and extensive vegetated roofs. Ecol Eng 35(3):369–380

Berryman A (1989) The conceptual foundations of ecological dynamics. Bull Ecol Soci Am 70:230–236

Bianchini F, Hewage K (2012) How green are the green roofs? Lifecycle analysis of green roof materials. Build Environ 48:57–65

Blandin P (2013) Ecology and biodiversity at the beginning of the twenty-first century: towards a new paradigm? In: Schwarz A, Jax K (eds) Ecology revisited reflecting on concepts, advancing science. Springer, New York. doi:10.1007/978-90-481-9744-617

Brenneisen S (2006) Space for urban wildlife: designing green roofs as habitats in Switzerland. Urban Habitats 4. http://www.urbanhabitats.org/v04n01/wildlife_full.html. Accessed 19 March 2015

Buist R (2008) Going for organics. Living Archit Monit 10(1):16–21

Butler C, Orians C (2011) Sedum cools soil and can improve neighboring plant performance during water deficit on a green roof. Ecol Eng 37(11):1796–1803

Carter T, Fowler L (2008) Establishing green roof infrastructure through environmental policy instruments. Environ Manage 42(3):151–164
Carter T, Keeler A (2008) Life-cycle cost-benefit analysis of extensive vegetated roof systems. J Environ Manage 87(3):350–363
Clewell A, Aronson J (2013) Ecological restoration, 2nd edn. Island Press, Washington DC
Coffman R, Waite T (2011) Vegetated roofs as reconciled habitats: rapid assays beyond mere species counts. Urban Habitats 6(1) http://www.urbanhabitats.org/v06n01/vegetatedroofs_full.html. Accessed 19 March 2015
Coleman D, Crossley D Jr, Hendrix P (2004) Fundamentals of soil ecology, 2nd edn. Elsevier, New York
Connelly M, Hodgson M (2008) Sound transmission loss of extensive green roofs-Field test results. Can Acoust 36(3):74–75
Cook-Patton S, Bauerle T (2012) Potential benefits of plant diversity on vegetated roofs: a literature review. J Environ Manage 106:85–92
Coulson S, Bullock J, Stevenson M, Pywell R (2001) Colonization of grassland by sown species dispersal versus microsite limitation in responses to management. J Appl Ecol 38(1):204–216
Cripps C, Eddington L (2005) Distribution of mycorrhizal types among alpine vascular plant families on the Beartooth Plateau, Rocky Mountains, USA, in reference to large-scale patterns in arctic-alpine habitats. Arct Antarct Alp Res 37(2):177–188
Cushman JC, Borland AM (2002) Induction of Crassulacean acid metabolism by water limitation. Plant Cell Environ 25(2):295–310
Daykin K, Benjamin L, Pantiel M (2013) The professional design guide to green roofs. Timber Press, Portland
Doran J, Zeiss M (2000) Soil health and sustainability: managing the biotic component of soil health. Appl Soil Ecol 15:3–11
Dunnett N, Nolan A (2004) The effect of substrate depth and supplementary watering on the growth of nine herbaceous perennials in a semi-extensive green roof. Acta Hortic 643:305–309
Dunnett N, Nagase A, Hallam A (2008a) The dynamics of planted and colonising species on a green roof over six growing seasons 2001–2006: influence of substrate depth. Urban Ecosyst 11(2008):373–384
Dunnett N, Nagase A, Booth R, Grime J (2008b) Influence of vegetation composition on runoff in two simulated green roof experiments. Urban Ecosyst 11:385–398
Dvorak BD (2011) Comparative analysis of green roof guidelines and standards in Europe and North America. J Green Build 6(2):170–191. doi:10.3992/jgb.6.2.170
Dvorak BD, Volder A (2010) Green roof vegetation for North American ecoregions: a literature review. Lands Urban Plan 96(4):197–213
Dvorak BD, Volder A (2012) Plant establishment on unirrigated green roof modules in a subtropical climate. AoB Plants. doi:10.1093/aobpla/pls049
Emilsson T, Berndtsson JC, et al (2007) Effect of using conventional and controlled release fertiliser on nutrient runoff from various vegetated roof systems. Ecol Eng 29(3):260–271
Enright C (2013) Green roofs are growing: standards support sustainability. Standardization News May/June. http://www.astm.org/sn/features/green-roofs-are-growing-mj13.html. Accessed 10 Feb 2014
Felson A, Pickett STA (2005) Designed experiments: new approaches to studying urban ecosystems. Front Ecol Environ 3:549–556
Feng C, Meng Q, Zhang YF (2010) Theoretical and experimental analysis of the energy balance of extensive green roofs. Energy Build 42:959–965
Forschungsgesellschaft Landschaftsentwicklung Landschaftsbau (Landscape, Research, Development and Construction Society) (2002) Guideline for the planning, execution, and Upkeep of Green-roof Sites. http://www.fll.de/shop/english-publications/green-roofing-guideline-2008-file-download.html. Accessed 19 March 2015
Forman RTT, Godron M (1986) Landscape ecology. Wiley, New York
Francis R, Lorimer J (2011) Urban reconciliation ecology: the potential of living roofs and walls. J Environ Manage 92:1429–1437

Friedrich C (2008) Don't call it dirt! Living Archit Monit 10(1):16–21
Gaffin SR, Rosenzweig C et al (2008) Variations in New York city's urban heat island strength over time and space. Theor Appl Climatol 94(1–2):1–11
Gaston K (ed) (2010) Urban ecology. Ecological reviews. Cambridge University Press, Cambridge
Getter K, Rowe DB (2006) The role of extensive green roofs in sustainable development. Hort Sci 41(5):1276–1285
Getter K, Rowe DB, Cregg B (2009) Solar radiation intensity influences extensive green roof plant communities. Urban For Urban Green 8(4):269–281
Gobster P, Barro S (2000) Negotiating nature: making restoration happen in an urban park context. In: Gobster P, Hull R (eds) Restoring nature: perspectives from the social sciences and humanities. Island Press, Washington DC, pp 185–207
Green Roofs for Healthy Cities (GRHC) (2006a) Green roof design 101 introductory course, 2nd edn. Cardinal Group, Toronto
Green Roofs for Healthy Cities (GRHC) (2006b) Case studies of green roof regulations in North America. Available via Docstoc. http://www.docstoc.com/docs/40655588/2006---Case-Studies-of-Green-Roof-Regulations-in-North-America. Accessed 1 Jan 2014
Gregoire B, Clausen J (2011) Effect of a modular extensive green roof on stormwater runoff and water quality. Ecol Eng 37(6):963–969
He H, Jim C (2010) Simulation of thermodynamic transmission in the green roof ecosystem. Ecol Model 221(24):2949–2958
Higgs E (2012) Changing nature: novel ecosystems, intervention, and knowing when to step back. In: Turner M, Weinstein R (eds) Sustainability science. Springer, New York, pp 383–398
Hobbs R (2004) Restoration ecology: the challenge of social values and expectations. Front Ecol Environ 2:43–48
Hobbs R (2007) Setting effective and realistic restoration goals: key directions for research. Restor Ecol 15(2):354–357
Hobbs R, Higgs E, Hall C (2013) Novel ecosystems: intervening in the new ecological world order. Wiley-Blackwell, New York
Hooper D, Chapin F III, Ewel J, Hector A, Inchausti P, Lavorel S, Lawton J, Lodge D, Loreau M, Naeem S, Schmid B, Setälä H, Symstad A, Vandermeer J, Wardle D (2005) Effects of biodiversity on ecosystem functioning: a consensus of current knowledge. Ecol Monogr 75(1):3–35
Jeffries P, Barea J (2001) Arbuscular Mycorrhizzae: a key component of sustainable plant-soil ecosytems. Mycota 9:95–113
Jim C, He H (2010) Coupling heat flux dynamics with meteorological conditions in the green roof ecosystem. Ecol Eng 36(8):1052–1063
John J (2013) Assessment of arbuscular Mycorrhizae fungi in a green roof system. MS Thesis, Dalhousie University, Halifax, Nova Scotia
Jorgensen S, Odum H, Brown M (2004) Emergy and exergy stored in genetic information. Ecol Model 178(1):11–16
Kahn P Jr (1999) The human relationship with nature: development and culture. MIT Press, Cambridge
Kaplan R, Kaplan S (eds) (1989) The experience of nature: a psychological perspective. Cambridge University Press, Cambridge
Kay J (2002) On complexity theory, exergy, and industrial ecology. In: Kibert CJ, Sendzimir J, Guy GB (eds) Construction ecology: nature as the basis for green buildings. Spon, London, pp 72–107
Kibert CJ, Sendzimir J, Guy GB (eds) (2002) Construction ecology: nature as the basis for green buildings. Spon, London
Köhler M, Keeley M (2005) The green roof tradition in Germany: the example of Berlin. In: Hoffman L, McDonough W (eds) Green roofs: ecological design and construction. Schiffer, New York, pp 108–112
Loreau M, Naeem S, Inchausti P, Bengtsson J, Grime J, Hector A, Wardle D (2001) Biodiversity and ecosystem functioning: current knowledge and future challenges. Science 294(5543):804–808

Louv R (2012) The nature principle: reconnecting with life in a virtual age. Algonquin Books, Chapel Hill, NC
Luckett K (2009) Green roof construction and maintenance. McGraw-Hill, New York
Lundholm J, Richardson P (2010) Habitat analogues for reconciliation ecology in urban and industrial environments. J Appl Ecol 47(5):966–499
Lundholm J, MacIvor J, MacDougall Z, Ranalli M (2010) Plant species and functional group combinations affect green roof ecosystem functions. PLoS One 5(3). doi: 10.1371/journal.pone.0009677
MacIvor J, Lundholm J (2011) Insect species composition and diversity on intensive green roofs and adjacent level-ground habitats. Urban Ecosyst 14(2):225–241
MacMahon J (1998) Empirical and theoretical ecology as a basis for restoration: an ecological success story. In: Groffman P, Pace M (eds) Successes, limitations and frontiers in ecosystem science. Springer-Verlag, New York, pp 220–226
Madre F, Vergnes A, Machon N, Clergeau P (2014) Green roofs as habitats for wild plant species in urban landscapes: first insights from a large-scale sampling. Landsc Urban Plan 122:100–107
Metselaar K (2012) Water retention and evapotranspiration of green roofs and possible natural vegetation types. Res Conserv Recyc 64:49–55
Millennium Ecosystem Assessment (MEA) (2005) Ecosystems and human well-being, vol 5. Island Press, Washington DC
Mitsch W, Jorgensen S (2004) Ecological engineering and ecosystem restoration. Wiley, New York
Molineux C, Fentiman C, Gange A (2009) Characterising alternative recycled waste materials for use as green roof growing media in the UK. Ecol Eng 35(10):1507–1513
Morgan S, Celik S, Retzlaff W (2012) Green roof storm-water runoff quantity and quality. J Environ Eng 139(4):471–478
Naveh Z (2000) The total human ecosystem: integrating ecology and economics. BioScience 50(4):357–361
Naveh Z, Lieberman A (1994) Landscape ecology: theory and application, 2nd edn. Springer, Berlin
Nuttle T, Hobbs R, Temperton V, Halle S (2004) Assembly rules and ecosystem restoration: where to from here? In: Temperton V, Hobbs R, Nuttle T, Halle S (eds) Search for ecological assembly rules and its relevance to restoration ecology. Island Press, Washington DC
Oberndorfer E, Lundholm J, Bass B, Coffman R, Doshi H, Dunnett N, Gaffin S, Köhler M, Liu K, Rowe DB (2007) Green roofs as urban ecosystems: ecological structures, functions, and services. BioScience 57(10):823–833
Odum H (1971) Environment, power and society. Wiley-Interscience, New York
Odum E (1992) Great ideas from ecology for the 1990's. BioScience 42:542–545
Odum H (2002) Material circulation, energy hierarchy, and building construction. In: Kibert CJ, Sendzimir J, Guy GB (eds) Construction ecology: nature as the basis for green buildings. pp 37–71, Spon Press, New York
Odum H, Nilsson P (eds) (1996) Environmental accounting: emergy and environmental decision making. Wiley, New York
Osmundson T (1999) Roof gardens: history, design, and construction. W. W. Norton and Company, New York
Ranalli M, Lundholm J (2008) Biodiversity and ecosystem function in constructed ecosystems. CAB Rev Perspect Agric Vet Sci Nutr Nat Resour 3(34):1–16
Rosenzweig M (2003) Reconciliation ecology and the future of species diversity. Oryx 37(2):194–205
Rowe DB (2011) Green roofs as a means of pollution abatement. Environ Pollut 159(8–9):2100–2110
Rowe DB, Getter K, Durhman A (2012) Effect of green roof media depth on Crassulacean plant succession over seven years. Landsc Urban Plan 104(3–4):310–319
Sayed OH et al (1994) Growth, water relations, and CAM induction in Sedum album in response to water stress. Biol Plant 36(3):383–388

Schrader S, Böning M (2006) Soil formation on green roofs and its contribution to urban biodiversity with emphasis on Collembolans. Pedobiologia 50:347–356
Schroll E, Lambrinos J, Righetti T, Sandrock D (2011) The role of vegetation in regulating stormwater runoff from green roofs in a winter rainfall climate. Ecol Eng 37(4):595–600
Simmons M, Gardiner B, Windhager S, Tinsle J (2008) Green roofs are not created equal: the hydrologic and thermal performance of six different extensive green roofs and reflective and non-reflective roofs in a sub-tropical climate. Urban Ecosyst 1(4):339–348
Simons R, Choi E, Simons D (2009) The effect of state and city green policies on the market penetration of green commercial buildings. J Sustain Real Estate 1(1):139–164
Single Ply Roofing Institute (SPRI) (2010) ANSI/SPRI RP-14 Wind design standard for vegetative roofing systems. www.spri.org Waltham, MA, 36 p
Smith K, Roebber P (2011) Green roof mitigation potential for a proxy future climate scenario in Chicago, Illinois. J Appl Meteorol Climatol 50(3):507–522
Snep R, Opdam P (2010) Integrating nature values in urban planning and design. In: Gaston K (ed) Urban ecology. Cambridge University Press, New York, pp 261–286
Snodgrass EL, MacIntyre L (2010) The green roof manual. Timber Press, Portland
Solecki WD, Leichenko RM (2006) Urbanization and the metropolitan environment -Lessons from New York and Shanghai. Environment 48(4):8–23
Suaris W, Irwin P (2010) Effect of roof-edge parapets on mitigating extreme roof suctions. J Wind Eng Ind Aerodyn 98(10):483–491
Sutton R (2013a) Seeding native grass on green roofs. J Living Archit 1(1):1–26
Sutton R (2013b) Rethinking extensive green roofs to lessen emphasis on biomass. J Living Archit 1(1):2
Sutton R (2014) Aesthetics for green roofs and green walls. J Living Archit 1(2):1–15
Sutton R, Harrington J, Skabelund L, MacDonagh P, Coffman R, Koch G (2012) Prairie-based green roofs: literature, templates and analogs. J Green Build 7(1):143–172. doi:10.3992/jgb.7.1.143
Teemusk A, Mander Ü (2010) Temperature regime of planted roofs compared with conventional roofing systems. Ecol Eng 36(1):91–95
Temperton V, Hobbs R (2004) Assembly rules and ecological restoration: where to from here? In: Temperton V, Hobbs R, Nuttle T, Halle S (eds) Search for ecological assembly rules and its relevance to restoration ecology. Island Press, Washington DC, pp 34–54
Temperton V, Hobbs R, Nuttle T, Halle S (eds) (2004) Search for ecological assembly rules and its relevance to restoration ecology. Island Press, Washington DC
Thayer R (1989) The experience of sustainable landscapes. Landsc J 8(2):101–110
Turner M (1989) Landscape ecology: the effect of pattern on process. Annu Rev Ecol Syst 20:171–197
Weiler S, Schloz-Barth K (2009) Green roof systems: a guide to the planning, design, and construction of landscapes over structure. Wiley, New York
White P, Jentsch A (2004) Disturbance, succession, and community assembly in terrestrial plant communities. In: Temperton V, Hobbs R, Nuttle T, Halle S (eds) Search for ecological assembly rules and its relevance to restoration ecology. Island Press, Washington DC, pp 342–366
Wiens J (2005) Toward a unified landscape ecology. In: Wiens J, Moss M (eds) Issues and perspectives in landscape ecology. Cambridge Press, New York, pp 148–151
World Commission on Environment and Development (WCED) (1987) Our common future, report of the world commission on environment and development. United Nations, Switzerland: Annex to UN General Assembly Document A/42/427
Wu J (2013) Landscape ecology. In Leemans R (ed) Ecological systems: selected entries from the encyclopedia of sustainability science and technology. Springer, New York, pp 179–200
Zobel M, Otsus M, Liira J, Moora M, Möls T (2000) Is small-scale species richness limited by seed availability or microsite availability? Ecology 81(12):3274–3282

Chapter 2
Monitoring Abiotic Inputs and Outputs

Lee R. Skabelund, Kimberly DiGiovanni and Olyssa Starry

Abstract Green roof monitoring is critical to understand and improve the design, implementation, and management of green roof ecosystems. Creating resilient, less resource intensive living roofs fitting their larger eco-regional context, specific local setting, and unique project objectives means understanding inputs and outputs. This chapter addresses monitoring abiotic inputs and outputs related to green roof hydrology (precipitation and irrigation, storage, outflow, and evapotranspiration), water quality, energy fluxes, temperatures, meteorological conditions (wind), and gas/carbon exchange. This chapter presents monitoring approaches and equipment needs from literature and researcher interviews detailing several relevant examples. Important design, educational, and management opportunities relating to effective monitoring programs are discussed.

Keywords Hydrology · Water quality · Energy fluxes · Temperatures · Meteorological conditions · Substrate characteristics · Gas/carbon exchanges

2.1 Introduction

Intensively managed ecosystems generally follow a high input, high output model requiring relatively large subsidies of time, substances, energy, and materials while frequently shedding stormwater and contributing various effluents directly and indirectly to the environment (Arvidson 2012).

L. R. Skabelund (✉)
Department of Landscape Architecture/Regional & Community Planning, The College of Architecture, Planning & Design, Kansas State University, 302 Seaton Hall, Manhattan, KS 66506-2909, USA
e-mail: lskab@k-state.edu

K. DiGiovanni
Department of Civil & Environmental Engineering, Villanova University, Villanova, PA 19085, USA

O. Starry
Urban Honors College, Portland State University, Portland, OR 56478, USA

R. Sutton (ed.), *Green Roof Ecosystems*, Ecological Studies 223,
DOI 10.1007/978-3-319-14983-7_2

Because any ecosystem leaks or exports nutrients, materials, and energy, excess inputs become outputs and potential wastes degrading or harming their surrounding surface water, air, and ambient temperatures (Odum 1969; Oke 1978; Spirn 1984). We should and can create "sustainable urban social-ecological systems" (Byrne and Grewal 2008, p. 1) including green roofs (Rowe 2011) and other living infrastructure.

What green roof ecosystem inputs and outputs relate to benefits and concerns? What is known about such inputs and outputs?

Stormwater outputs produce financial impacts so the quantity stored, and its timed release, must be reconciled with precipitation intensity, as well as substrate and vegetation characteristics. Many urban areas now closely monitor stormwater runoff (for one example, see Kurtz et al. 2010). How does one monitor and account for the constituents retained by or exported from a green roof? How do these flows compare amongst rooftops?

To assess heat attenuation on and within a living roof system, requires measuring insolation and heat flows. Yet how do substrate and vegetation type influence hydrology and microclimate? And, how do plant and root growth above and within a substrate, the production of new vegetation from seeds, and human management strategies influence hydrologic processes, energy flow, climatic conditions, and the creation of a living, supportive substrate? Teasing apart these complex interactions relies first on solid qualitative and quantitative data about each.

Researchers, designers, and managers must account for constituents retained or leaving a green roof. They need to ask: What current methods and equipment are being used to measure and analyze inputs and outputs? How are the data analyzed and then used to improve green roof design, monitoring, and management?

In this chapter, we first define green roof ecosystem inputs and outputs. Next, we provide a general overview of green roof monitoring. We focus primarily on monitoring the inputs and outputs associated with water and energy fluxes from green roof systems and gas/carbon exchange. We present examples demonstrating green roof monitoring applications related to research, design, and management goals—listing important challenges and lessons learned from green roof monitoring. Opportunities for the future of green roof monitoring and research are also discussed.

2.2 Defining Inputs and Outputs

Every ecosystem consists of many interconnected variables and researchers cannot feasibly monitor everything, so they must clearly define what is to be monitored and why. Monitoring specific abiotic inputs and outputs brings vital understanding to the interactions and functions associated with both biotic and abiotic conditions.

We define ***inputs*** as substances and energy added to a green roof (for example, water in the form of precipitation and irrigation, added nutrients, and energetic inputs like sunlight or solar radiation). We define ***outputs*** as substances and fluxes modified on *or* leaving a green roof (for example, the outflow of water nutrients in

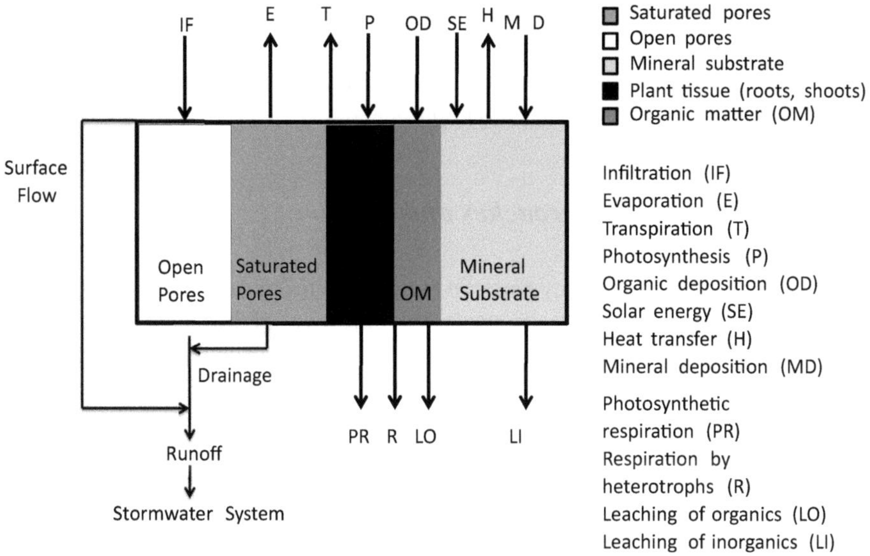

Fig. 2.1 Green roof inputs and outputs model (Olyssa Starry and Rich Pouyat, unpublished)

substrates or runoff, evapotranspiration, and heat energy). Inputs and outputs from green roof systems are generally conceptualized in Fig. 2.1.

Closely examining "green roof benefits" related to stormwater management, urban heat loads, energy use, and carbon sequestration helps researchers effectively monitor the dynamic conditions influencing important green roof attributes and functions. Thus, we monitor conditions and factors related to optimize water and energy inputs, and reduce or eliminate negative outputs—namely, outflow, excess heat, carbon dioxide, nutrients, and heavy metals. In doing so we better understand how to create ***low input*** (water and energy conserving), and ***low output*** (less runoff and non- or minimally-polluting) green roof ecosystems and can also enhance our ability to sequester carbon and achieve other project goals.

It is important to note that ways to monitor green roof inputs and outputs vary. "We direct some monitoring at the inputs and outputs themselves (i.e. the energy and material fluxes in the system), some at describing processes (such as evapotranspiration or microbial activity) that drive those fluxes, and others at physical conditions (such as temperature or wind speed) which may directly or indirectly influence fluxes and other aspects" (John Lambrinos 2014, pers. comm.).

2.3 Planning for Green Roof Monitoring

It is impossible to learn from green roof ecosystems without closely observing and understanding monitoring goals and objectives. This section discusses: (1) monitoring approaches and goals in light of the needs and demands associated with observational studies versus experimental monitoring designs; (2) expectations regarding equipment and maintenance; (3) data collection; (4) data analysis and the technical expertise needed to successfully undertake effective green roof system monitoring; and (5) monitoring precautions.

2.3.1 Monitoring Approaches and Goals

The most important part of any research project and accompanying monitoring process is articulating specific needs, goals, and objectives of the study. To do this requires a reasonable understanding of the available literature and the project context. This includes the type of site, regional and local setting, funding, expertise, personnel, equipment, and other necessary support systems.

Relating monitoring to the specific type of site(s) and study under consideration before deciding what type of monitoring to undertake is essential. Will monitoring be done on an existing green roof, on a new or proposed green roof, or on models, mock-ups, modules, or platforms? Will monitoring activities examine integrated or modular systems, or both?

It is also vital to understand conditions associated with the particular green roof study system under consideration. Precipitation, temperature, relative humidity, wind speed, wind direction, and solar radiation can change seasonally, and can be affected by surrounding buildings, structures, and vegetation. For building retrofits, "data collected before renovation can be a valuable measure of the new green roof's performance" (Onset Computer Corporation 2012, p. 3).

An example of the type of monitoring questions appropriate at this stage might be: How hot and cold does this location get? How do adjacent building masses influence sun/shade patterns and wind movements—and thus precipitation, relative humidity, temperatures, and other conditions on the roof?

Monitoring is tempered through identification of project objectives. For example, if our primary interest is to improve design and management, as opposed to understanding ecosystem functions, or if we look for trends over time and space or specifically try to address a narrower questions through controls, then our methods and analysis will likely be different. (Karban and Hunzinger 2006).

Based on project goals, level and intensity of monitoring activities varies. Welker et al. (2013) describes a three-tiered, low, medium, and high, approach to monitoring and provides a framework for balancing monitoring between project goals and monitoring costs. For example, a low level approach for monitoring the hydrology and ecology of a green roof might include visual inspections while a high level approach include sensor systems collecting continuous data (Welker et al. 2013).

Generally, two overarching approaches, observational and experimental describe green roof monitoring. Each approach may include qualitative and quantitative data collection and analysis, and these approaches and accompanying monitoring activities can be carried out simultaneously to address specific green roof research questions, hypotheses, and/or practical design and management issues.

Observational monitoring studies rely on systematic collection, recording, and analysis of relevant data over some period of time. Observation of green roof conditions in space and time are frequently and systematically recorded to document changes, dynamics, and particular conditions for selected variables addressing research questions of interest. Hand-written notes, quick counts or measurements, accompanied by photographs taken from the same locations through time can effectively supplement data collected from other monitoring devices.

Experimental research/monitoring focuses on theoretical or hypothesis testing and can be related to one or more topics (e.g.): vegetation types, substrate depths, substrate types, roof slopes, micro-climatic conditions, supplemental irrigation, nutrients, shading devices, etc. Experimental monitoring relies on systematic collection, recording, and analysis of relevant data over a period of time, and can include a controlled comparison targeting a specific research question and hypothesis. Standard statistical designs and protocols should lead to significant inference about the data and experimental designs require consultation with technical experts. Other researchers must be able to replicate methods. Thus, researchers must balance the need for replicated treatments in their design against the feasibility of including multiple roof-scale measurements. Depending on the research question, working with modules or experimental containers may be an option. As well, one or more control rooftops may be monitored so that comparisons can be made between the green roof (or green roof modules) and nearby black, dark-gray, and/or white or light-colored roofs.

Experimental research requires more statistical rigor (i.e., replication) than observational study. Observational study requires less replication but are also less generalizable. Neither approach is better than the other, but the distinction is important. Often the two approaches are integrated. Some—including Tilman 1989, and Havens and Aumen 2000—argue that these two approaches must be integrated.

Importantly, green roof monitoring and data collection can be done to support green roof management, to evaluate performance relative to particular green roof project goals or models, or to collect data as part of the process of testing specific hypotheses about how green roofs function.

2.3.2 Monitoring Equipment and Maintenance

Monitoring equipment can be simple, such as a hand-held thermometer, manual rain gauge, and a stormwater collection container, through complex, such as a series of tipping buckets, temperature sensors, and multiple flow sensors—all connected to data logger(s) and a satellite-operated wireless data distribution network. Value comes from using basic probes or sensors for repeated measurements of green roof systems over several seasons or years. Although single measurements of variables such as substrate temperature or moisture content give snapshots of systems constantly in flux, such samples taken at a regular intervals over a longer time discern overall rates, trends, and patterns (Tsiotsiopoulou et al. 2003). Spot measurements are especially useful so that comparisons between two different green roof treatments can be made. These measurements can be conducted in the context of field or classroom visits. They can also be incorporated into regular monitoring system maintenance and green roof management. For example, commercial systems can be used to assess green roof media water content, viewed in real time and used to manage green roof irrigation systems (e.g. Linda Tools green roof in Brooklyn, NY). Commercially available green roof monitoring systems are available and in some instances equipment including weather stations, sensors, and loggers can be ordered from a single established company.

Complex monitoring networks may combine educational and commercial systems to collect continuous data increasing replicates (Lea-Cox 2012). Sampling larger roof areas accounts for variation and accessing data remotely improves safety or security concerns and reduces visits. For example, remote access enables roof data to be visualized during storms. Transferring sensor information to a network, computer, or handheld device synchronously, makes data more accessible in the classroom or laboratory in real time (see Fig. 2.2). Green roof monitoring systems also document green roof performance to improve green roof adoption, design, and management. In a survey of architects and building managers in Chicago and Indiana, Hendricks and Calkins (2006) identified ways that designers can increase public understanding of the benefits associated with new or innovative green roof practices. They noted that early adopters key on recognition of the environmental services provided as revealed by monitoring. As such, monitoring is seen as vital to

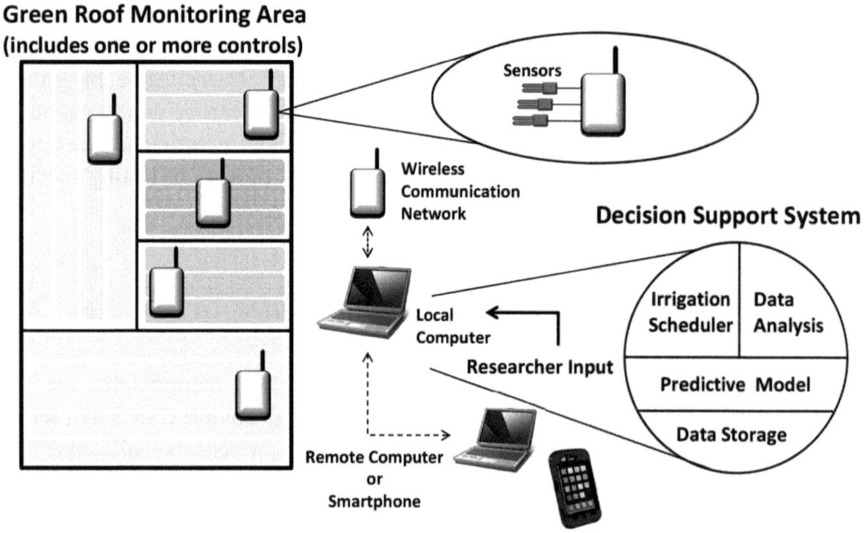

Fig. 2.2 A wireless sensor network system supports green roof monitoring and facilitates real time data collection and analysis. (Adapted with permission from Lea-Cox 2012)

the "green building" certification process, particularly since LEED™ (Leadership in Energy and Environmental Design) and other rating systems require pre- and post-implementation monitoring, and have the potential to improve design, implementation, and management while also deepening our collective knowledge about green roof dynamics and functioning.

2.3.3 Data Collection

Depending on the study, data collected on abiotic green roof conditions and processes may be documented directly by a researcher in a hand-written notebook, portable device, or computer—and/or the data may be wirelessly signal-fed to a data-logger/computer or physically transmitted via wires/cables linked to a data-logger/computer. A type of networked wireless sensor system can be developed and deployed to assist with data collection for green roof research (Fig. 2.2). Green roof monitoring activities should be dated and recorded as to their time and purpose in a logbook or recording device.

2.3.4 Data Analysis and Technical Expertise

Technical expertise needed for green roof monitoring and data analysis typically includes personnel familiar with scientific research methods and statistical analysis. Familiarity with rigorous, systematic data collection and analysis procedures and the ability to trouble-shoot equipment or device failures is vital. For example, in order to provide useful measurements, soil moisture sensors must be placed properly and calibrated appropriately to give accurate readings (Starry 2013). Some commercial organizations provide guidance specific to green roofs regarding weather station selection, logging capacity, configurations, setup, data download, and deployment options; sensor placement and positioning for different sensor types and purposes; and sensor cable protection, weather station grounding, battery maintenance, and sensor calibration (Onset Computer Corporation 2012).

2.3.5 Monitoring Precautions

In any monitoring scenario the unique nature of green roofs needs to be considered. Safety precautions include fall awareness and fall protection training for any person maintaining or otherwise walking on a green roof in situ (Omar et al. 2013). Additional care should also be taken not to disturb the green roof system when studying an actual green roof, especially the roof membrane or monitoring devices and equipment.

Table 2.1 Chapter 2 topics

Amount and frequency of precipitation and irrigation
Substrate moisture and interception
Water outflows (stormwater runoff)
Evapotranspiration (ET)
Water quality
Surface energy balance (including latent heat fluxes)
Temperature dynamics (surfaces and sub-surfaces)
Wind speed, direction, and dynamics
Gas exchange and carbon sequestration

2.4 Topics of Green Roof Monitoring

Different types of green roof monitoring and data collection approaches are used to quantify inputs, outputs, and the factors that control them. Abiotic data can be collected and evaluated on an established or newly implemented green roof (in situ) and/or on modules or microcosms located on top of a roof or constructed on platforms established as experimental prototypes on the ground or roof. In the remainder of this chapter, we discuss abiotic green roof monitoring related to the nine topics listed in Table 2.1 above.

2.5 Monitoring Hydrology

This section addresses: (1) monitoring of water inputs—amount, rate, and duration of precipitation and supplemental irrigation; and (2) monitoring of water outputs including quantity and quality of green roof runoff, and evapotranspiration rates. The water balance is important to consider in evaluating green roof hydrology and system performance. Represented schematically in Fig. 2.3, the green roof water balance is given as $P - ET - \Delta S = D$ where P is precipitation (and/or irrigation), ET is evapotranspiration, ΔS the change in storage, and D is drainage (outflow—assuming no surface runoff) all defined as volume flow rates.

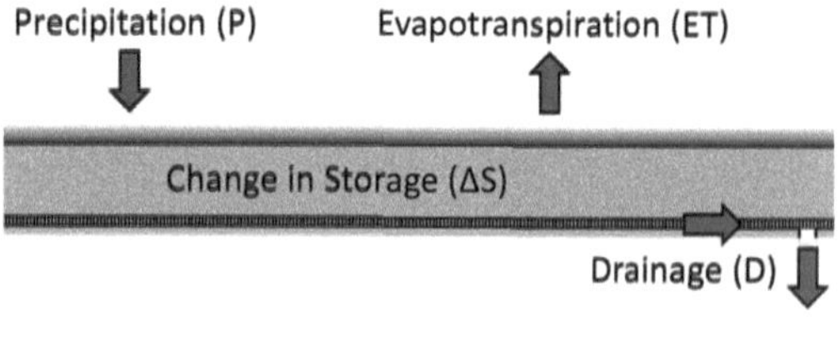

Fig. 2.3 Green roof water balance. Note: Water drainage is often referred to as runoff or outflow since it leaves the green roof ecosystem. (DiGiovanni 2013 with permission from ASCE)

2.5.1 Water Inputs: Precipitation and Irrigation

Water inputs to green roof systems include precipitation and irrigation (Chap. 4) and are important considerations for green roof performance related to the establishment and viability of live plants and seed (Nagase and Dunnett 2013; Skabelund et al. 2014). Geographically and temporally variable precipitation patterns and supplemental irrigation dictate the amount, rate, and duration of water inputs onto green roofs. Water inputs can be monitored through various means: tipping bucket rain gauges (heated or non-heated), non-recording (manually checked) rain gauges, plate gauges, weighing type rain gauges, radar rain data, and flow meters. For example, the ***amount and frequency of precipitation*** can be monitored using a manual rain gauge or by using a tipping bucket connected to a data-logger to record rainfall and snowmelt at some pre-selected interval (e.g. every 5–15 min). While a nearby weather station can be used to estimate onsite precipitation, spatial variability in weather patterns and structural differences (e.g. if the green roof falls in the lee of other buildings) warrant onsite measurement of precipitation.

Irrigation can be measured with flow meters, but this approach is challenging if more than one water delivery pipe is active. ***Irrigation needs and demands*** can be measured over time and an observational and/or experimental design employed to examine (as one example) water conservation practices—from rooftop water harvesting and re-use, drip and sub-surface irrigation, or completely eschewing irrigation (see Rowe et al. 2014 and Chap. 4).

Green roofs may be observed every few days, and when vegetation shows signs of wilting or browning, irrigation can be applied for a specified amount of time and/or quantity of water. In other cases, soil moisture sensors, linked to programmed irrigation systems, can provide supplemental water at specified rates.

Irrigation not only impacts green roof performance (biomass production, vegetative health, and summertime cooling potential), but it can strongly influence substrate moisture and stormwater drainage monitoring results (Getter and Rowe 2009; Fassman-Beck et al. 2013). For example, by taking up retention capacity and generating runoff, irrigation can become a disservice. Precipitation and irrigation must be accounted for when preparing green roof monitoring plans and analyzing collected data.

Observational research of the Upper Seaton Hall Green Roof (see Fig. 2.4) in Manhattan, Kansas (USA) focuses on monitoring vegetation, substrate temperatures, soil moisture, micro-meteorological conditions, and other factors—while seeking to understand the influences of irrigation or its lack, on an integrated living roof system in the central Great Plains. Along with plant growth and survival, precipitation, irrigation, stormwater runoff, and meteorological variables are monitored. The primary objective of the project is to determine how selected native grasses and forbs fare within varying substrate depths (7.5–17.5 cm), with and without supplemental irrigation (Skabelund et al. 2014). A Campbell Scientific data logger records air temperature, relative humidity, green roof surface and sub-surface temperatures, rainfall, and wind speed and direction every 5 min. Instruments installed in June

Fig. 2.4 **a** Kansas State University's Upper Seaton Hall Green Roof has been monitored since May 2009 using **b** a Campbell Scientific CR23X micro-logger (Lee R. Skabelund)

2009 on the green roof (Fig. 2.4) included a Campbell Scientific (CR23X micro-logger), BP solar panel (10 W 16.8 V); RM Young weather station; three surface temperature probes; six sub-surface temperature probes; and one Texas Electronics (TR-525I) tipping-bucket rain gauge. Hard-wired cables connect instruments and sensors to the data-logger. A manual, all-weather manual rain gauge (Productive Alternatives) was also placed at the south end of the green roof near the tipping bucket. In 2013, an additional 5TE Temperature/Soil Moisture Sensor (purchased from Decagon Devices) and a U23-002 Hobo Pro v2 Ext Temp/Relative Humidity Data Logger (Onset) were added to better understand stresses on vegetation following the complete elimination of irrigation in mid-August 2012.

2.5.2 Water Storage: Substrate Moisture and Interception

Water storage on a green roof is achieved in the substrate and drainage layers as well as on/in the green roof vegetation through canopy interception and internal plant storage. Monitoring water storage on a green roof provides important information regarding stormwater management benefits as well as water conservation/irrigation and plant survival.

Green roof stormwater retention can be quantified by monitoring water inputs and storage terms as well as by quantifying water inputs and outputs (as discussed in the following sections). On a per event basis, the stormwater retention of a green roof is dictated by the available water storage capacity, in the substrate, drainage layers, and vegetation (and also by intra-storm ET which restores water storage capacity during an event). Interception and intra-storm ET gained increased research focus and some researchers seek to quantify the interception capacity of green roof vegetation through laboratory techniques under simulated rainfall conditions (Fassman and Simcock 2012; Rostad et al. 2011).

Though various components of the green roof contribute to green roof water storage, the primary contribution is almost always substrate storage in the form of substrate moisture. Substrate storage on a green roof can be directly quantified through the use of volumetric water content (VWC) sensors.

Fig. 2.5 Decagon soil moisture sensor set to sense moisture top-to-bottom of 7–12 cm substrate profile (Olyssa Starry and Liz Ensz, unpublished)

The use of VWC sensors for quantification of substrate storage warrants various considerations regarding sensor selection, installation, calibration/validation, and substrate heterogeneity. A variety of VWC sensors with different capabilities are commercially available. Sensors that measure dielectric permittivity (the tendency of water to polarize in an electric field) to quantify VWC (e.g. Decagon Devices 5TE) are commonly used on green roof monitoring projects. Depending of the soil volume measured by sensors, some may be used for deeper substrate roofs and others more appropriately in shallower ones (John Buck 2014, pers. comm.).

Appropriate installation of VWC sensors is of particular importance to help avoid erroneous readings that can be attributable to faulty sensor placement (see Fig. 2.5 for the typical placement of a Decagon VWC sensor). VWC sensors are impacted by proximity to surface and drainage layers which can result in very wet or dry readings (John Buck 2014, pers. comm.), sensor spacing (e.g. recommended at 20 cm apart), and contact with substrate. An initial watering-in period of a few days after sensor placement makes readings more consistent by removing air pockets an embedded sensor (Griffin 2013). Substrate heterogeneity is also important and researchers attest that VWC readings are best taken in the field.

Soil moisture data helps with water balance calculations. Moisture sensors help detect field capacity, or excess water held after initial drainage. Outflow (and thus stormwater retention) from a green roof can be estimated by assuming that rain in excess of the field capacity of the green roof substrate flows out or off of the system. Researchers attest that it is best to take these measurements in the field using sensors since substrate disturbance can affect lab results.

The University of Maryland's green roof research program has used VWC sensors extensively in research evaluating critical green roof design factors related to stormwater runoff, substrate storage, and soil moisture by examining substrate composition and depths, vegetation species selection, vegetation metabolism (e.g. C3 vs. CAM), and regional environmental conditions especially rainfall frequency and intensity (John Lea-Cox 2014, pers. comm.). Maryland's monitoring and modeling work has primarily focused on: (1) quantifying substrate moisture content by sensing water flux between rainfall events; (2) modeling environmental fluxes using modified Penman-Monteith equations; and (3) assessing plant species effects, including differences in water use over time.

2.5.3 Water Outputs: Outflow Quantity

Green roof outflow (i.e., drainage and/or runoff) constitutes the liquid water leaving a green roof system during or following a storm event or irrigation. Outflow can be monitored directly by several techniques and used to understand green roof water retention in relation to water inputs (precipitation and supplemental irrigation) and outputs (evapotranspiration, through-flow or drainage and runoff). Some of the instruments utilized for the measurement of green roof outflow include in-line flow measurement devices and large capacity tipping bucket gauges, as well as flumes, weirs, and cisterns with accompanying water level or mass measurement devices. Sometimes, custom devices are utilized for monitoring green roof outflow, for example, the orifice restricted device (ORD) developed by Voyde et al. (2010).

For full-scale green roof systems, the total amount of green roof stormwater retention can be simply quantified through the water balance by subtracting the amount of outflow from the total estimated precipitation falling on the green roof. Roof-scale runoff or outflow is one of the most challenging aspects of green roof monitoring, so these systems need to be carefully designed and evaluated.

For example, the Portland, Oregon (USA) Hamilton Eco-Roof focuses on monitoring runoff and stormwater retention at the roof-scale. Two (7.62 and 12.7 cm depth), integrated eco-roofs (west 243 and east 234 m^2), each with a different substrate, were installed on the Hamilton apartment building in 1999 and monitored from 2001 to 2012. Fiberglass flumes (Fig. 2.6) measure outflow from each roof section while a rain gauge measures precipitation input (Kurtz et al. 2010, p. 18). A small, V-trapezoidal Plasti-Fab flume was installed adjacent to, and immediately upstream of, each primary roof drain. The primary roof drain is sealed and isolated to direct all flow through the flume prior to entering the drain. An American Sigma Model 950 bubbler-type flow meter is used to measure water level in each flume. Because of spillover and other physical challenges, "it is not unusual to measure more runoff coming from the east side than the total rainfall that fell on the roof. This makes the use of the east side data problematic…" (Kurtz et al. 2010, pp. 18–19). Nevertheless, this technique, when properly executed, yields valuable data and the flume setup has been verified to measure a range of flows accurately from inter-event outflow to medium and large events. Data from this roof has been used to improved policy and design guidance for new green roof construction and retrofits (Timothy Kurtz 2014, pers. comm.).

Fig. 2.6 Drain **a** and flume **b** setup on the Hamilton Ecoroof for stormwater outflow monitoring (City of Portland, BES)

Utilizing various techniques, the outflow and stormwater retention performance of green roofs is being evaluated by researchers worldwide. Of note, 66 percent of publications on green roofs are from USA and EU representing primarily the performance of green roofs in temperate environments (Blank et al. 2013). Extensive documentation of the stormwater retention performance of green roof systems supports that green roofs effectively reduce the volume of stormwater, attenuate peak flows and increase the time to peak in comparison to conventional roof surfaces (Charlesworth et al. 2013; Fioretti et al. 2010). Pioneering studies of green roof stormwater retention and runoff reduction in the US were completed by Berghage et al. (2009). Reviews of other early works are available in the literature e.g. Berndtsson (2010). Continued research in the field has yielded published works from various researchers including Carson et al. (2013), Fassman-Beck et al. (2013), Rosatto et al. (2013), Song et al. (2013), Palla et al. (2012), Stovin et al. (2012) and Voyde et al. (2010) (Chaps. 4 and 5).

2.5.4 *Water Outputs: Evapotranspiration (ET)*

Evapotranspiration (ET) describes the transfer of water to the atmosphere from the combined effect of evaporation from e.g. substrate and leaf surfaces and transpiration through vegetation. Monitoring ET is important to inform irrigation practices and because ET is linked to a variety of benefits that can be provided by green roofs. For example, ET restores the retention (water storage) capacity of a green roof and the ability to capture stormwater and also provides benefits linked to micro-meteorological regulation and carbon sequestration. ET achieved during dry days between storm events has the greatest influence on green roof stormwater retention (Voyde 2010). Intra-storm evapotranspiration (though often assumed negligible) can also be an important mechanism for stormwater retention particularly for low intensity long duration storm events (DiGiovanni et al. 2010).

Evapotranspiration can be monitored using weighing lysimeters (see Fig. 2.7). ET can also be modeled by capturing site specific wind, temperature, and other

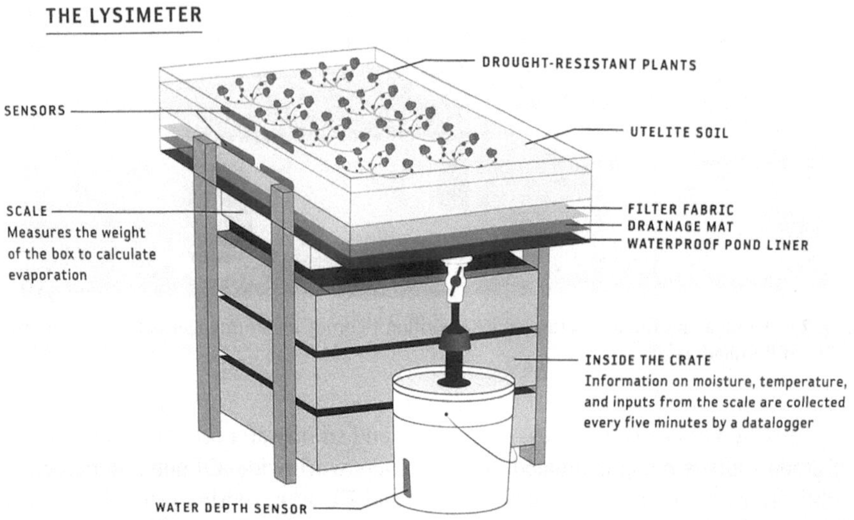

Fig. 2.7 Green roof lysimeter setup at the Utah Natural History Museum in Salt Lake City, Utah (Graphic by University of Utah & Natural History Museum of Utah)

micro-meteorological conditions—and then related to substrate and vegetation type. Simple or sophisticated analysis of ET patterns, dynamics, and changes through time are possible (refer to Chap. 3).

Additional to lysimeters, ET can be measured by a variety of methods (Jensen et al. 1990) including soil water depletion techniques and energy balance. Weighing lysimeters are widely considered the only directly quantitative means to measure ET (Tanner 1967; Jensen et al. 1990; Rana and Katerji 2000; Xu and Chen 2005). Other ET measurement techniques include eddy covariance and scintolometers. Nouri et al. (2012) reviews ET measurement techniques for urban landscape vegetation, including green roofs.

ET measurement is costly; therefore, estimates of ET are often used and come from various techniques including temperature-based, radiation-based, and combination-based equations. Widely recognized to yield the most accurate results, Penman-Monteith-based combination equations are widely applied, though other methods are recognized to be less data intensive. Data required to estimate ET varies depending on the method and can include solar radiation, temperature, relative humidity and wind speed monitored using pyranometers, temperature-and-relative humidity probes, and wind sensors respectively. Furthermore, the development of appropriate coefficients to adapt ET estimates from reference vegetation to green roof vegetation, from regionally available data sets to local conditions and to account for substrate moisture conditions are ongoing by various researchers (DiGiovanni et al. 2011; Schneider 2011; Sherrard and Jacobs 2012; Starry 2013; DiGiovanni et al. 2013; and DiGiovanni 2013).

Despite the importance of evaporative processes in managing stormwater and providing other valuable ecosystem services, comprehensive monitoring and measurement of ET from green roofs is rare. Reported studies measuring

Fig. 2.8 Green roof monitoring setup on two roofs in Salt Lake City, Utah: (**a**) Natural History Museum of Utah; (**b**) and (**c**) University of Utah J. Willard Marriott Library (Lee R. Skabelund)

evapotranspiration from green roofs are limited to a small body of literature including Berghage et al. (2009), Voyde et al. (2010), DiGiovanni et al. (2010), Feller et al. (2010), Sherrard and Jacobs (2012), DiGiovanni et al. (2013), DiGiovanni (2013), Starry (2013), Wadzuk et al. (2013), and Marasco et al. (2014).

University of Utah research is an example of a recent project employing innovative monitoring tools and techniques for ET monitoring on green roofs. Located in Salt Lake City, Utah (USA) researchers have been monitoring two in-situ green roofs since fall 2013 (Fig. 2.8). The Natural History Museum green roof covers 1115 m^2, and the Marriott Library roof 632 m^2. One of the major purposes of the study is to monitor ET, so researchers set up four weighing-lysimeter systems. ET data collected by the lysimeters will be compared to ET measurements made by a Campbell eddy covariance system. Eddy covariance is one of the most widely accepted micro-meteorological methods to directly measure fluxes such as water vapor and is based on the covariance between wind speed and humidity measured separately but simultaneously at high frequency. Researchers expect accurate green roof ET time series from a point scale, not easily achieved by most other means. Accounting for winds that generate errors for scale readings requires considerable time to calibrate sensors and validate results. Researchers recognize instrumentation limitations (e.g. a tipping bucket will miss some measurements and it is hard to apply depth sensors on green roofs to provide continually reliable outflow data). The same is true for irrigation monitoring. Utah's arid climate requires irrigation, yet researchers have not found good tools to measure the flows from irrigation tubes or sprinklers. (Burian and Feng 2014, pers. comm.). Estimating the amount of water applied is thus necessary based on designed and implemented water pressure and flow rates.

2.5.5 *Water Output: Outflow Quality*

The quality of water outflow from a green roof system is an important consideration of green roof performance. Green roofs can improve water quality through filtration and adsorption in the substrate (Wang et al. 2013), plant uptake of nutrients, and microbial action, though these processes are not well studied (Dietz 2007). Furthermore, the export of nutrients and other constituents based on rainfall intensity (Teemusk and Mander 2007) and fertilization (De Cuyper et al. 2005) can cause concentrations in green roof outflow to exceed standards and/or objectives set for receiving water bodies (Van Seters et al. 2009).

Water quality can be measured in relation to precipitation and rooftop runoff by taking one or more samples from a single integrated green roof, control roof, or from a series of experimental green roof modules set on platforms. Understanding the base nutrient conditions of the substrate is critical (Chap. 5).

Water quality samples from green roofs can be collected by grab sampling or with automated samplers. Care needs to be taken to intentionally sample different regions of the runoff hydrograph. Subsequent laboratory analyses can be performed for determining the concentration or presence of various constituents including nutrients and metals. Basic water chemistry parameters can also be monitored either in the laboratory or through data collection in the field. To quantify parameters such as conductivity, temperature, dissolved oxygen, pH and turbidity, a variety of probes, sondes, or other measuring devices can be utilized to collect discreet or time-series data sets. Some studies have evaluated the promising use of turbidity as a proxy for TSS concentration in green roof outflow (Al-Yaseri 2013).

Reviews of water quality studies from full-scale and laboratory green roofs including factors that impact green roof performance are presented in the literature, e.g. Berndtsson (2010) and various other studies have been reported (see for example Berghage et al. 2007; Teemusk and Mander 2007; Dunnett et al. 2008; Simmons et al. 2008; Berghage et al. 2009; Bliss et al. 2009; Van Seters et al. 2009; Gregoire and Clausen 2011; Schroll et al. 2011; Morgan et al. 2012; Toland et al. 2012; Alsup 2013; Clark and Zheng 2013; Gnecco 2013; Seidl et al. 2013; Wang et al. 2013 and Zapater-Pereyra 2013) (Chap. 5). A USEPA funded report (Culligan et al. 2014) discusses an analysis of water quality and quantity benefits for selected New York City green roofs, noting relevant ET and soil moisture monitoring and research needs.

2.6 Monitoring Energy Flows and Temperatures

Below, we address green roof surface energy balance, temperature dynamics (surface and subsurface), and thermal fluxes to and from buildings with green roof systems. We discusses approaches to monitoring (e.g. surface temperatures on a green roof system in comparison to conventional roof surfaces) and also address ways to monitor or otherwise understand the energy flows associated with a specific green roof ecosystem. As well, we discuss monitoring the relationships between green roof energy flows and specific types of green roof plant systems.

2.6.1 Surface Energy Balance

The surface energy balance of a green roof generally differs from that of conventional roofs. In comparison, green roofs can provide energy benefits to individual

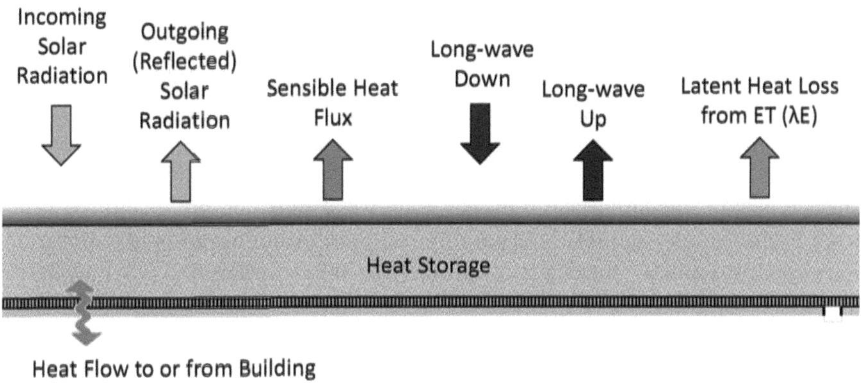

Fig. 2.9 A green roof energy budget consists of various measurable components. (DiGiovanni adapted from Gaffin et al. 2011)

buildings (as further discussed in Sect. 2.6.3) and impact ambient conditions. Energy benefits can be associated with green roofs by passive cooling through evaporative processes and latent heat fluxes as well as reflection of solar radiation (characterized by albedo) and reduction of sensible heat fluxes. With widespread adoption of green roofs, mitigation of the urban heat island (UHI) effect can be achieved reducing, "peak energy demand, air conditioning costs, air pollution and heat-related illness and mortality" (PlaNYC 2008).

The surface energy balance is represented as: $R_n = \lambda E + H + G$ where R_n is net radiation, H is the heat flux to the air also known as sensible heat flux, λ is the latent heat of vaporization, E is the rate of vaporization (evapotranspiration), and G is the heat flux to the soil also known as soil heat flux (Hanks 1992). Monitoring surface energy balance components (see Fig. 2.9) can be achieved through the use of various sensor technologies and estimation techniques.

Net radiation includes incoming and outgoing long-wave and short-wave radiation. Incoming solar radiation (often referred to as short-wave radiation) and outgoing (reflected) solar radiation are measured using pyranometers. Incoming and outgoing long-wave and/or infrared radiation are monitored using pyrgeometers. Data from monitoring surface and air temperatures as well as wind speed can be used to estimate sensible heat flux. Surface temperature can be monitored using direct contact thermocouple sensors and infrared radiometers. Air temperature can be measured using probes coupled with an appropriate solar radiation shield. Latent heat fluxes associated with evapotranspiration (Sect. 2.5.4, Chap. 3) can be monitored by various techniques or by backing out the term in the energy balance if all other parameters are known. Studies evaluating energy balance and latent heat fluxes related to green roofs include Jim and He (2010), Susca et al. (2011), Coutts et al. (2013), Kim and Park (2013), Nagengast (2013), Peng and Jim (2013) and Song et al. (2013).

One example of energy flux studies is that of Columbia University and City College of New York with ongoing studies through Drexel University. Researchers have deployed a high quality sensor network on a diverse array of green roofs in NYC. The sensors enable quantification of various components of the energy balance. Five monitored green roofs, presented in Gaffin et al. (2009), vary in their layout, materials, and structure. Monitoring stations use identical sensor selections paired with a monitored "control" roof. Back-to-back pyranometers (Kipp and Zonen CMP3) measure incoming and outgoing short-wave radiation and determine surface albedo. Incoming and outgoing long-wave radiation are quantified using surface temperature and relative humidity inputs, with surface infrared radiometers (Apogee Instruments SI-111, previously IRR-P) and temperature/relative humidity probes (Campbell Scientific CS215). The surface infrared radiometers are, "particularly useful for monitoring green roof leaf temperatures which have a complex geometry" (Gaffin et al. 2009, p. 2655). Gaffin et al. (2009, p. 2654) also noted that, "sensors are increasingly becoming available to measure all four SW and LW fluxes," including net radiometers like Kipp and Zonen CNR2 and Hukesflux NR01 with back-to-back pyranometers and pyrgeometers. The sensible heat calculation requires air temperature and wind speed. These are monitored with temperature/relative humidity probes (CS215) and wind sensors (RM Young 05013). Latent heat fluxes due to evapotranspiration (ET) are quantified using inputs of wind speed, temperature and relative humidity and by weighing lysimeter. Figure 2.10 shows one NYC green roof monitoring setup.

Fig. 2.10 Meteorological and hydrological monitoring at the Ethical Culture Fieldston School green roof (Kimberly DiGiovanni)

2.6.2 Temperature Dynamics

Green roof temperatures are related to cooling, reduction of the urban heat island (UHI) effect and building energy savings. ***Temperature dynamics and changes*** can be measured using surface and sub-surface temperature sensors. It is helpful to monitor different kinds of conventional or control roofs as each "conventional" roof type will likely perform differently in some respects—with the same being true for different green roof substrate types and depths including in different climatic conditions. For example, on the 6875 m^2 Walmart Green Roof in Chicago, Illinois (USA) the energy impact of an extensive (approximately 7.5 cm) green roof was analyzed and compared with an adjacent white roof based on 2006–2009 monitoring. The following parameters were measured at points distributed across the two rooftop types: (1) surface temperatures; (2) temperatures under the roof membrane; (3) temperatures below the roof deck; and (4) temperatures in the substrate profile for the green roof only. Heat flux (Q, in watts per square meter)—a measure of energy flowing in or out of the store through the roof—was also monitored. HVAC air intake temperatures were measured from July 2009 to July 2010 (Walmart et al. 2013). "To analyze energy impact of the green roof, the heat flux data collected from the roof was integrated into a simplified building model [then] into the full store energy model"—helping researchers "interpret field data [and] allot heat flux differences properly." The model translated temperature difference into "energy use difference" by the rooftop and air handling units by "modeling air temperature difference on the green side as precooling or preheating…" (Walmart et al. 2013, p. 21). Models were run for the Chicago store, then data were extrapolated to a model in Houston, Texas to gauge likely green roof energy impacts. Average annual conditions were studied as well as peak heating and cooling periods to determine the green roof's effect on store energy use (Walmart et al. 2013).

2.6.2.1 Surface Temperatures: Green Roofs and Control Roofs

Surface temperature can be monitored using direct contact thermocouple sensors and infrared radiometers. Researchers in diverse geographic regions have successfully used those techniques for measuring green roof surface temperature in comparison to conventional roof surfaces. Sidwell et al. (2008) evaluated a southern Illinois green roof and a black roofing membrane (ethylene propylene dieneterpolymer or EPDM) control using temperature sensors; the green roof fluctuated between 23.6 and 29.8 °C; the EPDM control roof 19.1 and 46.3 °C. Monitoring by Dvorak and Volder (2013) in central Texas (USA) found an un-irrigated modular *Sedum* green roof was 18.0 °C cooler at the surface and 27.5 °C cooler below the modules in comparison to a control roof during summer months. DeNardo et al. (2005) in Pennsylvania (USA) found an 8.9 cm green roof substrate surface to be 6 °C warmer in winter months and 19 °C cooler in summer months compared to a control. Wong et al. (2003) in Singapore found intensive rooftop garden temperatures 30 °C cooler than a control roof, and in Japan, Onmura et al. (2001) found temperatures to be 30–60 °C cooler on a green roof than a nearby control roof. Other researchers (including Blanusa et al. 2013) have considered the cooling effect of different types of green roof vegetation.

2.6.2.2 Sub-Surface Temperatures: Substrate

It is important to understand how warm or cold it gets beneath the substrate surface since green roofs can insulate buildings. In combination with substrate moisture levels substrate temperatures also strongly influence evapotranspiration rates, vegetative health, and microbe and invertebrate diversity (Chap. 7, Chap. 14).

Temperature profiles can be measured for different roof surface types using thermocouples. Pearlmutter and Rosenfeld (2008) applied thermocouples to different locations on small building "cell" replicates to compare various methods of roof-cooling including from mesh shading, soil, and gravel. A heat flux plate was also placed under the soil and simultaneous measurements made of global radiation (using a Kipp and Zonnen pyranometer), wind speed (via a LSI constant temperature hot-wire anemometer) and ambient air temperature (using a Campbell 21x datalogger). They found that though roof shading material provided more overall daily cooling, gravel-covered roofs optimize daily cooling potential which is important for the desert climate in which their work was conducted.

Dvorak and Volder (2013, p. 30) placed thermistors at multiple depths in green roof modules to compare the effects of irrigation on cooling. "Ambient air temperatures were collected on the rooftop with non-forced ventilation shielded air temperature instrumentation (Humidity and Temperature Probe HMP155 and 41005-5 radiation shield)." Substantial temperature reduction in unirrigated modules was noted (compared to standard roof surfaces).

Green roof cooling in relation to plants is the focus of several studies. Most use similar thermistor technology to measure temperature but different monitoring advice can be gleaned from each study. Jim (2012) noted for their study on plant effects in the tropics, for example, concerns arose about the effect of advection on adjacent plots and they recommend larger study plots with sensors placed in the middle of the plots. They found that grass plots cooled more effectively than groundcover or shrub. Butler and Orians (2011) used a Maxim iButton high capacity temperature logger DS1922L and found temperature regulation might be one mechanism that makes *Sedum* a nurse plant.

2.6.3 Building Thermal Fluxes (Insulating Properties)

Green roofs impact the heat gain and loss to and from buildings and influence the heating and cooling loads. Green roof substrate provides insulation and vegetation reflects solar radiation more effectively than most conventional roof surfaces preventing solar heat gain and increasing the thermal resistance of the building (Eumorfopoulou and Aravantinos 1998). Monitoring of thermal fluxes in and out of buildings can be achieved using a variety of sensors including thermocouples, thermistors, temperature probes (inside and ambient air temperature), and heat flux plates. Heating and cooling impacts of green roofs can be evaluated through energy usage data, typically metered through power-supply companies. Studies evaluating the thermal gain and thermal resistance of buildings with green roofs include Pierre (2010), Fioretti (2010), Becker and Wang (2011), Zhao and Srebric (2012), Chan and Chow (2013), Darkwa (2013), Moody and Sailor (2013) and Olivieri (2013).

2.7 Monitoring Wind Speed and Direction on Green Roofs

Wind speed and dynamics can influence the stability of the entire green roof system. Wind also has a major influence on the movement and drying (through ET) of substrates and the viability of green roof vegetation. Thus, green roof monitoring should document wind speed and direction by using wind sensors—in tandem with monitoring other relevant micro-meteorological, hydrologic, and substrate variables. Monitoring the impact of ***wind dynamics*** on vegetation and/or green roof system movement, dislodgement, breakage, and overall ***green roof stability*** can be accomplished by using a wind tunnel for modules, or cameras and observations for an entire green roof system. ***Wind scour or loss of substrate*** can be measured using devices placed in the substrate and observed over time and/or using one or more high-resolution video cameras to record movement of particles during windy periods. Simple or highly sophisticated wind scour modeling and analysis are possible (Laminack et al. 2014).

2.7.1 Stability of Substrate and the Entire Green Roof System

Zhang et al. (2007) indicate that green roof "soil erosion" induced by winds decreases with higher levels of plant cover. Roots bind plant masses to the substrate, thus providing a windbreak from erosive forces. To observe such phenomena, University of Central Florida researchers implemented two, full-scale green roofs to continuously monitor wind effects, using "a grid of very low differential pressure transducers and a high speed anemometer for wind speed and direction." A geosynthetic erosion control blanket was used on one roof, significantly reducing substrate loss (Wanielista et al. 2011, p. v). Field data from several monitoring stations with high wind velocities may better define design parameters for all green roof-building options. Cao et al. (2013) explored wind load characteristics for green roof modules. A series of wind tunnel experiments were carried out on a scaled-down module installed in different positions on two types of building models. They investigated peak force and moment coefficients of the model rooftop and the effects of parapets and other design parameters Retzlaff et al. (2010) employed a subsonic, recirculating wind tunnel to evaluate wind uplift and wind scour of partially and non-vegetated modular green roof systems.

2.7.2 Built Context Influences Wind Patterns and Dynamics

Wind and wind variability are strongly influenced by building mass, height and parapets thus influencing green roof systems. For example, visual observations of wind and snowfall on two Kansas State University green roofs indicate patterns of wind and precipitation respond to building mass, location, and height (Skabelund et al. 2014). Use of an anemometer to monitor wind speed and direction provides useful information on the dynamics associated with wind strength, direction, and patterns related to the urban context.

2.8 Monitoring Substrate Attributes

Without understanding the specific attributes of green roof substrate characteristics it is unlikely that we can create green roof ecosystems that are resilient and also minimize resource demands, especially supplemental irrigation and nutrients (Beattie and Berghage 2004). Below, we address ways to monitor substrate pH, nutrients, organic matter, metals, and other constituents that are seen as vital to plant systems (but potentially detrimental to downstream aquatic systems). We also examine ways to effectively assess changes in soil attributes over time, noting that once a green roof is installed, substrate properties can be sampled to inform maintenance decisions like fertilization frequency. Monitoring substrate attributes can inform balanced maintenance and management decisions (e.g. fertilization to secure system survival and health targeted to reduce nutrient and metal loads in green roof outflow).

Substrate chemical parameters and organic matter content vary greatly across time and space. Nutrient and pH studies are time and resource intensive, so it is important that they incorporate additional roof metadata so that findings can be generalized. The green roof research community also needs to agree on appropriate reporting units. For example, organic matter content is sometimes reported as volume per unit substrate and other times reported as mass per unit substrate. Despite these challenges, a few model studies are emerging. For example, one study in Germany using data spanning at least 20 years, and in some cases much longer, found that though substrate porosity of modern extensive roofs increases over time, the C/N ratio declined (Köhler and Poll 2010).

Zheng and Clark (2013) evaluated five different *Sedum* species under variable substrate pH conditions. By identifying species-specific characteristics and optimizing substrate pH, Zheng and Clark suggested that *Sedum* growth can be optimized. Panayiotis et al. (2003) studied four substrates for their capacity to sustain *Lantana camara L.* Physical and chemical evaluation included "weight determination at saturation and at field capacity, bulk density determination, water retention, air filled porosity at 40 cm, pH and electrical conductivity." Plant growth evaluated "shoot length, shoot number, main shoot diameter and the number of buds and flowers" (Panayiotis et al. 2003, p. 619). Studies of pH optimization and conductivity should be considered in regards to native plants and mixed vegetative systems on green roofs in North America.

2.9 Monitoring Gas Exchange and Carbon Sequestration

Carbon sequestration occurs on a green roof as photosynthesis creates biomass, especially root mass. By sequestering carbon, green roofs help to mitigate climate change. ***Carbon fluxes via gas exchange***, and ***carbon sequestration*** by changes in CO_2 through photosynthesis and respiration can be measured directly as atmospheric CO_2 exchange, or indirectly as changes in C stocks over time (i.e., collecting, drying and weighing substrate and root samples). Modeling and analysis of atmospheric carbon fluxes, net ecosystem productivity, and carbon sequestration through time is possible and sophisticated equipment exists to do so (Chap. 5). As noted below, several efforts are being made to construct carbon budgets for green roofs.

Foundational work conducted in Michigan documented changes in plant biomass and associated carbon content over time in order to assess green roof carbon sequestration. It is important to note that this approach ignored carbon losses from the system via respiration and leaching. Getter et al. (2009) monitored 13 green roofs (nine in Michigan and four in Maryland). For twelve roofs, plant material and substrate were harvested seven times across two growing seasons. Roofs ranged from 1 to 6 years in age and from 2.5 to 12.7 cm in substrate depth. Replicate samples of aboveground biomass were collected, dried, and ground. Carbon accumulation was determined by multiplying dry matter weight by total C concentration. This study documented an accumulation of carbon (above ground and below ground) of 377 g C m^{-2} over a two year period (Getter et al. 2009). This carbon data was used to support ecological observations about the different *Sedum* and compare carbon sequestration with carbon flows to and from the green roof. Whittinghill et al. (2014) took a similar approach in their study of different landscape areas (including green roofs). Carbon content analysis was performed on above-ground biomass, below-ground biomass (roots), and soil and substrate collected at the end of the 2010 and 2011 growing seasons (Whittinghill et al. 2014). Researchers in Vancouver are updating calculations to incorporate respiration using experimental chambers "Li-Cor LI-8100" (Gaumont-Gauy and Halsall 2013). In their 2012 study, five chambers, 314 cm^2 in area, measured CO_2 fluxes at the roof-atmosphere interface for five different *Sedum* species. One additional chamber measured CO_2 flux from an unplanted surface to assess respiration. The net ecosystem productivity (NEP) was determined as the balance between gross CO_2 assimilation through plant photosynthesis and CO_2 release through plant and decomposer respiration. Gaumont-Gauy and Halsall (2013) found that net C assimilation integrated across plant types was 440 g m^{-2} yr^{-1}. A range of uptakes were observed for different species whereby endemic species native to the region exhibited higher net carbon fixation compared to others; these findings are supported by Starry et al. 2014 who also noted a range in uptake for different *Sedum* species. Gaumont-Gauy and Halsall (2013) further note that their study does not include carbon lost from the system via leaching; future work may involve a more integrated study of all the different green roof carbon pathways (Chap. 5).

2.10 Synthesis of Green Roof Monitoring: Approaches, Costs, Challenges, and Lessons Learned

2.10.1 Green Roof Monitoring Approaches

Green roof monitoring ranges from simple to complex, and data extensive to data intensive. Common approaches and tools include observation, experimentation, computer modeling, and in situ sensors. For measurements of water quality and substrate attributes, samples require additional lab support. The importance of integrated green roof monitoring is highlighted in the following section.

2.10.2 Integrated Green Roof Monitoring

Integrated green roof monitoring brings together observation, experimentation, and data collection in a manner that enables researchers to understand complex interrelationships over an extended period of time. Early work by Carter and Rasmussen (2006) in Athens, Georgia (USA), Glass and Johnson (2009) in Washington, DC, Berghage et al. (2009) in Pennsylvania, and other researchers set the stage for more in-depth and integrated green roof monitoring now occurring in many locations in North America. Recent examination of old extensive green roofs in Germany by Thuring and Dunnett (2014) provides insights that can help guide integrated monitoring efforts, especially for roofs employing shallow mineral substrates. Following are five brief examples of integrated green roof monitoring in the United States and Canada:

University of Pittsburgh Researchers at University of Pittsburgh integrated monitoring and evaluated green roofs compared to conventional roof tops focusing on various interrelated areas including stormwater management, water quality, and thermal benefits. They utilized a suite of sensor systems including flumes, weir boxes, ultrasonic sensors, soil moisture sensors, rain gauges, thermocouples, temperature probes, net radiometers, laboratory water quality analyses, and data loggers connected by modem and electronic networks (Neufeld et al. 2009).

EPA Region 8 Green Roof The first large-scale extensive green roof in Colorado was created atop the ninth floor of the EPA building in Denver. Covered with *Sedum* species, cacti, and grasses, this 1858 m^2 roof is near a gravel ballast control roof. Both roofs have: (1) weather stations to measure temperature differences; (2) instruments to monitor stormwater runoff rates and quantities; and (3) water collectors for water quality analysis. In 2008–2009, Klett et al. (2012) evaluated green roof vegetation (biomass) in relation to different substrates, zeolite amendments, and irrigation regimes. They used digital image analysis (employing SigmaScan Pro 5.0 image analysis software) and manually collected two-dimensional data. To assess water-holding capacity of substrates, they collected volumetric moisture content data using a Delta-T ThetaProbe ML2X. Their analytical methods included multivariate analysis.

Vancouver Island University VIU researchers employed a sophisticated green roof monitoring design strategy for evaluating gas/vapor exchanges, vegetation, meteorological conditions, water and energy fluxes, and water quality from green roofs. Their integrated monitoring design strategy, initiated in January 2012, includes automated and portable CO_2 and H_2O exchange chambers, digital cameras, weather stations (precipitation, radiation, temperature, relative humidity), water level sensors, soil heat flux sensors, and sensors to monitor dissolved organic carbon in green roof runoff (CDOM/FDOM sensor). Data logging/acquisition systems (Campbell Scientific) and computational software (Matlab) are integral to their green roof monitoring (Gaumont-Guay 2014, pers. comm.).

Portland State University Researchers at the Green Building Research Lab are pursuing several monitoring objectives. Projects include "very simple monitoring setups (a weather station, soil moisture, soil temperature)" and also more complex systems "involving those same sensors as well as surface heat flux sensors, net radiometers, arrays of air temperature rakes, and HVAC monitoring" (Sailor 2014, pers. comm.). One of PSU's projects addressed reciprocal effects of solar panels and green roofs and for other integrated monitoring, data loggers interfaced with indoor environmental quality sensors and outdoor weather sensors to monitor air temperature, CO_2, occupancy, relative humidity, and equipment run time. PSU researchers currently monitor stormwater and heat loads associated with a 3716 m^2 green roof (Williams 2013).

University of Toronto Toronto's Green Roof Innovation Testing Laboratory (GRIT Lab) uses "real time data monitoring and ongoing field observation to study the metrics associated with [green roof] systems" (ASLA 2013). The 372 m^2 GRIT Lab section of roof is dedicated to conducting experimental research—with 33 (1.22 x 2.44 m) modules. Each module is instrumented with eight sensors—one soil moisture sensor, a rain gauge to measure runoff and flow rates from each module, and five thermal sensors along a vertical axis to generate a thermal profile. One infrared radiometer records the average surface temperature of a 0.914 m diameter circle. At least 12 researchers are involved in this green roof research project—intended to be holistic and integrated by evaluating interrelated processes, including meteorological conditions, heat and energy flows, gas exchanges, water quantity and quality, soils, vegetation, and fauna (ASLA 2013). MacIvor (2014 pers. comm.), noted that the GRIT Lab "green roof has been online since late 2010 but has only been fully instrumented (full array of stormwater and thermal sensors, data loggers and dedicated computer with macros scripts to recall and subset data from all sensors), calibrated, and fully automated since June 2013." Early monitoring efforts at this site focused on irrigation and plant success (MacIvor et al. 2013), but ongoing research is targeting a number of different questions, especially those related to pollination. Troubleshooting, calibration, and modification of monitoring sensors and equipment was seen as vital to the process (Hill et al. 2015).

2.10.3 Green Roof Monitoring Costs and Funding Sources

The cost of monitoring green roofs ranges widely. Basic observational monitoring can be conducted for little cost while complex monitoring operations require hundreds of thousands in equipment and personnel. Specific equipment costs are available from the manufacturers, but also range within the same monitoring parameters depending on the sensor specification, range, and accuracy.

At Michigan State University (MSU), Bradley Rowe (2014, pers. comm.) noted that Campbell Scientific data loggers were the most cost intensive equipment purchased for their green roof monitoring work. The MSU Plant and Soil Sciences green roof was instrumented with heat flux sensors, moisture sensors, thermocouples, and a weather station, costing approximately $10,000 (see Getter et al. 2011). The monitoring system was pieced together using either existing or purchased equipment. Most MSU funding for green roof monitoring came from green roof suppliers, the USEPA, and internal university grants—with the largest grants from Ford Motor Co. and numerous smaller grants from companies such as LiveRoof and XeroFlor America. As is the case with other monitoring projects, many donated green roof materials were contributed.

Retzlaff (2014, pers. comm.) described five green roofs monitored on five different campus buildings at Southern Illinois Edwardsville University (SIEU)—the largest 1486 m^2 and the smallest 28 m^2. SIEU also has four (4) green roof projects monitoring stormwater runoff and a green wall test area. SIEU has used Hobo data loggers and simple soil thermal devices to monitor temperatures of green roof systems. The loggers cost approximately $300 each and the thermal probes $65. For measuring stormwater runoff, SIEU has used inexpensive five-gallon plastic gas cans that they weigh to track runoff from each storm event. SIEU researchers kept costs down by obtaining meteorological data from a local reporting station. They received permission to use an available wind tunnel (costing others more than $275,000 to install) for their green roof wind research. Between 2004 and 2014 SIEU received approximately $100,000 in external funds for their green roof and green wall research projects. The largest grants were an EPA P3 award and direct funding from the National Roofing Contractors Association.

At the University of Maryland, Lea-Cox (2014, pers. comm.) noted that "research instrumentation costs are considerably higher than what would ultimately be installed on commercial green roofs." Monitoring costs inevitably depend on the size of the green roof, the complexity of the research, and available equipment.

In short, complex monitoring systems are costly. Most monitoring systems described in this chapter cost between $5000 and $20,000 (including in-kind loans or donations). They may also require the expertise of specialists or consultants to maintain them and troubleshoot any site-specific challenges. More research is necessary to quantify the benefits of these monitoring systems relative to costs.

2.10.4 Green Roof Monitoring Challenges and Lessons Learned

Monitoring green roofs can present a variety of challenges including collection of representative data sets, instrumentation and maintenance of monitoring systems, as well as management and interpretation of collected data. Researchers conducting green roof monitoring have experienced many challenges and offer lessons learned from these experiences in this section.

Researchers recognize that collecting data that is representative of the overall green roof system is a goal of green roof monitoring that can be a challenge to meet. For example, modules and smaller platforms may be quite constraining in regards to the growth demands or requirements of some grasses and herbaceous vegetation. This is fine if the goal is to test selected species growth and viability in these constrained systems and compare them with integrated (monolithic) green roof systems but does not necessarily reflect the functional characteristics of the larger, integrated systems. Furthermore, adjacent walls or structures can have a significant influence on wind patterns and thus rainfall events—concentrating more precipitation (rainfall or snowfall) on one portion of a green roof and reducing or eliminating rainfall/snowfall on another portion of the same roof.

Beyond creating monitoring systems to allow for representative data collection, there are limitations with monitoring instrumentation. For example monitoring of hydrologic inputs and outputs from green roofs can present potential challenges and limitations including the following: (1) Flumes and gauges may not capture low flows and are susceptible to debris blockages (2) Flow meters can also be blocked and disabled by particles and magnetic flow meters require full pipe flow for operation (3) Tipping buckets cannot capture precise precipitation rates that are very small or very large/rapid and are generally only good for small areas as they can be overwhelmed by large amounts of flow. (4) Cisterns, rain-barrels, and buckets collecting runoff may overtop in larger storm events making accurate runoff measurements impossible (Rowe 2014, pers. comm.).

Other researchers have also experienced equipment related limitations to green roof monitoring. For example, when asked about the pros and cons related to monitoring hydrology from 12 mock-up green roof panels and three mock-up control roof panels (1.524 × 1.524 m) in Fayetteville, Arkansas, Mark Boyer (Univ. of Arkansas) stated: "For me it was the tipping bucket capacity. I really wanted to be able to compare the lag time of runoff off of a green roof compared to a conventional membrane roof and I thought the tipping buckets could do that for me. We had attempted using weirs on the first green roof that installed, but there were problems associated with that, so I was hopeful that the tipping buckets would solve the problem. We tried using tipping buckets to measure the quantity and timing of runoff but their capacity was exceeded and so we had to resort to capturing all of the runoff and omitting the timing effect" (Apr. 2014, pers. comm.).

Data collection and data management present another monitoring challenge. Collecting soil moisture and other data with a data logger is helpful, but setting everything up and getting all equipment working the way is supposed to work can be time consuming and very challenging (Rowe 2014, pers. comm.). Downloading recorded data (especially for data recorded every 5–15 min) can also be quite time consuming. Some devices automatically save data with file names indicating the date and time data was downloaded. Correlation with daylight savings times may be needed for some devices. Some data may need to be collected using a USB or other direct cable connection. Ethernet or wireless connections may be able to speed this process up and costs may be minimal if wireless or Ethernet connections already exist. Otherwise, costs will increase. Linking data collection devices to the Internet can be very helpful and save time if done well (enabling ready access to multiple users and allowing for sharing of results from anywhere that a researcher can access the Internet). Quick and ready assessment of data is possible via networked monitoring and analytical equipment, but requires well-trained and funded personnel.

Interpretation and analysis of monitoring data from green roofs presents further challenges. Careful analysis and interpretation of monitoring results is required prevent conclusions that are incomplete, problematic, and misleading. For example, Berghage et al. (2009) note the importance of relating concentrations of green roof outflow to total volume. "Although the runoff concentrations (from the green roofs) were typically higher, the loading was not always higher" (pp. 4–16). Furthermore, Berghage et al. (2009) note that interpretation from green roof studies must recognize that green roofs are dynamic systems with living properties that impact the system outputs. They found that outflow from unplanted substrate sections was considerably higher in both concentration of tested water quality parameters and in total volume of outflow than planted green roofs, suggesting "that newly planted roofs are likely to have much higher runoff loading rates than established roofs" (pp. 4–17). The study also demonstrated seasonal variation in runoff for some (but not all) runoff constituents monitored, and this may be attributed to seasonal fluctuation in plant uptake.

Researchers also recognize that monitoring needs are tied to regional and site-specific conditions related to the location, design, and size of the installation. Furthermore, establishing and maintaining a green roof monitoring network requires sustained funding and appropriately trained personnel for the maintenance and upkeep of monitoring equipment and acquisition of quality data sets.

2.11 Future Directions for Green Roof Monitoring

The future of green roof monitoring holds many opportunities, particularly at the intersections relating hydrologic processes, evapotranspiration, energy transfer, vegetation, nutrient cycling and other services provided by green roofs. Integrated studies considering holistic and multi-faceted approaches to evaluating green roofs are increasingly needed. Such integrated studies reveal findings valuable for understanding various interrelated processes and concomitant benefits.

It is important to note that monitoring (including experimentation) is sometimes an afterthought in regards to green roof research. To improve monitoring outcomes, researchers need to collaborate with practitioners as part of the green roof planning and design process to create "designed experiments" (Felson et al. 2014; Sutton 2013) to address various research needs. Such monitoring can help us choose better materials for green roofs in the years to come (Friedrich 2005).

One step that would help to unite green roof researchers collecting monitoring data is a platform that would facilitate the comparison of national green roof datasets. Some national databases already exist, but these lack an option to search for monitoring data. Table 2.2 illustrates how such a database might be set up to include information relevant to researchers.

Various topics in need of research attention exist beyond those mentioned previously within the context of integrated green roof research. Given continued reliance on succulents, comparisons between *Sedum*-dominated, systems exclusively supporting native grasses, forbs and other indigenous species, and mixed vegetative systems need to be monitored in relation to long-term stormwater runoff and water quality trends. Furthermore, studies are needed to address the impacts of different types of green roofs on air quality, a topic of research that has received little focus as of yet.

Many new types of sensors, including fiber optic cables, provide opportunities to improve measurements of soil moisture. Many other possibilities exist or will soon present themselves and green roof researchers need to remain alert to the costs and benefits of emerging sensing technologies, tools, and communication devices.

The development of monitoring networks incorporating automated sensor/system technologies real-time, remote sensing networks and data management systems with low-cost sensor technology will aid in advancing green roof monitoring initiatives. Mooney-Bullock et al. (2012) provide an example of a low-cost sensor network using new technology to monitor green infrastructure including green roofs, revealing how real-time monitoring can be implemented in an affordable manner.

Table 2.2 Green roof database categories

Roof location	Size, slope, aspect, substrate depth, etc.	Hydrology-related data	Energy-related data	Microclimatic data	Biodiversity studies
Roof sample	10,000 ft.^{2}	Yes	Yes	Yes	Logbook observations
	929 m^2				

Further, the assessment of neighborhood scale impacts of green roof adoption, which has received limited attention, would expand the scope of green roofs benefits and dovetail into research related to city-wide planning and green infrastructure planning and networks (e.g. Green City, Clean Waters and NYC Green Infrastructure Plan).

Overall, the future of green roof monitoring presents many intriguing and practical research opportunities. What is of interest about many green roof monitoring projects is the time and expertise required to design, install, test, calibrate, and validate data generated by the instruments and equipment which can lead to the question: Is simpler better? That depends on the research questions being asked and the particular green roof types and contexts.

References

Alsup S, Ebbs S, Battaglia L, Retzlaff W (2013) Green roof systems as sources or sinks influencing heavy metal concentrations in runoff. J Environ Eng-ASCE 139(4):502–508

Al-Yaseri I, Morgan S, Retzlaff W (2013) Using turbidity to determine total suspended solids in stormwater runoff from green roofs. J Environ Eng-ASCE 139(6):822–828

American Society of Landscape Architects (ASLA) (2013) Award of excellence: green roof innovation testing (GRIT) laboratory. ASLA professional awards project narrative. http://www.asla.org/2013awards/394.html. Accessed 3 April 2015

Arvidson AR (2012) Greening the landscape. WW Norton, New York

Beattie D, Berghage R, (2004) Green roof media characteristics: the basics. In: Proc. of 2nd North American green roof conference: greening rooftops for sustainable communities, Portland, Oregon, 2–4 June 2004

Becker D, Wang D (2011) Green roof heat transfer and thermal performance analysis. Unpublished Carnegie Mellon University Report, Pittsburgh, PA

Berghage R, Jarrett A, Beattie D et al (2007) Quantifying evaporation and transpirational water losses from green roofs and green roof media capacity for neutralizing acid rain. Pennsylvania State University, State College, Pennsylvania

Berghage R, Beattie D, Jarrett A, Thuring C, Razaei F (2009) Green roofs for stormwater runoff control. T. P. S. University. USEPA, National Risk Mgmt Research Lab, Water Supply & Water Resources Division. EPA/600/R-09/026

Berndtsson J (2010) Green roof performance towards management of runoff water quantity and quality: a review. Ecol Eng 36(4):351–360

Blank L, Vasl A, Levy S, Grant G, Kadas G, Dafni A, Blaustein L (2013) Directions in green roof research: a bibliometric study. Build Environ 66:23–28

Blanusa T, Vaz Monteiro M, Fantozzi F, Vysini E, Li Y, Cameron R (2013) Alternatives to Sedum on green roofs: can broad leaf perennial plants offer better 'cooling service'? Build Environ 59:99–106

Bliss DJ, Neufeld RD, Ries RJ (2009) Stormwater runoff mitigation using a green roof. Environ Eng Sci 26(2):407–417

Butler C, Orians CM (2011) Sedum cools soil and can improve neighboring plant performance during water deficit on a green roof. Ecol Eng 37(11):1796–1803

Byrne LB, Grewal P (2008) Introduction to ecological landscaping: a holistic description and framework to guide the study and management of urban landscape parcels. Cities Environ 1(2):article3, p 1–19

Cao J, Tamura Y, Yoshida A (2013) Wind tunnel investigation of wind loads on rooftop model modules for green roofing systems. J Wind Eng Ind Aerodyn 118:20–34

Carson TB, Marasco DE, Culligan PJ, McGillis WR (2013) Hydrological performance of extensive green roofs in New York city: observations and multi-year modeling of three full-scale systems. Environ Res Lett 8(2):24–36
Carter TL, Rasmussen TC (2006) Hydrologic behavior of vegetated roofs. J Am Water Resour Assoc 42(5):1261–1274
Chan ALS, Chow TT (2013) Evaluation of overall thermal transfer value (ottv) for commercial buildings constructed with green roof. Appl Energy 107:10–24
Charlesworth SM, Perales-Momparler S, Lashford C, Warwick F (2013) The sustainable management of surface water at the building scale: preliminary results of case studies in the UK and Spain. J Water Supply: Research & Technology-AQUA 0003–7214) 62(8):534–544
Clark MJ, Zheng Y (2013) Plant nutrition requirements for an installed sedum-vegetated green roof module system: effects of fertilizer rate and type on plant growth and leachate nutrient content. HortScience 48(9):1173–1180
Coutts A, Daly E, Beringer J, Tapper N (2013) Assessing practical measures to reduce urban heat: green and cool roofs. Build Environ 70:266–276
Culligan P, Carson T, Gaffin S, Gibson R, Hakimdavar R, Hsueh D, Hunter N, Marasco D, McGillis W, O'Conner TP (2014) Evaluation of green roof water quantity and quality performance in an urban climate. USEPA National Risk Management Research Laboratory, Office of Research & Development, EPA/600/R-14/180
Darkwa J, Kokogiannakis G, Suba G (2013) Effectiveness of an intensive green roof in a subtropical region. Build Serv Eng Res Technol 34(4):417–432
De Cuyper K, Dinne K, Van De Vel L (2004) Rainwater discharge from green roofs. In Proc. of 30th CIB W062 International Symposium on Water Supply and Drainage for Buildings, Paris, France. 16–17 September 2004
De Nardo JC, Jarrett AR, Manbeck HB, Beattie DJ, Berghage RD (2005) Stormwater mitigation and surface temperature reduction by green roofs. Trans ASAE 48:1491–1496
Dietz M (2007) Low impact development practices: a review of current research and recommendations for future directions. Water Air Soil Pollut 186(1–4):351–363
DiGiovanni K (2013) Evapotranspiration from urban green spaces in a northeast United States' city. Dissertation, Drexel University
DiGiovanni K, Gaffin S, Montalto F (2010) Green roof hydrology: results from a small-scale lysimeter setup (Bronx, NY). Low Impact Dev 2010:1328–1341. doi:10.1061/41099(367)114
DiGiovanni K, Gaffin S, Montalto F, Rosenzweig C (2011) The applicability of classical predictive equations for the estimation of evapotranspiration from urban green spaces: green roof results. World Environmental and Water Resources Congress, Palm Springs, California. 22–26 May 2011
DiGiovanni K, Montalto F, Gaffin S, Rosenzweig C (2013) Applicability of classical predictive equations for the estimation of evapotranspiration from urban green spaces: green roof results. J Hydrol Eng 18(1):99–107. (Figure 4 with permission from ASCE)
Dunnett N, Nagase A, Booth R, Grime J (2008) Influence of vegetation composition on runoff in two simulated green roof experiments. Urb Ecosyst 11:385–398
Dvorak B, Volder A (2013) Rooftop temperature reduction from unirrigated modular green roofs in south-central Texas. Urb For Urb Gr 12(1):28–35
Eumorfopoulou E, Aravantinos D (1998) The contribution of a planted roof to the thermal protection of buildings in Greece. Energy Build 27(1):29–36
Fassman E, Simcock R (2012) Moisture measurements as performance criteria for extensive living roof substrates. J Environ Eng 138(8):841–851
Fassman-Beck E, Voyde E, Simcock R, Hong YS (2013) Four living roofs in three locations: does configuration affect runoff mitigation? J Hydrol 490:11–20
Feller M, Traver R, and Wadzuk B (2010) Estimation of green roof evapotranspiration: experimental results. In Proc. of 2010 International Low Impact Development Conference: Redefining Water in the City, San Francisco, CA. 11-14 April 2010. doi: 10.1061/41099(367)8. http://ascelibrary.org/doi/book/10.1061/9780784410998

Felson AJ, Pavao-Zuckerman M, Carter T, Montalto F, Shuster B, Springer N, Stander E, Starry O (2014) Mapping the design process for urban ecology researchers. BioScience 63(11):854–865
Fioretti R, Palla A, Lanza LG, Principi P (2010) Green roof energy and water related performance in the Mediterranean climate. Build Environ 45(8):1890–1904
Friedrich CR (2005) Principles for selecting the proper components for a green roof growing media. pp 262–273. In: Proc. of 3rd North American Green Roof Conference, Washington, DC. 4–6 May 2005
Gaffin SR, Khanbilvardi R, Rosenzweig C (2009) Development of a green roof environmental monitoring and meteorological network in New York City. Sensors 9(4):2647–2660
Gaffin SR, Rosenzweig C, Khanbilvardi R et al (2011) Stormwater retention for a modular green roof using energy balance data. Columbia University Press, New York
Gaumont-Guay D, Halsall R (2013) The carbon balance of a Pacific west coast green roof. In: Proc. of Cities Alive: 11th Annual Green Roof and Wall Conference. San Francisco. 23–26 Oct 2013
Getter KL, Rowe DB (2009) Substrate depth influences sedum plant community on a green roof. HortScience 44(2):401–407
Getter KL, Rowe DB, Robertson GP, Cregg BM, Andresen JA (2009) Carbon sequestration potential of extensive green roofs. Environ Sci Technol 43(19):7564–7570
Getter KL, Rowe DB, Andresen JA, Wichman IS (2011) Seasonal heat flux properties of an extensive green roof in a Midwestern U.S. climate. Energy Build 43:3548–3557
Glass CC, Johnson PA (2009) Monitoring of a new green roof for water quality and quantity. In: Proc. of Greening Rooftops for Sustainable Communities Conference, Awards & Trade Show. Baltimore, MD. 30 April to 2 May 2008
Gnecco I, Palla A, Lanza LG, LaBarbera P (2013) A green roof experimental site in the Mediterranean climate: the storm water quality issue. Water Sci Technol 68(6):1419–1424
Gregoire B, Clausen J (2011) Effect of a modular extensive green roof on stormwater runoff and water quality. Ecol Eng 37(6):963–969
Griffin W (2013) The effects of green roof substrate composition on plant growth and storm water retention of Mid-Atlantic green roofs. Diss. U of Maryland
Hanks R (1992) Applied soil physics: soil water and temperature applications, 2nd edn. Springer-Verlag, Berlin
Havens KE, Aumen NG (2000) Hypothesis-driven experimental research is necessary for natural resource management. Environ Manag 25(1):1–7
Hendricks J, Calkins M (2006) The adoption of an innovation: barriers to use of green roofs experienced by Midwest architects and building owners. J Gr Build 1:148–168
Hill J, Perotto M, Yoon C (2015) Processes of quantifying the hydrological performance of extensive green roofs. In Proc. of 2015 RCI Convention & Trade Show, San Antonio, TX. 5–10 March 2015
Jensen ME, Burman RD, Allen RG (1990) Evapotranspiration and irrigation water requirements. ASCE Manuals and reports on engineering practice No. 70. Am. Society of Civil Engineers, NY
Jim CY (2012) Effect of vegetation biomass structure on thermal performance of tropical green roof. Ecol Eng 8(2):173–187
Jim CY, He HM (2010) Coupling heat flux dynamics with meteorological conditions in the green roof ecosystem. Ecol Eng 36(8):1052–1063
Karban R, Hunzinger M (2006) How to do ecology: a concise handbook. Princeton University Press, New Jersey
Kim S-C, Park B-J (2013) Assessment of temperature reduction and heat budget of extensive modular green roof system. Korean J Hortic Sci Technol 31(4):503–511
Klett JE, Bousselot JM, Koski RD, O'Connor TP (2012) Evaluation of green roof plants and materials for semi-arid climates. USEPA National Risk Management Research Laboratory, Office of Research & Development, USEPA/600/R-12-592
Köhler M, Poll PH (2010) Long-term performance of selected old Berlin greenroofs in comparison to younger extensive greenroofs in Berlin. Ecol Eng 36:722–729

Kurtz T et al (2010) Stormwater management facilities monitoring report: December 2010. Bure au of Environmental Services, Sustainable Stormwater Mgmt Program, City of Portland, OR: p 172
Laminack KD, Dvorak BD, Volder A (2014) Substrate loss is minimal in vegetated and un-vegetated extensive roof modules over a 14-month period. J Living Archit 1(4):1–13
Lea-Cox JD (2012) Using wireless sensor networks for precision irrigation scheduling. In: Kumar M (ed) Problems, perspectives and challenges of agricultural water management. InTech Press, Rijeka, pp 233–258
MacIvor JS, Margolis L, Puncher CL, Matthews BJC, (2013) Decoupling factors affecting plant diversity and cover on extensive green roofs. J Environ Manag 130:297–305
Marasco DE, Hunter BN, Culligan PJ, Gaffin SR, McGillis WR (2014) Quantifying evapotranspiration from urban green roofs: a comparison of chamber measurements with commonly used predictive methods. Environ Sci Technol Lett 48(17):10273–10281
Moody S, Sailor DJ (2013) Development and application of a building energy performance metric for green roof systems. Energy Build 60:262–269
Mooney-Bullock R, Buffam I, Bolan M (2012) Urban learning laboratory features four educational, monitored green roofs. In: Proc. of Cities Alive: 10th Annual Green Roof & Wall Conference. Chicago. 17–20 October 2012
Morgan S, Celik S, Retzlaff W (2012) Green roof stormwater runoff quantity and quality. J Environ Eng 139(4):471–478
Nagase A, Dunnett N (2013) Establishment of an annual meadow on extensive green roofs in the UK. Landsc Urb Plan 112:50–62
Nagengast A (2013) Energy performance impacts from competing low-slope roofing choices and photovoltaic technologies. Diss. Carnegie Mellon University
Neufeld RD, Monnell J, Ries RJ (2009) Monitoring protocol and data collection: comparison of the runoff water quantity, quality and thermal performances of two green roof technologies: thin vs. thick. Sch. of Engineering, U Pittsburgh
Nouri H, Beecham S, Kazemi F, Hassanli AM (2012) A review of ET measurement techniques for estimating the water requirements of urban landscape vegetation. Urb Water 10(4):247–259
Odum EP (1969) The strategy of ecosystem development. Science 164:262–270
Oke TR (1978) Boundary layer climates. Methuen Company, New York
Olivieri F, Di Perna C, D'Orazio M, Olivieri L, Neila J (2013) Experimental measurements and numerical model for the summer performance assessment of extensive green roofs in a Mediterranean coastal climate. Energy Build 63:1–14
Omar MS, Quinn MM, Buchholz B, Geiser K (2013) Are green building features safe for preventive maintenance workers? examining the evidence. Am J Ind Med 56:410–423
Onmura S, Matsumoto M, Hokoi S (2001) Study on evaporative cooling effect of roof lawn gardens. Energy Build 33:653–666
Onset Computer Corporation (2012) Monitoring green roof performance with weather stations. MKT1017-0912, Bourne, MA
Palla A, Gnecco I, Lanza LG (2012) Compared performance of a conceptual and a mechanistic hydrologic models of a green roof. Hydrol Process 26(1):73–84
Panayiotis N, Tsiotsiopoulou P, Chronopoulos I (2003) Soil amendments reduce roof garden weight and influence the growth rate of Lantana. Hortscience 38(4):618–622
Pearlmutter D, Rosenfeld S (2008) Performance analysis of a simple roof cooling system with irrigated soil and two shading alternatives. Energy Build 40(5):855–864
Peng LLH, Jim CY (2013) Green-roof effects on neighborhood microclimate and human thermal sensation. Energies 6(2):598–618
Pierre J, Bisby L, Anderson B, MacDougall C (2010) Thermal performance of green roof panels in sub-zero temperatures. J Gr Build 5(2):91–104
PlaNYC (2008) Sustainable stormwater management plan. City of New York
Rana G, Katerji N (2000) Measurement and estimation of actual evapotranspiration in the field under Mediterranean climate: a review. Eur J Agron 13(2–3):125–153

Retzlaff W, Celik S, Morgan S, Graham M, Luckett K (2010) Into the wind—wind tunnel testing of green roof systems. In: Proc. of Cities Alive: 8th Annual Green Roof & Wall Conference. Vancouver, BC. 30 Nov–3 Dec 2010
Rosatto H, Meyer M, Laureda D et al (2013) Water retention efficiency of green roof systems in extensive and intensive type covers. Revista De La Facultad De Ciencias Agrarias 45(1):169–183
Rostad N, White S, DiGiovanni K, Montalto F (2011) Canopy interception in vegetated stormwater management features. In: Proc. of American Geophysical Union 2011 Meeting, San Francisco, California. 5–9 December 2011
Rowe DB (2011) Green roofs as a means of pollution abatement. Environ Pollut 159(8–9):2100–2110
Rowe DB, Kolp MR, Greer SE, Getter KL (2014) Comparison of irrigation efficiency and plant health of overhead, drip, and sub-irrigation for extensive green roofs. Ecol Eng 64:306–313
Schneider K (2011) Quantifying evapotranspiration from a green roof analytically. Master's Thesis, Villanova University
Schroll E, Lambrinos J, Righetti T, Sandrock D (2011) The role of vegetation in regulating stormwater runoff from green roofs in a winter rainfall climate. Ecol Eng 37(4):595–600
Seidl M, Gromaire M-C, Saad M, De Gouvello B (2013) Effect of substrate depth and rain-event history on the pollutant abatement of green roofs. Environ Pollut 183:195–203
Sherrard JA, Jacobs JM (2012) Vegetated roof water-balance model: experimental and model results. J Hydrol Eng 17(8):858–868
Sidwell A, Gibbs-Alley J, Forrester K, Jost V, Luckett K, Morgan S, Yan T, Noble B, Retzlaff W (2008) Evaluation of the thermal benefits of green roof systems. In: Proc. of 6th Annual Greening Rooftops for Sustainable Communities Conference, Baltimore, MD. 30 April to 2 May 2008
Simmons M, Gardiner B, Windhager S, Tinsley J (2008) Green roofs are not created equal: the hydrologic and thermal performance of six different extensive green roofs and reflective and non-reflective roofs in a sub-tropical climate. Urb Ecosyst 1(4):339–348
Skabelund LR, Blocksome C, Brokesh D, Kim HJ, Knapp M, Hamehkasi M (2014) Semi-arid green roof research 2009-2014: Resilience of native species. In Proc. of 12th Annual Cities Alive Green Roof and Wall Conference, Nashville, TN. 12–15 November 2014
Song U, Kim E, Bang JH, Son DJ, Waldman B, Lee EJ (2013) Wetlands are an effective green roof system. Build Environ 66:141–147
Spirn AW (1984) The granite garden: urban nature and human design. Basic Books, New York
Starry O (2013) The comparative effects of three sedum species on green roof stormwater retention. Dissertation, U of Maryland
Starry O, Lea-Cox JD, Kim J, van Iersel MW (2014) Photosynthesis and water use by two Sedum species in green roof substrate. Environ & Experimental Botany 107(2014):105–112
Stovin V, Vesuviano G, Kasmin H (2012) The hydrological performance of a green roof test bed under UK climatic conditions. J Hydrol 414:148–161
Susca T, Gaffin SR, Dell'Osso GR (2011) Positive effects of vegetation: urban heat island and green roofs. Environ Pollut 159(8–9):2119–2126
Sutton, RK (2013) Seeding green roofs with native grasses. J Living Archit 1(1):26
Tanner CB (1967) Measurement of evapotranspiration. In: Hagan RM, Haise HR, Edminster TW (eds) Irrigation of agricultural lands, Amer Soc Agron Monograph 11:534–574
Teemusk A, Mander U (2007) Rainwater runoff quantity and quality performance from a green roof: the effects of short-term events. Ecol Eng 30(3):271–277
Thuring CE, Dunnett N (2014) Vegetation composition of old extensive green roofs (from 1980s Germany). Ecol Process 3:4
Tilman D (1989) Ecological experimentation: strengths and conceptual problems. In: Likens GE (ed) Long-term studies in ecology: approaches and alternatives. Springer Verlag, Berlin, pp 136–157
Toland DC, Haggard BE, Boyer ME (2012) Evaluation of nutrient concentrations in runoff water from green roofs, conventional roofs and urban streams. Trans A S A & B E 55(1):99–106
Tsiotsiopoulou P, Nektarios A, Chronopoulos I (2003) Substrate temperature fluctuation and dry-weight partitioning of *Lantana* grown in four green roof growing media. J Hortic Sci Biotechnol 78(6):904–910

Van Seters T, Rocha L, Smith D, MacMillan G (2009) Evaluation of green roofs for runoff retention, runoff quality, and leachability. Water Qual Res J Can 44(1):33–47
Voyde E, Fassman E, Simcock R (2010) Hydrology of an extensive living roof under sub-tropical climate conditions in Auckland, New Zealand. J Hydrol 394(3–4):384–395
Voyde E, Fassman E, Simcock R, Wells J (2010) Quantifying evapotranspiration rates for New Zealand green roofs. J Hydrol Eng, Special Issue: Low Impact Development, Sustainability Science, and Hydrological Cycle 15:395–403
Wadzuk B, Schneider D, Feller M, Traver R (2013) Evapotranspiration from a green-roof stormwater control measure. J Irrigation Drain Eng 139(12):995–1003
Walmart et al (2013) Green roof performance: a cost-benefit analysis based on Walmart's Chicago store. 31 Jan 2013 report
Wang X, Zhao X, Peng C, Zhang X, Wang J (2013) A field study to evaluate the impact of different factors on the nutrient pollutant concentrations in green roof runoff. Water Sci Technol 68(12):2691–2697
Wanielista M, Minareci M, Catbas N, Hardin M (2011) Green roofs and wind loading. Florida Dept. of Environmental Protection and U of Florida Stormwater Management Academy, FDEP:934
Welker AL, Mandarano L et al (2013) Application of a monitoring plan for storm-water control measures in the philadelphia region. J Environ Eng 139(8):1108–1118
Whittinghill LJ, Rowe DB, Schutzki R, Cregg BM (2014) Quantifying carbon sequestration of various green roof and ornamental landscape systems. Landsc Urb Plan 123:41–48
Williams C (2013) Portland state will build green roof research site on Walmart's new North Portland store. 23 Oct 2013 News: Portland State U
Wong NH, Chen Y, Ong CL, Sia A (2003) Investigation of thermal benefits of rooftop garden in the tropical environment. Build Environ 38:261–270
Xu CY, Chen D (2005) Comparison of seven models for estimation of evapotranspiration and groundwater recharge using lysimeter measurement data in Germany. Hydrol Process 19(18):3717–3734
Zapater-Pereyra M, van Dien F, van Bruggen JJA, Lens PNL (2013) Material selection for a constructed wetroof receiving pre-treated high strength domestic wastewater. Water Sci Technol: J Int Assoc Water Pollut Res 68(10):2264–2270
Zhang CL, Zou X-Y, Yang P et al (2007) Wind tunnel test and cs tracing study on wind erosion of several soils in Tibet. Soil Tillage Res 94:269–82
Zhao MJ, Srebric J (2012) Assessment of green roof performance for sustainable buildings under winter weather conditions. J Cent South Univ Technol 19(3):639–644
Zheng, Y, Clark MJ (2013) Optimal growing substrate pH for five sedum species. Hortscience 48(4):448–452

Personal Communication

John Lambrinos (2014) Dept. of Horticulture, Oregon State University http://horticulture.oregonstate.edu/content/john-lambrinos
John Buck (2014) Civil & Environmental Consultants, Inc., Pittsburgh, Pennsylvania https://www.linkedin.com/in/johnbuckenvironmentalscientist
John Lea-Cox (2014) University of Maryland, College-Park https://www.psla.umd.edu/people/dr-john-lea-cox
Timothy Kurtz (2014) Bureau of Environmental Services, City of Portland, Oregon https://www.portlandoregon.gov/bes/44422
Steve Burian and Youcan Feng (2014) Dept. of Civil & Environmental Engineering, University of Utah http://www.civil.utah.edu/burian; http://urbanwater.utah.edu/people.html
David Gaumont-Guay (2014) Biology Dept., Vancouver Island University https://www2.viu.ca/biology/faculty/gaumont-guay.asp

David J. Sailor (2014) Dept. of Mechanical and Materials Engineering, Portland State University http://web.cecs.pdx.edu/~sailor/CV.pdf
Scott MacIvor(2014) Dept. of Biology, York University http://grit.daniels.utoronto.ca/people/students/
William Retzlaff (2014) College of Arts & Sciences, Southern Illinois University Edwardsville http://www.siue.edu/artsandsciences/biologicalsciences/faculty-staff/faculty-pages/retzlaff.shtml

Lee R. Skabelund ASLA is an associate professor of Landscape Architecture at Kansas State University and has overseen the installation and monitoring of the Seaton Hall green roof and a campus bio-retention treatment train.

Dr. Kimberly DiGiovanni is a post-doctoral research fellow and adjunct assistant professor in the Department of Civil and Environmental Engineering at Villanova University. She is a graduate of Drexel University's doctoral program in Environmental Engineering and is an alumna of the Sustainable Water Resource Engineering Laboratory at Drexel University. Dr. DiGiovanni's research interests and focus include the eco-hydrology of natural vegetated spaces and engineered urban green spaces, including green roofs.

Dr. Olyssa Starry is Assistant Professor at Portland State University's Honors College. Her research relates various aspects of green roof structure to function through both experimentation and modeling. Her doctoral work at the University of Maryland, College Park assessed the role of different Sedum species in the green roof water cycle utilizing controlled chambers for monitoring gas exchange as well as outdoor experimental green roof platforms.

Chapter 3
Climates and Microclimates: Challenges for Extensive Green Roof Design in Hot Climates

Mark T. Simmons

Abstract Green roof systems have been developed and adopted in the temperate and cool-temperate climates of Europe and North America. Although these regions can get extreme weather, they generally do not experience climatic extremes of high temperatures, prolonged drought, and intense rainfall events of tropical and subtropical regions. This presents challenges for green roof design to not only provide adequate growing conditions for plants, but also to improve roof performance with respect to intrinsic (e.g. cooling building, extension of roof membrane lifetime) and extrinsic (e.g. flash flood mitigation, building cooling, reduction of heat island effect) benefits. Therefore, the components of conventional green roof including plant palette, growing media composition and the other synthetic layers need to be modified. The characteristics of green roof water retention, plant water availability, plant selection, and thermal properties are all critical factors which need to be adapted to help address the harsher environmental conditions and performance demands of hot climates. If these problems can be overcome, the combined environmental, ecological and sociological benefits suggest green roofs could be an imperative technology for towns and cities in tropical and subtropical regions of the world.

Keywords Green roof · Tropical and subtropical climates · Growing substrate · Plant selection · Stormwater

M. T. Simmons (✉)
Lady Bird Johnson Wildflower Center, University of Texas, 4801 La Crosse Avenue, Austin, TX 78739, USA
e-mail: msimmons@wildflower.org

R. Sutton (ed.), *Green Roof Ecosystems*, Ecological Studies 223,
DOI 10.1007/978-3-319-14983-7_3

3.1 Introduction

3.1.1 The Characteristics of Hot versus Temperate Climates

Green roofs represent a synthesized ecosystem subject to environmental extremes for plants. Extensive green roofs are described as having a thin (<20 cm; 7.8 in) layer of growing media, and depending on elevation, subject to the extremes of high wind, high thermal load, varying air humidity and often limited plant availability (Oberndorfer 2007). In effect, surface weather and ground conditions are oftentimes poor predictors of green roof microclimate where air and soil conditions are exacerbated to such an extent that from a plant perspective the growing conditions are significantly compromised. These extreme stresses can be significantly amplified in warmer climates.

Historically, green roof systems have been recorded in different regions across Europe and Asia (Snodgrass and Snodgrass 2006), but the contemporary extensive green roof (henceforth "green roof") has largely been developed in the temperate and cool-temperate climates of Europe and North America (Aber and Melillo 1982; Williams et al. 2010). Although periods of heat and drought can impact temperate regions, compared to tropical and subtropical zones, temperate climates can generally be described as experiencing moderate rainfall spread across portions or most of the year (sporadic drought notwithstanding), cool or cold winters mild to warm summers and moderate diurnal temperate variation (Peel et al. 2007). By contrast, warm tropical and subtropical climates (henceforth "hot climates") have cool to warm winters and warm to hot summers with rain events distributed either through the year (e.g. wet tropical) or seasonally (e.g. hot arid or Mediterranean) depending on geographic location. In hot climates the conditions of increased water (too much and too little) stress and high temperatures govern most of the challenges of green roof design. These differences can have a direct effect on the ecological function of the green roof—heat stress (both above and below ground), periodic saturation, and periodic drought all dictate the ecological response and hence design of green roofs in warmer climates. In terms of plant ecology and plant selection perhaps the greatest consequence of hot climate environment is a broader ecological niche—the sedum-dominated roofs in temperate systems are characterized by high water use efficient, succulent plants capable of withstanding cold winters and warm summers on a shallow-well drained medium. Conversely, in many hot climates plants must with stand high leaf and root temperatures, prolonged drought and occasional prolonged media saturation. To be able to tolerate heat, drought and prolonged saturation suggest plants with a different ecophysiological niche. From a plant-selection perspective this may be overcome by paying less attention to the conventional, green roof, temperate-climate plant palette and selecting from regional florae adapted to these more stressful conditions. However, the characteristics of the growing media to mitigate extreme hydrological and thermal conditions may require significant redesign.

3.1.2 *Temperate Green Roof Design challenges*

Green roof design has traditionally focused on growing media composition and structural design optimized in terms of minimal cost and weight (roof load bearing) to achieve desired performance goals and to ameliorate soil microclimate and water availability to accommodate appropriate plants. Temperate climate extensive green roof design has thus been optimized so much so the ecological niche for green roof plants is very narrow (Snodgrass and Snodgrass 2006). This suggests that in warmer, non-temperate systems with greater climatic extremes (e.g. high daytime and night time temperatures, frequent flash flood events), green roof design may require revision. All green roofs potentially offer significant intrinsic (e.g. cooling building, extension of roof membrane lifetime) and extrinsic (e.g. flash flood mitigation, reduction of heat island effect) benefits. But all aspects of conventional green roof design—plant palette, substrate composition and profile design—may likely need to be modified to accommodate these different environmental conditions and performance expectations.

Plant selection for green roofs in temperature regions has focused mainly on shallow rooted, succulent plants which exhibit Crassulacean Acid Metabolism (CAM) in the family Crassulaceae and less commonly on a selection of herbaceous grasses and forbs native to temperature regions (Snodgrass and Snodgrass 2006). In terms of stormwater, thermal mitigation and habitat characteristics temperate green roofs, designed correctly, can perform well. However, translating this technology to warmer regions presents a challenging suite of climatic problems including: flash flooding, prolonged drought, high day and night-time air and soil temperatures and limited available water supply. Ironically, the benefits of green roofs in these warmer environments might hypothetically be more justified than in temperate climates, by providing mitigation performance for the very characteristics that challenge their design and implementation (Kaufman et al. 2007; Alexandri and Jones 2008; Simmons 2008).

In this chapter I identify the short- and long-term challenges and benefits of micro and meso-climate that affect green roof design in hot climates and describe evidence and propose theories to overcome them.

3.2 The Benefits and Problems of Green Roofs in Hot Climates

3.2.1 *Emergence of Research*

Until recently, efforts to successfully implement extensive green roofs in hot climates have been comparatively few. Williams et al. (2010) suggest the major barriers have been unfamiliarity with green roof technology and inexperience of the emerging green roof industry, lack of regionally relevant research and inappropriate carry-over of the design (substrates and drainage layers) and biology (species), from temperate regions directly to hot climates. However, over the last decade green roof hot climate research has been initiated in a few locations around the world including: Australia (Williams et al. 2010), Southeast Asia (Tan and Sia 2005), Southern (Mediterranean) Europe (Fioretti et al. 2010), Central America (Müller Garcia 2005), and in USA: Texas (Simmons et al. 2008; Volder and Dvorak 2014), Florida (Sonne 2006a; Wanielista et al. 2008), Georgia (Carter and Rasmussen 2006) and Hawaii (Cabugos et al. 2007).

The specific problems around hot climate green roof success include low species/individual plant survival rates, due to drought (Farrell et al. 2012) for other reasons to be discussed below, poor stormwater performance under high rainfall intensity (Simmons et al. 2008) or prolonged wet events and weediness (Williams 2010). Additionally, from an implementation perspective, the limited expertise of green roof technology and knowledge of realistic performance and absolute function among architects and landscape architects has inhibited broad adoption in hot climates (Williams 2010).

In many respects green roofs represent a novel technology, more so outside of temperate regions, and the lack of knowledge, records of failure and inevitable low implementation rates has dramatically inhibited further development of this technology in hot climates.

3.2.2 Water Retention and Plant Water Availability

The ability for roofs to retain stormwater can vary a lot among green roof types with some having little or no retentive performance despite manufacturers claims (Simmons et al. 2008). Media composition and depth (Monterusso and Rowe 2005), drainage and retention layers (Simmons 2008) and the growth form and physiology of the plant suite (Dunnett and Kingsbury 2004; Schroll et al. 2011) all can have a direct effect on water retention performance (FLL 2008). Paradoxically, to some extent green roof design has been driven by the need for the conflicting goals of good stormwater retention and adequate drainage (both in the media and immediately above the roof membrane), while at the same time leaving sufficient available water in the growing media for plant uptake storm water retention (FLL 2008). This requires water to be held in different states and/or in different component of the green roof system with plant available water held at field capacity or below in the growing media and storm water retained in the media and in other retention structures as absorbent mats or combined drainage-retention layers below the growing media. European green roof standards have focused on the provision or assumption that either plant selection or frequency of rainfall events can meet plant growth requirements while still maintaining good water retention qualities (FLL 2008). But these guidelines may fall short of the provision of performance requirements for hotter and wetter climates. Despite the recommendations for drainage and retention of water green roofs in hot climates have sometimes failed to perform (Williams et al.

2010; MacIvor et al. 2011). This may be due to inappropriate combination of specifications of media, drainage, plant selection etc. and it is difficult to tell whether or not guidelines have been closely adhered to (Dvorak 2011). For example, FLL (2008) guidelines suggest that the growing media, should exhibit a broad range of particle sizes where the larger fraction represented by a porous, mineral-based material such as expanded shale, expanded clay, recycled brick, tile, scoria or pumice depending on local availability works well for a variety of temperate green roof assemblages (Molineux et al. 2009). But this may not be ideal for all plants types on green roofs in other regions. For example, research has generally been in support of increased organic matter (greater than FLL recommendations) to aid both plant establishment and especially to improve plant available water (Molineux et al. 2009). The problem with excessively increasing organic matter this is that under warm and wet conditions organic matter in the growing media may rapidly decompose under increased bacterial and fungal biological activity, dramatically reducing effective root volume. Even though some organic matter is continually added by vegetative components, high levels of organic matter are unlikely to be maintained. This suggests that other stable components meet the positive water retention (and other characteristics of organic matter) be substituted, for example hydrophilic gels, perlite and vermiculite which hold water, air and have high cation exchange capacity for plant nutrient supply (Getter and Rowe 2006; Sutton et al. 2012).

The ability for green roofs to be able to pump (evapotranspire) water out of the green roof while at the same time maintaining adequate plant water in the growing availability is a conundrum (Chap. 4). Keeping water loss to a minimum is related to plant transpiration, media evaporation and water-holding capacity within the media. Transpiration is minimized using plants with high water use efficiency, which is one attraction of succulent CAM plants, characterized by low stomatal conductance (Korner et al. 1979) and minimized night-time transpiration. However, removal of water from the substrate is desirable to optimize long-term storm water retention during wet seasons: in wet seasons with high frequency rain events the faster the green roofs can remove water from the roof system the better it can absorb the next event. Therefore plants that can switch between low transpiration in dry periods and high transpiration in rain events i.e. facultative CAM, or equally broad soil water niche plants such as some prairie grasses and forbs would be ideal (Wolf and Lundholm 2008; Sutton et al. 2012).

But even plants with high water use efficiency, the plant available water can decline quickly following precipitation/irrigation events especially in shallow media (Van Woert et al. 2005). This implies that where supplemental irrigation is unavailable the need to use plants with very high drought tolerance regardless of succulence and photosynthetic pathway is mandatory (Farrell et al. 2012). One alternative is to design a roof that simulates other hot climate landscapes with annual seeds, bulbs or other cryptophytes (plants which maintain living tissue below ground and seasonally invisible) only emerge under favorable conditions. Such a 'brown' roof may not be most desirable aesthetically or even general performance but certainly suggests that they are worth investigation.

Evaporation from the surface is dependent on both air and soil conditions. Therefore optimizing canopy cover to shade the soil surface needs to be balanced by transpirational characteristics of the plant. In cool climates the effect of shading may be less important than other microclimate effects such as precipitation and media moisture properties (Wolf and Lundholm 2008). Conversely, in hot climates with exceptionally high surface temperatures up to 90 °C (Williams et al. 2010), canopy shading, particularly in dry season may be important in influencing media water availability.

Many commercial green roof manufacturers utilize additional water retention layers (porous/capillary blankets or 'egg carton' bucket layers Fig. 3.1) to improve storm water retention performance of the roof and can be very effective (e.g. Miller and Narejo 2005; Berghage and Gu 2009). Ironically, some of these drainage/retention layers are usually topped with a root barrier—making retained water effectively inaccessible to plant roots. In climates where water availability is at a premium this is an exceptional inefficient use of resources. Destruction of four-year experiment green roofs in Texas however showed that aggressive roots followed moisture gradients and often compromised these root barriers (Fig. 3.1). An alternative to this is to use hydroponic foam in place of a standard retention layer (Fig. 3.2). This

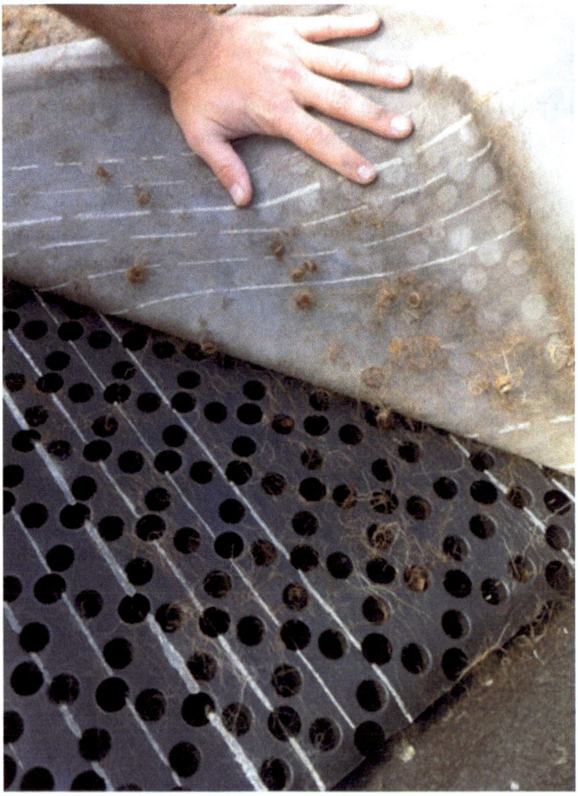

Fig. 3.1 Four-year old roots on a destructed green roof passing through root barriers into drainage/retention layer. (*Mark Simmons*)

Fig. 3.2 Experimental hydroponic foam layer used beneath the growing media to accommodate both stormwater retention and providing plant available water Note the roots both above and penetrating through foam layers (*Mark Simmons*)

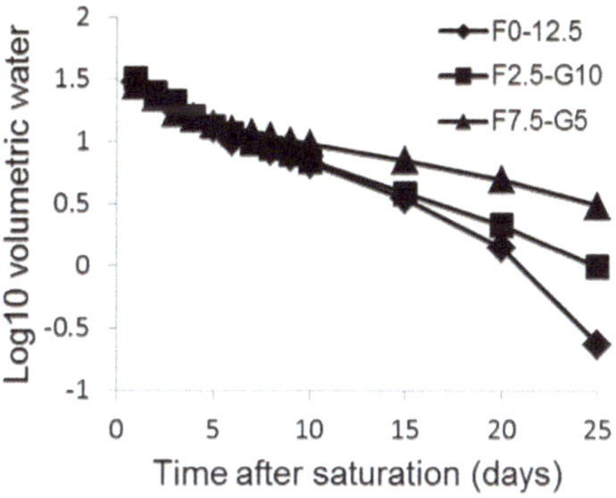

Fig. 3.3 Dry-down curves of vegetated trays containing different green roof growing media. (♦): No hydroponic foam layer, 12.5 cm growing media; (■): 2.5 cm hydroponic foam layer, 10 cm growing media; (▲): 7.5 cm hydroponic foam layer, 5 cm growing media (M.T. Simmons (2012) unpublished data)

provides for the retention of storm water retention while still simultaneously allowing accessibility to available water by roots. Trials in Texas indicate that hydroponic foam significantly prolongs the plant availability of water increases by reducing the rate of loss of total volumetric water content over time (Fig. 3.3). The wide range of commercial and potential products to aid water retention/availability is somewhat confusing and if performance is to be optimized then investigation and standardization (e.g. ASTM) is going to be essential to further green roof development in these harsher environments (Miller and Narejo 2005).

3.2.3 Root Temperature and Media Composition

Plant physiological processes are highly sensitive to temperature. Most vascular plant roots have a much narrower temperature envelope of performance compared to the aboveground stems and leaves. Although species specific, generally the operational temperature range of root physiological processes are from 4 °C to 30 °C. Above that upper temperature, respiration and other root processes decline rapidly and certain processes, particularly the synthesis of secondary materials slow down until above 48 °C where they stop and root mortality results (Xu and Huang 2000; Urban 2008; Sutton et al. 2012). Even in arid CAM plants these upper limits to root function still apply (Drennan and Nobel 1998).

Roof surface (waterproof membrane) temperatures in summer can easily exceed these critical temperatures. In Texas, roof temperatures have been recorded at 56 °C in early (spring) growing season (Simmons et al. 2008) and can exceed 70 °C in summer (Simmons et al. 2008), mid 50 s°C in Florida (Sonne 2006b) and up to 90 °C recorded in Australia (Williams et al. 2010). Simmons et al. (2008) recorded temperatures in weekly irrigated growing media (5 cm (2 in) below surface) ranging between 25 °C to 40 °C, similar to values recorded on green roofs in Singapore (Tan and Sia 2005) and Florida (Sonne 2006b) suggesting that there is sufficient heat flux through conduction, radiation and convection to limit root growth in at least the top layers of the media.

Excavated plants from extensive roofs exhibited low root density in the top 5 cm of the growing media suggesting that in some growing media the top layer may be redundant either due to temperature, high porosity or more water availability in these upper layers (Fig. 3.4). Collectively this evidence indicate that modification

Fig. 3.4 Four-year-old grass (Bouteloua dactyloides) grown on experimental green roof exhibiting low root density in upper layers of the growing media (10 cm total media depth) (Mark Simmons)

of media composition, specifically to alter the thermal conductivity (λ) and heat capacity, may help to improve the green roof environment in extreme climates. Media composition may also be critical to plant establishment. Any component that increases water retention will likely improve plant survival. MacIvor et al. (2011) examining a range of succulents, grasses and forbs on green roof modules in Toronto, Canada, concluded that plant cover and biomass declined on a media based on the FLL specifications of low organic matter. The coarse component of many commercial growing media can be naturally occurring (scoria, lava rock pumice), recycled (brick, tile) or processed (expanded shale or clay). These components often makes up the bulk of volume and are included to provide ballast, root anchor, and a site for plant available water and nutrients. However these materials can present a problem in hot climates as they may exhibit high thermal conductivity, transmitting heat through conduction (convection and radiation have a relatively small role in growing media thermal flux) down into the sensitive root-growth zone. One way to mitigate this to protect roots from high temperatures is to increase thermal insulation characteristics of the media by addition of organic or other non-coarse, lightweight materials like vermiculite that are known to have low thermal conductivity. Laboratory trials (Simmons, unpublished data) demonstrated that lightweight, porous organic and inorganic material added to media (50 % by volume) not only improved volumetric water content ($\theta = 0.248\ m^3.m^{-3}$; brick plus porous matter $\theta = 0.465\ m^3.m^{-3}$) but also reduced thermal conductivity across a range of soil water potential (ψ) (Fig. 3.5a). The trade-off of high organic volume is that, with a few exceptions, most commercially-available organic amendments used in green roof media break down over time—and with warmer climates this process is accelerated, reducing valuable root volume and causing plant decline or death. Additionally, these same laboratory trials revealed that one commercially available expanded clay-based material also demonstrated

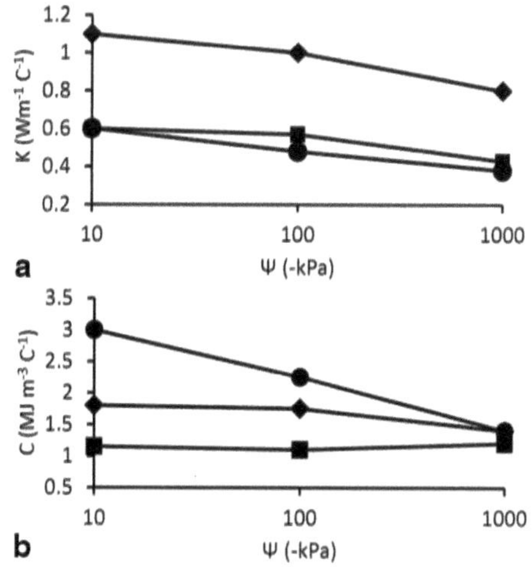

Fig. 3.5 A the relationship of thermal conductivity (κ) and B heat capacity (C) to growing-media water potential (Ψ) of three substrates: crushed brick (♦); crushed brick and porous organic/inorganic matter (■, 50:50 by volume) and a commercially available substrate (● expanded clay and organic/in-organic matter 50:50 by volume) (M. T. Simmons, T. Caldwell and M. R. Bright unpublished data)

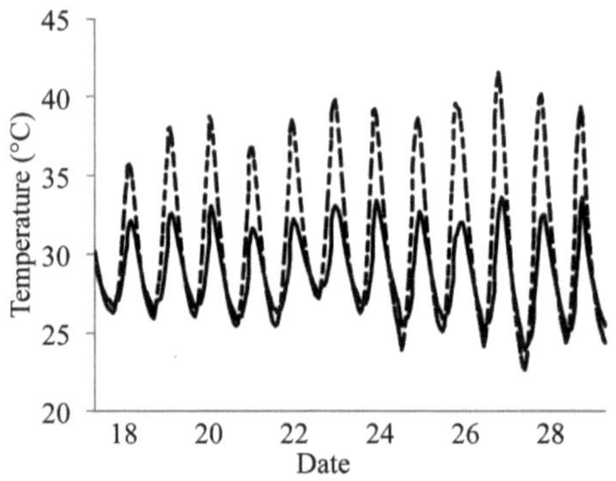

Fig. 3.6 Diurnal temperature flux of 10 cm deep growing media (2 types) on green roofs over 10 days (average high temperature 33 °C) in August in Texas. Solid line = media comprised of decomposed granite, perlite and organic matter; Dashed line = media comprised of expanded clay, expanded shale, sand and organic matter (M.T. Simmons unpublished data)

unusually high heat capacity (Fig. 3.5b). This could present a significant problem in hot climates where the cumulative effect of slow overnight cooling during warm months could lead to the build-up of excessively high temperatures in the media over time. Comparison of substrate temperature over an 10 day period in summer 2007 of test plots showed that a commercially available expanded clay/shale-based media did slowly reach higher maximums (consistently exceeding the critical 30 °C temperature where root function becomes impaired) compared to media containing decomposed granite and perlite (Fig. 3.6). While, without further investigation, it is not possible to determine the mechanism that drives this response, it does highlight the need for further investigation and specification of thermal properties of growing media for green roofs in hot climates.

3.2.4 Thermal Benefits in a Hot Climate

One key attribute of green roofs is their thermal benefits both to the building and immediate environment (Chap. 9). These characteristics of green roofs are no more important than in hot climates where daily maximum air temperature are higher, last longer through the day and persist over much or all of the year. Roof surface temperatures have been shown to be dramatically decreased in the presence of green roofs with deltas of 20 °C in Florida (Sonne 2006b), 38 °C in Texas (Simmons et al. 2008) and up to 60 °C in Japan (Wong 2003). This has mainly been attributed to the combination of shading (Wong 2003), solar reflectivity (Castleton et al. 2010), insulation (Barrio 1998), and evaporative cooling (Onmura et al. 2001) of all or some of the green roof components. This has several direct benefits. Firstly a damping of the diurnal temperature variations at the roof membrane combined with protection from ultra-violet radiation can extend the membrane integrity (Liu and Baskaran 2003). Secondly, it can reduce the energy budget of the building. The reduction of thermal flux through the building below the green roof can translate to savings in the build of up to 4 °C in temperate systems, to up to 15 °C in subtropical (Simmons et al. 2008). In a green roof test in Athens, Greece demonstrated that a building with a regular roof experienced internal air temperatures over 30 °C for 68 % of total time during a three-day test period in summer. Conversely, the green roofed building exceeded 30 °C only 15 % of the time. Whatever the mechanism this mitigation of thermal flux can amount to significant cost savings. Dunnett and Kingsbury (2004) suggest that there is an 8 % reduction in electricity use for air conditioning for every 0.5 °C decrease in internal temperature and if this model can be extrapolated to other regions, would represent a significant saving in hotter climates. In Florida, Sonne (2006b) estimated an energy reduction (cooling) of around 50 % for a two-story building with a 150 m^2 green roof. It has been argued however that green roofs for their thermal mitigation properties alone may not justify the resources as standard insulation is relatively inexpensive. According to one model on well- insulated buildings energy savings drop from 48 % for non-insulated to 2 % for well-insulated buildings.

Similarly, green roofs have been shown to cool ambient air temperatures that can translate to the larger scale especially in hot climates (Alexandri and Jones 2008). Microclimate modification by green roofs can affect both immediate local conditions by directly cooling air (Wong 2003), increasing reflectivity and by reducing long-wave radiation through the diurnal temperature cycle all of which can modify the urban heat-island effect (Getter and Rowe 2006; Oberndorfer et al. 2007, Santamouris In press). Even in the continental temperature climate of Toronto, Canada a study concluded that with only 6 % of total roof space dedicated to green roofs would result in a reduction of 1–2 °C in summer (Bass et al. 2003).

3.2.5 Plant Selection

Clearly, tolerance to drought, high temperatures (air and soil) and ability to tolerate media saturation for periods of time are desirable features. This suggests that plant selection for hot climates should examine those species with broader ecological niche and habitat generalists not specialists. A mix of growth forms for hot climate green roofs, may be the solution to optimize performance across all climate conditions through the year (MacIvor 2011; Wolf and Lundhom 2008). Succulence or CAM, although beneficial is not the only method of drought survival. Ability to reduce biomass through drought deciduousness, (Farrell et al. 2012) or avoidance as a seed (therophyte) or high water use efficiencies can all be successful drought survival strategies. Sedums of temperate climate origin although widely used for green roofs in Europe and North America, with a few exceptions, may not be suitable in hot climates as the exhibit relatively weak ability to fix CO_2 above 20 °C (Williams 2010; Livingston 2004). At night in hot climates, when gas exchange takes place stomata open in CAM plants, temperatures can easily exceed this through much of the growing season, and at high vapor pressure gradients and night time temperatures (>30 °C) CAM plants have been shown to exhibit significant decreases in net CO_2 gain (Herppich 1997; Livingston et al. 2004). While all the mechanisms that inhibit the use of the temperate-climate Sedums on green roofs in hot climates remain unidentified, evidence from these and other studies suggest that high day and/or night-time temperatures may be responsible. Some Sedums including non-European Sedums however have been shown to perform with some success in green roofs in Texas (Volder and Dvorak 2014) and under greenhouse conditions in the warm/temperate climate of Melbourne, Australia.

Farrell et al. (2012) had good drought survivorship of two Mexican and one Caucasian Sedum species that performed better than two succulent natives. In a Texas study survivorship on 18 extensive green roof units (Simmons et al. 2008) over 5-year period demonstrated that plant physiognomy or guild was not necessarily a prediction of plant survival (Fig. 3.7). Woody and non-woody forbs generally did less well than most graminoids and succulents. Some grasses did moderately well especially warm season bunch grasses, while a cool season grass, and more hydrologically-mesic graminoids did not. The three CAM species were better performers with up to 100 % survivorship (Fig. 3.7). With more supplemental irrigation

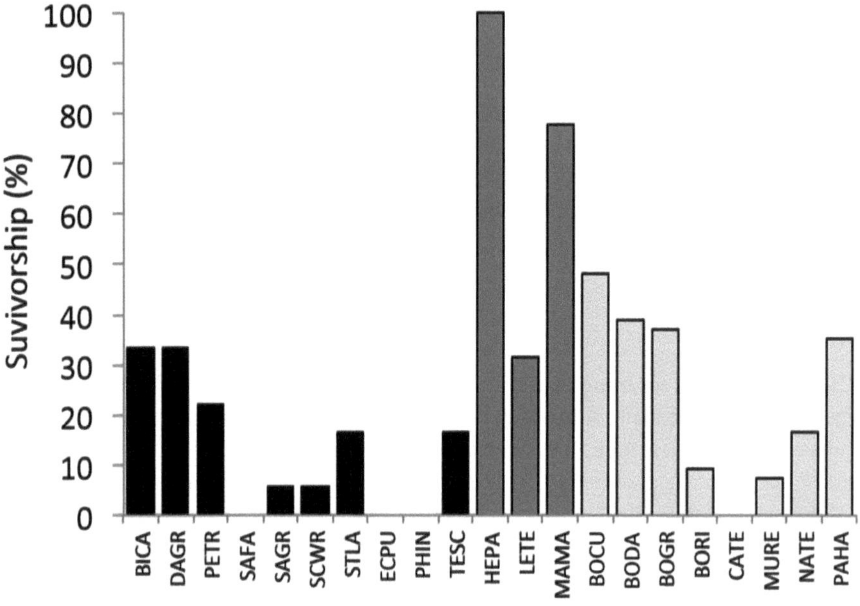

Fig. 3.7 The mean 5-year survivorship of 21 regionally native species on 18 green roof units in Austin Texas. Guilds: Black bar = forb; Dark grey bar = succulent/CAM; Light grey bar = graminoid. Species: BICA = Bignonia capreolata; DAGR = Dalea greggii; PETR = Penstemon triflorus; SAFA = Salvia farinacea; SAGR = Salvia greggii; SCWR = Scutellaria wrightii; STLA = Stemodia lanata; ECPU = Echinacea purpurea; PHIN = Phyla incisa; TESC = Tetraneuris scaposa; HEPA = Hesperaloe parviflora; LETE = Lenophyllum texanum; MAMA = Manfreda maculosa; BOCU = Bouteloua curtipendula; BODA = Bouteloua dactyloides; BOGR = Bouteloua gracilus; BORI = Bouteloua rigidiseta; CATE = Carex texensis; MURE = Muhlenbergia reverchonii; NATE = Nassella tenuissima; PAHA = Panicum hallii. Media depth = 100 mm. Irrigation regimen: minimum of 50 mm per month either by rainfall, irrigation or both (M T Simmons, unpublished data)

and improved media characteristics overall survivorship would likely increase. In another Texas study of fifteen different species of different geographic origins and different growth forms Dvorak and Volder (2012) found that only four of the fifteen species faired consistently well, demonstrating 100 % survival over three years and all were succulents.

Plant architecture may also have an influence on survivorship. Liu et al. (2012) examined the physiology and survival of thirty-one plants on green roofs in the humid subtropical climate of central Taiwan. The most successful species were those that exhibited succulent foliage, leaf hairs/spines, CAM and elevated plant height (up to 35 cm tall). Such physiological strategies to deal with drought stress are also common in grassland ecoregions and consequently lend themselves to green roof environments. Wolf and Lundholm (2008) in a study in cool temperate location (Nova Scotia, Canada) suggested that beyond the genus Sedum, some grasses were able to respond to water stress and lived longer-lived than succulents and woody plants and should be considered as further candidates for the green roof plant palette. Similarly, Sutton et al. (2012) reviewed grasses and forbs from North American prairie that had been used on green roofs under different irrigation regimens and

Fig. 3.8 Green roof on a residential building in Texas (Green Roof: Ecosystem Design Group, Lady Bird Johnson Wildflower Center, University of Texas at Austin Architects: Bercy Chen Studio LP)

Fig. 3.9 Green roof on a residential building in Texas (Green Roof: Ecosystem Design Group, Lady Bird Johnson Wildflower Center, University of Texas at Austin Architects: Bercy Chen Studio LP)

geographic locations and concluded that these could provide alternatives to Sedum species but stressed that more detailed studies are needed (e.g. Figures 3.8 & 3.9). One trait that may enhance the suitability of prairie grasses is that many grasses (and indeed many other species) are facultative mycorrhizal. This may help to improve performance by increasing effective root volume through the production of mycorrhizae, reducing water stress and nutrient uptake in a limited media depth of the green roof environment (Sutton et al. 2012).

Plant selection for some green roofs has mainly relied on tried successes of green roofs in temperate systems but more recently has examples from warmer climates (Dvorak and Volder 2013; Liu et al. 2012). This has led to some roof failure in hot climate and the basis for plant selection is undergoing an overhaul (e.g. Simmons et al. 2008; Liu et al. 2012; Sutton et al. 2012). While the issues of simple survival are obviously important, species selection would benefit from a fresh approach focusing on overall desired roof (e.g. storm water, thermal, and aesthetic characteristics), performance and letting that drive roof design and plant selection. That would mean selecting the most desired benefit(s), for example, drought tolerance, slow growth rate, then finding plant species or assemblages that meet these criteria. Finally, the extremes of conditions on hot climate green roofs suggest that plant selection screening should focus on species with a broader ecological niche selection - i.e. generalists (e.g. plants that can tolerate drought, yet endure occasional saturation) and not specialists (e.g. a species constantly requiring well-drained conditions). This is especially important with respect to tolerance range of the plant species to both soil water and soil temperature.

3.3 Conclusions

3.3.1 Imperative of Green Roofs in Hot Climates

Large proportion of global population lives in cities in subtropical and tropical regions around the world. In cities with high densities and high proportion of impervious surface and limited green infrastructure (Shanghai—Tokyo, New Delhi, Mumbai, Hong Kong, Sydney) green roofs may be the only green strategy to improve essential ecosystem services. However, for green roofs to be successful, a more holistic approach and understanding of all performance benefits have to be understood and quantified. It is not enough for justification of green roof technology to focus on one performance feature, say water retention, as cheaper methods to achieve the same goal (e.g. retention pond at grade), or thermal benefits (e.g. reflective white roof) may be available. In other words, taken individually, green roof performance attributes may exhibit incremental benefits at unjustifiably high costs. However, the environmental, ecological and sociological benefits taken together make a sound case for green roofs and even an imperative technology for the future of our cities. As described above green roofs vary regarding plant and media traits due to the local climate and microclimate (Table 3.1)

Table 3.1 The differences in in roof characteristics and plant and media traits for green roofs in temperate, hot-arid and hot-wet climates

		Temperate Climate	Hot Arid Climate	Hot Wet Climate
Roof conditions	Max leaf/root temperature	Moderate	High	High
	Min leaf/root temperature	Low	Moderate-Low	Moderate
	Sustained period of high temperatures	Infrequent	Frequent	Frequent
	Sustained period of low temperatures	Frequent	Infrequent	Never
	Sustained period of saturation	Frequent	Occasional	Frequent
	Sustained period of drought	Infrequent	Frequent	Infrequent
Desirable plant traits	Cool climate succulent	✓		
	Warm climate succulent		✓	✓
	Cool season herbaceous	✓		
	Warm season herbaceous		✓	✓
Desirable media traits	Drainage	Well-poor	Well-moderate	Well
	Retention	Low-moderate	High	Moderate-high
	Thermal conductivity	Moderate	Low	Low

3.3.2 *Research Questions for the Future*

Media composition affects abiotic and plan roof performance. The seemingly infinite potential compositions need continued systematic examination to meet the challenges of specific environments. Water retention and drainage, and thermal characteristics deserve particular attention.

Plant selection for hot climate green roofs needs a shift of focus away from plant survival only and more to ward desired roof performance. Once this is established—storm water, building cooling etc. - then the roof can be designed and appropriate species selected accordingly.

Although the climate conditions will somewhat drive plant selection where climates are more seasonal (wet-dry; cool warm) and as climate change theory predicts more climatic stochasticity, the selection procedure should examine species with broad ecological niches with respect to soil water and soil and air temperature conditions.

Justification for green roofs will rely on the quantification all the potential environmental (e.g. storm water retention, thermal moderation), ecological (e.g. habitat) and sociological (e.g. access to green space) benefits collectively (Chap. 9) especially as many are not mutually independent.

The FLL standards have been useful metrics around which green roofs can be designed, built and experimented with, and have been successful in a range of climates (Philippi 2005; Dvorak 2011). However, these standards should not limit the development of green roof growing media specifications for hot climates where innovative ways to improve thermal and hydrologic characteristics are needed. The investigation and current development of standards (e.g. FLL ASTM) of all green roof components will be essential for the adoption of green roofs as a major contributor to green infrastructure.

References

Aber JD, Melillo JM (1982) Nitrogen immobilization in decaying hardwood leaf litter as a function of initial nitrogen and lignin content. Can J Botany 60:2263–2269

Alexandri E, Jones P (2008) Temperature decreases in an urban canyon due to green walls and green roofs in diverse climate. Build Environ 43:4810–4493

Barrio EPD (1998) Analysis of the green roofs cooling potential in buildings. Energ Build 27:179–193

Bass B, Krayenhoff ES, Martilli A et al (2003) The impact of green roofs on Toronto's urban heat island. In: Proceedings of greening rooftops for sustainable communities. Chicago 2003

Berghage RD, Gu J (2009) Effect of drain layers on green roof stormwater performance. In: The seventh annual greening rooftops for sustainable communities conference. Atlanta.

Cabugos LA, Kaufman J, Cox LJ et al (2007) Feasibility of rooftop landscaping with native Hawaiian plants in Urban Distracts of Hawaii. In: The fifth greening rooftops for sustainable communities conference. Minneapolis

Carter TL, Rasmussen TC (2006) Hydrologic behavior of vegetated roofs. J Am Water Resour Assoc 42:1261–1274

Castleton HF, Stovin V, Beck SMB et al (2010) Green roofs; building energy savings and the potential for retrofit. Energ Build 42:1582–1891

Drennan PM, Nobel PS (1998) Root growth dependence on soil temperature for *Opuntia ficus-indica*: influences of air temperature and a doubled CO2 concentration. Funct Ecol 12:959–964

Dunnett N, Kingsbury N (2004) Planting green roofs and living walls. Timber Press Inc, Portland

Dvorak B (2011) Comparative analysis of green roof guidelines and standards in Europe and North America. J Green Build 6:170–191

Dvorak B, Volder A (2012) Plant establishment on unirrigated green roof modules in a subtropical climate. AoB Plants 5:1–10

Dvorak B, Volder A (2013) Rooftop temperature reduction from unirrigated modular green roofs in south-central Texas. Urban Urban Gr 12:28–35

Farrell C, Mitchell RE, Szota C. et al (2012) Green roofs for hot and dry climates: Interacting effects of plant water use, succulence and substrate. Ecol Eng 49:270–276

FLL (2008) Guidelines for the planning, construction and maintenance of green roofing. Forschungsgesellschaft Landschaftsentwicklung Landschaftsbau e. V., Bonn, p 119

Fioretti R, Palla A, Lanza LG et al (2010) Green roof energy and water related performance in the Mediterranean climate. Build Environ 45:1890–1904

Getter KL, Rowe DB (2006) The role of extensive green roofs in sustainable development. HortSci 41(5): 1276–1285

Herppich WB (1997) Stomatal responses to changes in air humidity are not necessarily linked to nocturnal CO2 uptake in the CAM plant *Plectranthus marrubioides* Benth. (Lamiaceae). Plant Cell Environ 20:393–399

Kaufman A, Cox LJ, Muira T, et al (2007) The potential for green roofs in Hawai'i, Report RM-15, University of Hawai'i at Monoa

Korner CH, Scheel JA, Bauer H (1979) Maximum leaf diffusive conductance in vascular plants Photosynthetica. 13:45–8

Livingston EH, Miller C, Lohr M (2004) Green roof design and implementation in Florida. In: The second annual greening rooftops for sustainable communities conference, Portland

Liu TC, Shyu GS, Fand WT et al (2012) Drought tolerance and thermal effect measurements for plants suitable for extensive green roof planting in humid subtropical climates. Energ Build 47:180–188

Liu K, Baskaran B (2003) Thermal performance of green roofs through field evaluation. In: Proceedings for the first North American green roof infrastructure conference, awards and trade show. Chicago

MacIvor JS, Ranalli MA, Lundholm JT (2011) Performance of dryland and wetland plant species on extensive green roofs. Ann Bot 107:671–679

Miller C, Narejo D (2005) State of the green roof industry in the United States. In: Geosynthetics research and development in progress (GRI 18). ASCE, Austin, pp 1–8

Molineux CJ, Fentiman CH, Gangea AC (2009) Characterising alternative recycled waste materials for use as green roof growing media in the U.K. Ecol Eng 35:1507–1513

Monterusso MA, Rowe DB (2005) Establishment and persistence of Sedum spp. and native taxa for green roof applications. HortSci 40:391–396

Müller GT (2005) Evaluation of the first green roof in the district of Xochimilco in Mexico City after 15 Months. In: World green roof congress. Basel, 2005

Oberndorfer E, Lundholm J, Bass B, et al (2007) Green roofs as urban ecosystems: ecological structures, functions, and services. Biosci 57:823–833

Onmura S, Matsumoto M, Hokoi S (2001) Study on evaporative cooling effect of roof lawn gardens. Energ Build 33:653–666

Peel MC, Finlayson BL, McMahon TA (2007) Updated world map of the Köppen-Geiger climate classification. Hydrol Earth Syst Sci 4:439–473

Philippi PM (2005) Introduction to the German FLL-guideline for the planning, execution, and upkeep of green-roof sites. In: The third annual greening rooftops for sustainable communities conference. Washington, 2005

Santamouris M (In press). Cooling the cities–a review of reflective and green roof mitigation technologies to fight heat island and improve comfort in urban environments. Sol Energy 103:682–703

Schroll E, Lambrinos J, Righetti T et al (2011) The role of vegetation in regulating stormwater runoff from green roofs in a winter rainfall climate. Ecol Eng 37:595–600

Simmons MT, Gardiner B, Windhager S, et al (2008) Green roofs are not created equal: the hydrologic and thermal performance of six different extensive green roofs and reflective and non-reflective roofs in a sub-tropical climate. Urban Ecosyst 11:339–348

Snodgrass EC, Snodgrass LL (2006) Green roof plants. Timber Press, Portland

Sonne J (2006a) Evaluating green roof energy performance. ASHRAE J 48:59–61

Sonne J (2006b) Energy performance aspects of a Florida green roof. In: 15th annual symposium on improving building energy systems efficiency in hot and humid climates. Orlando

Sutton R, Harrington JA Skabelund L et al (2012) Prairie-based green roofs: literature, templates and analogs. J Green Build 7:143–172

Tan PY, Sia A (2005) A pilot green roof research project in Singapore. In: Third annual greening rooftops for sustainable communities conference, awards and trade show. Washington

Urban J (2008) Up by roots. International society of arboriculture. Champaign, Illinois

VanWoert N, Rowe D, Andresen J et al (2005) Green roof stormwater retention effects of roof surface, slope, and media depth. J Environ Qual 34:1034–1044

Volder A, Dvorak B (2014) Event size, substrate water content and vegetation affect storm water retention efficiency of an un-irrigated extensive green roof system in Central Texas. Sustain Cities Soc 10:59–64

Wanielista MP, Hardin M, Kelly M (2008) A water quality assessment of two green roof stormwater treatment systems. In: ASCE, 2008 Honolulu. ASCE, pp 423–423

Williams NSG Rayner JP, Raynor KJ (2010) Green roofs for a wide brown land: opportunities and barriers for rooftop greening in Australia. Urban Urban Gr 9:245–251

Wolf D, Lundholm JT (2008) Water uptake in green roof microcosms: effects of plant species and water availability. Ecol Eng 33:179–186

Wong NH, Chen Y, Ong CL, et al (2003) Investigation of thermal benefits of rooftop garden in the tropical environment. Ecol Eng 38:261–270.

Xu Q, Huang B (2000) Growth and physiological responses of creeping bentgrass to changes in air and soil temperatures. Crop Sci 40:1363–1398

Dr. Mark T. Simmons is Director of Research at the Lady Bird Johnson Wildflower Center at the University of Texas. Restoration and design projects focus on the creation of urban prairies and riparian systems and engineered sustainable urban landscapes. Scientific research projects include: green roofs, green walls, native turfgrass, urban prairies and prescribed fire.

Chapter 4
Water Through Green Roofs

John G. Lambrinos

Abstract How green roofs process water is a critical component of their function and effective management. As green roof technology has spread from northern Europe's relatively cool and humid climate, green roof designs have had to adapt to regional variations in the timing and availability of water. The development of regionally appropriate designs requires a mechanistic understanding of green roof water relations, plant eco-physiology and irrigation technology. Emerging designs effectively match environmental conditions, substrate characteristics, plant physiological traits, plant community interactions, and expert systems for applying supplemental water.

Keywords Evapotranspiration · Hydrology · Irrigation · Plant selection · Plant water use strategies · Xeric climates

4.1 Green Roofs as Hydrological Systems

The vast majority of the water that lands on a conventional roof quickly flows off. In sharp contrast, water that lands on a green roof enters a complex hydrological system (Fig. 4.1). Stocks of water are held on and within plants, in substrate, and in various layered materials such as drainage and water retention fabrics. Water exits the system through evaporation from the substrate and plant surfaces, transpiration, and runoff. The magnitude of the various stocks of water as well as the flux of water between stocks and out of the system is governed by complex interactions among green roof components and the physical environment. This complexity makes green roof performance inherently dynamic and contingent on the details of system design and local conditions (Berndtsson 2010).

Nevertheless, most extensive green roofs share broadly similar hydrological characteristics. The bulk of the standing stock of water on a green roof is held in the substrate. The amount of water intercepted and held by vegetation is comparatively

J. G. Lambrinos (✉)
Department of Horticulture, Oregon State University, 4017 Agriculture and Life Sciences Building, Corvallis, OR 97331-7304, USA
e-mail: john.lambrinos@oregonstate.edu

R. Sutton (ed.), *Green Roof Ecosystems,* Ecological Studies 223,
DOI 10.1007/978-3-319-14983-7_4

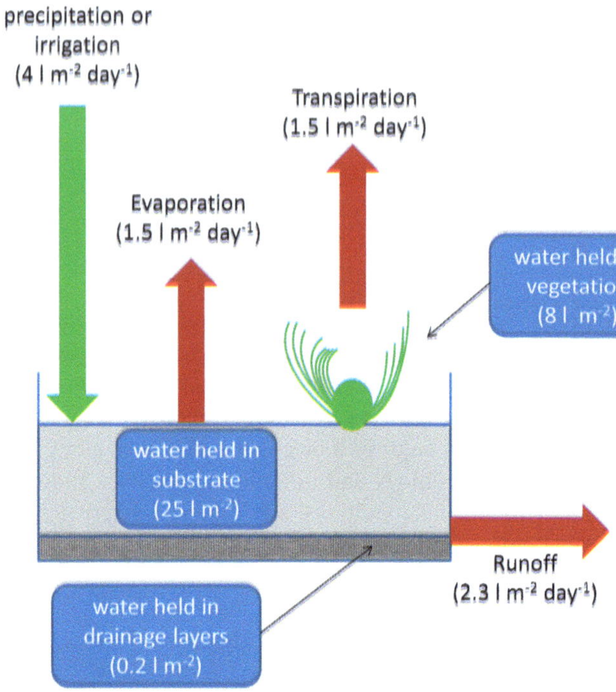

Fig. 4.1 Stocks and flows of water through a typical extensive green roof. The hypothetical scenario depicted is for an extensive green roof that has reached maximum water storage capacity during a spring rainstorm in Corvallis, OR U.S.A. The roof design is assumed to consist of 70 mm of pumice-based substrate, an 8 mm drainage layer, and *Sedum* sp. vegetation. Stock and flow estimates (*in parentheses*) are derived from published values for similar systems: precipitation (Schroll et al. 2011a), evaporation and transpiration (Voyde 2011), runoff as a function of % retention (Spolek 2008), substrate water storage capacity (Fassman and Simcock 2012), drainage layer water storage capacity (VanWoert et al. 2005a, Fassman and Simcock 2012), *Sedum* sp. water storage capacity (Berghage et al. 2007; Fassman and Simcock 2012)

small, at least on *Sedum* dominated roofs. Many studies have reported no significant difference in stormwater storage between vegetated and un-vegetated (substrate only) roofs, although results vary with storm size and season (Monterusso et al. 2004; VanWoert et al. 2005a; Dunnett et al. 2008; Lundholm et al. 2010; Buccola and Spolek 2011; Starry 2013). While the amount of water captured by most green roof vegetation is small relative to the substrate, differences in plant architecture and vegetation structure have been shown to significantly influence water capture. Roofs planted with grasses and forbs as well as roofs that combine different growth forms capture and retain significantly more water than sedum only roofs (Lundholm et al. 2010; Nagase and Dunnett 2012). Also, some potential green roof plant choices have exceptional water capturing properties. For example, the sponge like physical structure of many mosses allows them to hold 8–10 times their dry weight in water. In contrast to other vegetation types, moss covered roofs can often retain significantly more water than substrate only roofs (Anderson et al. 2010). The stock of water held within plant tissues can also be sizeable for some vegetation types.

For example, *Sedum* and other succulent species can be 80–90% water by weight under well-watered conditions (Berghage et al. 2007). While this stock of water does not directly influence broad hydrological properties such as stormwater retention to a significant degree, it is an important component of the drought tolerance mechanisms for many species.

The substrates used in extensive green roofs are designed to be highly permeable, but also have relatively high water holding capacity for their weight. Extensive substrates typically have maximum densities of around 1 g/cm^{-3} (62 lbs/ft^3) and target maximum water holding capacities (water storage at field capacity) that range from 25 to 65% by volume (FLL 2008). However, under field conditions, the actual maximum storage capacity of substrates is typically less than that estimated from laboratory techniques, likely as a result of structural changes to the substrate caused by plant root development and evapotranspiration (Fassman and Simcock 2012). In addition to substrate, extensive designs usually incorporate one or more layers designed to facilitate drainage, minimize erosion, or retain water for plant use. Few studies explicitly report water-holding capacities for these layers, but those that do report values less than 20% by volume (Miller 2003; VanWoert et al. 2005a).

Under most conditions, the water that enters an extensive green roof has a short residence time. Substrate profiles are only 20–150 mm (5–37 in.) deep and constrained by an impervious membrane. This limits the absolute storage capacity of extensive designs. Once maximum storage capacity is reached, the porous substrate and even more porous drainage layers quickly channel runoff off of the roof. In most designs runoff enters the municipal wastewater stream, but runoff can be captured by a gray water system and returned to the roof as irrigation (Compton and Whitlow 2006; Chang et al. 2011). In many studies the stormwater storage capacity of green roofs is defined in terms of the plant available water held in the substrate (storage at field capacity-storage at the wilting point of vegetation). This value represents the theoretical long-term average capacity of the roof to retain intercepted rainfall. Storage capacity varies considerably with substrate composition and age. Martin (2008) reported storage capacity values from eleven studies that ranged from 2 to 53%. When conditions are favorable for photosynthesis, green roof vegetation quickly depletes the stock of water in the substrate, restoring the storage capacity of the roof. A number of studies report that under ideal conditions the water storage capacity of vegetated roofs can be completely restored within about a week (VanWoert et al. 2005b; Durhman et al. 2006; Berghage et al. 2007; Voyde et al. 2010a).

The restoration of storage capacity through evapotranspiration is a key component of roof function, but it is a process that is strongly dependent on climatic conditions, the water status of the roof, and the composition of the vegetation. Evapotranspiration is extremely sensitive to a number of climatic drivers, principally temperature, humidity, and wind (Allen et al. 1998). In addition, the transpiration component of evapotranspiration is strongly tied to water availability. Plants maximize transpiration when water is readily available. Under well-watered conditions, succulent green roof vegetation contributed more than 50% of the evapotranspiration from extensive green roofs (Berghage et al. 2007; Voyde et al. 2010b). Transpiration declines as water availability declines until water stress reaches a plant specific threshold, at which point transpiration ceases. Many of the drought-adapted species commonly used in extensive green roof designs have relatively high transpiration rates when water is available. Sedum roofs have reported maximum evapotranspiration rates that range from 5 to 6 mm day^{-1} (VanWoert et al. 2005b; Durhman et al. 2006; Voyde 2011; Sherrard and Jacobs 2012; Chap. 2). Roofs planted with non-xerophytic species such as *Spartina alterniflora* and *Solidago canadensis* can attain evapotranspiration rates that are an order of magnitude greater (Compton and Whitlow 2006), and roofs planted with a combination of growth forms can exhibit significantly greater rates than monocultures (Lundholm et al. 2010).

4.2 Climatic Influence on Green Roof Water Dynamics

Extensive green roofs balance two inherently conflicting goals: lightweight and water storage. They achieve their relatively lightweight (even when saturated with water) by being shallow and using light porous substrates. Coarse substrate texture also minimizes the risk of ponding in the shallow profile even during extreme rain events. Despite the limitation on total water storage capacity set by this substrate design, the presence of substrate alone significantly retards the timing and reduces the amount of runoff compared to conventional roof designs (VanWoert et al. 2005a; Schroll et al. 2011a). The presence of plants can potentially significantly enhance this stormwater function by dynamically restoring storage capacity through transpiration. However, the typical substrate design creates a unique and often severe water environment for plants that constrains plant choice. The rapid flow of water through the system via runoff or evapotranspiration can quickly put plants under water stress. As a consequence, shallow-rooted, drought tolerant species have typically been used on extensive green roofs even in mesic climates. The most common growth form used are succulents, with a distinct fondness for members of the Crassulaceae, although a broader range of regionally matched species and growth forms are increasingly being used (Dvorak and Volder 2010; Sutton et al. 2012). This system can be remarkably effective at both attenuating runoff and maintaining healthy vegetation if individual rain events are small in total volume and spaced at moderate (1–2 week) intervals between periods during which conditions are favorable for photosynthesis (Berghage et al. 2007). Many temperate climates such as those in northern Europe and in northeastern North America meet these criteria for substantial parts of the year. Average yearly stormwater retention values for roofs in these climates typically range from 30 to 60 %, broadly approximating mass balance estimates of watershed evapotranspiration for their regions (Gregoire and Clausen 2011; Carson et al. 2013).

However, a range of regional climates have conditions that are challenging for green roofs (Fig. 4.2). Climates with low or strongly seasonal precipitation may not provide enough water in a green roof context to sustain many otherwise drought adapted species. Even if a green roof plant assemblage can survive under a particular regional water regime it can still suffer reduced plant cover, be constrained in

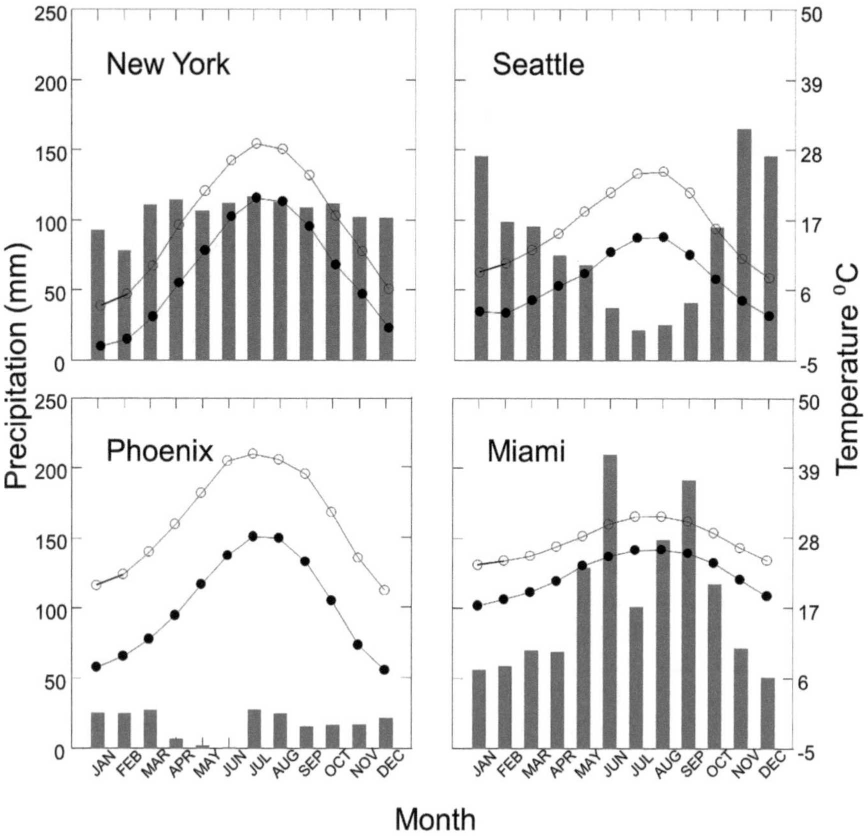

Fig. 4.2 Seasonal precipitation and temperature values for four North American cities. Values are thirty-year (1981–2010) averages for total monthly precipitation (*bars*), average monthly minimum temperature (*solid circles*) and average monthly maximum temperature (*open circles*). Data from NOAA National Climatic Data Center (NOAA 2014)

species or growth form diversity, or suffer reduced aesthetics (Nagase and Dunnett 2010; Schroll et al. 2011b; MacIvor et al. 2013). Other aspects of regional climate such as temperature extremes can also restrict plant choices. For example, many sedum species are intolerant of hard frosts or high temperatures (Chap. 3) (Boivin et al. 2001; Livingston et al. 2004; Simmons et al. 2008; Williams et al. 2010; Rowe et al. 2012).

The pattern of water availability imposed by regional climate can also influence other aspects of green roof performance. The thermal benefits of green roofs are influenced by the water content of the substrate and by evapotranspiration, and have been shown to vary by building type and location (Sailor et al. 2012). During extended dry periods, low levels of soil moisture and reduced evapotranspiration can lower the thermal benefit of a green roof (Sun et al. 2014). In the humid tropics, green roofs can become large heat sinks, partly because extreme daytime temperature can cause plants with crassulacean acid metabolism (CAM) physiology to stop transpiring. Much of the stored heat can later be transmitted into the building at rates greater than a conventional roof (Jim 2014). Green roof stormwater attenuation performance is also strongly influenced by climatic conditions. In general, the proportion of precipitation retained by green roofs declines as storm volume and frequency increase and during winter when the potential evapotranspiration is limited (Mentens et al. 2006; Villarreal 2007; DiGiovanni et al. 2010; Carson et al. 2013).

A general, lack of published long-term monitoring data (Chaps. 2 and 13) as well was design differences among roofs make it difficult to assess the impact that regional climate has on overall green roof performance. The few data that are available principally report stormwater performance. In the Pacific Northwest of North America the majority of the yearly precipitation falls during the winter when the potential for evapotranspiration is small. In addition, winter storms can come closely spaced in time creating prolonged periods of precipitation. Spolek (2008) reports that roofs in Portland, OR monitored over 2–3 years had total rainfall retention that ranged from 12 to 25 %. A Seattle roof monitored for a year had a total retention of 31 % (Berkompas et al. 2009). Both of those values are among the lowest reported from full-scale monitored green roofs (Carson et al. 2013). The overall, long-term retention seems to be driven by reduced retention performance during the winter. On the Portland roofs monitored by Spolek (2008), total rainfall retention was only 12 % during the winter compared with 42 % during the spring and summer. Similarly, test-bed scale roofs in Corvallis, OR retained less than 28 % of intercepted rainfall during the winter, which was less than half their retention capacity during the summer (Schroll et al. 2011a). In the Corvallis study, the seasonal differences were partly attributable to the higher frequency and volume of storm events during the winter as well as to the reduced recharge capacity provided by the vegetation.

In other regions, such as sub-tropical Florida, the timing and intensity of individual storm events can influence stormwater performance even if the bulk of rainfall occurs during favorable evapotranspiration conditions. Using a field validated mass balance model Hardin et al. (2012) predicted green roof stormwater retention values for several Florida locations ranging from 33 to 51 %. Locations with moderate predicted retention efficiencies seem to reflect the large, overall volume of annual precipitation at those sites. Similarly, green roofs in climates that experience large yearly variation in conditions can exhibit large variation in stormwater performance. The southern California climate is notable for its extreme inter-year variation in total precipitation. There Bennett et al. (2008) report that modeled rainfall retention efficiency for a typical extensive design varied from 21 to 64 % depending on yearly precipitation patterns. It is important to note that total stormwater retention may not be an appropriate performance metric for some goals. Delay in runoff as well as reductions in peak flow for individual storm events can be more relevant to stormwater management, and these aspects of attenuation performance can be high relative to conventional roofs across a range of storm size and timing (Fioretti et al. 2010). In general, high variability in rainfall patterns and weather conditions make it difficult to predict the hydrological performance of green roofs based on storm characteristics. Instead, mechanistic models of water flux through the green roof system provide the best predictive descriptor of stormwater performance (Stovin et al. 2012).

4.3 Plant Water Use Strategies for Green Roofs

Ideally, green roof vegetation should combine high transpiration capacity with the ability to tolerate extended periods of water deficit. This is perhaps not as conflicted a goal as it might seem. Plants possess an incredible diversity of water use strategies, many of which are appropriate in green roof contexts across a range of regional climatic conditions. Farrell et al. (2013a) have developed a conceptual model to screen potential plants for green roof applications based on their water use under mesic and xeric conditions as well as their ability to minimize water stress during periods of water deficit. Here I categorize a slightly broader (although still not comprehensive) list of water use strategies into syndromes that relate to green roof performance.

4.3.1 Water Loss Minimizers

Many species from xeric or seasonally dry climates are exceptionally good at conserving water. Structural adaptations in these species include waxy or hairy leaf coverings, leaf orientations that reduce insolation and heating, fewer or smaller leaves, and reduced stomatal density. Many species also have succulent stems or leaves and use internally stored water to buffer the effects of soil water deficit. Another principal physiological adaptation is crassulacean acid metabolism (CAM) photosynthesis. Plants with one or more of these adaptations can survive extended periods of water deficit. Drought tolerance screening experiments have identified a number of species that can survive on typical extensive substrates without water for more than 130 days (Durhman et al. 2006; Bousselot et al. 2011; Farrell et al. 2012). Many water loss minimizing species used in green roof applications are low-growing perennial succulents, particularly those in the family Crassulaceae, but perennial forbs (Fig. 4.3a) and woody shallow rooted shrubs are also common.

An inherent tradeoff of many of the adaptations that minimize water loss is reduced photosynthesis. Consequently many drought tolerant species have comparatively low photosynthetic rates per leaf area or biomass (Körner et al. 1979). Water loss minimizers emphasize consistent (although relatively low) photosynthetic capacity across a range of water conditions. Consequently, extreme water loss minimizers might not be the most appropriate plant choices for green roofs that experience modest periods of water deficit. In these cases, species with a greater peak capacity to transpire would be more desirable. Interestingly, great variation exists in the peak transpiration capacity among succulent species commonly used on green roofs, even within the same genus (Voyde et al. 2010b; Farrell et al. 2012; Starry 2013). Some of this variation potentially reflects dynamic responses to water availability in some species (see Sect. 4.3.2 *Water loss adapters*).

On the other hand, some water loss minimizers are robust choices in climates where green roof substrates are likely to experience more prolonged water deficit, or in climates where water availability is highly variable. Some xerophytic peren-

Fig. 4.3 Species representing different water use syndromes useful in green roof contexts. **a** Species like *Erigeron linearis* can withstand long periods of drought even on thin mineral soils, because they have consistently low transpiration rates. **b** Many species in the genus *Sedum* facultatively switch between C3 and CAM photosynthetic pathways allowing them to have exceptional drought tolerance but also achieve moderate transpiration rates when water is available. **c** Geophytes like *Camassia quamash* avoid water stress through dormancy. **d** Bryophytes like *Racomitrium canescens* tolerate long periods of desiccation. **e** Although not exceptionally drought tolerant, many warm season grasses like *Zoysia* sp. can withstand extreme temperature and light conditions. (Photograph credits: (a) Richard Martinson, (b, c), Erin Schroll, (d) John Lambrinos, (e) Alec Kowalewski)

nials have pronounced water loss adaptations, but are poor choices for extensive green roofs because they minimize water stress by accessing water stores from deep or spatially complex soil profiles (Ehleringer and Mooney 1983); the process of hydraulic redistribution is a tactic wholly unavailable on shallow, extensive green roofs.

4.3.2 Water Loss Adapters

Some species are particularly adept at adjusting their water use to the amount of water available, enabling them to have higher photosynthesis rates when water conditions are favorable. This is a desirable trait for a green roof plant. Woody perennials often adjust water use through gross morphological changes such as replacing photosynthetically efficient leaves with more water use efficient ones, or by shedding leaves altogether and entering a period of drought induced dormancy (Westman 1981). Many perennial temperate grasses achieve exceptional drought dormancy by combining leaf senescence with physiological dehydration tolerance mechanisms (Volaire and Norton 2006). Not all drought dimorphic species are desirable for extensive green roof applications. Many of these species (both woody and herbaceous) have deep or extensive root systems. In addition, drought induced changes can create undesirable aesthetics or pose fire safety concerns from the accumulation of dry biomass.

In some species, photosynthetic pathways are highly plastic and plants facultatively adjust photosynthetic metabolism based on water availability and other environmental conditions (Andrade et al. 2009). Many succulent species are known to switch between C3 and CAM photosynthetic pathways or to adjust the diurnal timing of gas exchange and CO_2 fixation in relation to water availability (Sayed 2001; Fig. 4.3b). These adjustments allow some succulent species to achieve moderate transpiration rates during periods of high water availability. However, the details of photosynthetic plasticity and its influence on water use patterns are highly species specific, nuanced, and dependent on a number of environmental factors (Herrera 2009). For example, although *Phedimus albus* (syn. *Sedum album*) and *Phedimus kamtschaticus* (syn. *S. kamtschaticum*) are both broadly known to switch between C3 and CAM metabolism, they display markedly different physiological performance in a green roof context (Starry et al. 2014). Starry found that *S. kamtschaticum* had significantly higher transpiration, higher daily carbon assimilation, and switched from C3 to CAM metabolism at a lower substrate water availability compared to *S. album*. As a consequence it used 35 % more water than *S. album*. However, perhaps partly because of its more parsimonious water use *S. album* was more drought tolerant than *S. kamtschaticum*. Rowe et al. (2014) found similar results with syn. *S. kamtschaticum* var. *floriferum* (trade name *S. floriferum*). In a greenhouse study using experimental roof modules, *S. album* survived 84 days without water, but *S. floriferum* did not.

4.3.3 Water Stress Avoiders

Some species take drought dormancy to the extreme and either complete their entire life cycles before water availability declines, or exist for extended periods as highly specialized drought survival structures. Many of these species have ruderal life history strategies or evolved under strongly seasonal water availability such as deserts or seasonal wetlands. Desert annuals that complete their life cycle within a brief few weeks are prime examples. Many annuals from a range of different habitats could be suitable for green roof contexts, although they have not often been used to date (Chap. 10). Nagase and Dunnett (2013) report that a diverse annual meadow can be easily and economically established on an extensive green roof in the central UK climate. A diverse assemblage of species provided abundant flowers throughout the summer and fall even without irrigation, and the system required very little maintenance apart from annual mowing. However, the long-term performance of annual plant based systems has not been investigated. Annual systems will likely require a tolerance for large dynamic changes in the species composition of the roof over time (Chaps. 12 and 13).

Species with perennating organs such as geophytes are another category of drought stress avoiders that are potentially suitable in a green roof context (Schroll et al. 2011b; Nagase and Dunnett 2013; Van Mechelen et al. 2014; Benvenuti 2014; Fig. 4.3c). Like annuals, these species often produce strikingly attractive flowers, however bloom times can be relatively brief and all above ground structures typically die back completely. However, in contrast to many annuals, the amount of senescent biomass is relatively small and mowing management is not required. Many geophytes have particularly early or late bloom times that can be a valuable trait in terms of pollinator resources as well as expanded aesthetics (Benvenuti 2014).

4.3.4 Water Loss Tolerators

Some species lack well-developed adaptations for conserving water, but instead have a remarkable ability to withstand desiccation (Hoekstra et al. 2001). Many of these species are non-vascular bryophytes and lichens, but some vascular *resurrection* plants have this ability as well (Gaff 1989). Mosses have most commonly been used on green roofs (usually in combination with sedum) in northern temperate climates (e.g. Bengtsson et al. 2005; Oberndorfer et al. 2007). However, their drought tolerance properties make them potentially suitable choices for a number of climates with extended periods of water scarcity (Anderson et al. 2010). Also, their lack of roots and prodigious water retention capacity also suggest that they could be used to develop extremely lightweight but still highly functional systems.

4.3.5 Water Loss Sensitive

Plants that do not have well developed drought tolerance mechanism can still be suitable choices for extensive and semi-intensive green roofs in some contexts. Because of their tolerance of extreme temperature and light conditions, warm season turf grasses have been used on green roofs in sub-tropical and tropical climates, although typically supplemental irrigation is provided (Jim 2012; Ju et al. 2012; Sutton et al. 2012; Chen 2013). Some wetland species have broad habitat tolerances or some ability to withstand short periods of dry conditions. They may be good choices for relatively wet climates. MacIvor et al. (2011) found that several wetland species were able to survive condition on an extensive green roof in maritime Nova Scotia over two growing seasons, although their overall cover was less than that of more dryland-adapted species.

It is important to note that the performance oriented syndromes described above don't necessary reflect evolutionary or ecological tradeoffs. Species can combine aspects and traits that span categories. Indeed, most plants exhibit some characteristics of each syndrome to varying degrees. Still, the syndromes provide a useful way of relating dominant species traits to functional green roof goals as well as to the abiotic constraints imposed by different regional or situational contexts. Closely related species or species that share similar life forms and life history can often have very different overall water use patterns (Wolf and Lundholm 2008). This makes it suspect to use growth form or simple morphological traits as screening tools for green roof plant assembly. Plant choice decisions need to be based on detailed species-specific functional traits that are evaluated in the context of specific climate profiles and performance goals (Chaps. 9 and 11).

4.4 Modifications to Green Roof Water Dynamics

The typical extensive green roof design can be modified in a number of ways that alter the dynamics of water through the system. These changes are often made to better match a particular green roof to local climatic conditions, or to enhance particular green roof functions such as aesthetics or habitat quality.

4.4.1 Plant Assemblage Design

The species and growth form composition of green roof vegetation can have a strong influence on water capture and retention, as well as the rate at which storage capacity is restored following a rain event (see Sect. 4.1). For many functional goals such as stormwater management and building thermal load reduction, vegetation designs must balance a tradeoff between high transpiration capacity and drought tolerance. The optimal balance between these two functional traits depends strongly on the specific climatic context of the roof. However, designing systems that are inflexibly tailored to a narrow regime is unwise. Climatic conditions vary within and between years, and spectacularly so in some climates. Incorporating species with an ability to facultatively adjust water use depending on water availability is one strategy for improving performance under variable conditions (Chap. 11). The water use plasticity of many *Sedum* and other succulent species make them good choices for extensive green roofs. However, there is considerable variation in water use patterns as well as climatic tolerances among succulent species (Voyde et al. 2010b; Farrell et al. 2012; Starry 2013). In addition, species from other growth forms and taxonomic groups can also exhibit high degrees of water use plasticity (Farrell et al. 2013a). Such species-specific functional traits are too rarely taken into account when making green roof plant selections, partly reflecting lack of accessible data on the functional traits of candidate green roof species.

Another strategy for designing functionally resilient green roof vegetation is to combine species with complementary water use patterns or functional traits. Several studies have reported a positive relationship between the species or growth form richness of green roof vegetation and water management performance as well as other functions (Lundholm et al. 2010; Nagase and Dunnett 2010; Cook-Patton and Bauerle 2012; Chap. 9). Although the mechanistic causes of these relationships are not well understood, one likely reason is trait complementarity. Specifically exploiting complementarity could be an effective design strategy. For example, in the Pacific Northwest of North America most of the precipitation falls during the cool winter when potential evapotranspiration is low. During these periods mosses can significantly increase water storage capacity above that of the substrate itself through water held in their complex physical structure (Anderson et al. 2010). During the spring when conditions are more favorable for photosynthesis, vascular plants can contribute significantly to recharge capacity through transpiration (Schroll et al. 2011a). Preliminary results suggest that combining both moss and sedum can significantly improve yearly stormwater retention over single species vegetation types (Van Hoosen pers. comm.).

Another potential cause of the positive relationship between performance and vegetation diversity is that some species facilitate the growth and survival of other species in the assemblage. One broad way that facilitation can happen is that species modify abiotic conditions, making them more favorable for themselves or other species (Hastings et al. 2007). Butler and Orians (2011) showed that the growth and overall health of the perennial forbs *Agastache rupestris* and *Asclepias verticillata* on green roofs were decreased by the presence of *Sedum* species during favorable conditions but were increased during summer water deficit. Similarly, Heim (2013) found that the presence of the moss *Polytrichum commune* increased the growth of the perennial forb *Solidago bicolor* under experimental green roof conditions. In both studies the cause of the observed facilitation is equivocal, but both the *Sedum* and moss decreased temperature and increased water availability in the substrate. Plant-microbial symbioses are another broad mechanism for facilitation. For example, the symbiotic relationship between most plants and mycorrhizae fungi can directly increase their ability to acquire and uptake water, particularly under drought conditions (Augé 2001; Chap. 7). Most members of the Crassulaceae do not form arbuscular mycorrhizal associations (Wang and Qiu 2006). However, the near absence of arbuscular mycorrhizal fungi within newly installed engineered substrates could be a significant factor limiting the range of plant species that are suitable for green roofs as well as the drought tolerance of diverse green roof systems (John 2013).

As with complementarity, facilitation could potentially be exploited to improve functions associated with plant water use. For example, moss could be used to facilitate the establishment of vascular species, reduce the frequency of extreme drought stress, or reduce the overall need for supplemental water. However, manipulating species interactions to achieve specific functional goals is complicated by the inherent dynamism of green roof vegetation (Chap. 12). Although few published long-term studies exist, those that do suggest that the composition and relative species abundance of green roof vegetation can change dramatically in the years following establishment (Chap. 12). In some cases, vegetation changes seem to reflect consistent successional trajectories that are constrained by substrate type and depth, and by water availability (Köhler 2006; Rowe et al. 2012; Bates et al. 2013; Thuring and Dunnett 2014). However, there can also be considerable year-to-year variability in community structure much of which correlates with variation in drought stress (Bates et al. 2013). Indeed, one aspect of developing drought resilient vegetation may be accepting a degree of un-planned variation in its composition, including the natural colonization of species from the regional species pool (Chaps. 10, 11 and 12).

4.4.2 Substrate Design

Since substrate is the largest single store of water in a green roof system unsurprisingly its composition, depth, and slope strongly influence patterns of water flow through the system (Li and Babcock 2014). Substrate design is therefore an important way of optimizing extensive green roofs to particular environmental constraints or functional goals. In some cases substituting natural soil profiles for highly engineered substrates may be a productive approach (Chap. 6). Also, some authors have argued that wetland-like systems could be a practical alternative under some conditions (Song et al. 2013). More commonly, a number of adjustments have been made to the basic engineered substrate design in order to attain specific performance goals or in response to environmental constrains. However, developing appropriate design criteria is complicated by the complex interactions between substrate, plants, and environmental conditions (Chap. 5). For example, in isolation the influence of substrate characteristics on water retention and runoff dynamics are straightforward to predict using existing mass balance and more mechanistic hydrological models. However, because hydrological performance reflects strong interactions between a number of highly variable system components and environmental conditions it is necessary to calibrate and validate models to each specific case, limiting their usefulness as design tools (Li and Babcock 2014).

Another example of this contextual complexity occurs with the relationship between substrate depth and plant performance. Increasing substrate depth increases the store of water available to plants (VanWoert et al. 2005a), buffers plant roots from cold stress (Boivin et al. 2001; Rowe et al. 2012), and can accommodate species with deeper rooting profiles (Sutton et al. 2012). A number of studies have documented a positive relationship between growth and survival and substrate depth for a number of potential green roof species. Dvorak and Volder (2010) reviewed this literature and concluded that without irrigation only the most drought adapted succulent species are able to tolerate the water stress conditions imposed by the shallowest (<10 cm) substrate profiles across a range of climates. Increasing substrate depth or providing supplemental irrigation greatly expands the diversity of species and functional types that a roof can support. However, the results are highly species specific and can vary over time (Dunnett et al. 2008; Getter and Rowe 2009; Rowe et al. 2012). For example, in the Rowe et al. (2012) study seven species performed well on 2.5–7.5 cm substrate depths when they were evaluated 2 years after establishment; yet by year seven only two of these species were still present on any media depth.

Most extensive green roof substrate designs are based on guidelines established by the German Landscape Research, Development, and Construction Society

(FLL). The guidelines set performance criteria for key parameters such as permeability, water storage capacity, and maximum load. Performance targets vary for different green roof configurations and contexts, but broadly specify that substrates should have high permeability (saturated water flow ≥ 0.001 cm s^{-1}), hold 35–65% v/v water at field capacity, contain $\leq 15\%$ w/w of fine (<63 μm) particles, and contain 10–20% v/v organic matter (FLL 2008). Typically designers have achieved these performance targets using substrates composed primarily of course inorganic aggregates such as expanded clay, pumice, and a number of recycled materials such as crushed brick. A number of modifications to substrate composition have been proposed to increase the amount of water available to plants or to dampen fluctuations in water availability. These include increasing the organic matter content or incorporating other water retention additives such as polymer gels. As with substrate depth, the efficacy of these strategies appears to be highly context dependent. For example, in a short-term greenhouse experiment Nagase and Dunnett (2011) evaluated the influence of substrate organic matter content on the growth of four forbs and grass species. All four species responded differently to the level of organic matter, and results depended on the watering regime. Under a dry regime increasing organic matter content above 10% by volume did not have any significant effect on plant growth. However, under a well-watered regime some species increased growth considerably with higher organic matter content. Nagase and Dunnett speculate that the lush growth might be a disadvantage under more natural conditions that include periodic drought. Papafotiou et al. (2013) tested the influence of substrate depth, organic matter type and content, and irrigation frequency on the growth of several drought adapted Mediterranean species. Similar to the Nagase and Dunnett (2011) study, they found significant interactions between treatments. Notably, however, some of the best plant performance was observed on the shallow (15 cm) compost amended substrate even under minimal irrigation. Other water retention amendments such as hydrophilic polymers (hydrogels) and silicate granules can increase overall as well as plant available water holding capacity of the substrate, although the magnitude of the effect depends on the type of additive and substrate (Farrell et al. 2013b). The incorporation of hydrogels into green roof media has been found to increase the growth of grasses and non-succulent forbs (Oschmann et al. 2007; Sutton 2008). Biochar is another potential amendment that could increase plant available water. Beck et al. (2011) report that green roof substrate amended with 7% biochar had significantly greater water retention than non-amended substrate. However, no published studies have evaluated the effect of biochar on plant available water or plant performance in a green roof context.

In addition to the substrate itself, most extensive green roof profiles include a number of plastic or woven geotextile layers, some of which are explicitly designed to retain and increase plant available water. However, few studies have directly evaluated how these layers influence the plant water relations of the system (Chap. 3). In the only experimental study that has been reported to date, Savi et al. (2013) report that 90% of the water retained by these layers is potentially available to plants, compared to 34% for the substrate itself. The presence of the layers also had a significantly positive effect on plant water status and survival. However, the transfer of water through the roof profile was strongly influenced by diurnal temperature patterns and the details of system design, suggesting that designs could be optimized to enhance plant water availability under specific environmental conditions.

The wide variety of roof designs, the complex interactions between design parameters and environmental conditions, and the high degree of species-specific responses make it nearly impossible to establish universal design prescriptions for green roof substrate. Instead, a promising approach is to develop regionalized and function specific design (Chap. 9) criteria based on local field testing. Fassman and Simcock (2012) used this approach to develop design specifications for extensive green roof media for Auckland, NZ that will maintain plants without irrigation under typical conditions and capture 100 % of runoff from storms that have less than 25 mm (1 in.) of precipitation.

Despite the complications described above, incorporating heterogeneous substrate depths as well as improving the plant available water capacity of the substrate are promising strategies for maintaining species or growth from diverse vegetation on extensive green roofs. In climates that experience extreme water deficit, adjustments within extensive design constraints may not be sufficient if aesthetics or plant diversity are important design goals. An example of such a system is the green roof installed on the Oregon Dental Service (ODS) building in Bend, OR (Fig. 4.4). The average annual precipitation in Bend is only 29.5 cm (NOAA National Climatic Data Center) and since the roof was designed as an accessible recreation area of

Fig. 4.4 The green roof on top of the Oregon Dental Service (*ODS*) Building in Bend, OR. In extreme water environments like Bend, designs may need to incorporate deeper substrate depths if diverse vegetation is a goal. Substrate depths here vary from 20 to 81 cm (8–32 in.). (Photo: Richard Martinson)

the building, diverse and aesthetically pleasing vegetation was a design criterion. To support the vegetation substrate was composed of 50 % mushroom compost and 50 % pumice and varied in depth from 20 to 81 cm. The roof was planted with 29 native western U.S. plant species that were matched to the specific substrate microhabitats.

The initial plantings established well. However, over the next several years maintenance crews removed the majority of the perennial grasses and forbs that were part of the initial installation. The crews were unfamiliar with the plant palette and removed as weeds anything they did not recognize. The company that initially designed and installed the roof has recently been hired to re-establish some of these plantings and to provide ongoing maintenance. This experience highlights the need for comprehensive management plans and properly trained maintenance personnel in order to ensure the long-term success of green roof systems (Chap. 13).

4.4.3 *Irrigation*

One of the fundamental appeals of extensive green roofs is that they can help solve a number of problems associated with urbanization at a low expenditure of resources such as energy, nutrients, or water. The use of resource inputs in their management therefore often receives skepticism. More practically, present and predicted freshwater scarcity will put increasing financial as well as legislative restrictions on commercial and residential water use (Falkenmark and Xia 2013). Although largely developed in northern Europe, the modern extensive design that combines shallow well-drained substrates with mats of low growing succulents adapts well across a range of climates even with no or minimal irrigation. Recent and ongoing research has identified regionally adapted vegetation and substrate designs (including increased depth) that can be used to develop more regionally tuned versions of this basic low input design. Examples include arid Australia (Razzaghmanesh et al. 2014; Farrell et al. 2012), subtropical New Zealand (Voyde et al. 2010b), Mediterranean Europe (Van Mechelon et al. 2014), and North America (Dvorak and Volder 2010; Sutton et al. 2012; Chaps. 3 and 11).

However, a number of potentially appropriate uses exist for irrigation on extensive green roofs depending on the context and functional goals. During the establishment period the growth and survivorship of even highly drought tolerant species is increased by supplemental water (Dunnett and Nolan 2004; Thuring et al. 2010; Sutton 2013). The development of high plant cover and health during establishment can reduce weed pressure and potentially influence other aspects of long term performance. After establishment, climatic variation can periodically create periods of extended water deficit even in relatively mesic climates. These periodic stresses may act as a stochastic species filter contributing to observed long term declines in species and growth form richness on un-irrigated green roofs (Köhler 2006; Rowe et al. 2012) or dramatic changes in species dominance (Chap. 12). Overall, irrigation can greatly expand the pool of plant species, growth forms, and functional types that are suitable for a particular green roof context, particularly on shallow substrates or in water limited climates (Monterusso et al. 2005; Price et al. 2011; Schroll et al. 2011b; MacIvor et al. 2013). Even if plants can survive a particular green roof water environment without irrigation, their growth and traits related to aesthetics such as flowering and canopy cover can be improved with irrigation (Nagase and Dunnett 2010; Schroll et al. 2011).

Irrigation can potentially indirectly enhance a number of other green roof functions via its effects on the diversity and composition of vegetation. As described above (*plant assemblage design*), a number of studies have documented a positive relationship between green roof vegetation diversity and stormwater management performance. In aggregate, diverse green roof species assemblages can be more resilient to environmental perturbations such as drought stress compared to less diverse assemblages (Nagase and Dunnett 2010; Chap. 10). In addition, some aspects of aesthetic preference are related to functional and structural diversity. In a survey of Australian office workers, the most preferred living roof type had taller, grassy, and flowering vegetation while low growing succulent vegetation was least preferred (Lee et al. 2014).

The degree to which green roofs moderate internal building temperature is partly influenced by the water content of the system and by rates of evapotranspiration (Barrio 1998). Irrigation could potentially increase evapotranspirative cooling and be a tool for reducing building cooling costs. Sun et al. (2014) modeled thermal performance of a green roof in Beijing, China and estimated that the value of the avoided building cooling costs related to irrigation was greater than the monetary costs of the irrigation itself. However, other studies in a Mediterranean and a subtropical climate have reported low cooling efficiencies associated with green roofs and minimal or no contribution to building cooling associated with irrigation (Jim and Peng 2012; Schweitzer and Erell 2014).

Given the wide range of functional benefits associated with irrigation it seems reasonable to expect that its use can be justified to meet a range of performance goals in a number of contexts. In these cases irrigation systems should be optimized to maximize water use efficiency relative to the functional goals. Unfortunately, few published studies have evaluated irrigation system design in a green roof context. In one of the few, Rowe et al. (2014) compared the performance of overhead, drip, and sub-irrigation systems. They found that the overhead system resulted in the highest substrate water content and wasted the least amount of water in the form of runoff. The overhead system also produced a more even distribution of water through the three-dimensional substrate profile. Sub-surface and drip systems created more heterogeneous distributions, likely because vertical and lateral capillary movement of water was limited in the porous substrate. As a result, plant growth and health were greatest under overhead irrigation. However, optimum system design is likely dependent on other system elements such as substrate composition and depth, water retention layers, the specific species composition of the vegetation, environmental conditions (e.g., diurnal winds, relative humidity, and shade patterns), and cost constraints. In addition, the use of irrigation can directly decrease substrate stormwater storage capacity under some climatic conditions (Schroll et al. 2011a). The design of the irrigation system as well as other roof components such as substrate water holding capacity will need to reflect a balance between potential tradeoffs such this. Irrigation management will also likely need to be periodically adjusted over time as the roof ages.

The development of expert irrigation systems for green roofs that match the timing and amount of applied water to actual plant water needs is still in its infancy. Mass balance equations can be used to estimate the water status of the substrate and therefore predict the need for irrigation if the relationship between water status and plant water stress is known. The most difficult parameter of water mass balance to directly measure is usually evapotranspiration (Chap. 2). Several predictive evapotranspiration models have been developed for green roofs based on empirically derived regression models as well as a range of more mechanistic standard agricultural models (Kasmin et al. 2010; Rezaei and Jarrett 2006; Voyde et al. 2010b; DiGiovanni et al. 2012; Sherrard and Jacobs 2011; Karanam et al. 2013; Starry 2013). Estimates of evapotranspiration for green roofs based on reference evapotranspiration estimates require the application of attenuating factors (DiGiovanni et al. 2012) or water stress coefficients for water-limited conditions. Furthermore, there are few published values for crop coefficients or plant factors specific for green roof plant species that are necessary to adapt physically based reference evapotranspiration estimates to green roof vegetation types (DiGiovanni et al. 2012; Sherrard and Jacobs 2011; Starry 2013). Alternatively, empirically derived estimates of evapotranspiration for green roof systems were found to be robust across water status for specified vegetation types (Voyde 2011). Voyde et al. (2010b) use such an approach to develop irrigation guidelines for succulent planted roofs in Auckland, New Zealand. Empirical approaches require the calibration and validation of transpiration models for the variety of roof designs and environmental contexts. Currently no consensus exists on the most appropriate modeling approach for green roof contexts. Monitoring approaches including the development of inexpensive wireless sensor networks could provide a practical way of directly measuring the real-time water status of a roof in high spatial and temporal resolution (Starry et al. 2011; Lea-Cox et al. 2013; Chap. 2).

Diverting runoff from green roofs into a gray-water system or using secondarily treated municipal wastewater for irrigation are two other approaches to minimizing the impact of green roof irrigation on regional water demand. In addition, the storage capacity of a gray water system can also greatly improve the stormwater performance of roofs in some climates. In central Florida, a green roof system that incorporated a cistern and irrigation retained 87 % of yearly runoff compared with only 43 % retention for a system without irrigation and a cistern (Hardin 2006; Wanielista and Hardin 2006). A number of potential problems arise with the use of gray water for irrigation (Maimon et al. 2010). Only one published study has evaluated gray water use in a green roof context. Moritani et al. (2013) exposed *S. kamtschaticum* to periodic irrigation water with elevated salinities typical of some (but not all) gray water. The irrigation water increased salt stress for the plants, which reduced evapotranspiration. This could have a positive impact on the overall irrigation requirements for the system, but of course could detrimentally influence long-term plant growth and survival as well as stormwater performance.

4.5 Future Research Needs and Questions

A great deal of progress has been made in understanding how green roofs use, store, and regulate the flow of water. However, there are still a number of areas in need of further research as well as a number of issues for which the broader community of green roof designers, managers, and users are still developing optimal solutions.

4.5.1 More Integrated and Regionalized System Designs

There has been progress in developing green roof designs that are better tuned to local climates or specific user goals such as native wildlife habitat. However, there is still great potential to develop more fully integrated system designs that coordinate substrate composition, vegetation composition, water management, and long term maintenance plans. Ideally, sets of region-specific design and management specifications should be developed to guide designers as well as those charged with maintaining the long-term functioning of green roofs. These specifications should be flexible enough to guide the design of roofs that vary in functional performance goals. For example, what is the best design for a roof in the U.S. Pacific Northwest whose main functional goal is stormwater management? Does the design need to change if wildlife habitat or aesthetics are prime performance goals? What if maintenance budgets or expertise are limited?

4.5.2 More Automated Systems for Tracking Water Status and Broader Performance

As green roofs are increasingly used in climates that might necessitate the use of irrigation there is a need for high spatial and temporal resolution data describing their water status. Automated systems that track water status coupled with efficient irrigation designs could significantly minimize the water use impact of green roofs. Even when green roofs are not irrigated, automated systems that track performance metrics such as runoff retention or thermal performance would facilitate adaptive management and could potentially be used to provide performance based credits or incentives within a payment for ecosystem services framework. The technical hurdles to the development of such systems are beginning to be overcome, but costs need to decline and more commercialized off-the-shelf systems need to be developed.

4.5.3 Policy and Design Guidelines for the Appropriate Use of Supplemental Water

Depending on the functional goal and the local climate, the use of irrigation as well as other inputs such as fertilizer may be unavoidable. We need better frameworks for assessing whether that resource use is appropriate, or whether alternative strategies such as *cool roofs* (Millstein and Menon 2011) might be more appropriate in some contexts. In regions such as the western U.S. specific policy guidelines and incentives regarding the use of water for green roofs might be helpful. For example, the City of Portland, OR U.S.A. offers building permit incentives for developers that incorporate green roofs into building designs. Importantly, the city also has set water use standards and provides technical assistance to meeting the standard (City of Portland 2009).

References

Allen RG, Pereira LS, Raes D, Smith M (1998) Crop evapotranspiration-guidelines for computing crop water requirements-FAO Irrigation and drainage paper 56. FAO, Rome, 300, 6541

Anderson M, Lambrinos JG, Schroll E (2010) The potential value of mosses for stormwater management in urban environments. Urban Ecosyst 13(3):319–332

Andrade JL, Cervera JC, Graham EA (2009) Microenvironments, water relations, and productivity of CAM plants. In: De La Barrera E, Smith WK (eds) Perspectives in biophysical plant ecophysiology: a tribute to Park S. Nobel. Universidad Nacional Autonoma de Mexico, Mexico, pp 95–120

Augé RM (2001) Water relations, drought and vesicular-arbuscular mycorrhizal symbiosis. Mycorrhiza 11(1):3–42

Barrio EPD (1998) Analysis of the green roofs cooling potential in buildings. Energy Build 27(2):179–193

Bates AJ, Sadler JP, Mackay R (2013) Vegetation development over four years on two green roofs in the UK. Urban For Urban Green 12(1):98–108

Beck DA, Johnson GR, Spolek GA (2011) Amending greenroof soil with biochar to affect runoff water quantity and quality. Environ Pollut 159(8):2111–2118

Bengtsson L, Grahn L, Olsson J (2005) Hydrological function of a thin extensive green roof in southern Sweden. Nord Hydrol 36:259–268

Bennett S, Conrad-Saydah A, McCormick S, Morris D, Protteau M (2008) Vegetated roofs in the Los Angeles river watershed: the environmental and economic impacts of stormwater runoff and management. A group project submitted in partial satisfaction of the requirements for the degree of Master's of environmental science and management. University of California, Santa Barbara

Benvenuti S (2014) Wildflower green roofs for urban landscaping, ecological sustainability and biodiversity. Landsc Urban Plan 124:151–161

Berghage R, Jarrett A, Beattie D, Kelley K, Husain S, Rezai F, Long B, Negassi A, Cameron R (2007) Quantifying evaporation and transpirational water losses from green roofs and green roof media capacity for neutralizing acid rain. Report, National Decentralized Water Resources (NDWRCP) Research Project. Pennsylvania State University

Berkompas B, Marx KW, Wachter HM, Beyerlein D, Spencer B (2009) A study of green roof hydrologic performance in the Cascadia region. In: She N, Char M (eds) Low impact development for urban ecosystem and habitat protection. American Society of Civil Engineers, pp 1–10

Berndtsson JC (2010) Green roof performance towards management of runoff water quantity and quality: a review. Ecol Eng 36(4):351–360
Boivin M, Lamy M, Gosselin A, Dansereau B (2001) Effect of artificial substrate depth on freezing injury of six herbaceous perennials grown in a green roofsystem. HortTech 11:409–412
Bousselot JM, Klett JE, Koski RD (2011) Moisture content of extensive green roof substrate and growth response of 15 temperate plant species during dry down. HortScience 46(3):518–522
Buccola N, Spolek G (2011) A pilot-scale evaluation of greenroof runoff retention, detention, and quality. Water, Air, Soil Pollut 216(1–4):83–92
Butler C, Orians CM (2011) Sedum cools soil and can improve neighboring plant performance during water deficit on a green roof. Ecol Eng 37(11):1796–1803
Carson TB, Marasco DE, Culligan PJ, McGillis WR (2013) Hydrological performance of extensive green roofs in New York City: observations and multi-year modeling of three full-scale systems. Environ Res Lett 8(2):024036
Chang NB, Rivera BJ, Wanielista MP (2011) Optimal design for water conservation and energy savings using green roofs in a green building under mixed uncertainties. J Clean Prod 19(11):1180–1188
Chen CF (2013) Performance evaluation and development strategies for green roofs in Taiwan: a review. Ecol Eng 52:51–58
City of Portland (2009) Ecoroof handbook. Department of Environmental Servces. City of Portland, Oregon. https://www.portlandoregon.gov/bes/article/259381. Accessed 5 Aug 2014
Compton JS, Whitlow TH (2006) A zero discharge green roof system and species selection to optimize evapotranspiration and water retention. Proceedings of the 4th North American green roof conference: greening rooftops for sustainable communities, Boston, MA. The Cardinal Group, Toronto
Cook-Patton SC, Bauerle TL (2012) Potential benefits of plant diversity on vegetated roofs: a literature review. J Environ Manage 106:85–92
DiGiovanni K, Gaffin S, Montalto F (2010) Green roof hydrology: results from a small-scale lysimeter setup (Bronx, NY). Low Impact Dev 2010:1328–1341
DiGiovanni K, Montalto F, Gaffin S, Rosenzweig C (2012) Applicability of classical predictive equations for the estimation of evapotranspiration from urban green spaces: green roof results. J Hydrol Eng 18(1):99–107
Dunnett N, Nolan A (2004) Effect of substrate depth and supplementary watering on the growth of nine herbaceous perennials in a semi-extensive green roof. Acta Hort (ISHS) 643:305–309
Dunnett N, Nagase A, Booth R, Grime P (2008) Influence of vegetation composition on runoff in two simulated green roof experiments. Urban Ecosyst 11:385–398
Durhman AK, Rowe DB, Rugh CL (2006) Effect of watering regimen on chlorophyll fluorescence and growth of selected green roof plant taxa. HortScience 41(7):1623–1628
Dvorak B, Volder A (2010) Green roof vegetation for North American ecoregions: a literature review. Landsc Urban Plan 96(4):197–213
Ehleringer J, Mooney HA (1983) Productivity of desert and Mediterranean-climate plants. In: Lange OL, Nobel PS, Osmond CB, Ziegler H (eds) Physiological plant ecology IV. Springer, Heidelberg, pp 205–231
Falkenmark M, Xia J (2013) Urban water supply under expanding water scarcity. In: Larsen TA, Udert KM, Lienert J (eds) Source separation and decentralization for wastewater management. IWA Publishing, London, pp 59–70
Farrell C, Mitchell RE, Szota C, Rayner JP, Williams NSG (2012) Green roofs for hot and dry climates: interacting effects of plant water use, succulence and substrate. Ecol Eng 49:270–276
Farrell C, Szota C, Williams NS, Arndt SK (2013a) High water users can be drought tolerant: using physiological traits for green roof plant selection. Plant Soil 372(1–2):177–193
Farrell C, Ang XQ, Rayner JP (2013b) Water-retention additives increase plant available water in green roof substrates. Ecol Eng 52:112–118
Fassman E, Simcock R (2012) Moisture measurements as performance criteria for extensive living roof substrates. J Environ Eng 138(8):841–851

Fioretti R, Palla A, Lanza LG, Principi P (2010) Green roof energy and water related performance in the Mediterranean climate. Build Environ 45(8):1890–1904
FLL (2008) Guideline for the planning, execution and upkeep of green-roof sites. English edition. Forschungsgesellschaft Landschaftsentwicklung Landschaftsbau, Bonn
Gaff DF (1989) Responses of desiccation tolerant 'resurrection'plants to water stress. In: Kreeb KH, Richter H, Hinckley TM (eds) Structural and functional responses to environmental stress. SPB Academic Publishing, The Hague, pp 255–268
Getter KL, Rowe DB (2009) Substrate depth influences sedum plant community non a green roof. HortScience 44(2):401–407
Gregoire BG, Clausen JC (2011) Effect of a modular extensive green roof on stormwater runoff and water quality. Ecol Eng 37(6):963–969
Hardin MD (2006) The rffectiveness of a specifically designed green roof stormwater treatment system irrigated with recycled stormwater runoff to achieve pollutant removal and stormwater volume reduction. Masters thesis, University of Central Florida
Hardin M, Wanielista M, Chopra M (2012) A mass balance model for designing green roof systems that incorporate a cistern for re-use. Water 4(4):914–931
Hastings A, Byers JE, Crooks JA, Cuddington K, Jones CG, Lambrinos JG, Talley TS, Wilson WG (2007) Ecosystem engineering in space and time. Ecol Lett 10(2):153–164
Heim A (2013) The effects of soil depth, competition and facilitation on plant growth, soil temperature and water loss on an extensive green roof. Masters thesis, St. Mary's University
Herrera A (2009) Crassulacean acid metabolism and fitness under water deficit stress: if not for carbon gain, what is facultative CAM good for? Ann Bot 103(4):645–653
Hoekstra FA, Golovina EA, Buitink J (2001) Mechanisms of plant desiccation tolerance. Trends Plant Sci 6(9):431–438
Jim CY (2012) Effect of vegetation biomass structure on thermal performance of tropical green roof. Landsc Ecol Eng 8(2):173–187
Jim CY (2014) Heat-sink effect and indoor warming imposed by tropical extensive green roof. Ecol Eng 62:1–12
Jim CY, Peng LL (2012) Substrate moisture effect on water balance and thermal regime of a tropical extensive green roof. Ecol Eng 47:9–23
John J (2013) Assessment of arbuscular mycorrhizal fungi in a green roof system. Masters thesis, Dalhousie University
Ju JH, Bae GT, Kim WT, Yoon YH (2012) Computation of irrigation interval and amount as affected by growing substrate and soil depth planted with Zoysia japonica in green roof during a dry summer. J Environ Sci Int 21(3):289–296
Karanam H, Retzlaff W, Morgan S, Celik S (2013) Experimental and theoretical analysis of evapotranspiration in green roof systems. In: Dincer I, Colpan CO, Kadioglu F (eds) Causes, impacts and solutions to global warming. Springer, New York, pp 1055–1066
Kasmin H, Stovin VR, Hathway EA (2010) Towards a generic rainfall-runoff model for green roofs. Water Sci Technol 62:898–905
Köhler M (2006) Long-term vegetation research on two extensive green roofs in Berlin. Urban Habitat 4(1):3–25
Körner C, Scheel JA, Bauer H (1979) Maximum leaf diffusive conductance in vascular plants (Review). Photosynthetica 13:45–82
Lea-Cox JD, Bauerle WL, Van Iersel MW, Kantor GF, Bauerle TL, Lichtenberg E, King DM, Crawford L (2013) Advancing wireless sensor networks for irrigation management of ornamental crops: an overview. HortTechnology 23(6):717–724
Lee KE, Williams KJ, Sargent LD, Farrell C, Williams NS (2014) Living roof preference is influenced by plant characteristics and diversity. Landsc Urban Plan 122:152–159
Li Y, Babcock RW Jr (2014) Green roof hydrologic performance and modeling: a review. Water Sci Technol 69(4):727–738
Livingston EH, Miller C, Lohr M (2004) Green roof design and implementation in Florida. In: Proceedings of the second annual greening rooftops for sustainable communities conference. The Cardinal Group, Toronto, pp 353–364

Lundholm J, MacIvor JS, MacDougall Z, Ranalli M (2010) Plant species and functional group combinations affect green roof ecosystem functions. PLOS One 5(3):e9677
MacIvor JS, Ranalli MA, Lundholm JT (2011) Performance of dryland and wetland plant species on extensive green roofs. Ann Bot 107(4):671–679
MacIvor JS, Margolis L, Puncher CL, Carver Matthews BJ (2013) Decoupling factors affecting plant diversity and cover on extensive green roofs. J Environ Manage 130:297–305
Maimon A, Tal A, Friedler E, Gross A (2010) Safe on-site reuse of greywater for irrigation-a critical review of current guidelines. Environ Sci Technol 44(9):3213–3220
Martin BK (2008) The dynamic stormwater response of a green roof. Dissertation, The University of Guelph
Mentens J, Raes D, Hermy M (2006) Green roofs as a tool for solving the rainwater runoff problem in the urbanized 21st century? Landsc Urban Plan 77(3):217–226
Miller C (2003) Moisture management in green roofs. Proceedings of the 1st greening rooftops for sustainable communities conferene. The Cardinal Group, Toronto, pp 177–182
Millstein D, Menon S (2011) Regional climate consequences of large-scale cool roof and photovoltaic array deployment. Environ Res Lett 6(3):034001
Monterusso MA, Rowe DB, Russell DK, Rugh CL (2004) Runoff water quantity and quality from green roof systems. Acta Hort (ISHS) 639:369–376
Monterusso MA, Rowe BD, Rugh CL (2005) Establishment and persistence of Sedum spp. and native taxa for green roof applications. HortScience 40(2):391–396
Moritani S, Yamamoto T, Andry H, Inoue M, Kato K, Saito H (2013) Effect of combined water and salinity stress factors on evapotranspiration of Sedum kamtschaticum Fischer in relation to green roof irrigation. Urban For Urban Green 12(3):338–343
Nagase A, Dunnett N (2010) Drought tolerance in different vegetation types for extensive green roofs: effects of watering and diversity. Landsc Urban Plan 97(4):318–327
Nagase A, Dunnett N (2011) The relationship between percentage of organic matter in substrate and plant growth in extensive green roofs. Landsc Urban Plan 103(2):230–236
Nagase A, Dunnett N (2012) Amount of water runoff from different vegetation types on extensive green roofs: effects of plant species, diversity and plant structure. Landsc Urban Plan 104(3):356–363
Nagase A, Dunnett N (2013) Establishment of an annual meadow on extensive green roofs in the UK. Landsc Urban Plan 112:50–62
NOAA (2014) National Climatic Data Center. Climate normals. http://www.ncdc.noaa.gov/data-access/land-based-station-data/land-based-datasets/climate-normals. Accessed 5 June 2014
Oberndorfer E, Lundholm J, Bass B, Coffman RR, Doshi H, Dunnett N, Gaffin S, Köhler M, Liu KKY, Rowe B (2007) Green roofs as urban ecosystems: ecological structures, functions, and services. BioScience 57(10):823–833
Oschmann C, Kobayashi N, Perkuhn C, Grüneberg H, Wissemeier A (2007) Study to expand the range of wild plants for extensive roof greening systems using superabsorbent polymers (SAP). Acta Hort (ISHS) 813:421–426
Papafotiou M, Pergialioti N, Tassoula L, Massas I, Kargas G (2013) Growth of native aromatic xerophytes in an extensive Mediterranean green roof as affected by substrate type and depth and irrigation frequency. HortScience 48(10):1327–1333
Price JG, Watts SA, Wright AN, Peters RW, Kirby JT (2011) Irrigation lowers substrate temperature and enhances survival of plants on green roofs in the Southeastern United States. HortTechnology 21(5):586–592
Razzaghmanesh M, Beecham S, Kazemi F (2014) The growth and survival of plants in urban green roofs in a dry climate. Sci Total Environ 476:288–297
Rezaei F, Jarrett AR (2006) Measure and predict evapotranspiration rate from green roof plant species. Proceedings Penn State College of Engineering research symposium. Penn State University
Rowe DB, Getter KL, Durhman AK (2012) Effect of green roof media depth on Crassulacean plant succession over seven years. Landsc Urban Plan 104(3):310–319

Rowe DB, Kolp MR, Greer SE, Getter KL (2014) Comparison of irrigation efficiency and plant health of overhead, drip, and sub-irrigation for extensive green roofs. Ecol Eng 64:306–313
Sailor DJ, Elley TB, Gibson M (2012) Exploring the building energy impacts of green roof design decisions–a modeling study of buildings in four distinct climates. J Build Phys 35(4):372–391
Savi T, Andri S, Nardini A (2013) Impact of different green roof layering on plant water status and drought survival. Ecol Eng 57:188–196
Sayed OH (2001) Crassulacean acid metabolism 1975–2000, a check list. Photosynthetica 39(3):339–352
Schroll E, Lambrinos JG, Righetti T, Sandrock D (2011a) The role of vegetation in regulating stormwater runoff from green roofs in a winter rainfall climate. Ecol Eng 37(4):595–600
Schroll E, Lambrinos JG, Sandrock D (2011b) An evaluation of plant selections and irrigation requirements for extensive green roofs in the Pacific northwestern United States. HortTechnology 21(3):314–322
Schweitzer O, Erell E (2014) Evaluation of the energy performance and irrigation requirements of extensive green roofs in a water-scarce Mediterranean climate. Energy Build 68:25–32
Sherrard JA Jr, Jacobs JM (2012) Vegetated roof water-balance model: experimental and model results. J Hydrol Eng 17(8):858–868
Simmons M, Gardiner B, Windhager S, Tinsley J (2008) Green roofs are not created equal: the hydrologic and thermal performance of six different extensive green roofs and reflective and non-reflective roofs in a sub-tropical climate. Urban Ecosyst 11:339–348
Song U, Kim E, Bang JH, Son DJ, Waldman B, Lee EJ (2013) Wetlands are an effective green roof system. Build Environ 66:141–147
Spolek G (2008) Performance monitoring of three ecoroofs in Portland, Oregon. Urban Ecosyst 11(4):349–359
Starry O (2013) The comparative effects of three Sedum species on green roof stormwater retention. Dissertation, University of Maryland
Starry O, Lea-Cox J, Ristvey A, Cohan S, House G (2011) Utilizing sensor networks to assess evapotranspiration by greenroofs. Paper presented at 2011 American Society of Agricultural and Biological Engineers meeting, Galt House Louisville, Kentucky, 7–10 August, 2011
Starry O, Lea-Cox JD, Kim J, van Iersel MW (2014) Photosynthesis and water use by two Sedum species in green roof substrate. Environ Exp Bot 107:105–112
Stovin V, Vesuviano G, Kasmin H (2012) The hydrological performance of a green roof test bed under UK climatic conditions. J Hydrol 414:148–161
Sun T, Bou-Zeid E, Ni GH (2014) To irrigate or not to irrigate: analysis of green roof performance via a vertically-resolved hygrothermal model. Build Environ 73:127–137
Sutton RK (2008) Media modifications for native plant assemblages on green roofs. University of Nebraska Lincoln, Landscape Architecture Program: faculty scholarly and creative activity, paper 16
Sutton RK (2013) Seeding green roofs with native grasses. J Liv Archit 1(1):1–20
Sutton RK, Harrington JA, Skabelund L, MacDonagh P, Coffman RR, Koch G (2012) Prairie-based green roofs: literature, templates, and analogs. J Gr Build 7(1):143–172
Thuring CE, Dunnett N (2014) Vegetation composition of old extensive green roofs (from 1980s Germany). Ecol Process 2014(3):4
Thuring CE, Berghage RD, Beattie DJ (2010) Green roof plant responses to different substrate types and depths under various drought conditions. HortTechnology 20(2):395–401
Van Mechelen C, Dutoit T, Hermy M (2014) Mediterranean open habitat vegetation offers great potential for extensive green roof design. Landsc Urban Plan 121:81–91
VanWoert EL, Rowe DB, Andersen JA, Rugh CL, Fernandez RT, Xio L (2005a) Green roof stormwater retention: effects of roof surface, slope and media depth. J Environ Qual 34:1036–1044
VanWoert ND, Rowe DB, Andresen JA, Rugh CL, Xiao L (2005b) Watering regime and green roof substrate design affect Sedum plant growth. HortScience 40(3):659–664
Villarreal EL (2007) Runoff detention effect of a sedum green-roof. Nord Hydrol 38(1):99–105
Volaire F, Norton M (2006) Summer dormancy in perennial temperate grasses. Ann Bot 98(5):927–933

Voyde EA (2011) Quantifying the complete hydrologic budget for an extensive living roof. Dissertation, The University of Aukland

Voyde E, Fassman E, Simcock R (2010a) Hydrology of an extensive living roof under sub-tropical climate conditions in Auckland, New Zealand. J Hydrol 394(3):384–395

Voyde E, Fassman E, Simcock R, Wells J (2010b) Quantifying evapotranspiration rates for New Zealand green roofs. J Hydrol Eng 15(6):395–403

Wang B, Qiu YL (2006) Phylogenetic distribution and evolution of mycorrhizas in land plants. Mycorrhiza 16(5):299–363

Wanielista M, Hardin M (2006) A stormwater management assessment of green roofs with irrigation. In: 2nd biennal stormwater management research symposium, 4 May 2006, pp 153–164

Westman WE (1981) Seasonal dimorphism of foliage in Californian coastal sage scrub. Oecologia 51(3):385–388

Williams NS, Rayner JP, Raynor KJ (2010) Green roofs for a wide brown land: opportunities and barriers for rooftop greening in Australia. Urban For Urban Green 9(3):245–251

Wolf D, Lundholm JT (2008) Water uptake in green roof microcosms: effects of plant species and water availability. Ecol Eng 33(2):179–186

Dr. John G. Lambrinos evaluates the design and function of green roofs in the Pacific Northwest. He has published regionally focused studies on appropriate green roof plant species and irrigation regimes.

Chapter 5
Nutrient Cycling in Green Roof Ecosystems

Ishi Buffam and Mark E. Mitchell

Abstract In this chapter we consider the cycling of Carbon (C), Nitrogen (N) and Phosphorus (P) in green roof ecosystems. The focus is placed primarily on N and P because these are the nutrients most often limiting to plant growth in terrestrial ecosystems, and because leaching of these elements to downstream aquatic ecosystems is a concern due to their potential to contribute to eutrophication. Extensive green roofs are commonly sources of phosphorus and dissolved organic carbon in runoff, while they may be either a source or a sink for nitrogen. Plant communities, substrate characteristics, substrate depth, and roof age all play a role in regulating nutrient export. Seasonal variation in runoff nutrient concentrations suggests the importance of temperature and light-mediated processes. Nitrogen leaching may drop off rapidly with the age of the ecosystem and vary with new inputs (atmospheric deposition of N, new fertilizer additions), while roofs leach out P for years or decades under current construction regimes, likely resulting from mineralization of P-rich organic matter in the roof substrate. Conceptual models of nutrient cycling developed from natural terrestrial ecosystems provide a useful starting point for interpreting the important nutrient cycling processes on green roofs. However, the engineered nature of green roof ecosystems, often with a high-nutrient substrate coupled to plants adapted to low-nutrient, extreme environments, gives rise to unique characteristics. There is still little known of the dynamics of important processes for recycling of nutrients within green roof ecosystems, and more studies which include modeling, full roof-scale experiments, and long-term monitoring are needed for improved understanding of these ecosystems.

Keywords Nitrogen · Phosphorus · Carbon · Nutrient fluxes · Nutrient dynamics · Organic matter · Mass-balance ecosystem models

I. Buffam (✉)
Department of Biological Sciences and Department of Geography, University of Cincinnati, 731H Rieveschl Hall, 312 College Drive, Cincinnati, OH 45221, USA
e-mail: ishi.buffam@uc.edu

M. E. Mitchell
Department of Biological Sciences, University of Cincinnati, 312 College Drive, Cincinnati, OH 45221, USA

R. Sutton (ed.), *Green Roof Ecosystems*, Ecological Studies 223,
DOI 10.1007/978-3-319-14983-7_5

5.1 Organization and Scope of this Chapter

We begin with a general overview of nutrient cycling processes and rationale behind their study in green roof ecosystems and then highlight the unique characteristics of green roofs relevant to nutrient cycling. We review the current state of knowledge with respect to the cycling of C, N and P in green roof ecosystems, most of which is based on observations of dissolved nutrient concentrations in roof runoff. We further examine observations of temporal dynamics in runoff nutrient concentrations, as a window into process understanding. The applicability of a simple terrestrial ecosystems nitrogen cycling model to green roofs is discussed. We highlight gaps in knowledge, and finish with a series of open questions relevant to green roof nutrient cycling, which we hope will provide a springboard for future research studies.

5.2 Rationale for Studying Nutrient Cycling in Green Roof Ecosystems

Our motivation for studying N and P derives from an interest in balancing the healthy functioning of green roof ecosystems (particularly the vegetation) with concerns for pollution and eutrophication of downstream aquatic ecosystems (eg. Carpenter et al. 1998). Carbon (C) is the currency of chemical energy flow within ecosystems, and its cycling couples with the cycling of N and P through biomass production and decomposition. There is also a general interest in the C sequestration potential of different ecosystems, related to efforts to slow atmospheric greenhouse gas accumulation.

Improved runoff water quality, including reduction of nutrients in runoff, has been touted as one of the benefits of green roofs, but it is not clear under what conditions this can be expected. As engineered ecosystems, green roofs are designed to have a sufficient availability of potentially limiting nutrients (especially N and P) to support a healthy, thriving plant community. As a result, there is often an excess of N and P, some of which is leached out during runoff events. This is particularly true for new roofs or roofs which have been newly fertilized. An understanding of the processes underlying C, N and P dynamics in green roof ecosystems would help in predicting the response of these systems to changes in environmental conditions over time, and in predicting and understanding the effects of varying system design.

5.3 Context: Nutrient Cycling in Terrestrial Ecosystems

5.3.1 Overview and Description of Elements

Nutrient cycling and availability to biota are factors of central importance in regulating the structure and function of ecosystems. As defined here, nutrient cycling involves the movement and transformation of bioactive elements into, out of, and within a given ecosystem (Fig. 5.1). Inputs typically involve atmospheric deposition or weathering of minerals from the geosphere, while exports include hydrological leaching losses. Microbially-mediated gas exchange may contribute either inputs to the system from the atmosphere (e.g. photosynthesis, nitrogen fixation) or exports from the system to the atmosphere (e.g. respiration, denitrification). Internal recycling involves physical, chemical and biochemical transformations and movement within and between different living and non-living components of the ecosystem.

Carbon (C) is the most abundant element in living matter, and carbon-carbon bonds in organic matter represent the common currency of chemical energy in ecosystems. Net ecosystem carbon balance (NECB; Chapin et al. 2006), describes one of the most fundamental characteristics of any ecosystem. Major fluxes in and out of terrestrial ecosystems are typically CO_2 in (associated with primary production) and CO_2 out (associated with autotrophic and heterotrophic respiration), with the balance of these two terms defining Net Ecosystem Production (NEP). Other

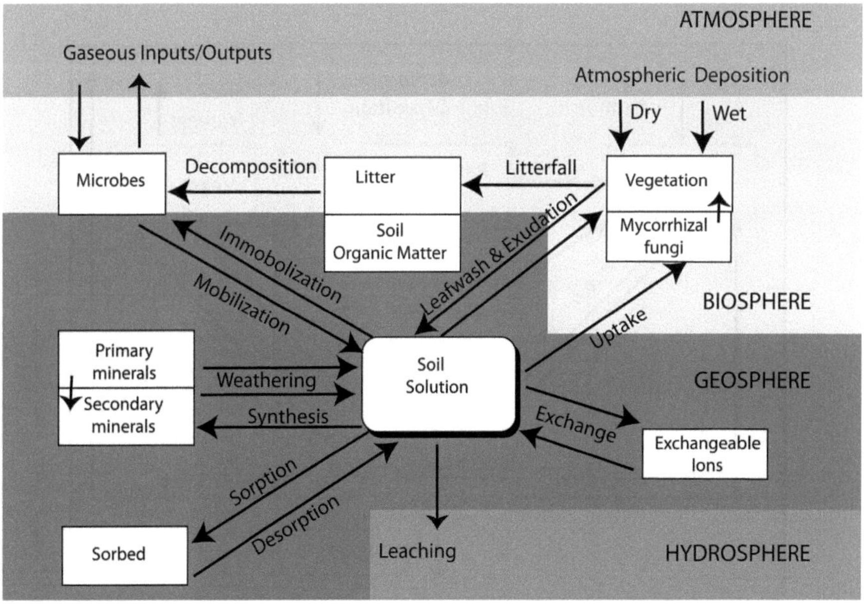

Fig. 5.1 A generalized schematic illustrating major pools and pathways for nutrient storage, transformation, and movement in ecosystems. (From Dahlgren 1998)

fluxes include hydrologic export (runoff) of dissolved organic carbon (DOC) and dissolved inorganic carbon (DIC; dissolved CO_2 and bicarbonate) from the leaching and weathering of soils; gaseous fluxes of CH_4, CO, and volatile organic compounds (VOC); and soot and CO_2 loss in the event of fire (Chapin et al. 2006).

Nitrogen (N) is a key building block for amino acids, proteins, and the nitrogenous bases of DNA. It is commonly the productivity-limiting nutrient in terrestrial ecosystems (Chapin et al. 2011). In unpolluted terrestrial ecosystems, microbially-mediated fixation of atmospheric N_2 provides the main source of reactive nitrogen (N_r, Galloway et al. 2003) while atmospheric deposition of the inorganic forms Ammonium (NH_4^+) and primarily Nitrate (NO_3^-) provide an additional source (Fig. 5.2). Both NH_4^+ and NO_3^- are accessible to plants and microbes and can be assimilated or immobilized into organic pools of N by plants and microbes, respectively. Microbial communities mineralize these organic forms of N into NH_4^+, by decomposing organic matter. The inorganic forms of N have differential mobility in soils: NH_4^+ binds on cation-exchange soil surfaces, thus experiences slow diffusion; NO_3^-, however, diffuses rapidly through soils and can readily leach out if present during periods of hydrologic flushing (Chapin et al. 2011). Importantly, several redox transformations involving N occur in soils. Nitrification is a process in which specialized microbes use NH_4^+ as an energy source, oxidizing it to NO_3^-. In low oxygen environments, denitrifying microbes use NO_3^- as a terminal electron acceptor in the process of denitrification, producing N_2O and N_2 as byproducts.

Phosphorus (P) is a necessary macronutrient, required for biosynthesis of key compounds including ATP, DNA and phospholipids. In natural systems, the major

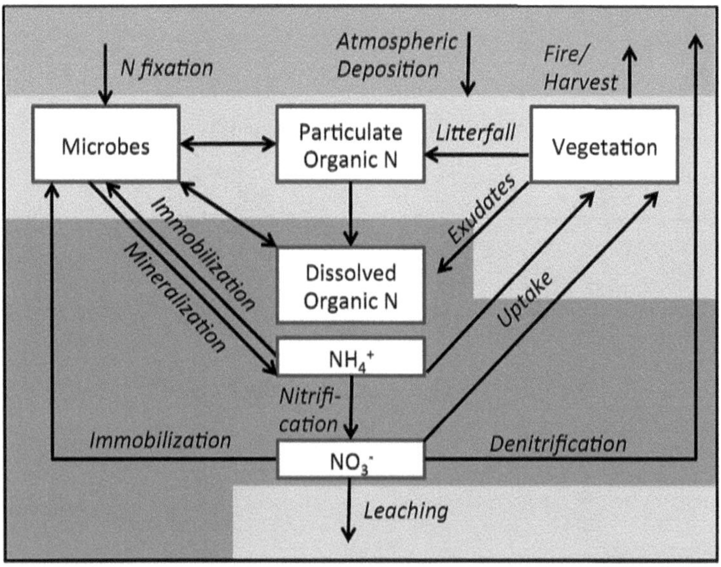

Fig. 5.2 Simplified diagram of the terrestrial nitrogen cycle with major pools (*boxes*) and fluxes (*arrows*) represented for a model ecosystem. (Modeled after Dahlgren 1998)

sources of P are from the weathering of rocks and the mineralization of organic material by microbes. Both of these processes release the water-soluble and biologically-accessible form of P, i.e. phosphate (PO_4^{3-}). Phosphate can either be taken up by plants or microbes, adsorbed to substrate, precipitated out of solution, or lost from the system via runoff. Phosphate is chemically active, commonly forms precipitates in the soil, and thus tends to experience slow diffusion (Chapin et al. 2011). Unlike N, very little P is introduced to systems via atmospheric sources.

5.3.2 Nutrient Cycling Dynamics in Natural Ecosystems

As a consequence of their high concentration in living tissue relative to environmental sources, the availability of Nitrogen (N) and Phosphorus (P) limit primary productivity in many ecosystems, with N limitation particularly common in terrestrial ecosystems (Vitousek and Howarth 1991; Crews 1999). Co-limitation by N and P occurs at a fairly consistent mass ratio of 15:1 for available N:P, with higher ratios leading to P limitation, and lower ratios leading to N limitation (Chapin et al. 2011). Labile organic C supply may also limit secondary production (e.g., microbial decomposition rates)(Marschner and Kalbitz 2003).

In terrestrial ecosystems, nutrient (N and P) cycling involves highly localized exchanges between plants, microbes, and their physical environment (Chapin et al. 2011). Unmanaged ecosystems tend to be nearly closed systems with respect to limiting nutrients, where internal recycling of nutrients is very high relative to inputs and outputs. On an annual basis, more than 90 % of N and P taken up by plants in natural terrestrial ecosystems commonly comes from recycled nutrients, i.e. from soils that store nutrients derived from the previous years' plant material (Likens et al. 1977). The macronutrients N, P, and K are typically required by plants in excess of that obtained through mass flow (water uptake by roots), thus diffusion and saturated flow in soils are important sources of these nutrients. Plant associations with mycorrhizal fungi are common especially in low-nutrient environments, and enable substantially increased uptake rates by increasing the effective root surface area and capacity to hold and store nutrients (Smith and Read 2008). Mycorrhizal fungi are most helpful in increasing plant access to slowly-diffusing nutrients, i.e. PO_4^{3-} and NH_4^+.

Internal nutrient cycling in terrestrial ecosystems consists of negative feedback loops that, over time, can lead to homeostasis within the system, where nutrient availability and plant uptake are balanced so as to maintain steady state. For instance, high nutrient uptake and retention (high nutrient use efficiency) by plants leads to nutrient depleted soils, which leads in turn to slower decomposition, which leads to lower nutrient availability, which slows plant growth. Conversely, high nutrient availability typically leads to high nutrient losses from a system, ultimately decreasing availability (Shaver and Melillo 1984).

5.3.3 Nutrient Cycling in Managed or Otherwise Heavily Impacted Ecosystems

Managed systems with substantial human influence can have considerably different characteristics than unmanaged ecosystems. This distinction is related both to changes (typically increases) in nutrient inputs, as well as changes in internal cycling processes. For instance, a nitrogen budget study of the city of Phoenix, AZ, USA revealed that commercial fertilizer and combustion were two major sources of nitrogen in this urban ecosystem. Atmospheric NO_X export, denitrification, and accumulation within the system (in large part as buried rubbish) were the major fates for nitrogen (Baker et al. 2001). This contrasts with unmanaged temperate ecosystems, where N-fixation is typically the major input, and major fates are denitrification, hydrologic runoff export, or accumulation in biomass for aggrading systems (Chapin et al. 2011).

5.4 Distinctive Characteristics of Green Roofs Relevant to Nutrient Cycling

Like many other managed or human-impacted systems, green roofs tend to have high inputs of N and P, in this case mostly due to the integration of nutrient-rich organic and inorganic fertilizers into the substrate. Their location within urban areas also means that atmospheric deposition fluxes will tend to be high, containing NO_X from automobile exhaust and other combustion processes, and possibly higher concentrations of P in mobilized dust (Pett-Ridge 2009).

In the context of nutrient cycling, the distinction between intensive and extensive green roofs is important. Intensive green roofs or "roof gardens" have deep substrate (>20 cm), can be quite heavy, and may have a wide variety of plant types (even shrubs or trees), requiring substantial management. Extensive green roofs are a modern modification of the roof-garden concept with shallower substrates (often <10 cm), and are more strictly functional in purpose than intensive roofs, requiring less maintenance (Oberndorfer et al. 2007). In this chapter we primarily consider extensive green roofs.

Extensive green roofs are engineered ecosystems with distinctive characteristics that may give rise to unique patterns of nutrient cycling. Although they include interacting biotic and abiotic components like natural ecosystems, the components have been selected/designed for specific purposes rather than co-developing over time. This contrasts with natural ecosystems, in which soil and plants have developed in tandem, and are linked by local climate and geology, such that plant communities and soil types "match". Thus, while in natural ecosystems high nutrient soils will typically be vegetated with nutrient-loving plants, green roofs commonly contain high-nutrient substrate coupled with plants adapted to low-nutrient environments. The match or mismatch between plant and substrate characteristics is a function of the design and relatively young age of these ecosystems.

Important constraints on green roof system design (e.g., FLL 2008) include climate and exposure conditions of the roof environment; weight limitations due to underlying building structure; cost; and limited body of available knowledge about these ecosystems. These conditions combine to severely restrict the available plant palette for extensive green roofs to drought-tolerant, wind-tolerant, low-profile, typically slow-growing plants, and the succulent CAM-photosynthesizing *Sedums* have been commonly used (Though, see other chapters in this book for examples of diverse green roof plant communities).

Green roof substrate is an engineered soil analog that has been designed to provide structure and nutrition to support a healthy, thriving plant community. Substrates have typically not been designed with nutrient retention, or runoff water quality, in mind. In service of the plant community, the substrate is typically nutrient-rich (often containing compost, which tends to be enriched in P), and may be supplemented with slow-release inorganic fertilizers containing N, P, and K. Integration of P-rich material commonly leads to substrate with about equal amounts of N and P, in spite of 15-fold higher plant demand for N (e.g., Chapin et al. 2011). Likely due to cost restrictions and to simplify design, there are no distinct soil horizons in typical extensive green roof substrate. Instead, organic matter is combined with inorganic aggregate material in a homogeneous mixture. This homogeneous mixture contrasts to natural soils, where distinct horizons develop and organic matter content decreases from the surface downward. Notably, some green roofs now contain a dual-layer construction, with a top organic nutrient-rich layer and a mineral sub-layer designed in part to retain leaching nutrients (e.g., Wang et al. 2013), more analogous to the physical arrangement of natural soil profiles. The homogenous and nutrient-rich nature of green roof substrate presumably influences nutrient cycling processes.

Green roofs are catchments, in that there is a fixed, measurable influx and outflow of water with a solvable water balance—but they have unique characteristics relative to most natural watersheds. Green roof substrate and underlying layers are designed to drain freely, in order to avoid extended periods of standing water around plant roots, anoxia and associated difficulties for the plant communities. Designers attempt to balance free drainage with runoff reduction by including substrate materials with high water holding capacity, and in some systems by including drain board to retain standing water below the substrate in a separate layer (e.g., FLL 2008). As a consequence, hydrologic residence times can be expected to vary greatly among roofs depending on roof design and climate conditions, and within a given roof over time due to changes in season, precipitation patterns, and evapotranspiration rates. Green roofs can experience extended periods of complete water retention, when precipitation events are relatively small and evapotranspirative demand is high; on the other hand, during larger storm or melt events, hydrologic residence times shorten once the system is saturated. This unique hydrology has important ramifications for nutrient cycling within green roofs, but at this point there have been few detailed water balance studies that would shed light on the interaction between hydrology and the cycling of C, N and P in these systems.

5.5 Current State of Knowledge on Nutrient Cycling in Green Roofs

5.5.1 Overview

Most of the current knowledge about nutrient cycling in green roofs involves the recording of patterns (mainly, nutrient concentrations in runoff) rather than the direct measurement of processes. Most green roof research, at least that published in the English-language literature involves relatively short-term (< 1 year) monitoring studies, and there is a shortage of experimental and modeling studies. As such, the review of the current state of the knowledge presented here is biased towards relatively young roofs, and towards extensive roofs more common in North America and Northern Europe, and the subject of most of the published research studies. However, enough of a body of research to synthesize information about the general patterns of C, N and P in green roof runoff is available to offer some conjecture about important contributing mechanisms. In the following section we detail the patterns observed in the literature to date with respect to each nutrient in green roofs, with a focus on the role of substrate variation. The reader is also referred to Berndtsson 2010, who provide an excellent review of green roof impacts on runoff quantity and quality including fluxes of metals, and Rowe (2011), who provides an informative overview of environmental impacts of green roofs.

5.5.2 Carbon Cycling in Green Roofs: Patterns Observed, and Implications for Processes

Like any vegetated space, green roofs have the potential for biomass and soil accumulation that can represent a carbon sink if inputs exceed exports. For green roofs, the major fluxes in and out of the ecosystem are presumably atmospheric CO_2 in (associated with primary production), atmospheric CO_2 out (associated with autotrophic and heterotrophic respiration), and hydrologic runoff of dissolved organic carbon (DOC) from the leaching of soil organic matter.

There is very little known about the dynamics of atmospheric CO_2 exchange in green roof ecosystems, which would give insight into short-term changes in carbon stocks. However, Gaumont-Guay and Halsall (2013) carried out detailed, year-long measurements of CO_2 exchange for a newly-established extensive green roof in British Columbia, Canada. These measurements were used to model rates of photosynthesis, autotrophic respiration, and heterotrophic respiration, enabling calculations of net ecosystem productivity (NEP). The NEP for the monitored green roof was 2 g C m^{-2} yr^{-1}, equivalent to only about 1 % of the value for a growing temperate forest (Aber and Melillo 2001). Based on these measurements, this particular roof was near steady state with respect to atmospheric carbon exchange, i.e. not serving as either a strong sink or source.

Although extensive green roofs have the potential to rapidly accumulate carbon over the short-term especially if they are established using seeds or small cuttings, the total biomass plateaus at a low value relative to most natural vegetated ecosystems. Using newly-established extensive green roof plots seeded with *Sedum spp.*, Getter et al. (2009) measured a net carbon storage of 275 g C m^{-2} in biomass, and 100 g C m^{-2} in substrate, over the course of 2 years. This net storage rate is on par with storage rates in aggrading temperate forests, which typically have NEP ranging from 200 to 400 g C m^{-2} yr^{-1} (Aber and Melillo 2001). However, a survey of 12 existing extensive green roofs in Michigan and Maryland, USA revealed that C storage in aboveground biomass averaged only 162 g C m^{-2} (Getter et al. 2009), with total biomass estimated at 260 g C m^{-2}. This is slightly lower than average biomass storage in the arctic tundra, desert, or temperate grassland biomes (325–375 g C m^{-2}), and much lower than the temperate forest biome (13,350 g C m^{-2}) (Saugier et al. 2001).

Green roofs are a source of dissolved organic carbon (DOC), which leaches out in runoff, often reaching concentrations above 50 mg/L. This range of DOC is comparable to that in streams draining peatlands (Mulholland 2003), and gives runoff from many green roofs a brownish tint typical of humic-rich waters. The variation in DOC effluent concentrations is primarily related to organic matter content in the substrate, with higher organic matter content typically giving rise to higher effluent DOC (Berndtsson et al. 2009). Runoff DOC concentrations are also sensitive to vegetation type and cover (Table 5.1) and substrate depth (Seidl et al. 2013). Anecdotally, an older, intensive green roof in Japan had runoff DOC concentrations of < 15 mg C L^{-1}, lower than most reported values for extensive roofs (Berndtsson et al. 2009). Few studies have focused on DOC, but amendment of substrate with biochar was found to decrease runoff DOC levels in one plot-scale study (Beck et al. 2011).

Hydrologic export of DOC can represent a substantial proportion of a green roof's net annual carbon exchange. Assuming atmospheric deposition of DOC averaging 1–5 mg C L^{-1}, and average runoff concentrations of DOC ranging from 20 to 80 mg C L^{-1}, an extensive green roof receiving of 1000 mm yr^{-1} of precipitation and exporting 400 mm yr^{-1} of runoff (i.e., 60 % runoff reduction, Gregoire and Clausen 2011), would have a net DOC export of 3–31 g C m^{-2} yr^{-1}. This represents a large range of uncertainty, but even the lowest estimate is greater than the annual NEP measured by Gaumont-Guay and Halsall (2013), and the highest estimate is on the same order as the short-term substrate sequestration rate measured by Getter et al. (2009). This implies that hydrologic export is of importance to the net carbon budget of green roofs, and should be included in any study concerned with carbon balance in these systems.

Because of the low potential for long-term plant biomass accumulation on extensive roofs, appreciable long-term carbon storage depends mainly on substrate accumulation of organic matter. Substrate organic matter content on newly constructed extensive roofs can be up to 65 g/L (~6.5 % by weight, assuming bulk density ~1 g/cm^3) based on widely-used guidelines (FLL 2008), but is typically lower. This translates to an initial C stock of up to ~3250 g C m^{-2} for a 10 cm thick substrate

Table 5.1 Controlled studies investigating the effects of variation in plant species or cover on green roof runoff nutrient concentrations (units = mg/L of C, N or P). Studies marked with an asterisk (*) also present nutrient fluxes, exhibiting the same general patterns as concentrations

Reference	Vegetation	TN	NO_3^-	NH_4^+	TP	PO_4^{3-}	DOC
Aitkenhead-Peterson et al. 2011	*S. kamtschaticum*		≈6.5	≈0.10		≈0.38	≈40
	D. cooperi		<0.5	≈0.10		≈0.25	≈38
	T. calycinum		<0.5	≈0.10		≈0.35	≈42
	Bare		≈6.0	≈0.14		≈0.40	≈33
Beck et al. 2011*	*L. perenne*	79.2	63.4		17.4	14.8	73.6
	S. hispanicus	No data	17.9		10.3	7.3	78.8
	Bare	No data	178		22.1	19.8	139.8
Emilsson et al. 2007*	Pre-vegetated mat (*S. album, S. acre, P. spurius*)	≈3	<1	≈0.03	≈0.3	≈0.08	
	Shoot established (*S. album, S. acre, P. spurius*)	≈25	≈20	≈0.02	≈0.6	≈0.02	
	Bare	≈35	≈25	≈0.06	≈0.5	≈0.04	
Vijayaraghavan et al. 2012	*S. mexicanum*		<2.5			≈25	
	Bare		≈25			≈50	
Wang et al. 2013	*S. lineare* (64 plants/m^2)	1.76		0.32	0.57		
	S. lineare (32 plants/m^2)	2.87		1.01	0.69		

TN total nitrogen, *TP* total phosphate, *DOC* dissolved organic carbon, *S.* Sedum, *D.* Delosperma, *T.* Talinum, *L.* Lolium, *P.* Phedimus layer. Where measurements on older roofs have been made, there is evidence for an increase in substrate organic matter content over time (Schrader and Böning 2006; Köhler and Poll 2010), particularly for shallow, single layer extensive green roofs (Thuring and Dunnett 2014). Organic matter accumulation has been observed in new green roof plots planted from seed, in which substrate organic matter content increased from 2.3 to 4.3 % over 5 years (Getter et al. 2007), and for established green roofs which increased from ~4 to ~6 % organic matter over the course of about 20 years (Köler and Poll 2010). Each 1 % increase in substrate organic matter content would amount to a net C storage of about 500 g C m^{-2} for a 10 cm thick substrate layer (again, assuming bulk density ~1 g/cm^3). Although it is not known how generalizable these observations are, the patterns suggest appreciable rates of potential C storage in substrate, with total storage slightly lower than that in forest soils, on a per area basis (e.g., Lal 2005).

Of note, increased organic matter also brings the practical concern of increased weight to the roof, mainly due to the high water holding capacity of the organic matter (discussed in Thuring and Dunnett 2014). The patterns observed suggest variation in carbon sequestration potential among different extensive green roofs. Some studies have noted that extensive green roofs can be a sink for carbon if they are planted from seed (Getter et al. 2009); however many green roofs are constructed with vegetation in place, sometimes at already high coverage, as in the case of pre-vegetated mats. In that case, there is less potential for C sequestration in plant biomass. Gaumont-Guay and Halsall's (2013) study showing atmospheric CO_2 exchange essentially in balance for a new extensive green roof in British Columbia, coupled with the fact that there is appreciable runoff of DOC from most green roof ecosystems, suggests that some roofs are actually net sources for carbon. However, other studies have indicated that organic matter content in green roof substrate can rise over time, indicating longer-term C storage (Schrader and Böning 2006; Köhler and Poll 2010). The actual dynamic will depend on the initial conditions (substrate organic matter content and type, and vegetation type and coverage), as well as local climate and roof hydrology. The change over time in substrate organic matter content will be a good indication of whether green roof ecosystems are gaining or losing carbon over the longer term.

5.5.3 Nitrogen Cycling in Green Roofs: Patterns Observed, and Implications for Processes

Most research on N in green roofs has focused on the dissolved phase; this includes the concentrations or fluxes of N into the system via atmospheric deposition (primarily NO_3^-) and out of the system via roof runoff (total nitrogen (TN), NO_3^-, and NH_4^+). In general, most roofs appear to operate as sinks for NH_4^+ (Berndtsson et al. 2006). Nitrate and total N, however, may be higher or lower in green roof runoff than precipitation (Teemusk and Mander 2007; Hathaway et al. 2008; Mendez et al. 2011). These inconsistent patterns may be attributable to variations in climate, vegetation, substrate type, fertilizer regime, substrate depth, and age of the green roof.

Table 5.2 Controlled studies investigating the effects of fertilizer amendments on green roof runoff nutrient concentrations (units=mg/L of C, N or P). Studies marked with an asterisk (*) also present nutrient fluxes, exhibiting the same general patterns as concentrations

Citation	Fertilizer treatment	TN	NO_3^-	NH_4^+	TP	PO_4^{3-}	DOC
Clark and Zheng 2013*	High fertilization (60 g/m² N; 9.8 g/m² P)		≈7.0		≈2.0		
	Low fertilization (10 g/m² N; 1.6 g/m² P)		≈0.5		≈1.3		
	No added fertilizer (0 g/m² N; 0 g/m² P)		≈0.0		≈1.1		
Emilsson et al. 2007*	High fertilization (10 g/m² N; 3.56 g/m² P)	≈116	≈75	≈20.0	≈7.0	≈6.0	
	Medium fertilization (5.0 g/m² N; 1.5 g/m² P)	≈12	≈11	≈0.04	≈0.3	≈0.05	
	Low fertilization (2.5 g/m² N; 0.73 g/m² P)	≈12	≈8	≈0.03	≈0.4	≈0.05	

TN total nitrogen, *TP* total phosphate, *DOC* dissolved organic carbon

Green roof vegetation may impact N runoff flux by several mechanisms: runoff volume reduction due to evapotranspiration, plant uptake and assimilation of N, and the release of N from litter and root exudates. The presence of typical green roof plants, compared to substrate only green roofs, usually reduces concentrations and fluxes of TN, NO_3^-, and NH_4^+ in green roof runoff (Table 5.1). Other studies have shown NO_3^- concentrations and fluxes in green roof runoff may vary with plant species (Table 5.1).

Substrate composition and fertilizer regime impact green roof N dynamics (Tables 5.2 and 5.3). If N levels in the substrate exceed biological (plant and microbe) requirements, either due to the initial substrate conditions or subsequent fertilization, the green roof will likely act as a N source (via runoff). Substrate mixes and fertilizers that include high NO_3^- levels will result in particularly high losses of N in runoff, due to the leachability of NO_3^- (Tables 5.2 and 5.3). For example, a sod roof in Estonia exposed to a fertilizer amendment resulted in runoff concentrations of over 30 mg/L TN, made up primarily of NO_3^- (Teemusk and Mander 2011). Ammonium additions may also result in NO_3^- runoff, due to the transformation of NH_4^+ to NO_3^- through the microbially-mediated process of nitrification (Fig. 5.2). This process may be responsible for the high levels of NO_3^- observed in runoff from green roof plots amended with NH_4^+ (Emilsson et al. 2007). A *Sedum* extensive green roof in Ohio (see Civic Garden Center green roof case study (Chap. 16) exposed to fertilization with corn gluten, resulted in a more than 8-fold increase in NO_3^- runoff concentrations (Buffam et al. Submitted). This was presumably due to

Table 5.3 Controlled studies investigating the effects of substrate type on green roof runoff nutrient concentrations (units = mg/L of C, N or P). Studies marked with an asterisk (*) also present nutrient fluxes, exhibiting the same general patterns as concentrations

Reference	Substrate type	Vegetation	TN	NO_3^-	NH_4^+	TP	PO_4^{3-}	DOC
Beck et al. 2011*	Commercial substrate (gravel, sand, silt, clay, pumice, compost, and paper fiber)	*S. hispanicus*	No Data	17.9		10.3	7.3	78.8
		L. perenne	79.2	63.4		17.4	14.8	73.6
	Commercial substrate with 7 % by weight biochar	*S. hispanicus*	No Data	22.5		8.3	7.3	25.7
		L. perenne	10.1	2			9.2	21.6
Vijayaraghavan et al. 2012	Garden Soil (peat, clay, organic matter (22 %))	*S. mexicanum*		<2.5			25	
	Commercial substrate (volcanic material, compost, organic and inorganic fertilizers)	*S. mexicanum*		≈13			≈29	
Wang et al. 2013	Commercial substrate with adsorption layer	*S. lineare*	1.76		0.32	0.57		
	Grass charcoal soil with adsorption layer	*S. lineare*	2.28		0.58	1.6		

TN total nitrogen, *TP* total phosphate, *DOC* dissolved organic carbon, *S.* Sedum, *L.* Lolium

mineralization of organic N in the corn gluten to NH_4^+ and subsequent nitrification, but requires further controlled study. Similarly, increasing substrate compost levels resulted in higher concentrations of TN in runoff (Hathaway et al. 2008). Substrates with high organic matter content sometimes, but not always, have increased N leaching (Van Seters et al. 2009). The inconsistencies may be due to different microbial communities or environmental conditions, as well as interactions with other substrate components. For example, the ion exchange capacity of green roof substrates will depend on the concentration and total volume of charged components, such as organic material and/or clay.

Substrate depth may influence the N cycling dynamics of green roofs by altering green roof hydrology, interaction times, substrate moisture and temperature, microbial habitat, binding and exchange sites, and the amount of leachable material. Only a few controlled studies have been performed, and these found that NO_3^- leaching increased with substrate thickness (Seidl et al. 2013; Table 5.4). Even with increases in the water holding capacity of the substrate at increased substrate depths, fluxes of NO_3^- can be higher (Seidl et al. 2013). In a study employing a dual substrate layer, with a basal layer for adsorbing nutrient runoff and an overlying

Table 5.4 Controlled studies investigating the effects of substrate depth on green roof runoff nutrient concentrations (units = mg/L of C, N or P). Studies marked with an asterisk (*) also present nutrient fluxes, exhibiting the same general patterns as concentrations

Citation	Substrate depth	TN	NO_3^-	NH_4^+	TP	PO_4^{3-}	DOC
Seidl et al. 2013*	6 cm		1.1			3.8	50
	16 cm		5			6	93
Wang et al. 2013	5 cm commercial substrate layer, 5 cm adsorption layer	≈2.07		≈0.86	≈0.69		
	5 cm commercial substrate layer, 10 cm adsorption layer	1.76		≈0.32	0.57		
	5 cm commercial substrate layer, 20 cm adsorption layer	≈1.5		≈0.42	≈0.37		
	5 cm commercial substrate layer, 30 cm adsorption layer	≈1.4		≈0.41	≈0.29		

TN total nitrogen; *TP* total phosphate; *DOC* dissolved organic carbon

organic layer to supply nutrients to green roof vegetation, N leaching was reduced when the depth of the adsorption layer was increased (Wang et al. 2013; Table 5.4).

Implications for N Cycling Processes: Patterns in NO_3^- and NH_4^+ found in runoff suggest that N mineralization and nitrification are likely occurring at appreciable rates in the substrate of many extensive green roofs, but direct process measurements have not been made. Other N transformations (Fig. 5.2) may be important but their rates have not been measured. Changes in the N sources (fertilizer amendments, substrate composition, vegetation), as well as reservoirs (vegetation presence and species, substrate), influence the export of N. Substrate depth may also impact N cycling, but this requires further controlled study.

5.5.4 *Phosphorus Cycling in Green Roofs: Patterns Observed, and Implications for Processes*

Research has consistently shown that green roofs act as sources of P for the first several years following installation, predominantly in the form of PO_4^{3-} (Monterusso et al. 2004; Berndtsson et al. 2006), with concentrations ranging from very low (ex. 0.025; Gregoire and Claussen 2011) to very high (ex. 29 mg/L; Vijayaraghavan et al. 2012). Concentrations of PO_4^{3-} in green roof runoff may reach and exceed levels comparable to wastewater (3–10 mg/L) (Metcalf and Eddy 1991). As P is typically the limiting nutrient in freshwater ecosystems, and P enrichment can lead to eutrophication (Carpenter et al. 1998), runoff from green roofs may pose a threat to aquatic ecosystems. Understanding how green roofs function as ecosystems may allow for improved green roof designs that limit or at least slow the export of P in runoff from green roofs. The export of P in green roof runoff under different experimental conditions and over time can provide clues as to the dominant processes controlling green roof P cycling.

Green roof vegetation may impact green roof P dynamics in a number of ways including: uptake of PO_4^{3-} and assimilation into organic tissues; reduced erosion of sediments containing bound P; and hydrologic and moisture changes due to evapotranspiration. One might therefore expect a negative relationship between plant presence and P runoff. Some studies do indeed indicate that plant presence reduces both TP and PO_4^{3-} concentrations and fluxes (Table 5.1) and that plant species identity may (Beck et al. 2011) or may not (Monterusso et al. 2004; Aitkenhead-Peterson et al. 2011) be important in altering P in runoff. However, it is still unclear what plant-mediated processes are responsible for observed patterns.

The composition of the green roof substrate and substrate amendments has been implicated as one of the most important determinants of P in green roof runoff. If P levels exceed the binding and uptake capacities of the substrate and biota, then P will be leached from the system. Imbalances are commonly attributed to P from the organic matter component of the substrate (Hathaway et al. 2008; Van Seters et al. 2009), presumably released by microbial mineralization of organic P to produce PO_4^{3-}. Compost, which typically has very high P content, has been suggested as the source of PO_4^{3-} to green roof runoff in several studies (Fig. 5.3; Berndtsson et al. 2009). Likewise, amendments of conventional fertilizers at high levels, in contrast to controlled release fertilizers (CRF), result in higher concentrations of TP and PO_4^{3-} in runoff (Emilsson et al. 2007) (Table 5.2). Conventional fertilizers release nutrients faster than controlled release fertilizers, likely overwhelming binding and exchange sites and exceeding the P requirements of plants and microbes. Substrate amended with Biochar (pyrolyzed biomass, with high binding capacity and surface area), showed a minimal ability to bind PO_4^{3-}, slightly reducing export via runoff (Beck et al. 2011; Table 5.3).

The depth of green roof substrate may play a role in green roof P dynamics by altering green roof hydrology, biology, and the physicochemical processes. In a controlled study, increases in substrate depth for extensive green roofs resulted

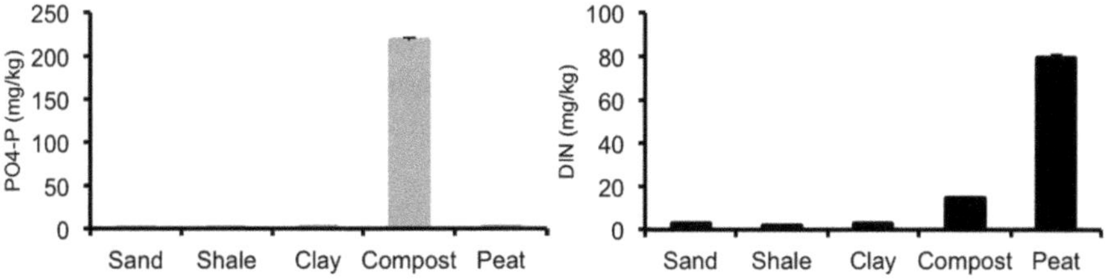

Fig. 5.3 Leachable PO_4^{3-} and Dissolved Inorganic Nitrogen ($DIN = NO_3\text{-}N + NH_4\text{-}N$) for commonly used green roof substrate components. Results are expressed in terms of mg of nutrient released per kg of dry material, for a 1:10 mixture of the respective component and either deionized water (for phosphate) or 2 M KCl solution (for DIN). Error bars are standard error of the mean for 3 replicates. (Buffam and Licardi, unpublished data)

in higher runoff concentrations of PO_4^{3-} (Seidl et al. 2013). On the other hand, for dual substrate layer green roofs (a nutrient substrate layer over an adsorption layer), increasing the thickness of the adsorption layer reduced TP concentrations in runoff (Wang et al. 2013). In contrast to most extensive green roofs, an older, intensive green roof in Japan was found to release little or no P in runoff (Berndtsson et al. 2009). These mixed results indicate the need for further study of the impacts of substrate depth on P dynamics.

Green roof age may alter green roof P cycling due to changes in the sources of P within the substrate, vegetation establishment and growth, and the water holding capacity of the substrate. Studies of green roof runoff water quality have found decreases in P leaching over time (Berndtsson et al. 2006; Buffam et al. Submitted). Köhler et al. (2002) found that the ability of a green roof to retain P increased from 26.1 % in the first year of installation to 79.9 % retention after 4 years. The authors concluded that this trend was likely due to plant establishment. Similar trends in other studies were attributed to the gradual leaching of available P in the green roof substrate over time (Van Seters et al. 2009; Buffam et al. Submitted).

Implications for P Cycling Processes: P is exported from green roof systems via runoff, especially for recently installed green roofs. It appears that for many green roofs, the major source of P is the organic component of the substrate, especially compost when present. PO_4^{3-} released by mineralization or substrate weathering may overwhelm the uptake and binding capacities of the biological and physico-chemical pools within the system. Amending green roofs with fertilizers, especially conventional as opposed to controlled release fertilizers, either worsens this imbalance or, for older roofs, reestablishes it. However, in the absence of fertilizer amendments, since the primary sources of P are within the substrate, the export of P in runoff should and does appear to decline with the age of the roof.

5.5.5 Temporal Dynamics Observed in Green Roof Nutrient Runoff

Although most available data on nutrient patterns from green roofs is for relatively few points in time for a given system, a small number of studies have examined temporal variation, and these dynamics can be analyzed to reveal information about potentially important nutrient cycling processes. The temporal dynamics can be partitioned into four time scales: within event variation, among-event variation, seasonal variation, and longer-term variation related to aging of the roof over years-decades.

Within-event green roof nutrient dynamics studies have mostly focused on contrasting the first runoff from the green roof with a sample taken later in the precipitation event. The first-flush effect is the term used to describe observations of higher pollutant loads in the initial runoff water due to the flushing of materials from dry and/or wet deposition or otherwise accumulated in the system (Zobrist et al. 2000; Berndtsson et al. 2006). A first flush effect would suggest the importance of new atmospheric deposition or between-event internal nutrient cycling for dynamic runoff concentrations, while a lack of first flush (steady concentrations during a runoff event) would suggest that substrate leachable nutrient concentrations are stable over short (hours-days) timescales. Few measurements of this type have been made for green roofs, but there are examples of a first flush effect with higher concentrations of TP, NH_4^+, and NO_3^- (Berndtsson et al. 2008) in initial runoff samples compared to samples taken later in the event.

Among event variation in nutrient cycling may be expected due to variations in precipitation event intensity, duration, and antecedent conditions such as roof moisture and temperature. These changes may alter the physicochemical (sorption and weathering) and biological (mineralization and plant or microbial uptake) process rates within green roofs. Additionally, a higher precipitation volume will most often result in more runoff, potentially leading to a dilution of dissolved compounds in the runoff. Studies have found lower specific conductivity in green roof runoff following larger precipitation events (Berghage et al. 2009) and higher concentrations of total P and PO_4^{3-} and lower NO_3^- in green roof runoff during large events (Teemusk and Mander 2007). Other studies, however, have observed no effect of event size on total P concentrations (Gregoire and Clausen 2011). These inconsistencies suggest the importance of climatic and roof conditions, which vary between studies and through time.

Seasonally-driven green roof nutrient dynamics may be expected due to variation in plant productivity and microbial activity responding to variation in temperature and light within the green roof ecosystem. Researchers have observed increased runoff NO_3^- concentrations in the summer months compared to fall and winter (Van Seters et al. 2009; Buffam et al. Submitted), and higher levels of P in green roof snowmelt vs. rain-driven runoff (Gregoire and Clausen 2011). Frequent temporal sampling has revealed contrasting seasonal patterns for different systems (Fig. 5.4). A 2-year study of runoff water quality on a 1–3 year old sloped extensive green roof with a pre-vegetated *Sedum* mat revealed strong seasonal patterns, with most nutrient and base cation concentrations reaching their maximum concentrations in the warmest months (Buffam et al. Submitted). In contrast, monthly measurements of N and P concentrations and fluxes in runoff from newly-established tray-based extensive green roof plots showed a winter maximum for phosphate, and no clear seasonal pattern in nitrate (Fig. 5.4). Phosphate runoff fluxes were not measurably affected by the presence or absence of plants in this plot-scale study, while nitrate runoff fluxes were decreased in the presence of plants (Fig. 5.4, Johnson 2014). These results suggest that nutrient dynamics in newly-established green roofs are influenced both by plant uptake of N, and by temperature-mediated substrate processes such as weathering or organic matter mineralization.

Long-term aging of a green roof ecosystem may influence its water holding capacity and runoff chemistry. For example, substrate on a 5 year old roof had more than 3 times the water holding capacity of the original substrate, nearly twice the organic matter and pore space, and increased free airspace (Getter et al. 2007). While it appears that runoff of P, and potentially N, decreases with age (eg., Berndtsson et al. 2006, Köhler et al. 2002), few studies have directly investigated changes in

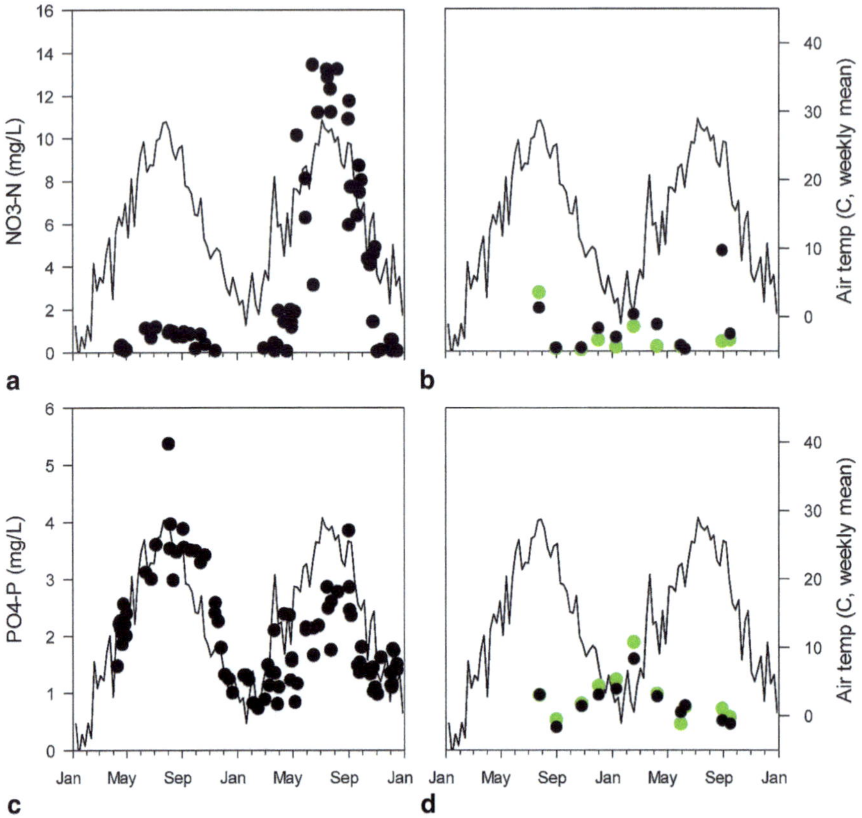

Fig. 5.4 **a, c** (leftmost panels): Seasonal dynamics for nitrate (NO_3^-) and phosphate (PO_4^{3-}) runoff concentrations from the Civic Garden Center cottage green roof in Cincinnati, OH, for the majority of the runoff events for 2 years (For roof description see Case Study, Chap. 16). *Dots* indicate concentration; the *solid black line* indicates weekly average air temperature. Note the increase in NO_3^- in the second year likely due to a May fertilization event, and the decline in PO_4^{3-} from 1 year to the next, attributed to aging of the roof (From Buffam et al. submitted). **b, d** (rightmost panels) Seasonal dynamics for NO_3^- and PO_4^{3-} runoff concentrations from newly established tray-based extensive green roof plots, either with (*light green dots*) or without (*black dots*) vegetation. (From Johnson 2014)

C, N or P cycling in green roofs over longer (decadal) timescales. Changes in the plant community, microbial community, substrate makeup, and water retention with roof age are all likely to impact green roof nutrient dynamics and thus runoff water quality; however these dynamics are not well understood.

Taken together, these observations of temporal variation indicate that green roofs are dynamic systems, which are strongly influenced by environmental conditions (weather, climate) and time. Future controlled studies may help to isolate which variables are most responsible for these patterns and in doing so, uncover the dominant green roof nutrient cycling processes.

5.6 Conceptual Model of Long Term Green Roof Nutrient Cycling Dynamics

The long-term (decadal-scale) changes in nutrient cycling in green roofs are of relevance to managers and researchers alike. Although little is currently known about these long-term dynamics from direct measurements, paradigms and models developed for natural terrestrial ecosystems provide a useful framework for generating hypotheses about green roofs. The "Nutrient Retention Hypothesis", developed to describe nutrient cycling over successional time (centuries-millennia) in N-limited forest ecosystems (Vitousek and Reiners 1975) offers one such starting point. In this conceptual model, there are several predictable phases during succession, with different nutrient retention characteristics. For primary succession, there is a long time period during which the system is nutrient limited, while vegetation grows and soils build up. During this "aggradation" period, nutrients are held tightly within the system, and export of the limiting nutrient will be essentially zero, while export of other essential nutrients will also be reduced relative to inputs. After vegetation and soils have built up, the system experiences a period of transition where nutrient exports increase, until reaching steady state (inputs = outputs). Secondary succession generally follows the same sequence as primary succession, but begins with a disturbance, resulting in high loss of nutrients during a reorganization period, after which the aggradation period starts again as vegetation regrows. The ecosystem then proceeds as with primary succession, towards steady state (Vitousek and Reiners 1975; Aber and Melillo 2001).

This basic pattern of changing nutrient retention vs. export over successional time can be envisioned with a 3-compartment model (Fig. 5.5; Vitousek et al. 1998), which is a simplified version of a generalized nutrient cycle (Fig. 5.1). In this simplified model, the eventual steady state occurs as a consequence of internal feedbacks resulting in equilibrium between the rate of formation of organic nutrient pools (vegetation, soil) which hold nutrients within the system, and the rate of decomposition of those pools into soluble, bioavailable inorganic nutrients which can be lost through leaching (Vitousek et al. 1998).

Applying the concepts from this model allows us to construct a basic set of hypotheses for the long-term trajectory of N and P outputs from green roof ecosystems (Fig. 5.6). These hypotheses are informed by observations of nutrient runoff concentrations from green roof ecosystems (Sect. 5.5, and summarized below), and acknowledge the unique characteristics of extensive green roofs as engineered ecosystems (Sect. 5.4).

Summary of five relevant and commonly observed nutrient patterns from extensive green roofs:

- Organic matter integrated into substrate, and fertilizers either initially integrated or added later, provide a large nutrient supply for green roof ecosystems.
- Green roofs are a source of high P and N in runoff, for new roofs and following fertilization events.
- Substrate typically contains very low N:P, i.e. is enriched in P relative to N supply, and relative to plant or microbial demand.

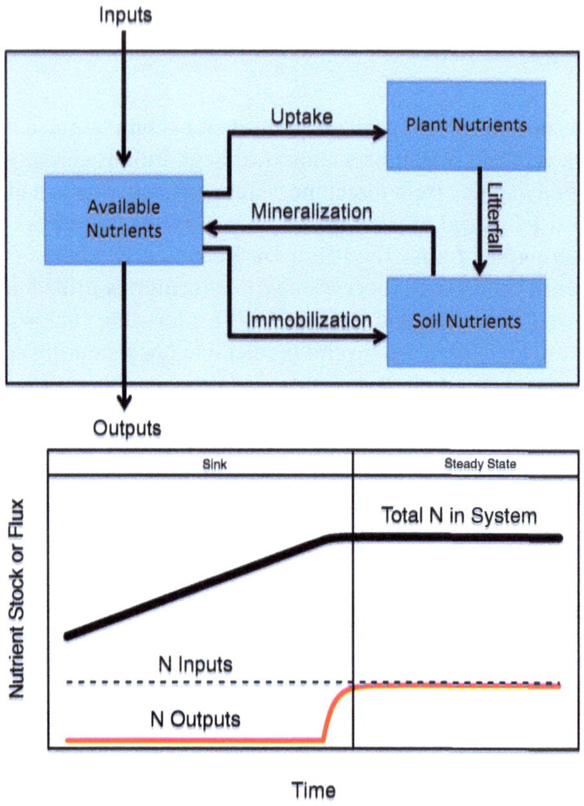

Fig. 5.5 Nutrient cycling within a terrestrial ecosystem can be modeled in a simplified way with 3 compartments: available inorganic nutrients, plant nutrients, and soil nutrients (including microbially bound nutrients). (*Top*) The 3-compartment model, developed to represent changes in N cycling during succession in a N-limited forest ecosystem. (*Bottom*) Model predictions for variation in outputs (hydrologic runoff), and total system N storage during succession, in the absence of disturbance and with constant, moderate inorganic nitrogen inputs (atmospheric deposition). Under this scenario, a N-limited ecosystem will be a sink for N for a period of time (many centuries, in the model) while vegetation and soil pools grow. Internal feedbacks between the rates of plant uptake, litterfall, and net mineralization result in the system tending towards steady state over time, where N outputs are equal to inputs. (Adapted from Vitousek et al. 1998)

- After an initial period of leaching, green roofs may be either a source or a sink for N.
- Green roofs are consistently a source for P in runoff, but runoff concentrations may be low for older roofs.

There are a few ways in which the development of green roofs over time is clearly different than the development of forest ecosystems as proposed in the 3-compartment model (Vitousek et al. 1998), and in general models of terrestrial nutrient dynamics over successional time (Chapin et al. 2011; Vitousek and Reiners 1975). One difference is that green roofs as currently constructed start with an abundance of nutrients in the system, relative to plant needs. As a result, green roofs can be expected

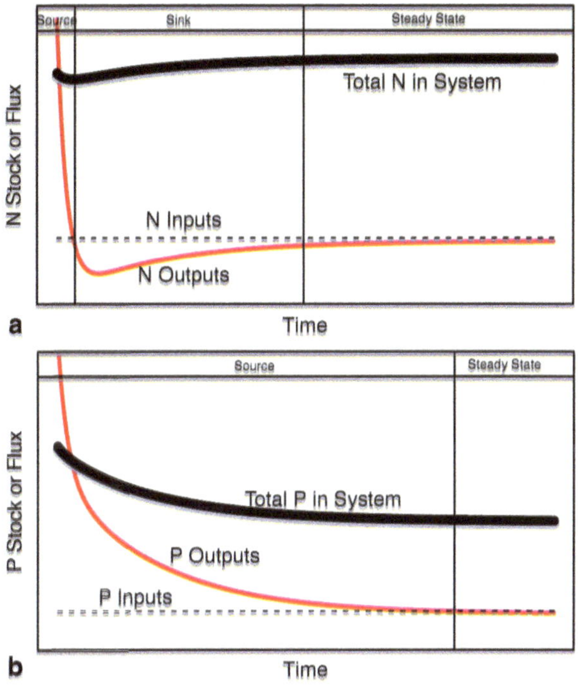

Fig. 5.6 Different hypothesized trajectories for long-term N (panel a, above) and P (panel b, below) dynamics in an extensive green roof, based on application of a simplified 3-compartment model for nutrient cycling (Vitousek et al. 1998). Both nutrients begin with high rates of leaching outputs (*thin, solid line*) due to excess initial nutrient stocks in substrate relative to plant demands. Both are also predicted to eventually approach steady state where plant uptake is matched by the substrate mineralization rate, and nutrient outputs equal inputs (*thin, dotted line*). **a** Total system N (*thick line*) may start at levels relatively close to the steady state system N content, and thus may approach steady state fairly rapidly. During the intervening transition period, the roof may serve as a sink or a source for N. The dynamics of this transition depend on the initial nutrient supply, how long plants take to reach their maximum biomass, and the rates of internal processes (plant uptake, net mineralization, and litterfall production). **b** For P, the system is consistently a source and the approach to steady state is delayed. This is mainly because P is typically present in great excess relative to N (thus non-limiting). P (whether in organic or inorganic form) is also "stickier" than NO_3^- in terms of binding to soil organic and mineral particles, thus will have a greater tendency to resist being flushed quickly out of the system

to leach out high levels of nutrients when they are newly constructed (Fig. 5.6), and this has been broadly observed (e.g., Tables 5.1, 5.2, 5.3 and 5.4). This contrasts with the models of primary succession in forested ecosystems (Fig. 5.5), in which nutrients are in short supply, thus tightly held within the system during early succession. The creation of a green roof ecosystem may instead be analogous to a disturbance in a natural ecosystem, which is typically followed by a brief period of high nutrient leaching.

In the absence of disturbance, based on the simple ecosystem model the system is predicted to gradually approach steady state, where nutrient inputs = outputs (e.g., Vitousek and Reiners 1975; Fig. 5.5). The intervening transition phase could take on different characteristics, depending on the initial nutrient stocks and on the balance between inorganic nutrient consumption (plant uptake and soil organic matter formation) and production (net mineralization) over time (e.g., Fig. 5.6). If initial nutrient substrate stocks are not too high, and plants continue to grow over an extended period of time, the ecosystem may experience an "aggradation" phase where outputs < inputs, i.e. the system holds tightly to nutrients, particularly the limiting nutrient. This may be the case with N for some green roofs. However, if initial nutrient stocks in the substrate are very high relative to demand, and the nutrient in question is in organic form or otherwise bound in the substrate, then the system may continue to leach out the nutrient over a long period of time, and never experience a phase where outputs < inputs. This appears to be the case with P for most green roofs as currently constructed.

This model has interesting implications when applied to green roof ecosystems. Assuming that basic models and paradigms from natural forested ecosystems apply, green roofs over time will tend to approach steady state with respect to nutrients—though, it should be noted that even in forested ecosystems the real situation is typically more complex than the model would predict (Vitousek et al. 1998; Aber and Melillo 2001). Even if the systems do generally approach steady state, a key question is how long this takes. Observations to date suggest that this may happen relatively rapidly (within a few years) for N, but may take many years/decades for P.

There are, additionally, several factors that may prevent green roofs from conforming to predictions based on the 3-compartment model. Foremost, the model is an extreme simplification of a complex and dynamic ecosystem, and for many green roofs it may be necessary to include other processes to accurately characterize the system. For example, in extensive green roof ecosystems, productivity may be limited by factors other than nutrient availability, such as moisture availability, wind stress, the low growth potential of the plants, or management choices. Also, the plant species, which by design are typically slow-growing and lack an overstory layer, have lower potential productivity and biomass accumulation than a forest ecosystem. Any subsequent fertilization of the roof would likely lead to additional pulses of nutrient export, slowing the approach to steady state. The lifespan of these ecosystems are also limited, as green roofs are likely to be replaced every 50 years or so. These factors contribute to the potential for a green roof ecosystem to be restricted to the "reorganization" phase of development for much of their lifespan, especially for P.

The simple 3-compartment model allows us to generate hypotheses about potential nutrient behavior of green roofs over a long (decadal) time frame (e.g., Fig. 5.6). However, although this approach does provide a useful starting point, there is not yet enough available field data to determine whether model predictions are borne out. Detailed long-term monitoring of green roofs is needed to fill this knowledge gap. As we continue to learn more about these ecosystems, it will be interesting to see where their long-term trajectories lie relative to those of non-engineered ecosystems.

5.7 Areas of Greatest Uncertainty and Suggestions for Future Research Directions

5.7.1 Overview

We are at an exciting juncture in green roof research, where there is a growing interest by policy-makers and managers in green roof implementation, and by researchers from different disciplines in green roofs as a study system. In spite of the growing interest and body of research, there remain a number of limitations to our understanding of nutrient cycling in green roofs (Table 5.5). These are in part related to the relatively narrow scope of published research studies. The majority of current knowledge of C, N and P cycling in green roofs consists of observations of patterns of concentrations (and occasionally fluxes) of these elements in runoff. Hydrologic fluxes are presumably the largest N and P loss pathway for most green roof systems, and runoff fluxes are of environmental concern because of their potential impact on downstream water quality. However, there are a number of other important transformations and fluxes influencing nutrient cycling and movement through ecosystems (Fig. 5.1), ultimately impacting both plant health within the ecosystem and nutrient runoff. Most of these internal pathways have not been quantified in green roof ecosystems, so their current degree of uncertainty is very high (Table 5.5).

Ecosystem analysis benefits from complementary approaches including the "four legs" supporting ecosystem science: theory, long-term study, cross-ecosystem comparison, and experiments (Carpenter 1998). To date, studies of green roofs in general, and certainly nutrient cycling in green roofs, have suffered from a relative lack of experimental and modeling studies; and most of our knowledge is based on short-term monitoring only, with a few comparative studies examining different types of roofs. Furthermore, full-scale studies are unreplicated, and the wide range of initial roof designs involving different substrate types and depths, different plant palettes, different climates, and different roof ages complicates among-roof comparisons. Controlled experimental treatments have tended to be small scale studies, typically using plots of $< 1\ m^2$ and often in a greenhouse setting. It is not clear how the small scale of these studies, and the restricted greenhouse climate (with higher and more consistent humidity than most natural systems), influences the nutrient cycling dynamics relative to full-scale roofs in a real climate setting. Finally, many of the studies to date have only included measurements of nutrient concentrations in runoff, rather than also measuring water balance and calculating nutrient fluxes. To move forward, an expanded scope of approaches is needed.

It is relevant to note that this assessment is based on the English-language scientific literature only. Broad implementation of extensive green roofs in German cities, especially during the mid-late twentieth century, has led to a rich history of green roof design and associated research in that country. Thus, there is an existing deep knowledge base, only some of which has been accessed by the scientific community writing in English. There is good potential for expanding partnerships to

Table 5.5 Ecosystem nutrient fluxes and transformations, their estimated potential magnitude in green roofs (relative to other processes for the same element), and current degree of uncertainty in green roofs

Flux or process	Movement or transformation	Potential Magnitude			Degree of uncertainty
		Carbon	Nitrogen	Phosphorus	
Atmospheric deposition	Input	L	M	L	Low
N-fixation	Input	-	M	-	High
Fertilizer addition	Input	L	H	H	Moderate
Mineral weathering	Input/transformation	L	L	M	High
Substrate OM accumulation	Transformation	M	M	M	Moderate
Microbial mineralization of OM	Transformation	H	H	H	High
Microbial immobilization	Transformation	L	M	M	High
Nitrification	Transformation	-	H	-	High
Sorption/desorption	Transformation	M	M	M	High
Plant net uptake[a]	Transformation	M	H	M	Moderate
Denitrification	Transformation/loss	-	L	-	High
Runoff	Loss	H	H	H	Moderate
Emission of N_2O	Loss	-	L	-	High
Aeolian transport	Loss	L	L	L	High

[a] Net plant uptake = plant uptake–(litterfall + root exudates and other losses)
OM organic matter, *L* low, *M* moderate, *H* high

build rapidly on this knowledge base—by making existing studies available across different language platforms, by international research collaborations, and by carrying out new studies on older, well-studied roofs in Germany (see for instance Köhler and Poll 2010; Thuring and Dunnett 2014).

5.7.2 Questions for Future Research

As an aid in thinking about the future of nutrient cycling research in green roofs, we present several pressing research questions below, ranging from the theoretical to the practical:

Q1: How can green roof nutrient cycles best be modeled, conceptually and mathematically? Development of models is important for increased understanding of green roof ecosystems, and for prediction of the impacts of variation in roof design, or in environmental conditions including climate. In this chapter we presented a rough conceptual schematic with associated expectations based on a simple 3-compartment model from N-limited terrestrial ecosystem over successional time (Vitousek et al. 1998). This enabled discussion of similarities and differences compared to the (relatively) well-understood nutrient-cycling processes in natural forested ecosystems, but there are many other types of models that could be useful in describing green roof ecosystem dynamics (e.g., Jorgensen 2011).

Q2: How does varying nutrient stoichiometry impact the nutrient cycling within green roof ecosystems, and specifically the capacity of these systems to support thriving plant communities while minimizing hydrologic loss of dissolved N and P? Currently, most green roofs have an abundance of P relative to N, based on the very low N:P ratios observed in effluent water. This suggests that P sources could be reduced without stressing plants, and this might result in more efficient nutrient retention within the ecosystem.

Q3: What role do soil microbial communities, including bacteria, saprotrophic fungi, and mycorrhizal associations play in processing C, N and P in green roof substrate, and in regulating the runoff of these nutrients? Very little is known on this topic, though McGuire et al. (2013) found that green roofs can be home to diverse fungal communities and that these communities differ among roofs. Mycorrhizal associations could be of importance for nutrient delivery to plants, particularly if engineering of green roofs shifts to lower nutrient substrates (See also Chap. 7 in this book).

Q4: Is Nitrogen-fixation an important source of N to green roof ecosystems? N-fixation can provide an important source of fixed nitrogen to ecosystems, especially in high light, N-limited environments (Chapin et al. 1991). Currently, N-fixation is probably not of major importance in most extensive green roofs, since *Sedums* are not among the families of plants commonly associated with N-fixing bacteria (Paul and Clark 1996). However, free-living soil microbes can perform

N-fixation under appropriate environmental conditions, and there are already examples of plants associated with N-fixation on green roofs (Chap. 16). The use of N-fixing plant associations could provide an important contribution of N to green roof ecosystems under some engineering designs, and would allow for reduced use of N fertilizers.

Q5: What is the role of green roof ecosystems in surface-atmosphere exchange of the trace greenhouse gases (GHG) methane (CH_4) and Nitrous Oxide (N_2O)? Although overall carbon exchange over long time frames may be relatively small in green roofs, CH_4 and N_2O exchange could still be relevant to urban GHG balances. It is not known whether green roofs are a source or a sink for these gases, nor how substrate characteristics affect CH_4 or N_2O exchange. Green roofs are unlikely to be sources of CH_4 because they are designed to be well-drained and thus presumably lack anoxic zones for CH_4 genesis. Green roofs could however be a sink for CH_4, as has been found for other vegetated ecosystems (Groffman and Pouyat 2009). Green roofs could be a source of N_2O, which is a byproduct of both nitrification and denitrification processes (Groffman et al. 1998; Bateman and Baggs 2005). N_2O emissions often increase after fertilization events (Aber and Melillo 2001), due to an increase in the rates of those processes. Denitrification is unlikely to be important for extensive green roofs that are well-drained, as this process requires anoxic microsites typical of waterlogged soils. However, denitrification could be important in some roofs that have thicker substrate or natural soils that drain less freely, and nitrification may contribute to N_2O emissions.

Q6: How can substrate amendments increase the capacity of green roof substrate to reduce hydrologic losses and bind nutrients? Early experiments with the integration of biochar, an agricultural soil amendment, have shown promising results in terms of increasing water-holding capacity and reducing leaching of DOC and sometimes N and P (Beck et al. 2011).

Q7: How can fertilizer applications best be matched with plant nutrient demands to reduce expensive fertilizer requirements and reduce impacts of pollution downstream? Unlike agro-ecosystems in which high productivity and harvest are the main goals, green roof ecosystems are designed for stability, sustainability, and minimal management requirements. Inherent slow growth and low nutrient requirements are characteristics of plants like *Sedum*, which are adapted to low-fertility environments (Chapin et al. 1986). These characteristics may persist even when they are placed in a high-nutrient environment, like a fertilized green roof. Therefore, the nutrient supply could be kept relatively low, with the goal of minimizing fertilizer nutrient inputs and designing a system that holds and recycles nutrients.

5.8 Synthesis and Implications

5.8.1 Summary of Current Knowledge

Most of the current knowledge about green roof nutrient cycling involves observation of patterns of runoff concentrations. As most extensive green roofs are currently constructed, both available N and especially P substrate pools begin high relative to the plant community's capacity to uptake nutrients. In spite of providing a substantial reduction in runoff volume, the runoff is often enriched in nutrients, particularly in younger roofs. Organic carbon and inorganic phosphorus are consistently high in runoff; Ammonium is typically reduced in runoff relative to the incoming atmospheric deposition flux, while nitrate may be either reduced or enriched. Although phosphorus leaching may be a transient effect, it represents a substantial disservice for at least several years after construction for many green roofs, with P concentrations in effluent water from many green roofs as high as in wastewater (Metcalf and Eddy 1991). It is also largely unnecessary since most of the systems have excess P relative to plant demand, and relative to availability of N and other nutrients.

A number of characteristics of green roof systems influence nutrient cycling and nutrient efflux in runoff. These include plant density or coverage, plant identity and functional characteristics, substrate type, and substrate depth. Substrate characteristics have been found to be a primary control on runoff nutrient fluxes from green roof systems. The organic component of the substrate has been implicated as the primary source for N and P leaching out of some green roof ecosystems (Hathaway et al. 2008; Fig. 5.3), while slow-release mineral fertilizers have been suggested as the main source of leaching N and P in other green roofs (Berndtsson et al. 2006). Plants also play an important role in sequestering N in biomass, reducing N losses in runoff (Table 5.1). Carbon sequestration in substrate or biomass in green roofs may occur but is likely a minor factor relative to the initial energy costs of building the system, and other important functions provided by the system, including long-term energy savings (Getter et al. 2009).

5.8.2 Looking Forward

In any complex ecological system, the observed hydrologic nutrient efflux is the result of the balance between other ecosystem processes (Fig. 5.1), many of which remain poorly studied in green roofs (Table 5.5). An understanding of the internal process dynamics is important for predicting responses to environmental variation (e.g., weather variability and climate shifts) and impacts of varying system characteristics (e.g., substrate organic matter content, substrate N:P ratio, substrate depth, plant assemblage). Studies of temporal dynamics, on time-scales ranging from minutes to years, provide useful information on potentially important processes. Studies taking a mass-balance approach to following C, N, and P dynamics through green roof ecosystems would provide a valuable addition to the knowledge base. Future studies should emphasize controlled experimental manipulations at the full-roof scale, direct process measurements, mathematical process-based models linking green roof hydrology and biogeochemistry, and long-term observations spanning decades, which is a relevant time-scale for these ecosystems.

5.8.3 Management Implications

At this point, there is enough information available to suggest next-generation substrate designs that would minimize nutrient losses while sustaining plants and providing the other services for which green roofs are designed. To reduce the output of nutrients, particularly inorganic N and P, the keys are minimizing inputs and/or retaining nutrients more tightly within the system. Lowering nutrient inputs in precipitation is not under control of local managers, but the initial nutrient pool sizes can be lowered by decreasing the amount of N and (especially) P rich fertilizers and organic matter in the substrate, by encouraging biological N-fixation rather than fertilization as a source of N, and by matching fertilization regime carefully to measured system needs. This may mean elimination of P in added fertilizers, since most systems as currently constructed have an abundance of P relative to N.

Nutrient retention is a challenge because of the high potential for nutrient leaching loss in systems like green roofs that are well-drained and rapidly flushed during large precipitation and melt events. Strategies to increase retention (and decrease runoff) could include increased stability of the substrate nutrient pool, and increased uptake rates by, and storage in, vegetation. Increased retention within the substrate could be achieved by amendments which bind nutrients tightly and thereby lower rates of loss by mineralization, desorption, and weathering, or by any enhancement to water retention within the ecosystem. For example, initial experiments with biochar have shown some promise in both water and nutrient retention (Beck et al. 2011). Increased storage by vegetation could either be sustained by harvesting vegetation periodically, or by using vegetation that would decompose slowly and thus contribute to the long-term accumulation of organic matter in the substrate. Finally, runoff from green roofs can be re-used/re-cycled such that the excess P is used for fertilizing other vegetation, or at least prevented from entering local waterways.

References

Aber J, Melillo J (2001) Terrestrial ecosystems. Harcourt Academic Press, San Diego

Aitkenhead-Peterson J, Dvorak B, Voider A, Stanley N (2011) Chemistry of growth medium and leachate from green roof systems in south-central Texas. Urban Ecosyst 14(1):17–33

Baker L, Hope D, Xu Y, Edmonds J, Lauver L (2001) Nitrogen balance for the Central Arizona-Phoenix (CAP) ecosystem. Ecosystems 4(6):582–602

Bateman E, Baggs E (2005) Contributions of nitrification and denitrification to N2O emissions from soils at different water-filled pore space. Biol Fertil Soils 41(6):379–388

Beck D, Johnson G, Spolek G (2011) Amending greenroof soil with biochar to affect runoff water quantity and quality. Environ Pollut 159(8–9):2111–2118
Berghage R, Beattie D, Jarrett A, Thuring C, Razei F, O'Connor T (2009) Green roofs for stormwater runoff control. Cincinnati, OH. EPA/600/R-09/026
Berndtsson J (2010) Green roof performance towards management of runoff water quantity and quality: A review. Ecological Engineering 36:351–360
Berndtsson J, Emilsson T, Bengtsson L (2006) The influence of extensive vegetated roofs on runoff water quality. Sci Total Environ 355(1–3):48–63
Berndtsson J, Bengtsson L, Jinno K (2008) First flush effect from vegetated roofs during simulated rain events. Hydrol Res 39(3):171–179
Berndtsson J, Bengtsson L, Jinno K (2009) Runoff water quality from intensive and extensive vegetated roofs. Ecol Eng 35(3):369–380
Buffam I, Mitchell M, Durtsche RD (Submitted) Seasonal variation in green roof runoff water quality
Carpenter, S (1998), The need for large-scale experiments to assess and predict the response of ecosystems to perturbation. In: Pace ML, Groffman PM (eds) Successes, Limitations, and Frontiers in Ecosystem Science. Springer, New York, pp 287–312
Carpenter S, Caraco N, Correll D, Howarth R, Sharpley A, Smith V (1998) Nonpoint pollution of surface waters with phosphorus and nitrogen. Ecol Appl 8(3):559–568
Chapin F, Shaver G, Kedrowski R (1986) Environmental controls over carbon, nitrogen and phosphorus fractions in Eriophorum-vaginatum in Alaskan tussock tundra. J Ecol 74(1):167–195
Chapin F, Bliss L, Bledsoe L (1991) Environmental regulation of nitrogen fixation in a high arctic lowland ecosystem. Can J Bot 69:2744–2755
Chapin F, Woodwell G, Randerson J, Rastetter E, Lovett G, Baldocchi D, Clark D, Harmon M, Schimel D, Valentini R, Wirth C, Aber J, Cole J, Goulden M, Harden J, Heimann M, Howarth R, Matson P, McGuire A, Melillo J, H Mooney, Neff J, Houghton R, Pace M, Ryan M, Running S, Sala O, Schlesinger W, Schulze E (2006) Reconciling carbon-cycle concepts, terminology, and methods. Ecosystems 9(7):1041–1050
Chapin F, Matson P, Vitousek P (2011) Fundamentals of terrestrial ecosystem ecology. Springer, New York
Clark M, Zheng Y (2013) Plant nutrition requirements for an installed Sedum–vegetated green roof module system:effects of fertilizer rate and type on plant growth and leachate nutrient content. HortScience 48:1173–1180
Crews T (1999) The presence of nitrogen fixing legumes in terrestrial communities: evolutionary vs ecological considerations. Biogeochemistry 46(1–3):233–246
Dahlgren R (1998) Effects of forest harvest on stream-water quality and nitrogen cycling in the Caspar Creek Watershed. USDA Forest Service Gen Tech Rep PSW-GTR 168:45–53
Emilsson T, Berndtsson J, Mattsson J, Rolf K (2007) Effect of using conventional and controlled release fertiliser on nutrient runoff from various vegetated roof systems. Ecol Eng 29(3):260–271
FLL (2008) Guidelines for the planning, construction, and maintenance of green roofing- green roofing guideline. Forschungsgesellschaft Landschaftsentwicklung Landschaftsbau, Bonn
Galloway J, Aber J, Erisman J, Seitzinger S, Howarth R, Cowling E, Cosby B (2003) The nitrogen cascade. Bioscience 53(4):341–356
Gaumont-Guay D, Halsall R (2013) What's the carbon sequestration potential in green roofs? 11th Annual Cities Alive Green Roof and Wall Conference, San Francisco, CA
Getter K, Rowe D, Andresen J (2007) Quantifying the effect of slope on extensive green roof stormwater retention. Ecol Eng 31(4):225–231
Getter K, Rowe D, Robertson G, Cregg B, Andresen J (2009) Carbon sequestration potential of extensive green roofs. Environ Sci Technol 43(19):7564–7570
Gregoire B, Clausen J (2011) Effect of a modular extensive green roof on stormwater runoff and water quality. Ecol Eng 37(6):963–969
Groffman P, Pouyat R (2009) Methane uptake in urban forests and lawns. Environ Sci Technol 43(14):5229–5235

Groffman P, Gold A, Jacinthe P (1998) Nitrous oxide production in riparian zones and groundwater. Nutr Cycl Agroecosyst 52(2–3):179–186
Hathaway A, Hunt W, Jennings G (2008) A field study of green roof hydrologic and water quality performance. Trans Asabe 51(1):37–44
Johnson C (2014) Role of plant species richness on the quantity and quality of stormwater runoff in green roof plots. M.S. Thesis, Department of Biological Sciences, University of Cincinnati
Jorgensen S (ed) (2011) Handbook of ecological models used in ecosystem and environmental management. Taylor and Francis, Boca Raton
Köhler M, Poll PH (2010) Long-term performance of selected old Berlin green roofs in comparison to younger extensive green roofs in Berlin. Ecol Eng 36:722–729
Köhler M, Schmidt M, Grimme FW, Laar M, Assunção Paiva VL, Tavares S (2002) Green roofs in temperate climates and in the hot-humid tropics-far beyond the aesthetics. Environ Manage Health 13(4):382–391
Lal R (2005) Forest soils and carbon sequestration. Forest Ecol Manage 220:242–258
Likens G, Bormann F, Pierce R, Eaton J, Johnson N (1977) Biogeochemistry of a forested ecosystem. Springer-Verlag, New York
Marschner B, Kalbitz K (2003) Controls of bioavailability and biodegradability of dissolved organic matter in soils. Geoderma 113:211–235
McGuire K, Payne S, Palmer M, Gillikin C, Keefe D, Kim S, Gedallovich S, Discenza J, Rangamannar R, Koshner J, Massmann A, Orazi G, Essene A, Leff J, Fierer N (2013) Digging the New York City Skyline: soil fungal communities in green roofs and city parks. PLoS One 8(3):e58020
Mendez C, Klenzendorf J, Afshar B, Simmons M, Barrett M, Kinney K, Kirisits M (2011) The effect of roofing material on the quality of harvested rainwater. Water Res 45(5):2049–2059
Metcalf and Eddy (1991) Wastewater engineering: treatment, disposal, and reuse. McGraw-Hill, New York
Monterusso M, Rowe D, Rugh C, Russell D (2004) Runoff water quantity and quality from green roof systems. Expanding roles for horticulture in improving human well-being and life quality. Acta Hortic 639:369–376
Mulholland P (2003) Large-scale patterns in dissolved organic carbon concentration, flux, and sources. In: Findlay S, Sinsabaugh R (eds) Aquatic ecosystems: interactivity of dissolved organic matter. Academic Press/Elsevier Science, San Diego, pp 139–159
Oberndorfer E, Lundholm J, Bass B, Coffman R, Doshi H, Dunnett N, Gaffin S, Koehler M, Liu K, Rowe B (2007) Green roofs as urban ecosystems: ecological structures, functions, and services. Bioscience 57(10):823–833
Paul E, Clark F (1996) Soil microbiology and biochemistry, 2nd edn. Academic Press, San Diego
Pett-Ridge J (2009) Contributions of dust to phosphorus cycling in tropical forests of the Luquillo Mountains, Puerto Rico. Biogeochemistry 94(1):63–80
Rowe D (2011) Green roofs as a means of pollution abatement. Environ Pollut 159(8–9):2100–2110
Saugier B, Roy J, Mooney H (2001) Estimations of global terrestrial productivity: converging toward a single number? In: Roy J, Saugier B, Mooney H (eds) Terrestrial global productivity. Academic Press, San Diego, pp 543–557
Schrader S, Böning M (2006) Soil formation on green roofs and its contribution to urban biodiversity with emphasis on Collembolans. Pedobiologia 50:347–356
Seidl M, Gromaire M, Saad M, De Gouvello B (2013) Effect of substrate depth and rain-event history on the pollutant abatement of green roofs. Environ Pollut 183:195–203
Shaver G, Melillo J (1984) Nutrient budgets of marsh plants- efficiency concepts and relation to availability. Ecology 65(5):1491–1510
Smith S, Read D (2008) Mycorrhizal symbiosis, 3rd edn. Academic Press, London
Teemusk A, Mander U (2007) Rainwater runoff quantity and quality performance from a greenroof: the effects of short-term events. Ecol Eng 30(3):271–277
Teemusk A, Mander U (2011) The influence of green roofs on runoff water quality: a case study from Estonia. Water Res Manage 25(14):3699–3713

Thuring CE, Dunnett N (2014) Vegetation composition of old extensive green roofs (from 1980s Germany). Ecol Processes 3:4

Van Seters T, Rocha L, Smith D, MacMillan G (2009) Evaluation of green roofs for runoff retention, runoff quality, and leachability. Water Qual Res J Can 44(1):33–47

Vijayaraghavan K, Joshi U, Balasubramanian R (2012) A field study to evaluate runoff quality from green roofs. Water Res 46(4):1337–1345

Vitousek P, Howarth R (1991) Nitrogen limitation on land and in the sea-how can it occur. Biogeochemistry 13(2):87–115

Vitousek PM, Reiners WA (1975) Ecosystem succession and nutrient retention: a hypothesis. BioScience 25(6):376–381

Vitousek P, Hedin L, Matson P, Fownes J, Neff J (1998) Within-system element cycles, input-output budgets, and nutrient limitation. In: Pace M, Groffman P (eds) Successes, limitations, and frontiers in ecosystem science. Springer, New York, pp 432–451

Wang X, Zhao X, Peng C, Zhang X, Wang J (2013) A field study to evaluate the impact of different factors on the nutrient pollutant concentrations in green roof runoff. Water Sci Technol 68(12):2691–2697

Zobrist J, Muller S, Ammann A, Buchelli T, Mottier V, Ochs M, Schoenenberger R, Eugster J, Boller M (2000) Quality of roof runoff for groundwater infiltration. Water Res 34(5):1455–1462

Asst. Prof. Dr. Ishi Buffam University of Cincinnati, is an ecosystem ecologist and aquatic biogeochemist focusing on carbon and nitrogen transformations and hydrologic transport within watersheds and streams. His recent research explores the influence of vegetated roofs on urban surface water quality.

Mark E. Mitchell University of Cincinnati, is an ecologist studying nitrogen and phosphorus cycling dynamics in vegetated roof ecosystems. He received his M.S. from Loyola University Chicago and is working towards his Ph.D. in biology.

Chapter 6
Soil-Based Green Roofs

Brooke Byerley Best, Rebecca K. Swadek and Tony L. Burgess

Abstract Although typically eschewed in favor of engineered substrate, natural soils can provide an important ecological benefit for green roof systems in terms of jump-starting a viable habitat. They can act as fungal and microbial inoculants and can serve as an additional source of plants and insects (via seed banks, eggs, and larvae), presumably of species that naturally coexist. Even when no longer biologically active, natural soils can still benefit roof systems by mimicking the mineral-based properties of the natural habitat of a particular plant palette. However, concerns of fine particle illuviation, increased roof loading, and unpredictable biological activity dampen use of natural soils on green roofs. This chapter discusses the pros and cons of natural soils versus engineered substrates and how their properties can affect green roof systems. A single case study is also presented of the soil-based green roof at the Botanical Research Institute of Texas in Fort Worth (USA) that used a mixed engineered substrate–natural living soil system to model a local short-grass limestone prairie barrens ecosystem.

Keywords Biomimicry · Engineered substrate · Habitat template · Living soil · Natural soil · Prairie barrens · Seedbank · Soil inoculant · Soil-based roof

6.1 Introduction

Years of German research and experience taught the entire green roof industry that two main factors drive substrate choice for extensive green roof systems: weight and drainage. Current best practices (FLL 2002; Dunnett and Kingsbury 2008; Luckett 2009) typically encourage use of lightweight substrate with fairly large particle and

B. B. Best (✉) · T. L. Burgess
Botanical Research Institute of Texas, Fort Worth, TX 76107, USA
e-mail: bbest@brit.org

R. K. Swadek
Dept. of Parks & Rec. New York City Dept of Parks & Rec, New York, NY 10029, USA

T. L. Burgess
Inst. of Environmental Studies, Texas Christian University, Fort Worth, TX 76129, USA

R. Sutton (ed.), *Green Roof Ecosystems*, Ecological Studies 223,
DOI 10.1007/978-3-319-14983-7_6

pore sizes, and as a result of attempting to control for or optimize these, artificially produced mineral-based substrate mixes (hereafter, "engineered substrate" or ES) have dominated the commercial green roof industry. However, various limitations of engineered substrate often mandate the use of supplements such as compost or fertilizer to achieve a healthy roof ecosystem. If low maintenance and minimal inputs are among the primary goals of an extensive green roof, then designs should bias toward roof components, including the growth substrate, that promote a self-sustaining system.

So what about natural soils on green roofs? Intuition infers that soil from an intact, mature, natural plant habitat must already contain the components necessary to sustain its biotic inhabitants to a greater extent than artificially engineered roof substrate. Indeed, some of the earliest adopters of low-maintenance vegetated roofs—the Nordic peoples populating the blustery hillsides of Scandinavia around 900 AD—regularly used natural soils as they simply transplanted sod to their roofs from the surrounding environment (Hopkins and Goodwin 2011). But natural soils are being used less frequently in new roof systems. The authors know of very few documented large-scale, in situ extensive green roofs using a natural soil component (see Earth Pledge 2005 and Cantor 2008 for large-scale examples), though there are many examples of smaller private green roofs with natural soil (Dunnett et al. 2011).

Today we are more likely to find the same basic roof substrate guidelines across myriad reference works that all (1) derive or borrow from guidelines proposed for the German green roof industry (FLL 2002) and (2) discourage use of natural soils for the same handful of reasons (discussed below). Industry professionals can be adamant in their prejudice against its use (Friedrich 2005; Friedrich 2008), yet we have tangible proof of soil-based roofs still functioning, resulting in an incoherency between what is preached and what is evident. Somehow, in the years separating modern green roof designers and their Nordic predecessors, we learned the limitations of natural soils—including some undesirable maintenance chores that may arise from their use on a roof—but forgot their benefits. Engineered substrate can be profitably mass-produced and readily specified, whereas natural soils inherently require tailored analysis and amendment, which is usually less profitable for businesses, a fact that should not be overlooked in an industry that is fundamentally economic in its goals. Rather than an economic one, this chapter offers an ecosystem perspective of the advantages and disadvantages of both engineered substrate and natural soils on roofs and suggests avenues for future research.

6.2 Engineered Substrate (ES) vs. Natural Soils (NS)

Green roof design depends on the desired outcomes and intended purposes. If a garden-like setting for strolling is the owner's intent and finances are not limited, then builders usually follow guidelines for *intensive* roof systems, relatively deep-soiled, high-maintenance landscapes that, other than being located on a rooftop, resemble ground-level landscapes. If stormwater abatement, energy savings, or low maintenance are the intended goals or if budgets are limited or if the project involves retrofitting an old roof, then an *extensive* green roof is likely to be designed where shallow soils and a hardy plant palette create a functional vegetated surface that might achieve some level of self-sustenance. Because of the heavier weight-loading requirements of an intensive roof, these types of roofs can often be designed to meet user specifications without limit. Extensive roofs, by contrast, often are constrained by weight restrictions—soil depth, plant choice, and climate further circumscribing the system's ultimate configuration. Most ecosystem functions provided by an extensive green roof are linked to substrate as it relates directly to water retention and relates indirectly to plant performance. Even unvegetated gravel ballast or substrate layers will briefly detain stormwater and trap particulates (VanWoert et al. 2005), yet the addition of plants typically improves the system's retention and filtration abilities as well as providing a natural aesthetic (Sutton 2014). To function as something more than either mere ballast or rooting medium, both engineered substrate (ES) and natural soils (NS) must meet a handful of basic requirements.

6.2.1 Composition Overview

A number of convenient resources focus on the ins and outs of different green roof substrates, including the variety and properties of myriad components currently available (e.g., Miller 2003; Beattie and Berghage 2004; Friedrich 2005; Getter and Rowe 2006; Dunnett and Kingsbury 2008; Ampim et al. 2010; Dunnett et al. 2011; Nagase and Dunnett 2011). Most agree that rooftop substrates or soils are evaluated primarily based on gradation (texture) and organic composition. Both in turn are affected by physical, chemical, and biological soil properties that collectively correlate with the capacity for water retention, drainage rate, and nutrient availability, arguably the most important factors affecting plant success on green roofs. Typical composition of modern roof growing substrate (ES- or NS-based) is approximately 70–95 % minerals, the remainder being organics and fertilizer. Although recycled waste, plastic fillers, and absorptive additives are sometimes included among ES supplements to attain site-relevant goals, a stable, non-degradable mineral component with both good water and air capacity constitutes the bulk of most substrate while the remaining components mostly serve to nourish plants.

6.2.1.1 Mineral Components

The mineral base can take diverse forms: naturally sourced clay, sand, gravel, or vesicular volcanic rocks such as scoria; or artificial or modified minerals such as perlite, vermiculite, rockwool, or expanded clay, slate, or shale (Beattie and Berghage 2004; Friedrich 2005; Ampim et al. 2010). Commercially produced ES is often some proprietary blend of these materials and must be mixed in specified proportions with prescribed organics before planting can occur. By contrast, NS must be analyzed for existing amount and variability of relative proportions of sand, silt, clay, organics, plant nutrients, pH, salinity, and toxic elements. Then, extracted NS is often screened and amended with additional minerals to achieve desired weight, texture, and chemistry. The individual physical and chemical properties of the different mineral components are discussed later, but one could generalize that balancing a specific ES to the needs of a specific roof project usually involves increasing nutrient holding capacity, whereas NS-balancing aims to decrease both bulk density and fine particles and increase porosity. Most roof substrates are within the particle size range of 0.32–1.27 cm (1/8–1/2 inches) with minimal fines (Luckett 2009), but recent research shows that greater proportions of fines can be tolerable in specific systems (e.g. very shallow, 1-inch roofs) and can improve plant performance by increasing overall moisture and nutrient retention (Olszewski and Young 2011).

6.2.1.2 Organic Components

The organic ingredients of ES are typically some combination of compost, peat, coconut coir, or decomposed sawdust or bark (Friedrich 2005; Ampim et al. 2010) while the organic component of NS is typically a mixture of plant- and animal-derived detritus in diverse stages of decomposition to refractile humus. Organic matter in a roof soil helps provide plant nutrients, buffer against pH change, and retain moisture. For ES, the choice of a particular organic component must be considered in the context of each roof project as all components have advantages and disadvantages. For example, peat holds a considerable amount of moisture and remains saturated for long periods but is extremely hard to rewet when completely dry (Ampim et al. 2010), an unfortunate trait in a thin-soiled rooftop system that often experiences drought. Wood-based materials such as bark and sawdust have varying proportions of cellulose to lignin (depending on whether they come from soft- or hardwoods); more cellulose accelerates decomposition, whereas high lignin slows decomposition (Friedrich 2005). Compost added to an ES should be carefully sourced and tested to ensure it is free from residual herbicides, salts, undesirable seeds, or fertilizer by-products which can accumulate and linger in landscape or manure-derived composts (Friedrich 2005; Ampim et al. 2010). Compost should also be stabilized and fully mature because microorganisms in unfinished compost can compete with plants for nutrients (Beattie and Berghage 2004; Friedrich 2005; Ampim et al. 2010).

No matter their origination, the quantity of organics specified for either an ES or NS roof system should carefully be tailored to the local climate since temperature and humidity affect rate of decomposition (Wagai et al. 2008). For example, using a high proportion of organic matter in a tropical climate where decomposition may outpace replenishment can result in a reduction of overall roof substrate volume (Chap. 3), though some organic components (often those with high lignin content) are more stable than others. Conversely, in roof systems where total organic content increased over time, such as the doubling seen after just 5 years by Getter et al.

(2007), one can assume that new biomass production must be greater than the rate of decomposition (Chap. 5). Absent the creation or deposition of new organic content, initial organic content is completely reduced within 3–5 years (Luckett 2009).

A final issue is the potential for organic materials to decompose into finer particles that inhibit drainage (Friedrich 2008), although reports of this appear anecdotal with little verified evidence. However, assuming this is a possibility, a well-designed filter layer and careful selection of the organic amendment can ameliorate risk; for example, certain bark-based products with crystalline structure have drainage rates approaching those of sand, and fibrous buffered coir and similar materials can bind and help retain overall structure (Buist 2008). NS, on the other hand, with its inherently diverse organic component, brings a certain level of risk associated with fine particles.

6.2.1.3 Environmentally Appropriate Components

From an ecosystem perspective, it is important to consider the provenance of substrate components and the opportunity to use recyclables. New construction and retrofits alike can shrink their carbon footprints and cut costs by using locally-available components and recycled materials instead of purchasing off-the-shelf commercial mixes that have unknown provenance or contain unsustainably harvested materials such as peat (Campeau and Rochefort 1996; Graf et al. 2012; Caron and Rochefort 2013). Peat *can* be sustainably harvested, but you have to know the source. Like purchasing FSC-certified wood or composted bark, peat can be purchased as certified and used responsibly in a roof substrate, though with the number of alternatives now available, it could be avoided altogether. If absolutely required, appropriate peat sources are those associated with organizations such as the Canadian Sphagnum Peat Moss Association, which enforce preservation and reclamation policies.

Many construction and industrial wastes can be recycled as roof components including crushed brick, crushed roof tiles, aerated concrete, subsoil, and paper ash (Bengtsson et al. 2005; Emilsson and Rolf 2005; Cresswell and Sims 2007; Dunnett and Kingsbury 2008; Molineux et al. 2009). Incorporating these has the added benefit of local sourcing—thereby reducing transportation costs—as well as the green-minded reuse of otherwise low-value materials (Molineux et al. 2009). Harvested topsoil, if it comes from a reasonably proximal source, is also an eco-friendly choice since there are no energy expenditures from a manufacturing perspective. Of course, one should ensure restoration or recovery of the harvest site if the soil is not salvaged from land development. In the case study described below, for example, the topsoil was harvested from a planned quarry site. Thus, onsite soil stockpiling for eventual revegetation was never planned.

Regardless of the final components, whether built up from an ES base or thinned down from a NS base, a roof substrate will represent a compromise among cost, availability, system requirements, and design goals. It is this last, though, that should ideally come first. The key to a successful, long-lasting, and functional green roof is to determine clear design goals and manipulate composition to that end.

6.2.2 *Physical Properties*

There are a host of important physical properties to consider relative to a green roof substrate, including bulk density, aerated and total pore space, available water holding capacity, field capacity, permanent wilting point, capillary water rise, and desorption (Beattie and Berghage 2004). Conveniently these properties are all strongly influenced by soil texture and organic matter content, allowing designers to hone their specifications. Porosity, infiltration, shrink-swell capacity, erodability factor, available water-holding capacity, and permeability in particular are highly correlated with soil texture (Miller and Gardiner 2007). Another crucial physical factor is moisture retention, as two of the primary functions of green roofs are stormwater retention and sustaining vegetation (Francis and Lorimer 2011). This section will discuss the optimal physical characteristics for green roof substrates and the pros and cons of ES and NS.

6.2.2.1 Bulk Density

Bulk density, defined as the dry weight or mass per unit volume of a soil, is one of the most important features of a green roof system since weight is a main limiting factor (Beattie and Berghage 2004). Guidelines for maximum bulk density, such as those found in the German *Guidelines for Planning, Execution, and Upkeep of Green-Roof Sites* (FLL 2002), are set relatively low (around 1 g/cm^3 or 62 lb/ft^3) and therefore preclude the use of a single-component roof system (e.g. sand, silt, or clay). So while sand has high porosity and infiltration—ideal traits of a roof substrate—it is heavy with a high bulk density (Ampim et al. 2010). Clay on the other hand is relatively lightweight with a low bulk density but has low infiltration and high saturated weight, making it a poor roof substrate on its own as well (Miller 2003; Miller and Gardiner 2007; Ampim et al. 2010). As mentioned previously, this is why other lightweight supplements that facilitate adsorption and desorption are often necessary for substrate mixes. Soils with a high bulk density may also have a higher thermal conductivity, resulting in greater plant heat stress in the summer months (Brady and Weil 2008; Olszewski and Young 2011).

The general need for heavily engineered substrate places use of NS at a disadvantage because it lacks the same flexibility as ES for compatible blending. One option for bringing NS up to bulk density standards is combining it with ES materials, including many lightweight recycled aggregates (e.g. carbon8 and clay pellets, brick, and paper ash) that have a loose bulk density ranging from 0.83 to 0.91 g/cm^3 (51.5–56.4 lb/ft^3) (Molineux et al. 2009). However, if the design objective is for a specialized habitat requiring specific edaphic properties, such as wetlands, dunes, deserts, savannahs, or other dryland ecosystems, FLL standards simply are not appropriate substrate guidelines.

6.2.2.2 Porosity, Drainage, and Moisture Retention

High porosity of green roof substrate is favorable as it facilitates rapid desorption or drainage and reduces excessive moisture retention which minimizes potential for waterlogging (Chap. 4) (FLL 2002; Olszewski and Young 2011) and excessive roof loading. Sand, lava, pumice, perlite, ESCS (expanded slate, clay, or shale), rock-wool, and to a lesser degree vermiculite are all very porous or contain intra-particle pores (Bunt 1988; Handreck and Black 1994; Dunnett and Kingsbury 2008). These coarse components have large pore spaces facilitating rapid drainage but causing poor moisture retention (Bunt 1988). Finer components such as silt and clay have more total pore space but smaller pores that drain more slowly due to capillary forces exerted on water within these micropores (Miller and Gardiner 2007). This is something to consider relative to roof irrigation, if such activities are planned; the general lack of fines and thus capillary action in typical ES mean sub-irrigation systems won't work well (Beattie and Berghage 2004; Rowe et al. 2014, Chap. 4). NS-based substrates or any growing substrate admixture with a greater proportion of fines, on the other hand, should respond well to sub-irrigation if the lines can remain unclogged.

Organic amendments can also affect drainage properties. Nektarios et al. (2003) reported that addition of peat and perlite to sandy loam increased both porosity and moisture retention relative to unamended controls. Peat and coir are both old standards with regard to soil amendments for improved moisture control (Miller 2003) and can be useful in extremely shallow (2 cm) ES roofs that can completely dry out in as little as one day after irrigation (VanWoert et al. 2005; Olszewski and Young 2011).

The suite of combined drainage properties should be balanced to achieve proper roof function but should also match both the local climate and the design objectives. For example, green roofs in hot, arid regions (Chap. 3) will likely require substrates with enhanced moisture retention as those roofs are likely focused on building temperature reductions through plant shading rather than transpiration. Coastal areas with ample rainfall, however, might rather choose roof substrates with both rapid infiltration and high volume retention in order to better attenuate stormwater runoff (Chap. 4). And of course, substrate depth, slope, and antecedent moisture will all also affect drainage and retention to varying degrees (see Chap. 4; Monterusso et al. 2004; VanWoert et al. 2005; Getter et al. 2007).

Just as pedogenic processes of eluviation and illuviation occur in natural soils, green roof soils will also change over time. Getter et al. (2007) reported cases where overall hydraulic conductivity increased with roof age as porosity, organic content, and water holding capacity all increased in mature substrates, as did free air space (from roots and insects increasing macropore space and increasing initial infiltration capacity). And though age has not proven to be a factor in overall moisture retention rate, it may affect retention patterns as the roof settles (Mentens et al. 2006; Getter et al. 2007). Further, original organic inputs will eventually break down in mature green roof soils (Beattie and Berghage 2004) but should cycle back up as roof biomass is processed (Sutton 2013b), which could cause fluxes between the initial moisture retention of ES and later, more stabilized retention rates.

6.2.2.3 Available Water Holding Capacity, Field Capacity, and Permanent Wilting Point

Rainfall on intact natural soils percolates through the soil via gravitational forces. If rainfall rate exceeds drainage rate, then the soil eventually becomes saturated; all of its pore space is filled with water, and any additional water does not penetrate but rather sits atop or runs over the soil surface. Once rains subside, however, and all gravitational water has freely drained from the system, then a state called "field capacity" is reached (Miller and Gardiner 2007). The field capacity of soils or any growth substrate, including that found on green roofs, is important to plant health since this is the point at which most plants can begin extracting from the soil water held tightly in those small pores mentioned in Sect. 6.2.2.2. The amount of water in the soil at this time actually available to plants is called "available water capacity" and is largely dependent on physical properties such as soil texture and porosity. Thus, available water capacity ranges from field capacity at the upper, most saturated end and permanent wilting point at the lowest, driest end. Unavailable, hygroscopic water is tightly bound to soil particles, and therefore plants can neither extract nor make use of it.

Available water holding capacity, field capacity (FC), and permanent wilting point (PWP) are soil metrics that were historically defined relative to the performance capabilities of commonly cultivated plants. However, not all plants are bound by the limits of FC and PWP. Plants adapted to saturated conditions or succulents that must absorb water rapidly following a rainfall can uptake water in saturated or near saturated conditions before FC is technically reached. In contrast, plants adapted to aridity may continue to extract soil moisture well after the soil has reached a PWP that would threaten, for instance, a corn or tomato plant. Nevertheless, when available water is gone and only unavailable water remains, the permanent wilting point is reached. Beyond this, plants risk irreversible tissue damage and death.

Water holding capacity is therefore critical to consider during roof design, especially for thinner extensive roofs. Several recycled components such as crushed brick and aerated concrete can improve available water capacity of ES blends (Dunnett and Kingsbury 2008), while a compost supplement to a coarser textured NS can also increase plant available moisture (Handreck and Black 1994). The scenario above, though, highlights how too many small pores in growing substrate with hefty proportions of fine organic matter (humus) and clays—especially shrink-swell (smectite) clays such as montmorillonite—can store large volumes of water at field capacity, which is good for plants (ignoring weight), but also store large volumes at permanent wilting point (Miller and Gardiner 2007), which is unusable and *only* serves to increase loading weight.

Clays in particular, however, also have a steep sorption curve, meaning they gradually decline in moisture content under drying conditions. This could prove beneficial in the right NS-based roof system, allowing plants more time to prepare for drought conditions as soils slowly dry out (Miller 2003). Such an effect may be responsible for results seen by Olszewski and Young (2011) where non-succulent plant performance on an extremely shallow ES roof was positively correlated with greater proportions of fines in the blend, which exhibited greater water holding capacity and probably greater CEC for nutrients.

When it comes down to it, once you account for weight, how all of the above water-related properties affect permanent wilting point may actually be all that matters for plant performance. We can gather information, then, from natural systems that push plants (especially non-succulents) to water extremes, where plants undoubtedly have greater tolerance for wilting point conditions than those from wetter regions. For example, Burgess (1995) described how the moisture regimes of clays in semiarid grasslands of Southwestern North America cause an amplification of wet and dry weather sequences. Similarly, plants growing in hyperseasonal hillslope seeps in Texas are adapted to alternating periods of extreme saturation and desiccation and compensate with rapid annual growth and regeneration after each dieback. Llado (2011) and Jue (2011) demonstrated that some plants in these hillslope seeps could survive beyond permanent wilting point, as described by agricultural standards. These species are often warm season grasses or annuals that germinate in the spring months when water is plentiful and then undergo a period of dormancy during the summer months when the soil is desiccated, resuming growth once rains begin again (Swadek, unpubl. data). Such plant systems should be investigated further for candidate green roof species. Swadek and Burgess (2012) observed this as well in prairie habitats with shallow undeveloped soils. This ecosystem has been used as a model for green roofs as described below (Kinder 2009; Williams 2008; BRIT 2014).

6.2.3 Chemical Properties

Optimal chemical properties of roof soils include high cation exchange capacity (CEC) and appropriate ranges of plant available nutrients (Ampim et al. 2010). These are a delicate balance between optimal physical and chemical properties (see summaries in Emilsson 2008 and Ampim et al. 2010). While physical properties contribute to much of the drainage and water retention in a green roof medium, chemical properties such as pH, CEC, buffering capacity, and nutrient availability all play crucial roles in determining plant success by facilitating nutrient uptake and buffering the soil from chemical changes (Ampim et al. 2010). Very little peer-reviewed, repeatable research has been published on this topic as much of this knowledge is accumulated within the commercial substrate industry and is therefore somewhat proprietary, but for the most part ES components tend to be less broadly based in terms of desirable chemical properties compared to NS and often require amendments to correct deficiencies.

6.2.3.1 pH and Buffering Capacity

In general, soil pH can vary among regions and among differing plant palettes, but it must be within a range that allows plants to uptake nutrients from the soil, typically between 5.5 and 7.0 (FLL 2002; Ampim et al. 2010). Although pH out of this range won't kill plants directly, extreme values will severely restrict performance. Amendments to optimize pH are numerous and of variable utility according to the roof system in question. Molineux et al. (2009) found that addition of organic matter reduced pH for a variety of recycled alkaline roof substrates (crushed brick, clay pellets, paper ash pellets, and carbon8 pellets), bringing each component closer to FLL guidelines.

However, apart from the initial substrate pH at the time of roof installation, part of every design goal should also include promotion of a pH-stable roof over time, which we know is possible based on reports of mid-nineteenth century gravel-based roofs that still maintain relatively neutral pH (Emilsson 2008). This long-term stability may not always be possible without human intervention, especially in areas with high incidence of acid rain such as the northeast US where slightly alkaline growing substrate is often chosen as it is able to buffer or neutralize acid rain to some extent (Beattie and Berghage 2004). For example, the pH of ESCS is generally neutral to slightly alkaline (8.5), varying slightly with the exact raw materials and fuels used to create it (Friedrich 2005). But eventually the buffering capacity of this and other basic ES blends will be expended, and both soil pH and roof runoff pH will eventually decrease (Berghage et al. 2007), resulting in the need for periodic lime treatments over the long term (Friedrich 2005).

The initial pH of NS will depend on its geologic provenance as soils with high limestone content usually have pH buffered around 7.5 while soils from spodosols under coniferous boreal forest will be slightly acidic. Though these may require additional lime dust to increase or sulfur to decrease overall pH (Friedrich 2005), NS that includes fine organic matter should theoretically be better suited for long-term buffering capacity than unamended ES.

6.2.3.2 Nutrient Availability, CEC, and AEC

Plants require a variety of nutrients to live, including three macronutrients (nitrogen, phosphorus, and potassium), secondary nutrients (such as calcium, magnesium, and sulfur), and a suite of micronutrients required in very small amounts (boron, manganese, copper, iron, zinc, molybdenum, nickel, and chlorine). Each nutrient contributes in some vital way to the growth and development of plants—either through the formation of proteins and DNA or the development of cell walls, flowering structures, or general pollination (Miller and Gardiner 2007)—and they *must* be present in roof soils for plants to thrive. Because roots absorb most nutrients in soluble ionic forms, the ability of roof soils to retain these ions and slowly release them to plants is critical to vegetation success.

Cation exchange capacity (CEC) and anion exchange capacity (AEC) are the amounts of positive and negative ions a soil can retain, respectively. Positively charged nutrients are attracted to negatively charged soil particles—common in clays, organic materials, and ESCS aggregates—which reduces leaching during irrigation or precipitation (Miller and Gardiner 2007). Of the 14 essential mineral nutrients listed above, nine can be taken up in cation form while the other five enter as anions, three of which are micronutrients and only needed in very small amounts. Thus CEC is more important than AEC. The natural organic content and fine particulates in NS usually provide adequate CEC and AEC to sustain a plant community (Handreck and Black 1994), whereas many ES components carry a neutral charge that retains neither cations nor anions. Therefore clay-based ES and vermiculite amendments are often used to improve CEC on roofs where organic amendments or NS are undesirable (Beattie and Berghage 2004, Ampim et al. 2010). Roofs constructed with ES dominated by sand or perlite often suffer from low CEC (Ampim et al. 2010), and these may require regular fertilizing to correct nutrient losses from leaching.

In lieu of or in addition to organics or other nutrient sources, substrate blends often include slow release fertilizers to boost fertility (Beattie and Berghage 2004; Ampim et al. 2010). Care must be taken, however, to consider soil CEC and AEC in tandem with both initial plant uptake rates and lag time between soil installation and plant installation. Most nutrient leaching occurs following installation when over-fertilization adds more nutrients than can be held either by the unplanted soil or by the ecosystem when uptake by newly transplanted or germinating vegetation is limited (Friedrich 2005; Emilsson et al. 2007).

Sparse information on different roof substrate components and blends and their relevance to specific macro- and micronutrients is scattered throughout the literature, usually as small asides in investigations of plant performance. We offer findings from selected reports that relate to comparisons of ES versus NS. For example, use of zeolite in roof mixes can improve AEC (Ampim et al. 2009) and thus help retain nitrate and other anions for plant use, as can the incorporation of NS. Emilsson (2008) found that NS-based substrate retained more total and dissolved nitrogen over a 3.5 year period than a substrate based on crushed tile. Where phosphate leaching is a concern, addition of vermiculite ES can improve retention (Bunt 1988), as can mycorrhizal fungi (Miller and Gardiner 2007), suggesting that incorporation of NS with live fungal biomass or mycorrhizal inoculants could address this issue. Potassium is typically readily available in NS and can be specifically added to ES in the form of potassium-rich organic matter such as coir (Handreck and Black 1994). Sulfur can be deficient in highly sandy roof soils with low organic content. However, given that airborne sulfur can enter soils in highly industrial areas (Miller and Gardiner 2007), there is potential for urban green roofs to exploit air pollution as an anthropogenic buffer to sulfur deficiencies, but more research is needed to confirm this.

There is still ample need for research related to CEC, nutrient availability, and roof soils, including better methods for assessing and describing CEC on roofs. The current methods, derived from agricultural testing protocols, are weight-based, which may not be appropriate for lightweight green roof substrate (Beattie and Berghage 2004).

6.2.4 *Biological Properties*

Beyond best practice guidelines mandating use of weed-free substrates (FLL 2002, 9.2.14), there is a dearth of information on their biological properties, though biological activity of the substrate is essential to recycle roof biomass (Beattie and Berghage 2004). Perhaps this passive championing of sterility points to a larger trend. It is not difficult to find case studies of intensive roofs that used topsoil, but it *is* difficult to find examples using unsterilized or living topsoil since it is often assumed that all biological activity must be removed from a substrate before installing on a roof. Traditional practice, personal experience, perspective, and design intent can all paint the same subject with vastly different strokes. For commercial landscapers, natural soils are sources of unknown organisms and uncertainty. For an ecological designer, however, the unknown organisms are a source of ecosystem resilience and evolutionary potential. While both ES and sterilized NS by design have no biological properties to begin with and must start from scratch to build up soil biota, sterilized NS is more readily colonized than ES (Emilsson 2008). And neither can compare to living NS, which already possesses the microorganisms that can help establish and sustain a roof plant community. So what are the biological components of soil and how do they affect a rooftop ecosystem?

6.2.4.1 Soil Macro- and Microorganisms

Ignoring megafauna such as birds and squirrels, a multitude of macro-, meso-, and microfauna shape roof soils or substrates (see also Chap. 14). These include most of the same organisms found in soils at-grade: decomposers, nitrogen fixers, mineralizers, bioturbators, and the predators that keep the former in check. All of these have been found on green roofs, old and new, around the world, and would eventually colonize any rooftop habitat without anthropogenic help even with minimal substrate present.

The larger macro- and mesofauna of roofs include spiders, beetles, isopods, ants, snails, woodlice, mites, and springtails. When not feeding on each other, some of these organisms consume organic matter, fragmenting it and creating detritus, and some structurally modify the soil through burrowing and other activities, thereby improving soil aeration and drainage. Their activities stimulate increased microbial action which can alter soil chemistry (Moore et al. 1988).

Though relatively few investigations deal with mesofauna, a host of them occur on both young and old roof soils. Schrader and Boning (2006) reported that roofs of differing ages using ES (expanded clay or shale pellets) exhibited soil communities similar to other early succession soils such as those described in mining reclamation. This soil biotic maturation included the spontaneous colonization of Collembolans (Hexapoda; "springtails") that increased in density and richness with roof age, as well as changes in abiotic soil properties over time (e.g. decreased pH, increased carbon and nitrogen content). Collembolans are one of the first species to colonize virgin anthropogenic substrates (Madre et al. 2013) and are crucial to soil health. It is suspected that their population size within a roof system can become greater than what might be found at-grade because roof soils seldom have earthworms (presumably due to lack of colonization opportunities and low moisture conditions). Rumble and Gange (2013) report that microarthropod communities on extensive green roofs, though abundant, are "impoverished" in their composition due to the dry, hot conditions and that more thought should be given to soil biota during the design phase to help ensure plant success (Chap. 14) What happens below-ground affects what we see above-ground.

At a microbial level, healthy soils contain bacteria, protozoa, algae, arbuscular mycorrhizae, and other fungi (Chap. 7). Most of these are responsible for the fine-scale degradation of dead plant and animal matter, including each other. Their biological activity is crucial for regenerating carbon, nitrogen, sulfate, phosphate, and other inorganic elemental sources for plant use (Moore et al. 1988, Chap. 5). They release plant nutrients that have been immobilized in live and dead tissues. If this process of nutrient cycling is dependably happening in roof soils, it can replace the need for any fertilizer application subsequent to the initial loading during installation.

Mycorrhizal fungi, in particular, are important to soil biology and the health of plants, but they are initially missing from synthetic and sterilized roof substrate. Fungal introduction therefore must occur via airborne dispersal, purposeful inoculum, or accidental introduction via plug soil (John 2013; McGuire et al. 2013). The importance of mycorrhizal fungi as symbionts in community succession is clearly documented in restoration ecology(Bever et al. 2003; Renker et al. 2004; Vogelsang and Bever 2009). A trend of increasing inoculant use in roof habitats is evident, despite the fact that *Sedum* spp. and other Crassulaceae species do not benefit from symbiotic nutrient exchanges with these microbes (John 2013). For example, several studies demonstrate that external application of mycorrhizal fungi to ES roofs hastens early plant establishment and fosters long term plant performance by improving stress tolerance (Meyer 2004; John 2013). Essentially this is putting life back into an inert substrate that is naturally present in a healthy living soil. Roofs using ES and sterile NS, therefore, can benefit from fungal inoculations during installation, while living NS that remains biologically active presumably already contains the "correct" complement of microbes to benefit a matching suite of plants. The benefits of fungal complementarity are likely to hold true even if inoculant is chosen instead of living NS. Studies show that densities of mutualistic mycorrhizal fungi and native plant species are positively correlated, while invasive plant species co-occur with more generalist mycorrhizal fungi and fewer mutualistic fungi (Kulmatiski et al. 2006; Vogelsang and Bever 2009). Thus obtaining fungal inoculant tailored to the plants of a particular roof (1) could be useful in helping a roof system be more weed-resistant and (2) should be easier if the plant palette represents species that actually co-occur in nature and thus one source of all relevant species-specific mycorrhizal fungi.

Timing is also a factor that must be mentioned in relation to biological properties of roof soils, especially as it pertains to longevity of soil biota ex situ. Though seeds can persist in stockpiled soils for years, microbial viability rapidly diminishes almost as soon as topsoil is removed from its native environment. Numerous studies from mining reclamation document the decline in viable fungi and bacteria of soils that are stockpiled on site (Ross and Cairns 1981; McQueen and Ross 1982; Widdowson et al. 1982; Visser et al. 1984; Williamson and Johnson 1990; Johnson et al. 1991; Strohmayer 1999). These soils are scraped into massive piles (separating topsoil and subsoil) and left alone until needed for site reclamation, sometimes for many years. Although the outermost layers of topsoil stockpiles tend to retain biological activity—often evidenced by growth of an impressive population of ruderal plant species—fungi and bacteria deeper within the pile quickly go dormant and eventually become completely nonviable. When stockpiles are later respread, older piles take longer to attain plant coverage, presumably due to a lack of microbiota, requiring extended periods of erosion prevention measures. This information has led to mining and civil engineering recommendations of greater numbers of shallow stockpiles rather than fewer large piles to promote faster establishment and decrease the need for erosion controls. The green roof industry can use this information to develop best practices for living NS implementation. These ideally would include guidelines for handling stockpiles during extracting and transporting, as well as recommending harvest-to-installation timeframes to improve logistics coordination during roof construction and in the event of substrate removal during membrane repair or replacement.

6.2.4.2 Seedbank & "Weeds"

Some roofs are designed for growing only a specific suite of chosen plants and nothing more. Any plant found growing not on the list of approved species invites removal. These roofs function better with a highly controlled, ES-based substrate. Other roofs, however, are planned at the outset to accept eventual, inevitable colonization of other plants and make allowances for certain volunteer species (Nagase et al. 2013), while still other roofs are wholly designed for spontaneous colonization and self-organization (Chap. 12), such as the biodiverse roofs (brownfield roofs) of Europe (Ishimatsu and Ito 2013, Chap. 10). And then there are the roofs (usually of the "meadow" type) that incorporate living topsoil replete with microbes and a seedbank, a repository of plant seeds and diaspores that designers hope will germinate and augment the roof flora (case studies in Cantor 2008; Earth Pledge 2005, and Dunnett et al. 2011; Dvorak et al. 2013; BRIT 2014).

Natural topsoils almost assuredly contain some viable seeds if they have not been screened or heat-sterilized. The species present in the seedbank, though, may not match the current species at the donor or harvest site as community structures shift over time (Coffin and Laurenroth 1989). Therefore procurement of topsoil is ideally done with some knowledge of the extended history of the site. To the NS seedbank, green roof construction approximates a major above-ground disturbance event such as fire or flood, and the germination response of the seedbank tends to match typical patterns of primary succession with annuals coming up first followed by other less responsive species (Faist et al. 2013; BRIT 2014). Thus having seeds with differing germination responses allows NS seedbanks to buffer roof communities against perturbations during establishment and enhance overall resilience in an otherwise stressful environment. This can be advantageous for energy expenditures and labor inputs that might otherwise be invested in maintaining a controlled appearance with predetermined species and the efforts of constantly fighting a dynamic environment (Oberndorfer et al. 2007; Cantor 2008).

Of course, even the most aseptic ES will have to contend with opportunistic ruderals, arriving at the roof via wind, birds, contaminated tools, seed-mixes, work boots, or plug soils (Emilsson 2008; Nagase et al. 2013). Their elimination takes time and effort that translates to cost and eats away at savings gained elsewhere from the roof (e.g. reduced energy demands or stormwater inputs). Rather than avoid or eliminate all organisms except those chosen by the designer, building owners might rather embrace the novel habitat that is a vegetated roof—something natural and yet not, a "recombinant ecology" (Keurlartz and Van der Weele 2008) or "novel ecosystem" (Hobbs et al. 2006) that would not have existed without human intervention—and consider a range of allowances regarding compositional dynamics of species over time (see also Chap. 14). Dunnett et al. (2011) seem to agree that "[natural colonization] is the most cost-effective and ecological approach" to green roofs and that arrival of colonizing species are good for overall biodiversity. Pro-colonization roof designs are a nod to the roots of the green roof industry in Germany, purportedly sparked by observations of plants naturally colonizing sandy fire-retardant ballast layers over bitumen rooftops (Dunnett et al. 2011). If nothing else, roof industry stakeholders should find comfort in the fact that as hard as designers may try to concoct the perfect substrate, something will undoubtedly grow in whatever is put up there, be it gravel, finely crafted ES-blends, or topsoil. We simply have to adjust our acceptance level, educate the user, and learn to be patient as succession elaborates.

6.3 In Defense of Living Soil

Committing to an NS approach involves trade-offs. Probable increases in ecosystem services and decreases in direct and indirect costs (via local sourcing and recycling, fewer amendments, less long-term maintenance, etc.) are traded for potential increased structural cost to accommodate more weight, more clogging risk with fine particulates, more uncertainty in vegetation dynamics, and requirements for different kinds of expertise. Yet in the interest of promoting self-organizing roof systems and increased roof biodiversity, this is a trade-off we would like to see more designers and building owners choose and more researchers explore. Garden-style roofs and meadow-style roofs are fundamentally different. We propose that returns on investment in meadows are better than gardens from a wider, ecosystem perspective. They can sustain higher biodiversity and greater community resilience for equivalent maintenance costs, and they provide a better platform for ecosystem evolution in urban contexts (Lundholm et al. 2010; Cook-Patton and Bauerle 2012; Sutton 2013a). We also propose that adding viable NS communities to substrate of meadow green roofs is a simple means to promote rapid plant establishment and soil functional integrity (Table 6.1).

Restoration ecology research indicates that inoculation of damaged systems with healthy soils can expedite recovery of those systems (Middleton and Bever 2012), essentially leap-frogging over early successional stages in community maturation. Such evidence indicates that one should always prioritize building healthy, self-sustaining soil communities. If these function well, the vegetation and macrofauna will come. Rooftops, too, can be viewed as disturbed sites primed for restoration or reconciliation (Rosenzweig 2003; Oberndorfer et al. 2007; Francis and Lorimer 2011), using diverse designs focused on any number of healthier ecological goals.

If ES is equated with soil, it is *new* soil (entisol) or *young* soil (inceptisol) at best; however, it can be designed to mimic mature soil properties. As with any resilient ecosystem, green roofs are dynamic, and continuing human intervention is required to confine a plant palette within an initial design. But if one wants to hedge environmental bets and attempt something closer to a late succession community that requires fewer inputs to stabilize, use of living NS confers advantages. Restoration research has shown that inoculating just a small portion of new plug transplants with old/native soils can benefit performance of late succession species (the inoculated ones as well as their neighbors) and hinder the performance of early succession species (Middleton and Bever 2012). Theoretically, on a rooftop NS would promote establishment of the plants that form more stable vegetation instead of starting with colonizing species that proliferate and eventually decline while the new soil is developing. Of course, one could potentially derive similar advantages from a mycorrhizal/microbial inoculant (Meyer 2004) and not have to deal with the weight and fine particulate pitfalls of NS, but development of a fungal inoculant (often sold commercially in liquid form) ostensibly requires more effort than whole soil inoculant. Furthermore, whole soil/living NS inoculant methods might require very little inoculant volume if we subscribe to a "nucleation model" of succession whereby small pockets of soil inoculant become sources of colonization for neighboring uninoculated areas (Yarranton and Morrison 1974; Sutton 2008). This method would not be useful for seed recruitment, but it could facilitate microbe dispersal into the roof environment.

Living NS can also help address the issue of excessive lushness. Traditional ES green roofs typically add organics and fertilizer at initial installation to boost establishment and keep plants vigorous until nutrient cycling kicks in (Ampim et al. 2010; Sutton 2013b). However the amount of fertilizing amendment should be carefully calculated, not only to prevent post-installation leaching into nearby streams, but also to discourage excessive early growth, which may make the roof more susceptible subsequently to abiotic and biotic stresses (Getter and Rowe 2006; Dunnett and Kingsbury 2008; Ampim et al. 2010). Thus living NS is a solution to more problems than mere fertilizer addresses. Many shallow soil ecosystems that could be green roof models are oligotrophic, suggesting that plants adapted to green roofs can tolerate lower macronutrient levels. As mentioned above, an ap-

Table 6.1 Physical, chemical and biological properties of engineered substrates (ES) and natural soils (NS)

Property	Guidelines	ES	NS
Mineral components	Should comprise 70–95 % of total; particle size ratios may differ based on plant palette and weight and drainage restrictions	Combination of natural and modified minerals in appropriate ratios (e.g. expanded shale, clay, or slate) for drainage and nutrient retention; sterile	Predominantly natural minerals with ratios to maintain function; may be harvested soil; may dictate plant palette; not sterile
Organic components	Should comprise <30 % of total substrate	Via amendments (e.g. compost, peat, coconut coir, decomposed sawdust & bark)	Plant & animal- detritus in various decomposition stages; refractile humus
Environmentally appropriate components	Reuse of materials is desirable	Crushed brick, aerated concrete, paper ash	Topsoil, subsoil, compost
Bulk density (BD)	Low BD preferred to reduce weight loading; FLL (2002) suggests 1 g/cm^3	Easily blended to adhere to suggested values	Sand:clay ratio balances BD & porosity; combine w/low BD ES materials
Porosity (PO)	High PO preferred to encourage drainage & reduce weight loading through excessive moisture retention	Usually high PO and good drainage; low moisture retention problematic; low capillary action often means sub-irrigation not possible	PO decreases as fines increase giving slower drainage & higher moisture retention; allows sub-irrigation via capillary action
Available water holding capacity	FLL (2002) suggests 35 % (v/v) with air 10 %	FLL (2002) suggests 35 % (v/v) with air 10 %	Smaller fines may increase water holding capacity, requiring set plant palette
pH	pH 5.5–7.0 allows nutrient uptake; buffering capacity stabilizes pH over time	High pH (e.g. concrete, expanded shale/clay); buffers acid rain effects	Better suited for long term buffering due to fine organic matter
Cation exchange capacity (CEC)	Traditional CEC may not apply to thin roof substrates; few set standards	Poor CEC unless clay or vermiculite is present; sand and perlite are CEC-poor; amend to reduce leachates	Usually good through the binding action of positively charged organic matter and clays
Soil macro- and microorganisms	Most guidelines suggest sterile substrate	Colonized post installation; colonization of ES slower than sterilized NS	Builds up biota faster than ES; living NS contains fungi, bacteria, other organisms
Seedbank & "weeds"	Free of all propagative material	Absent by design; built-up post installation or through natural colonization	Absent by design if sterilized; monitor seedbank over time

propriate balance of soil microbial components (including pathogens) in living NS could reduce outbursts of weeds (i.e., early succession species) while also being stressful enough (via pathogens and moderately low nutrients) to encourage native plant growth without rampant overcrowding (McGuire et al. 2013), reducing risk for early failure if weather conditions are unpredictable.

Even if no longer biologically active, one might prefer NS where specific plant communities are being replicated. Logically, a shallow limestone glade community would perform best on a calcareous substrate (Chap. 11). Instead of constructing it anew from ES and amendments in an attempt to simulate the complex soil chemistry these plants are adapted to, why not incorporate a topsoil component from the actual model ecosystem? These glade or rock outcrop communities have been used as models for many green roof systems (Lundholm and Richardson 2010; Cook-Patton and Bauerle 2012; Sutton et al. 2012) because the plants there are adapted to high wind, drought, and poorly developed shallow soils, with only A horizons that are low in organic matter over bedrock. Although the plant communities from shallow soils are being modeled more frequently in green roofs, the natural soils of these ecosystems are not being simulated at the same rate. The case study at the end of this chapter (Sect. 6.5) documents one biomimicry attempt, and that roof has so far been considered a great success (BRIT 2014).

A cultural shift away from garden-style roofs will take time, just as it is taking time to transition from exotic-based, at-grade landscaping, including the standard English lawn that persists throughout suburbia (Jenkins 1994; Steinberg 2006). A highly biodiverse rooftop that mimics a natural, regionally appropriate plant community will always provide more ecosystem services with less cost than a high-maintenance roof garden, no matter how much pleasure the latter may bring to human visitors. Yet we as a community of green roof beneficiaries continue to accept what amounts to an un-ecologically-inspired norm. Many current roof design practices work hard to engineer nature out while expecting a natural result with a limited palette of plants colonizing a disturbed, artificial environment. In natural systems, colonizers often alter a disturbed site in ways that allow additional species to establish. Assisting in this process—by using local soils, by inoculating, by introducing living seedbanks—seems like a better use of our talents as "green" designers than demanding the removal of every vestige of wildness and struggling to arrest succession within a predetermined, static endpoint. Logic dictates that when trying to force organisms to establish in a novel habitat, one would attempt to replicate native conditions—growing tropical orchids in a humid greenhouse or growing bog plants in acidic, organic soil. We continue to design unnatural assemblages on artificial substrate and expect natural results probably because the urban populace and the landscaping professionals who serve them are unaware of natural processes and thus more comfortable with simpler, more controlled environments.

Definitely a roof with ES and a *Sedum* mat is a substantial improvement over tar and gravel and deserves some celebration. However, standardization and convenient installation, though more profitable in the short term, will reinforce stagnation within landscape design and management communities, just as lawns have. Fundamentally a living roof expresses a new, more mindful relationship between urban people and their watershed's ecosystem. The use of biomimicry design and NS will inevitably reinforce this new relationship, leading to a more integrated, healthier co-evolution within urban ecosystems. Designs using NS promote both more ecological reconciliation within urban environments and more sophistication in perception and techniques among practitioners of green design. Currently few commercial labs can perform biological soil tests, and relatively few consultants can write meaningful pedological interpretations for a specific context. Effective use of NS requires more ecological and pedological expertise than ES; thus using NS reinforces the development of experts who understand and value local, native soils. Such a trend carries the risk that greater commercial value of NS from shallow-soil habitats such as prairie barrens will lead to extracting these resources faster than they can be replenished, similar to mining horticultural peat moss or topsoil for urban landscaping, thereby threatening limited habitats. This issue should be anticipated with research on how to regenerate appropriate topsoil communities rapidly.

This is not to say that conservation risks and lack of professional expertise should inhibit the use of NS. Sod roofs with NS have a long history of successful use. When roofs are engineered for cold climates and accommodate snow loads, NS topping is straightforward, as in intensive-style green roofs. ES by contrast originally developed to solve problems with retrofitting extensive green roofs on older buildings with limited load capacity. Solving this problem with ES has allowed extensive green roofs to proliferate without adding structural costs. ES fits commercial construction culture; therefore developing living roof technology that combines advantages of NS and ES will inform this culture, which already has substantial influence on an urbanite's experience of ecological community.

6.4 Avenues for Further Research

Roof soil formation and appropriate habitat templates differ geographically. Precipitation and temperature regimes affect shrink-swell responses. Acid rain alters pH in heavily populated and industrial regions, and moisture retention and nutrient cycling vary widely between temperate and desert ecosystems. For these reasons and so many more, the green roof research community can only benefit from expanding the current list of roof substrate components, seeking innovative means for aligning an ever-increasing assemblage of extrinsic and intrinsic variables. Doing this means addressing perceived limitations and investigating means for their circumvention or exploitation.

One such avenue of inquiry involves reevaluating old industry norms. For example, the notion that fine particulates in substrate mixes will illuviate and clog filter fabric is prevalent throughout the literature (both academic and trade), but supporting data is almost never provided to back up such assertions. Whether this knowledge has been carried over from horticulture science or is based on early-industry experimentation, the green roof community does itself a disservice by not examining standards of substrate behavior using current materials and methods. Longitudinal analyses of in situ roofs, comparisons of different sources of fines (organic, clay, etc.), and particulate settling relative to slope are just a few topics in need of exploration. Further, researchers in the materials industry should be investigating the potential for improvement to filter layers. A non-clogging or self-cleaning filter layer or some as yet unimagined product may one day eliminate our concerns over fines, widening possibilities for substrate composition.

Another area that could benefit from more thorough investigation is slope and aspect. Though the effects of roof slope and aspect might seem somewhat intuitive (faster drainage, inconsistent soil moisture and retention, patchy vegetation, etc.) and anecdotal evidence abounds, there is little published research on how they affect intact roof systems, particularly with respect to soil biota (but see VanWoert et al. 2005). One can hypothesize that since slope affects soil moisture (Cantor 2008) and soil moisture affects both nutrient cycling (Miller and Gardiner 2007) and meso/microfaunal densities (Schrader and Boning 2006; McGuire et al. 2013), sloped green roofs should therefore exhibit non-uniform soil communities and ideally mimic hillslope plant communities, which would in-turn increase biodiversity. And since slope and aspect are inextricably coupled relative to their influence on radiant loading, effects of both should be investigated in tandem. Slope- and aspect-induced roof heterogeneity as a whole—from soils to plants—needs further long-term monitoring especially as it relates to ecosystem function. Where the effects of topographical variation on green roofs are well documented, so should be the above- and below-ground effects of slope and aspect.

Much effort has been expended in the process of finely crafting both plant palettes and soils to achieve uniform, low maintenance, and sustainable roofs at a cost to biodiversity or many of the ecosystem services associated with a more natural system. Modeling natural ecosystems can help promote these ends, but many of the ideal candidate ecosystems such as rock outcrop communities may require inclusion of native soils as well as native plants in order to best mimic the system and derive maximum benefits. The concerted use of habitat templates (Chap. 11) for green roof design is still a relatively new field and thus is in need of a deeper body of research, especially regarding long-term management or viability.

And while there has been much research on water retention and primary stormwater treatment via green roofs (Chap. 3, Chap. 4, Chap. 5), the long-term trends are still not well understood (Chap. 13). Like many wetland treatment systems, green roofs have the potential to become toxic systems in highly urbanized settings, acting as cumulative sinks for heavy metals and other airborne pollutants. Consideration for a plant palette that can withstand toxicity may be required, such as in mining reclamation and curbside bioswales. Further research is needed to document the efficacy of green roofs as mechanisms for primary stormwater treatment, explicitly with regard for the remediation potential of soil biota. This has largely been overlooked as a research focus, but at least one recent study found that fungal species composition of several highly urban green roofs closely resembled fungal suites seen in disturbed environments, some of which are resistant to major soil contaminants such as hydrocarbons and heavy metals (McGuire et al. 2013). This speaks to the capacity for "non-plant" green roof communities to participate in overall provisioning of ecosystem services.

There is also a need for careful accounting comparisons among green roofs using financial, ecological, and client value criteria, both after installation and after several years. Such research would indicate whether the potential extra financial costs of more intensive testing, extraction logistics, and site-specific blending of NS confer a resilience premium that offsets the convenience and standardization of ES.

There is obviously still much to be learned regarding the potential of roof soils and their components from an ecosystem perspective. When it comes to basic research on roof soils, we are steadily learning that the whole is indeed greater than the sum of its parts, that the interplay of the physics, chemistry, and biology of in situ roof soils is still largely unknown. Continued evaluation and monitoring across both timescales and landscapes is required to fill knowledge gaps, especially in the diverse environment found in North America. In comparison to the information already compiled from roof systems in northern Europe, large climatic differences across North America, Australia, and other continents make ongoing soil research even more necessary if green roof construction is to truly become established outside temperate regions (Chap. 3).

6.5 Case Study of a Living NS Green Roof

When the designers of the headquarters building for the Botanical Research Institute of Texas (BRIT) (Fort Worth, Texas, USA) decided to create a green roof, there were few precedents in the region. The idea of mimicking a local prairie habitat was put forth by community shareholders with hopes of the roof becoming a form of "reconciliation ecology" (Rosenzweig 2003), allowing habitat expansion for native species within a built environment. The landscape architects agreed to a biomimicry approach, and collaborating scientists and students started investigating possible template ecosystems.

The biomimicry strategy, also called habitat template strategy (Lundholm 2006), depends upon comparisons between a self-organizing, natural ecosystem and a model designed to simulate desirable aspects of that system. The sequence was to (1) explore and describe the wild system, (2) create a model system, (3) compare performance of the model system with the wild prototype, and (4) progressively refine the model to optimize desired functions. The advantage of this process brought learning about a poorly understood ecosystem in tandem with developing a regionally appropriate design for living roofs.

6.5.1 *The Model Ecosystems: Fort Worth Prairie Barrens and Glades*

Exploration refined project participants' perception of the plant communities associated with limestone, leading to a fundamental distinction between "barrens" and "glades" (Quarterman 1989; Homoya 1994; Baskin and Baskin 2003; Swadek and Burgess 2012). Barrens are habitats where shallow soil over bedrock restricts plant growth, vegetation usually occurring on soils 5–25cm (2–10 inches) deep over weathered limestone. Glades by contrast have extensive areas of exposed bedrock, with distinctive plant communities restricted either to soils in deeper crevices or soils less than 5 cm (2 inches) deep. In general, both of these habitats are dominated by short perennial bunchgrasses, prickly pear cactus, and yucca, which are hardy, drought-tolerant species, and annual grasses and forbs, which can complete their life cycle with seasonal rains. Patches of bare soil with cryptogamic crusts are also common.

6.5.1.1 Ecological Overview

Barrens and glades are constrained by limited soil moisture and have vegetation similar to both arid landscapes farther west and xeric glades to the east (Chap. 11). Plant cover is more open, vegetation strata are lower, and plant growth form spectra are more typical of drought-structured habitats than tallgrass prairie. The barrens and glades, together with the living roofs derived from them, can be understood as arid ecosystems with "pulse and reserve" dynamics (Noy-Meir 1973) adapted to exploit erratic soil moisture inputs. Soil moisture, the limiting resource, is available as discrete pulses after each rainfall event. Many rainfall events are relatively light; fewer saturate the soil deeply. Each soil moisture pulse is distinguished by the amount of precipitation, which determines depth of infiltration, and by the duration of availability, determined by temperature-dependent evapotranspiration. Each organism must use the pulsed regime of soil moisture to generate some form of reserve that survives a period of drought until the next moisture pulse. As in desert ecosystems, plants with different growth forms and phenological rhythms coexist by partitioning the soil moisture (Shmida and Burgess 1988); no single species can consistently dominate, because no single strategy can consistently sequester the erratically available resource and accumulate enough biomass to exclude the others. With less competitive exclusion, alpha-biodiversity of barrens is usually noticeably higher than adjacent grasslands. Quarterman (1989) described similar limiting factors in Tennessee limestone barrens.

6.5.1.2 Walnut Formation Barrens and Glades as A Living Roof Model

The project's first attempt to model barrens vegetation focused on the Walnut Formation along the western margin of the Fort Worth Prairie. The Walnut Formation consists of alternating layers of soft marls (calcareous clays) and hard limestone shell agglomerates (coquinites) of fossil *Texigryphaea* oysters (Scott et al. 2003) (Fig. 6.1). The Walnut shell agglomerate strata are massive, usually about 0.5 to 1 m thick, and form the most extensive glades in the region.

The most commonly mapped soils of Walnut barrens and glades are Maloterre, Aledo, Bolar, Purves, and Brackett series (Greenwade et al. 1977; Colburn 1978; Ressel 1981), with deeper mollisols adjacent in most landscape catenas. The Maloterre is a lithic ustorthent; the Brackett, a shallow typic haplustept; and all other soils are lithic calciustolls (SSS 1989a, 1989b, 1992, 1997, 2010). The soils have relatively high proportions of rock fragments, high levels of calcium carbonate, and given sufficient time will form dark, organic surface horizons (mollic epipedons) with basic pH, characteristic of semiarid grasslands and steppes.

Soils of Walnut barrens and glades are in various stages of profile development, depending upon rates of horizon formation in relation to erosion, which in limestone landscapes involves both mechanical transport and chemical dissolution. Many sites have soils that are too immature to be classified; essentially plants are growing in slightly weathered bedrock. Most marl strata are so weakly consolidated that they constitute a paralithic medium where plants can root directly into unweathered geological material. Maloterre soils are immature, with weakly devel-

Fig. 6.1 Example of typical Walnut Limestone with embedded oyster fossils. (*Photo Rebecca Swadek*)

oped horizons. Brackett soils have more differentiated horizons but lack the mollic epipedon characteristic of more mature soils in this landscape. Soil textures vary over short distances, depending upon what proportions of weathered limestone and clayey marl comprise the local parent materials. Loamy Aledo soils derive from weathered limestone; whereas smectitic clayey Purves soils develop from marls.

We concluded that a roof soil simulating a Walnut ecosystem should have a texture ranging from clay to sandy loam, with substantial amounts of organic matter and calcium carbonate, though the lime is not essential for all types of native barrens vegetation.

Expanses of bare limestone define the Walnut glades (Fig. 6.2). A distinctive plant association grows in extremely thin soils less than 5 cm (2 inches) deep, characterized by the low forbs, geophytes, succulents, and cool-season and warm-season annuals (Swadek and Burgess 2012). Unanchored poikilohydric, gelatinous fragments of the cyanobacterium *Nostoc commune* (star jelly) are also often seen on these glades. This habitat is similar to xeric limestone glades in Tennessee described by Quarterman (1989). Though Walnut glades are the most arid habitat in the region, they did not seem the most appropriate model for a living roof as the stature and cover of the vegetation appeared too sparse.

The barrens on deeper soils over Walnut strata, however, have greater plant cover and stature, reflecting more available soil moisture, and richer diversity, consistent with the greater diversity of soil types and opportunities to partition soil moisture. Often the vegetation is low, open grassland under 0.5 m (1.6 ft) tall, with

Fig. 6.2 Woodland (background), Walnut barrens with Indian paintbrush (*Castilleja indivisa*), and a Walnut glade (foreground) in the spring. In the late summer this habitat is typically very dry due to minimal, shallow soils. (*Photo Rebecca Swadek*)

scattered cactus emerging above the grass. Many species found on shallow glades soils also occur in barrens, together with characteristic perennial bunchgrasses. The richer plant diversity and higher vegetative cover of the barrens became the basis for choosing this ecosystem as a model for the first living roof tests.

Early experiments compared native plant performance between a commercial heat expanded shale aggregate substrate (Haydite) and natural barrens clay loam soils with different surface mulch treatments for moisture retention (Williams 2008; Kinder 2009). Most native barrens species grew poorly on the commercial mix, probably due to lack of lime and other nutrients; however during intense summer drought a higher proportion of established plants survived in the commercial Haydite mix. The native clay loam soils desiccated in summer, and few perennial species survived aside from one cactus, which thrived and eventually dominated many test boxes. Cool-season annuals persisted in most boxes, showing greatest density in boxes with gravel and tile mulches. Autumn seedling establishment also was noticeably greater in dead grass clumps. It was clear that, if a living roof were to have native species other than cactus and annuals, transplanted perennials would have to be irrigated for at least a year after installation. Also, a surface mulch of gravel or lightweight concrete tiles would enhance the diversity of a living roof (Williams 2008; Kinder 2009).

6.5.1.3 Goodland Formation Barrens and A Living Roof Model

The limited diversity of plants that survived in test boxes with clayey Walnut soil, together with the relatively heavy weight of saturated clays, led the project team to extend their focus to nearby barrens on the Goodland Formation, which offered outcrops up-slope and east of the Walnut. The Goodland Formation overlies the Walnut, and both formations possess similar vegetation structure with many overlapping species (Fig. 6.3).

Similar to the Walnut Formation, the Goodland consists of alternating strata of limestones and soft marls or shales (Hill 1901; Scott et al. 2003). However, Goodland Limestone differs from the harder, more impervious shell hash of the Walnut Formation; it usually has a nodular fabric that allows water to seep into networks of minor crevices, weathering the rock into cobble- to gravel-sized fragments. Thus Goodland Limestone seldom forms glades comparable with the extensive flat, bare surfaces of Walnut Limestone, but it does have extensive gravelly barrens with paralithic and immature soils. On eroding slopes, the Goodland Limestone strata are more resistant than interbedded marls and form prominent chalky-white bluffs and benches.

Goodland soils have been mapped with the same soil series as Walnut. Often areas with barrens are mapped as Aledo-Bolar complex (Ressel 1981). Although typical Bolar soils are too deep for barrens (SSS 1989b), shallow or truncated soils similar to Bolar calciustolls have been observed with barrens vegetation (Jue 2011; Llado 2011). Most Goodland barrens sites that we described had shallow soils that fit descriptions of Aledo and Maloterre series or were too undeveloped to be clas-

Fig. 6.3 Goodland Limestone habitat in summer. Soil depth is greater in the background and supports a denser assemblage of plants, including *Opuntia phaeacantha* (tulip pricklypear). (*Photo David M. Fisk*)

sified as a soil series. Excavations of the soils over Goodland Limestone showed a deeper weathering profile than over Walnut Limestone. The greater rooting depth and protection from direct evaporation may be a factor that creates differences in plant associations between Goodland and Walnut barrens. The lower horizons of Goodland-derived soils were up to 75 % rock fragments, so the actual volume of soil for roots may be relatively small. By eliminating the deeper rock fragments and increasing the depth of surface gravel mulch, the Goodland soil environment could be simulated with depth and weight more appropriate for a rooftop. Laboratory analyses of several samples indicated many barrens soils were sandy loams, with similar proportions of sand, silt, and clay on both Goodland and Walnut Formations (Williams 2008); however some samples had textures of clay loam, sandy clay loam, and loam. Apparently our first test boxes were filled from a site with atypical amounts of clay, perhaps a truncated B-horizon of a paleosol. This inconsistency highlights the importance of careful soil sampling of proposed extraction sites. The analyses also revealed that soils from both formations had very low levels of nitrogen and phosphorus, with pH ranging from 7.7 to 8.2. The only notable difference between the two types of barrens was calcium, which was two or three times higher in Goodland soils (15,304–16,660 ppm) than Walnut (5,519–8,214 ppm) (Williams 2008), perhaps due to a higher proportion of sand-sized limestone fragments in Goodland soils.

Experimental test boxes were filled with either Goodland barrens soil (sieved to remove gravel) or specific combinations of sieved barrens soil and a commercial green roof substrate dominated by Haydite. Results after a growing season reinforced the importance of supplemental irrigation and surface mulch for survival of transplanted perennials as large succulent perennials once again survived better than other transplants.

More importantly, though, the seedbank from test substrates containing some portion of native soil allowed those boxes to reestablish plant cover after a summer drought had killed most perennials. Although not tested, the contribution of soil microbes to this resilience was inferred from seedbank viability, allowing project members to conclude that starting with as much soil diversity as possible would allow maximum evolution of a biota adapted to rooftops. The intent was that some part of the native biodiversity destroyed by urbanization would find a new context in urban habitats. Whether it is a human gut or a biosphere, a fundamental challenge for managing a healthy, resilient ecosystem is promoting the evolutionary potential of the microbial community (Chap. 7). Therefore using healthy native soil for this roof project was preferable to an entirely synthetic, engineered substrate for the purpose of simulating a native prairie barrens ecosystem.

6.5.2 Challenges Posed by Using Natural Soil

Using native soil as the growing medium for a large roof area posed problems for conventional designers and contractors. Commercial green roof growing substrates are designed to be very lightweight. The engineered substrate sold by the company contracted for this project had a substantial amount of Haydite and very little silt or clay, creating a medium with very large pore sizes, promoting rapid infiltration and less water retention. The smaller particle sizes of natural soil create a relatively large volume of very small pores, retain water, and result in a much greater saturated weight than commercial media. Using methods and equipment specified by ASTM (2005), Williams (2008) found that a 2.54 cm (1 inch) depth of Goodland barrens soil had an approximate saturated weight of 34.2 kg/m^2 (7 lb/ft^2), while that of the commercial medium was 29.3 kg/m^2 (6 lb/ft^2). Later, a professional soil analysis lab analyzed similar materials by using an engineered substrate protocol. These results showed heavier saturated weights: 40.0–44.4 kg/m^2 (8.2–9.1 lb/ft^2) for Goodland barrens soil and 42.5 kg/m^2 (8.7 lb/ft^2) for topsoil from a nearby quarry. A blend of 67 % commercial medium with 33 % topsoil was 36.6 kg/m^2 (7.5 lb/ft^2). The commercial contractor's specification for maximum weight range of their growing substrate is 28.3–36.6 kg/m^2 (5.8–7.5 lb/ft^2). After several iterations of the design, the roof was re-engineered for a maximum load of about 244 kg/m^2 (50 lb/ft^2) above the metal deck.

Weights of mature plants also had to be considered, especially for succulents that store water internally. Mature barrens plants were collected and weighed several days after a rain, so the plants would be fully hydrated. Plants selected for weight analysis represented the largest and heaviest species likely to establish on the rooftop (two succulents and a perennial grass), thus estimating the upper range of load expected from a mature barrens community (Table 6.2).

Table 6.2 Weights of wild, mature barrens plants

Scientific name	Common name	Average kg/m² (lb/ft²) [n]	Range kg/m² (lb/ft²)
Opuntia phaeacantha	Tulip pricklypear	33.2 (6.8) [5]	24.9–50.8 (5.1–10.4)
Yucca pallida	Twistleaf yucca	10.3 (2.1) [6]	8.3–15.1 (1.7–3.1)
Muhlenbergia reverchonii	Seep muhly	23.4 (4.8) [3]	22.5–25.9 (4.6–5.3)

The commercial contractor had concerns about the amount of silt and clay in barrens soils, which could plug the fabric layer beneath the growing medium and retard drainage in their green roof system. The company's specifications recommend less than 10% combined clay and silt fractions. Purves soils are 35–55% clay (SSS 1997), and Aledo loams and clay loams typically have combined silt-clay fractions of 50% or more (SSS 1992). After several experiments with mixtures, a compromise was negotiated where the company would provide water-tightness and material integrity warranties for the lower components of their roof system but not for the growing medium and plants.

The utility of viable native topsoil depends upon keeping soil organisms and plant propagules alive. Research with restoration projects indicates that the viability of stockpiled soil biota and propagules decay over time (Strohmayer 1999; Norman and Koch 2005). Thus minimizing the interval between topsoil extraction and installation on the roof was highly encouraged, knowing fully that any biodiversity loss consequences of delays could not possibly be assessed. Further, considerable expense and time could be devoted to assessing the diversity and health of a soil's biota and seedbank. Such assessments were impractical given the roof construction schedule; therefore, surface vegetation was used as a proxy, assuming that healthy barrens vegetation indicated a suitable soil biota but realizing that this assumption involved considerable uncertainty.

Finding an appropriate topsoil extraction site also required thoughtful consideration, including not only soil health and plant composition but also the ethics of disturbing native habitats for urban landscaping. A future quarry site was chosen, thus the soil extraction became salvage of biota that would otherwise have been destroyed. The site had desirable vegetation structure and diversity, with a relatively low cover of the exotic species, growing on a thin Aledo soil over Goodland bedrock. The soil itself and the equipment required for removal and transport were donated to the project, so cost estimates unfortunately cannot be provided. However, in other cases time and materials for excavation and transportation could potentially increase the price of soil above ES. Approximately 15–25 cm (6–10 inches) of topsoil was extracted, screened, and installed with plants into modular trays in less than 3 weeks. After plant establishment, the trays were placed onto the rooftop by conveyor belt. At installation, the trays contained 6.3 cm (2.5 inches) of substrate with an additional 5 cm (2 inches) of substrate below the trays and a 1.2 cm (0.5 inches) gravel mulch layer on top, yielding an overall substrate depth of approximately 12.5 cm (5 inches) (Dvorak et al. 2013).

6.5.3 Results of Using Natural Soils

Initial maintenance included monthly walk-throughs, establishment watering, and hand removal of tree seedlings and specific invasive species—e.g., *Sorghum halepense* (johnsongrass), *Bothriochloa ischaemum* var. *songarica* (KR bluestem), *Torilis arvensis* (beggar's ticks). Stormwater harvested from the campus grounds was used to irrigate the roof during the first 2 years. At the end of the first growing season, weekly irrigation tapered to every few weeks, then ceased completely until the beginning of the following dry season, when semi-regular but minimal water was provided (e.g. 7.85 mm (0.31 inches) every 2–4 weeks). This irrigation regime tapered off at the end of the second growing season and a post-establishment irrigation plan adopted. This involved limiting the roof to emergency irrigation only during extended periods of both drought and extremely high temperatures (mostly as a means to reduce soil temperature and curb root die-off).

Plant performance was assessed after two growing seasons (Dvorak et al. 2013). From a list of 30 initial Walnut/Goodland test species, 15 were classified as successful. These include 7 of 17 forbs (41 %), 4 of 9 graminoids (44 %), and all 4 succulents (100 %). Several species (3 forbs, 1 graminoid) did not survive in the form of the original transplants but persisted in small numbers from the seedbank. Four species were identified as exceptional performers in terms of establishment speed and general proliferation (both vegetative and reproductive): *Bouteloua curtipendula* var. *curtipendula* (sideoats grama), *Bouteloua dactyloides* (buffalograss), *Muhlenbergia reverchonii* (seep muhly), and *Opuntia phaeacantha* (tulip pricklypear). Several unsuccessful species—e.g., *Aristida purpurea* (purple threeawn), *Asclepias asperula* (spider milkweed), *Convolvulus equitans* (Texas bindweed), and *Oenothera macrocarpa* (bigfruit evening primrose)—were expected to perform well due to their presence at the topsoil extraction site and their pervasiveness throughout Walnut and Goodland habitats (Swadek and Burgess 2012). Failure due to transplant shock is possible, but this speculation does not explain why these species also did not come up from the seedbank.

Aside from the 30 species purposefully transplanted or seeded, an additional 90+ species were observed growing on the roof. Most are species from the seedbank or from bird- or wind-aided dispersal, although some are suspected to have come from plug soil. Table 6.3 shows the breakdown of volunteer species into different functional groups.

Table 6.3 Breakdown of volunteer plant species observed within the first 2 years

Category	Proportion of total (N=93) (%)
Dicots	68
Annuals[a]	67
Grasses	75
Non-natives	25
Seedbank	70

[a] including species that can persist beyond one season if conditions are adequate

Fig. 6.4 The BRIT living roof in spring of its third year. All three planting assemblages, divided by gravel walking paths, can be seen here: yucca-dominated assemblage (foreground), cactus-dominated (midground), grass-dominated (background). (*Photo Rebecca Swadek*)

We considered 55 % of the volunteer/colonizing species established, including 31 desirable species suspected to have come from the seedbank. Enriched by these colonizing species, plant diversity is now higher on the roof compared to its barrens template, similar to biodiversity enrichment described by Sax and Gaines (2003). Slope and an original planting strategy that involved distinct species assemblages both likely played a role in encouraging heterogeneity and allowing different species to establish in different regions of the roof. The end result is now a novel ecosystem resembling local prairie barrens (Figs. 6.4), but with a unique, evolving vegetative cover serving as functional greenspace within the urban environment (as evidenced by its attractiveness to ground-nesting birds) (Fig. 6.5). Future planned research on this roof includes (1) examining the effects of slope on plant diversity, (2) measuring stormwater retention and water quality, and (3) comparing the roof arthropod community to the immediate grounds and a prairie template. Preliminary results from the arthropod study suggest that diversity after 2 years is similar to that found on older green roofs, while early findings from the slope study suggest slope-induced moisture gradients created roof microhabitats that differ in plant composition. This niche partitioning was aided by species from the seedbank that are now established in roof areas where most transplants failed. Thus, observations indicate that the NS was critical for sustaining roof health and accelerating establishment and coverage. Before the topsoil was extracted from the donor site—perhaps long before—seeds accumulated in the soil and thus had time to weather, break dorman-

Fig. 6.5 Ground-nesting birds began visiting the roof soon after installation, with frequency of nest discovery increasing each year. These are eggs of *Zenaida macroura* (mourning dove), the most common bird observed on the roof. (*Photo Brooke Byerley Best*)

cy, become scarified, and engage with microbiota, and otherwise experience the criteria necessary for germination. Seedbank contribution to vegetation must therefore be different than simply broadcasting purchased seed during initial planting.

References

Ampim P, Sloan J, Vempati R, Wu Y (2009) The impact of nanophase iron-oxide coated clinoptilolite zeolite on nutrient retention in a sand based medium. Agronomy abstracts. American Society of Agronomy, Madison

Ampim PAY, Sloan JJ, Cabrera RI, Harp DA, Jaber FH (2010) Green roof growing substrates: types, ingredients, composition and properties. J Environ Hortc 28(4):244–252

ASTM Standard E2399 (2005) Standard test method for maximum media density for dead load analysis of green roof systems. ASTM International, West Conshohocken. doi:10.1520/E2399-11. http: www.astm.org. Accessed 30 March 2014

Baskin JM, Baskin CC (2003) The vascular flora of cedar glades of the southeastern United States and its phytogeographical relationships. J Torrey Bot Soc 130(2):101–118

Beattie D, Berghage R (2004) Understanding the importance of growing media. Paper presented at Greening Rooftops for Sustainable Communities, Portland, OR, 2–4 June 2004

Bengtsson L, Grahn L, Olsson J (2005) Hydrological function of a thin extensive green roof in Southern Sweden. Nordic Hydrol 36(3):259–268

Berghage R, Beattie D, Jarrett A, O'Connor T (2007) Greenroof runoff water quality. Paper presented at Greening Rooftops for Sustainable Communities, Minneapolis, MN, 19 April–1 May 2007

Bever JD, Schultz PA, Miller RM, Gades L, Jastrow JD (2003) Prairie mycorrhizal fungi inoculant may increase native plant diversity on restored sites (Illinois). Ecol Restor 21:311–312

Brady NC, Weil RR (2008) The nature and properties of soils. Prentice Hall/Pearson Education, Upper Saddle River

BRIT (2014) Living roof monitoring report: plant performance, volunteer species, plant assemblage effects, and slope effects. Botanical Research Institute of Texas, Fort Worth

Buist R (2008) Going for organics. Liv Arch Monit 10(1):16–21

Bunt AC (1988) Media and mixes for container-grown plants: a manual on the preparation and use of growing media for pot plants, 2nd ed. Unwin Hyman, London

Burgess TL (1995) Desert grassland, mixed shrub savanna, shrub steppe, or semidesert scrub?: The dilemma of coexisting growth forms. In: McClaran MP, Van Devender TR (eds) The desert grassland. University of Arizona Press, Tucson, pp 31–67

Campeau S, Rochefort L (1996) Sphagnum regeneration on bare peat surfaces: field and greenhouse experiments. J Appl Ecol 33(3):599–608

Cantor SL (2008) Green roofs in sustainable landscape design. W.W. Norton & Company, New York & London

Caron J, Rochefort L (2013) Use of peat in growing media: state of the art on industrial and scientific efforts envisioning sustainability. Acta Hort 992:15–22

Coffin D, Lauenroth W (1989) Spatial and temporal variation in the seed bank of a semiarid grassland. Amer J Bot 76(1):53–58

Colburn WC (1978) Soil survey of Hood and Somervell counties, Texas. United States Department of Agriculture Soil Conservation Service in cooperation with Texas Agricultural Experiment Station

Cook-Patton SC, Bauerle TL (2012) Potential benefits of plant diversity on vegetated roofs: a literature review. J Environ Manage 106: 85–92

Cresswell D, Sims V (2007) Characterisation of mineral wastes, resources and processing technologies—Integrated waste management for the production of construction material. Case Study: Green roof substrate using aerated concrete waste. Report WRT 177/WR0115 (October) issued by University of Leeds, Akristos, and Mineral Industry Research Organization. DEFRA (Department for Environment, Food, and Rural Affairs), UK

Dunnett N, Kingsbury N (2008) Planting green roofs and living walls. Timber Press, Portland & London

Dunnett N, Gedge D, Little J, Snodgrass EC (2011) Small green roofs: low-tech options for greener living. Timber Press, Portland & London

Dvorak B, Byerley B, Volder A (2013) Plant species survival on three water conserving green roofs in a hot humid subtropical climate. J Liv Arch 1(1):39–53

Earth Pledge Foundation (2005) Green roofs: ecological design and construction. Schiffer Publishing, Atglen

Emilsson T (2008) Vegetation development on extensive vegetated green roofs: influence of substrate composition, establishment method and species mix. Ecol Eng 33:265–277

Emilsson T, Rolf K (2005) Comparison of establishment methods for extensive green roofs in southern Sweden. Urban Forest Urban Green 3:103–111

Emilsson T, Czemiel BJ, Mattsson JE, Rolf K (2007) Effect of using conventional and controlled release fertiliser on nutrient runoff from various vegetated roof systems. Ecol Eng 29:260–271

Faist AM, Ferrenberg S, Collinge SK (2013) Banking on the past: seed banks as a reservoir for rare and native species in restored vernal pools. Ann Bot 5:1–11

FLL (2002) Guidelines for the planning, execution and upkeep of green-roof sites. Forschungsgesellschaft Landschaftsentwicklung Landschaftsbau, Bonn

Francis R, Lorimer J (2011) Urban reconciliation ecology: the potential of living roofs and walls. J Environ Manage 92:1429–1437

Friedrich C (2005) Principles for selecting the proper components for a green roof growing media. Paper presented at Greening Rooftops for Sustainable Communities, Washington DC, 4–6 May 2005

Friedrich C (2008) Don't call it dirt! Living Arch Monit 10(1):16–21

Getter KL, Rowe DB (2006) The role of extensive green roofs in sustainable development. HortSci 41(5):1276–1285
Getter KL, Rowe DB, Andresen JA (2007) Quantifying the effect of slope on extensive green roof stormwater retention. Ecol Eng 31:225–231
Graf MD, Bérubé V, Rochefort L (2012) Restoration of peatlands after peat extraction. In: Vitt D, Bhatti, J (eds) Restoration and reclamation of boreal ecosystems: attaining sustainable development. Cambridge University Press, New York, pp 259–280
Greenwade JD, Kelley JD, Hyde HW (1977) Soil survey of Parker County, Texas. United States Department of Agriculture Soil Conservation Service in cooperation with Texas Agricultural Experiment Station
Handreck KA, Black ND (1994) Growing media for ornamental plants and turf. New South Wales University Press, Randwick
Hill RT (1901) Geography and geology of the Black and Grand Prairies, Texas, 21st Annual Report of the United States Geological Survey. Part VII. Texas. Vol 7. United States Geological Survey, Washington, DC
Hobbs RJ, Arico S, Aronson J, Baron JS, Bridgewater P, Cramer VA, Epstein PR, Ewel JJ, Klink CA, Lugo AE, Norton D, Ojima D, Richardson DM, Sanderson EW, Valladares F, Vila M, Zamora R, Zobel M (2006) Novel ecosystems: theoretical and management aspects of the new ecological world order. Global Ecol Biogeog 15(1):1–7
Homoya MA (1994) Barrens as an ecological term: an overview of usage in the scientific literature. In: Fralish JS, Anderson RC, Ebinger JE, Szafoni R (eds) 1994 proceedings of North American conference on savannas and barrens. Illinois State University, Normal. http://www.epa.gov/grtlakes/ecopage/upland/oak/oak94/Proceedings/Homoya.html. Accessed 23 Mar 2014
Hopkins G, Goodwin C (2011) Living architecture: green roofs and walls. CSIRO Publishing, Collingwood
Ishimatsu K, Ito K (2013) Brown/biodiverse roofs: a conservation action for threatened brownfields to support urban biodiversity. Landsc Ecol Eng 9:299–304
Jenkins VS (1994) The lawn: a history of an American obsession. Smithsonian Institution Press, Washington, DC
John J (2013) Assessment of arbuscular mycorrhizal fungi in a green roof system. MS thesis, Dalhousie University, Nova Scotia
Johnson DB, Williamson JC, Bailey AJ (1991) Microbiology of soils at opencast coal sites. I. Short-and long-term transformations in stockpiled soils. J Soil Sci 42:1–8. doi:10.1111/j.1365–2389.1991.tb00085.x
Jue ML (2011) Vegetative analysis of muhly hillslope seeps in North Central Texas. MS thesis, Texas Christian University, Texas
Keulartz J, Van der Weele C (2008) Framing and reframing invasion biology. Configurations 16(1):93–115
Kinder JW (2009) Appropriate design elements and native plant selection for living roofs in North Central Texas. MS Thesis, Texas Christian University, Texas
Kulmatiski A, Beard KH, Stark JM (2006) Soil history as a primary control on plant invasion in abandoned agricultural fields. J Appl Ecol 43:868–876
Llado LE (2011) Soil moisture dynamics of Muhly Seeps in a hillslope hollow during low flow and storm conditions. MS Thesis, Texas Christian University, Texas
Luckett K (2009) Green roof construction and maintenance. McGraw Hill, New York
Lundholm JT (2006) Green roofs and facades: a habitat template approach. Urban Habitat 4(1):87–101
Lundholm JT, Richardson PJ (2010) Habitat analogues for reconciliation ecology in urban and industrial environments. J Appl Ecol 47:966–975
Lundholm J, MacIvor JS, MacDougall Z, Ranalli M (2010) Plant species and functional group combinations affect green roof ecosystem functions. PLoS ONE 5(3):e9677. doi:10.1371/journal.pone.0009677
Madre F, Vergnes A, Machon N, Clergeau P (2013) A comparison of 3 types of green roof as habitats for arthropods. Ecol Eng 57:109–117

McGuire KL, Payne SG, Palmer MI, Gillikin CM, Keefe D, Kim SJ, Gedallovich SM, Discenza J, Rangamannar R, Koshner JA, Massmann AL, Orazi G, Essene A, Leff JW, Fierer N (2013) Digging the New York City skyline: soil fungal communities in green roofs and city parks. PLoS ONE 8(3):e58020. doi:10.1371/journal.pone.0058020

McQueen DJ, Ross CW (1982) Effects of stockpiling top soils associated with opencast mining 2: physical properties. N Z J Sci 25(3):295–302

Mentens J, Raes D, Hermy M (2006) Green roofs as a tool for solving the rainwater runoff problem in the urbanized 21st century. Landsc Urban Plan 77:217–226

Meyer (2004) Use of endotrophic mycorrhiza and soil microorgnisims in vegetation establishment on mineral greenroof substrates. Paper presented at Greening Rooftops for Sustainable Communities, Portland, OR, 2–4 June 2004

Middleton EL, Bever JD (2012) Inoculation with a native soil community advances succession in a grassland restoration. Restor Ecol 20(2):218–226

Miller, C (2003) Moisture management in green roofs. Paper presented at Greening Rooftops for Sustainable Communities, Chicago, IL, 29–30 May 2003

Miller R, Gardiner D (2007) Soils in our environment, 11th ed. Prentice Hall, Upper Saddle River

Molineux C, Fentiman C, Gange A (2009) Characterising alternative recycled waste materials for use as green roof growing media in the U.K. Ecol Eng 35:1507–1513

Monterusso MA, Rowe DB, Rugh CL, Russell DK (2004) Runoff water quantity and quality from green roof systems. Acta Hort 639:369–376

Moore JC, Walter DE, Hunt HW (1988) Arthropod regulation of micro- and mesobiota in below-ground detrital food webs. Annu Rev Entomol 33:419–435

Nagase A, Dunnett N (2011) The relationship between percentage of organic matter in substrate and plant growth in extensive green roofs. Landsc Urban Plan 103(2):230–236

Nagase A, Dunnet N, Choi M (2013) Investigation of weed phenology in an establishing semi-extensive green roof. Ecol Eng 58:156–164

Nektarios PA, Tsiotsiopoulou P, Chronopoulos I (2003) Soil amendments reduce roof garden weight and influence the growth rate of Lantana. HortSci 38:618–622

Norman MA, Koch JM (2005) Differences in species abundance between sites rehabilitated with direct return and stockpiled soil (Environmental department research bulletin no. 33). Alcoa World Alumina Australia. http://www.alcoa.com/australia/en/pdf/Research_Bulletin_ No_33.pdf. Accessed 9 April 2014

Noy-Meir I (1973) Desert ecosystems: environment and producers. Ann Rev Ecol Syst 4:25–51

Oberndorfer E, Lundholm J, Bass B, Coffman R., Doshi H, Dunnet N, Gaffin S, Kohler M, Liu K, Rowe B (2007) Green roofs as urban ecosystems: ecological structures, functions, and services. BioScience 57(10):823–833

Olszewski MW, Young CA (2011) Physical and chemical properties of green roof media and their effect on plant establishment. J Environ Hort 29(2):81–86

Quarterman E (1989) Structure and dynamics of the limestone cedar glade communities in Tennessee. J Tenn Acad Sci 64(3):155–158

Renker C, Zobel M, Opik M, Allen MF, Allen EB, Vosatka M, Rydlova J, Buscot F (2004) Structure, dynamics, and restoration of plant communities: do arbuscular mycorrhizae matter? In: Temperton VM, Hobbs RJ, Nuttle T (eds) Assembly rules and restoration ecology: bridging the gap between theory and practice. Island Press, Washington, DC, pp 189–229

Ressel DD (1981) Soil survey of Tarrant County, Texas, United States Department of Agriculture, Soil Conservation Service, in cooperation with Texas Agricultural Experiment Station

Rosenzweig (2003) Reconciliation ecology and the future of species diversity. Oryx 37(2):194–205. doi:10.1017/S0030605303000371

Ross DJ, Cairns A (1981) Nitrogen availability and microbial biomass in stockpiled topsoil in Southland. N Z J Sci 24(2):137–143

Rowe DB, Kolp MR, Greer SE, Getter, KL (2014) Comparison of irrigation efficiency and plant health of overhead, drip, and sub-irrigation for extensive green roofs. Ecol Eng 64:306–313

Rumble H, Gange AC (2013) Soil microarthropod community dynamics in extensive green roofs. Ecol Eng 57:197–204

Sax DF, Gaines SD (2003) Species diversity: from global decreases to local increases. Trends Ecol Evol 18(11):561–566

Schrader S, Boning M (2006) Soil formation on green roofs and its contribution to urban biodiversity with emphasis on Collembolans. Pedobiologia 50:347–356

Scott RW, Benson DG, Morin RW, Shaffer BL, Oboh-Ikuenobe FE (2003) Integrated Albian-Lower Cenomanian chronostratigraphy standard, Trinity River Section, Texas. In: Scott RW (ed) Cretaceous stratigraphy and paleoecology, Texas and Mexico. Perkins Memorial Volume, GCSSEPM Foundation Special Publications in Geology No. 1. Gulf Coast Section, Society of Economic Paleontologists and Mineralogists Foundation, Houston, pp 277–334

Shmida A, Burgess TL (1988) Plant growth-form strategies and vegetation types in arid environments. In: Werger NJA, Van der Aart PJM, During HJ, Verhoeven JTA (eds) Plant form and vegetation structure. S.P.B. Publishers, The Hague, pp 211–241

SSS (Soil Survey Staff) (1989a) Maloterre series. Official soil series descriptions, Natural resource conservation service, United States Department of Agriculture. https://soilseries.sc.egov.usda.gov/OSD_Docs/M/MALOTERRE.html. Accessed 27 Mar 2014

SSS (Soil Survey Staff) (1989b) Bolar series. Official soil series descriptions, Natural resource conservation service, United States Department of Agriculture. https://soilseries.sc.egov.usda.gov/OSD_Docs/B/BOLAR.html. Accessed 27 Mar 2014

SSS (Soil Survey Staff) (1992) Aledo series. Official soil series descriptions, Natural resource conservation service, United States Department of Agriculture. https://soilseries.sc.egov.usda.gov/OSD_Docs/A/ALEDO.html. Accessed 27 Mar 2014

SSS (Soil Survey Staff) (1997) Purves series. Official soil series descriptions, Natural resource conservation service, United States Department of Agriculture. https://soilseries.sc.egov.usda.gov/OSD_Docs/P/PURVES.html. Accessed 27 Mar 2014

SSS (Soil Survey Staff) (2010) Brackett series. Official soil series descriptions, Natural resource conservation service, United States Department of Agriculture. https://soilseries.sc.egov.usda.gov/OSD_Docs/B/BRACKETT.html. Accessed 27 Mar 2014

Steinberg T (2006) American green: the obsessive quest for the perfect lawn. W.W. Norton & Company, New York

Strohmayer P (1999) Soil stockpiling for reclamation and restoration activities after mining and construction. Restor Reclam Rev 4(7):1–6

Sutton R (2008) Media modifications for native plant assemblages on green roofs. Paper presented at Greening Rooftops for Sustainable Communities, Baltimore, MD, 30 April–2 May 2008

Sutton R (2013a) Seeding green roofs with native grasses. J Living Arch 1(1):15–35

Sutton R (2013b) Rethinking extensive green roofs to lessen emphasis on biomass. J Living Arch 1(1):36–38

Sutton R (2014) Aesthetics for green roofs and green walls. J Living Arch 1(2):1–20

Sutton R, Harrington JA, Skabelund L, MacDonagh P, Coffman RR, Koch G (2012) Prairie-based green roofs: literature, templates, and analogs. J Green Build 7(1):143–172

Swadek RK, Burgess TL (2012) The vascular flora of the North Central Texas walnut formation. J Bot Res Inst Texas 6(2):725–752

VanWoert N, Rowe DB, Andresen JA, Rugh CL, Fernandez RT, Xiao L (2005) Green roof storm water retention: effects of roof surface, slope, and media depth. J Environ Qual 34:1036–1044

Visser S, Fujikawa, Griffiths CL, Parkinson D (1984) Effect of topsoil storage on microbial activity, primary production and decomposition potential. Plant Soil 82(1):41–50

Vogelsang K, Bever J (2009) Mycorrhizal densities decline in association with nonnative plants and contribute to invasion. Ecology 90:399–407

Wagai R, Mayer LM, Kitayama K, Knicker H (2008) Climate and parent material controls on organic matter storage in surface soils: a three-pool, density-separation approach. Geoderma 147:23–33

Widdowson JP, Gibson EJ, Healy WB (1982) Effects of stockpiling top soils associated with opencast mining 1. Chemical properties and the growth of rye grass and white clover. N Z J Sci 25(3):287–294

Williams DA (2008) Appropriate design elements and soil selection for green roofs in North Central Texas. MS Thesis, Texas Christian University, Texas
Williamson JC, Johnson DB (1990) Determination of the activity of soil microbial populations in stored and restored soils at opencast coal sites. Soil Biol Biochem 22(5):671–675
Yarranton GA, Morrison RG (1974) Spatial dynamics of a primary succession: nucleation. J Ecol 62:417–428

Dr. Brooke Byerley Best serves as botanist and editor at the Botanical Research Institute of Texas (BRIT), a 501(c)3 non-profit organization dedicated to botanical conservation and science education. Her main research focus is performance monitoring of the building's biomimicry-based living roof.

Rebecca K. Swadek is a plant ecologist with the New York City Department of Parks & Recreation's Natural Resources Group and a research associate with the Botanical Research Institute of Texas (BRIT). Rebecca works primarily on wetland restoration projects for the City of New York while continuing research on green roofs and prairies in her spare time.

Dr. Tony L. Burgess is a research associate with the Botanical Research Institute of Texas (BRIT) and a retired Professor of Professional Practice in the Department of Environmental Science at Texas Christian University (TCU). Dr. Burgess has been studying biomimicry-based design since the early 1980s when he designed and helped construct the Caravan of Dreams Desert Dome in Fort Worth, Texas, and the Desert Biome habitat for the Biosphere II Project in Tucson, Arizona.

Chapter 7
Bacteria and Fungi in Green Roof Ecosystems

Krista L. McGuire, Sara G. Payne, Giulia Orazi and Matthew I. Palmer

Abstract Green roofs are one way by which cities are attempting to alleviate some of the problems associated with impervious surfaces in urban environments such as the urban heat island effect and stormwater runoff. In addition, green roofs provide a number of ecosystem services such as the provision of habitats for organisms residing in and migrating through the city that have only recently been studied and documented. Microorganisms such as fungi and bacteria have been found to be diverse and abundant components of green roof growing substrate and may contribute to some of the other benefits green roofs provide such as the removal of organic pollutants from precipitation. Here, we review several functional groups of microbes that may be useful for understanding in terms of green roof design and maintenance: mycorrhizal fungi, decomposer fungi, endophytes, N-fixing bacteria, and pathogens. These microbes interact with plant species and growing substrate in complex ways that require further investigation. The ecology of these microbial groups should also be considered, including their dispersal rates and how they respond to regional differences such as climate and seasonality. We highlight several research priorities for this area of work, which may ultimately facilitate greater functionality in green roof systems.

Keywords Soil · Microbes · Fungi · Bacteria · Urban parks · Mycorrhizae

K. L. McGuire (✉)
1305 Altschul Hall, Barnard College, Columbia University, 3009 Broadway, 10027 New York, NY, 10027, USA
e-mail: kmcguire@barnard.edu

S. G. Payne
44 Arden St. 5M NYC, NY, 10040, USA

G. Orazi
9 Sunset Rd., Rye, NY 10580, USA

M. I. Palmer
Columbia University MC 5557, 1200 Amsterdam Ave., NYC, NY 10027, USA

R. Sutton (ed.), *Green Roof Ecosystems*, Ecological Studies 223,
DOI 10.1007/978-3-319-14983-7_7

7.1 Introduction

Green roofs are so named for the vegetation that covers the otherwise impervious roof of a building. However, the fabric of the vegetated surface is supported by the growing medium in which an abundant and diverse community of microbes resides (Fig. 7.1). These microbes regulate a variety of the ecosystem services for which green roofs are valued such as the retention of water following precipitation events (Gaffin et al. 2009), the removal of air pollution (Yang et al. 2008), and the cycling of nutrients that support plant growth (Kremen 2005). Plants in non-engineered ecosystems cohabit with a variety of different microbes in their leaves, on their leaf surfaces, in their roots, and in the soil surrounding their roots.

Fig. 7.1 Microbial communities of green roofs can be studied and visualized with a variety of techniques. Microscopy can be used to assess mycorrhizal colonization (a), culturing can be used for assaying nutrient preferences and physiological capabilities for some microbes (b) and molecular techniques can be used for DNA and/or RNA sequences from bulk soil cores (c). In panel a, the plant roots were cleared and the mycorrhizal fungal tissue was selectively stained *blue*; the *tree-like structures* are arbuscules and the globular structures are vesicles of the AM fungi. In panel b, fungal colonies were grown on selective media assessing for heavy metal tolerances of difference species cultured from *green* roof soils in New York City. Panel c depicts a soil core (0–10 cm) from a *green* roof located on the Barnard College campus that was subsequently sequenced for both bacterial and fungal DNA. (*Photo credits: Krista McGuire (a and b); Sara Payne (c)*

These plant-associated microbes have been considered an extension of a plant's phenotype (Kristin and Miranda 2013), and complex feedbacks occur that may even shape plant traits (Friesen et al. 2011). In addition to engaging in biotic interactions, microorganisms and plants on green roofs must also cope with extreme abiotic conditions such as aridity, high winds, ultraviolet light exposure, and high temperatures. These harsh conditions may interfere with some symbioses and cause a reduction in microbial abundance. However, very few studies to date have evaluated the composition and function of green roof microbes, despite their anticipated importance to the survival and performance of green roof plants.

Initial research has shown that the microbes most prevalent on green roofs are bacteria and fungi, which are also globally ubiquitous and the most diverse and abundant components of terrestrial soils (Hawksworth 2001; Curtis et al. 2006; Pace 1997; Fierer and Jackson 2006). These microbes shape terrestrial ecosystems in particular by performing critical roles in the biogeochemical cycling of N, P, and C, including degrading soil organic matter into compounds required for plant survival and growth (Swift et al. 1979; Wardle et al. 2004). Microbes can also influence plant diversity and productivity (Klironomos 2002; van der Heijden et al. 1998; van der Heijden et al. 2008; Schnitzer et al. 2011). It is becoming increasingly evident that in order to understand the functioning of ecosystems it is paramount to characterize the assemblages of fungi and bacteria in soils. Numerous studies have examined microbial community composition and associated ecosystem services in non-built environments (Bru et al. 2011; Bell et al. 2005); however, the identities and functions of the urban microbiota are only beginning to be uncovered (Xu et al. 2014; McGuire et al. 2013).

Green roofs are constructed environments representing 'novel ecosystems' that often contain species assemblages not observed in the absence of human intervention (Hobbs et al. 2006). Nonetheless, the same biotic and abiotic factors that operate in unconstructed environments will also likely be operating in green roof communities. Here, we provide a review of the information that exists on green roof microbial communities and give recommendations on future research priorities. We also review the role of specific functional groups of microbes in non-engineered ecosystems to inform how microbes might be functioning in engineered roof communities (Table 7.1).

Table 7.1 Microbial groups likely to be important in green roof ecosystems

Microbial group	Specific taxa	Function
Arbuscular mycor- rhizal fungi (AM)	500–1500 species of fungi from the *Glomeromycota* phylum	Mutualistic with plants to facilitate soil nutrient uptake
Decomposers	Bacteria and fungi from numerous phyla	Nutrient cycling; organic contaminant degradation
Endophytes	Fungi from *Ascomycota* phylum	Diverse, mostly unknown, but some are protective against plant herbivores and pathogens
N-fixing bacteria	Plant-associated and free-living bacteria and cyanobacteria	Convert atmospheric nitrogen (N_2) to ammonia (NH_3)

7.2 Microbial Groups in Green Roofs

7.2.1 *Mycorrhizal Fungi*

One of the most important groups of plant-associated microbes that are likely to play an important role in green roof plant communities are mycorrhizal fungi. Broadly, mycorrhizae are mutualistic associations between plant roots and soil fungi in which photosynthetically derived carbon (C) from the plant is exchanged for limiting nutrients that the fungi take up from soils.

There are seven different types of mycorrhizal associations that are classified according to their anatomical structures and the groups of fungi that engage in the partnerships (Smith and Read 2008). However, the herbaceous plants that are cultivated on green roofs (notably those from the *Crassulaceae, Asteraceae, Poaceae, Fabaceae*) will almost exclusively form arbuscular mycorrhizal (AM) associations (Fig. 7.1a). Arbuscular mycorrhizal (AM) fungi (phylum *Glomeromycota*) are the oldest mycorrhizal association that evolved approximately 460 million years ago with the migration of plants from aquatic habitats onto land (Redecker et al. 2000; Schussler et al. 2001; Wang and Qiu 2006). As such, most plants have retained the capacity to form AM associations and they are currently estimated to be present in >90% of all plant species (Schussler et al. 2001; Wang and Qiu 2006). In non-engineered systems, mycorrhizal fungi enhance plant survival and performance in harsh environmental conditions that are similar to what are experienced on roof top environments, such as frequent soil drying, shallow soils for root nutrient foraging, and low nutrient conditions. AM fungi have long been known to aid plants in drought tolerance (Auge 2001). Mycorrhizal fungi also increase the volume of soil that a plant has access to for nutrient foraging, which would be beneficial on roofs that have minimal fertilizer inputs (Schwartz and Hoeksema 1998).

While only a few studies to date have looked at mycorrhizal fungi in green roof systems, the evidence so far indicate that AM fungi are diverse and abundant in both plant roots and growing substrate. In one study that evaluated AM fungal colonization in green roof plant roots in the UK, it was found that *Sedum* and moss both had high colonization levels averaging 50% or more (Rumble and Gange 2013). Another study evaluating AM colonization across green roofs in Nova Scotia found that three plant species (*Solidago, Poa*, and *Danthonia*) had high levels of colonization, but that *Sedum acre* had low colonization (John et al. 2014). These findings suggest that individual species of *Sedum* vary in their degree of AM fungal colonization, although the extent to which the degree of mycorrhizal dependency relates to long-term viability and stress-tolerance in the plants is not known. In another study from New York City that sequenced fungal DNA in green roof growing substrate containing native grassland communities, the second most abundant group of fungi was the *Glomeromycota*, which accounted for 20% of the total fungal community (McGuire et al. 2013). There were a total of 154 OTUs (operational taxonomic units) of AM fungi detected across the ten roofs sampled in the latter study. While the next-generation sequencing used in that study could not separate out AM fungi by species, the genera *Glomus* and *Rhizophagus* were the most abundant AM fungi, which are widespread and associate with a variety of plants. Future studies should evaluate whether or not these particular AM fungal taxa are better suited for tolerating the urban environment and to what extent they are benefiting their associate plants. Mycorrhizal functioning can exist along a continuum from mutualism to parasitism, and in disturbed ecosystems, the reversal of mutualisms to more parasitic relationships has been observed (Kiers et al. 2010). However, it is also possible that the abiotic stresses experienced on green roofs may result in greater symbiont reliance due to poor environmental quality (Schwartz and Hoeksema 1998).

7.2.2 Decomposers

Another group of microbes likely to play a significant role in green roof ecosystems are decomposers or saprotrophs. Free-living bacteria and fungi that decompose organic material are responsible for the majority of nutrient cycling in soils, and their activity influences soil-atmospheric gas exchanges and soil C storage (Conrad 1996; Canfield et al. 2010; Six et al. 2006; Trivedi et al. 2013). On established green roofs, the senescent leaves of the perennial vegetation, root turnover, root exudates, and dead microbial biomass will be the dominant inputs driving decomposer activity. Compost mixed with growing substrate prior to green roof construction will also provide substrate for microbial decomposers, but eventually those organic food sources will be exhausted unless further compost is added. Immediately following a green roof installation, when compost volume is high, there will likely be an abundance of nutrients available for decomposers, and their degradation capacity may be saturated. If so, there is the chance that excess nutrients will run off of buildings following precipitation events (Gregoire and Clausen 2011; Chen 2013). This leakage of nutrients may contribute to eutrophication and could be more detrimental to the ecosystem than having a gray roof (Chap. 5). For this reason, understanding the decomposition capacity on a roof should be a key research priority to inform the quantity of compost that should be added to the growing substrate. This information could also aid in minimizing the loss of effective soil volume that results from imbalances of organic matter inputs with decomposition rates. To date, no studies to our knowledge have evaluated decomposition rates on green roofs.

The high temperatures of rooftop environments and the mechanical disturbance of precipitation falling directly onto the shallow growing substrate are also likely to impact microbial decomposer composition and activity (Davidson and Janssens 2006). Fungal decomposers may be particularly important on green roofs, as they are less sensitive to water stress than bacteria (Manzoni et al. 2012). A recent study found that green roofs in New York City had higher fungal to bacterial ratios than park soils (McGuire et al. 2013), which may be due to the aforementioned drought tolerance of fungi. However, in mechanically disturbed soils, bacteria become more prevalent, as hyphal networks of fungi become damaged, so these ratios may change in regions that experience high levels of precipitation or foot traffic on shallow substrate. Bacteria and fungi have differing physiological capacities (de Boer et al. 2005; Waring et al. 2013), so if decomposer abundance is fungal rather than bacterial-dominated, there will be biogeochemical consequences that can affect C and N cycling.

7.2.3 Nitrogen-Fixing Bacteria

Nitrogen (N)-fixing bacteria are a group of microbes that may be crucial to the survival of certain groups of plants on green roofs. Nitrogen is an essential limiting nutrient for plant growth, namely since it serves as a building block for chlorophyll, as well as proteins, DNA, and RNA. Atmospheric N is one of the most abundant elements, however is rendered unusable for ecosystem use until the bacteria can convert atmospheric N_2 to ammonia, a readily usable form of nitrogen (Berthrong et al. 2014). The majority of reactive nitrogen is produced during N-fixation by bacteria, and is estimated to amount to nearly 100–300 Tg of nitrogen per year on land (Fields 2004). Generally, nitrogen fixing bacteria are characterized as a type of plant growth- promoting rhizobacteria (PGPR), which can be defined as free-living bacteria capable of colonizing plant roots and providing benefits to the host plant (Nadeem et al. 2014). There are three broad categories of N-fixing bacteria based on their photosynthesis abilities and associations with plant roots: root-associated, free-living photosynthetic, and free-living non-photosynthetic N-fixing bacteria. Symbiotic N-fixing microbes require compounds derived from host plant rhizospheres, whereas free-living photosynthetic nitrogen-fixing bacteria can utilize self-produced sugars, and free-living non-photosynthetic bacteria must acquire energy from decomposing organic matter.

Herbaceous plants found on green roofs form many of these associations with the N-fixing bacteria due to the wide range of benefits that the N-fixing bacteria provide to plants. For instance, two herbaceous plant families commonly cultivated on green roofs, Poaceae and Fabaceae, are able to form close associations with N-fixing bacteria *Azospirillum* and *Bradyrhizobium*, respectively (Saikia et al. 2014; Sanchez-Pardo and Zornoza 2014). N-fixing bacteria provide benefits to these plants such as: increased plant growth (Prabha et al. 2013), improved water and nutrient uptake (Bertrand et al. 2000; Kraiser et al. 2011; Mishra et al. 2014), and suppressed pathogen attack (Ji et al. 2014). Additionally, N-fixing bacteria exhibit a diverse tolerance to varying soil pH and aluminium concentrations, which enable plant survival in acidic soils, commonly experienced on green roofs. By inoculating green roof substrates with N-fixing bacteria, it is likely that green roof vegetation will exhibit increased survival.

7.2.4 Endophytic Fungi

Endophytic fungi are another diverse group of plant-associated microbes that can be found in the leaf, stem, and root tissues of most plant species, and may assist with plant survival on green roofs (Rodriguez et al. 2009). Some endophytes protect plants against herbivores and pathogens, as most of them produce protective alkaloid compounds (Clay and Schardl 2002). There are other endophytes that have been shown to confer tolerance of plant hosts to stressful environments (Rodriguez et al. 2008). In addition to endophytes, other bacteria and fungi that have been detected in the phyllosphere of plants (i.e., microbes residing in and on leaves) may or may not be endophytic, but may also contribute to plant survival and environmental tolerance in roof environments. However, phyllosphere microbial communities in trees have been found to be sensitive to urbanization, so it is unclear what their abundance might be or role they play in green roof ecosystems. In one study of the oak (*Quercus*) phyllosphere in urban and non-urban environments it was found that urban phyllosphere microbial communities were distinct and less diverse than phyllosphere communities in nonurban environments (Jumpponen and Jones 2010). In another study evaluating endophytes in rural and suburban forests of Japan corroborated these results and found fewer endophytes in suburban ecosystems (Matsumura and Fukuda 2013). Thus, while endophytes and other phyllosphere microbes have the potential to be beneficial in green roof communities, their abilities to tolerate urban environments need further investigation.

7.2.5 Pathogens

Plant pathogens are the most detrimental microbes for the maintenance and longevity of green roof plant communities. Pathogens may be particularly problematic on roofs that are planted with only a few species of plants, as monocultures of plants have long been known to be susceptible to pathogen outbreaks because they will accrue specialized plant pathogens that can easily spread to conspecific neighbors (Shipton 1977). However, these pathogens are somewhat difficult to detect prior to attack, as most soil-borne pathogens grow saprophytically in the rhizosphere in order to increase in numbers and outcompete the established beneficial microbes to access the host plant (Berendsen et al. 2012). There have been no published studies to date on pathogen dynamics in green roof communities to our knowledge, although one study observed pathogens in moss panels that were planted with single clones (Akita et al. 2011). Future research may uncover ways by which microbial inoculum can be managed and added to green roof plant communities to effectively reduce pathogenic outbreaks (Gopal et al. 2013).

7.2.6 *Microbial Interactions*

The various functional groups of microbes in green roof ecosystems are not self-contained and there are numerous examples of how these groups engage in antagonistic, commensal, and mutualistic relationships with each other in soils. For example, N-fixing bacteria may have indirect negative effects on plant pathogens because when N is high and not a limiting factor for plant growth, plants will synthesize and store high levels of nitrogen-rich compounds to aid in future defense mechanisms. Such defense mechanisms include biosynthetic enzymes, proteinase inhibitors, chitinases, alkaloids, and glucosinolates (Schultz et al. 2013; Friesen et al. 2011). When plants are under attack, photosynthesis is suppressed, thus forcing the plant to rely on these nitrogen-compound stores (Gomez et al. 2010; Schultz et al. 2013). Mutualistic microbes such as N-fixing bacteria and mycorrhizal fungi may also negatively impact pathogens, as they contribute to plant defense by producing antagonistic molecules on the plant interior and can modify the expression of plant defense pathways (Fravel 1988, Liu 2013). There are also synergies between decomposer bacteria and mycorrhizal fungi. Some bacteria in the rhizosphere actually facilitate mycorrhizal colonization of plant roots and are appropriately called, 'mycorrhiza helper bacteria' (Garbaye 1994). Decomposer bacteria in the genus *Pseudomonas* have also been studied extensively for their antagonistic effects on root pathogens (Weller et al. 2002; Fravel 1988). *Pseudomonas fluorescens* bacteria can actually enhance the upregulation of certain transcription factors involved in plant disease resistance (Van der Ent et al. 2009). These dynamic interactions are complex and difficult to study, but they are important to understand, as they may ultimately impact green roof functioning and may be useful for inoculum-based management strategies.

7.3 Plant-Soil-Microbial Feedbacks: Considerations for Green Roof Design

Plant choice on green roofs will impact the communities of resident fungi and bacteria, which may ultimately affect roof function. For example, the chemical constituents of plant tissue (including root tissue), root exudates, and plant residues can affect microbial biomass, microbial species composition, and microbial activity rates (Philippot et al. 2013; Bardgett and Shine 1999). Plant genomes also help mold the structure and functioning of their associated microbiomes; in turn, these microbiomes contribute to plant fitness (Turner et al. 2013). The plants chosen for cultivation on green roofs usually require low maintenance and are selected based on their abilities to tolerate the harsh roof environment. At present, the majority of green roofs worldwide contain European species belonging to the genus *Sedum* (Crassulaceae), which are hardy, succulent plants that can tolerate the rooftop environment in temperate climates. Recently, however, there has been an interest in experimenting with plant communities native to the regions where green roofs are being built, to facilitate habitat provisioning for associated native biodiversity in the urban environment and to increase ecosystem services (Lundholm et al. 2010). While there have been many studies examining how microbes benefit plants and vice versa, there is still much to uncover about how extreme abiotic conditions experienced on the green roof affect microbe-plant interactions.

The choice of growing medium will also have a significant effect on the composition and function of green roof microbial communities, as microbes strongly respond to their biochemical environment (Fierer et al. 2009). Soil organic matter (SOM) in particular exerts a significant influence on microbial communities, especially in terms of microbial biomass, community structure and function, and substrate utilization (Wardle 1992). Available SOM is thought to promote the production of polysaccharides, which allow better uptake and release of water and fosters the aggregation of soil particles, leading to improved soil structure. Microbes are essential in facilitating the micro-aggregation of soil particles (Duchicela et al. 2012). Certain groups of microbes, notably the AM fungi, are additionally crucial in promoting macro-aggregate formation and durability. In general, natural soil systems have a limited nutrient supply, and as a consequence, microbial biomass is tightly and positively linked to SOM, which greatly impacts microbial function, including microbial activities in carbon and nitrogen cycling (Booth et al. 2005; Cookson et al. 2006). Fluctuating amounts of SOM may also lead to alterations in microbial community composition, and since microbial communities vary widely in their ability to break down organic compounds, changing levels of SOM could promote the survival of certain microbes and hinder the persistence of others (Degens and Harris 1997). Soil pH, which is also linked to SOM, also strongly influences the incorporation of soil organic carbon and nitrogen into the microbial biomass and is one of the most significant predictors of bacterial community composition worldwide (Lauber et al. 2009).

The structural constituents of the growing substrate will also influence resident microbes, as soil texture is recognized to be a critical factor in shaping microbial community structure by influencing the availability of water and SOM in soils (Bossio et al. 1998; Wardle 1992; Wakelin et al. 2008). High silt and clay content positively correlate with SOM and microbial biomass; however, high clay content negatively impacts nitrogen mineralization by shielding organic nitrogen from microbial degradation (Strong et al. 1999). The physical organization of soil particles also exerts a strong influence on the growth and function of fungal hyphae. Highly compacted soils may have narrow sand pores, which prevent hyphae from extending throughout the soil matrix and limit hyphal diameter, especially of AM fungi (Drew et al. 2003; Wakelin et al. 2008). In disturbed soils aggregates are disrupted and as a result, fungi are unable to form extended hyphae (Wardle 1992). In green roofs, the growing substrate is often porous, which may be conducive to fungal growth. However, the porosity will also facilitate substrate drying, which may prevent certain species from establishing, and may select for taxa that can tolerate frequent wetting and drying.

In order to maintain local biodiversity in a green roof habitat, it would presumably be beneficial to select local substrates and their indigenous microbial communities as a component of the growing substrate. Developers of green roofs have been looking to use locally derived granular wastes as green roof starting materials (Oberndorfer et al. 2007), but substrates are often obtained from many sources, and each source harbors its own resident microbial population. When this mixture of foreign substrates and microbes is introduced into a new environment, the species present initially can prevent colonization by later species and change the overall community structure (Dickie et al. 2012). These effects, referred to as priority effects, can be deleterious to efforts of promoting local biodiversity if growing substrate is sourced from non-local materials.

7.4 Environmental and Regional Differences Affecting Rooftop Microbes

Urban green roofs are exposed to elevated levels of organic pollutants, which may be an additional selecting factor for microorganisms that can survive in these habitats. Heavy metals such as lead, arsenic, and copper and other organic contaminants are particularly of interest based on their prevalence and toxicity in urban atmosphere, soils, and groundwater (Clark et al. 2008; Srogi 2007). Microorganisms can tolerate these contaminants and high metal concentrations by utilizing a variety of physicochemical mechanisms to efficiently capture dissolved metal ions and convert metals from toxic to non-toxic forms. Other microbes can adapt to polluted urban areas by developing metal-resistance or utilizing contaminants as substrates through natural means of detoxification (Nikel et al. 2013; Hanif et al. 2010; Vullo et al. 2008). In a study of green roof fungal communities in New York City, the most abundant taxa were identified as fungi capable of degrading organic contaminants and tolerating heavy metal contamination such as *Pseudallescheria, Peyronellaea,* and *Thielavia* (McGuire et al. 2013).

Regional differences among green roof communities must also consider dispersal dynamics of microbes across the fragmented landscapes both within cities and across local urban to rural gradients, as local and regional wind patterns are likely to shape the community of fungi dispersing from green spaces on to green roofs and vice versa. Green roof ecosystems can be compared to island habitats residing within the 'ocean' of the urban environment. In actual island communities, two key processes that maintain species diversity over time are immigration and extinction. For green roof microbial communities, the ecological processes underpinning community assembly and the maintenance of diversity through time may be similar in some ways to island habitats, although the stress tolerance needed for immigrant propagules to survive and establish may cause higher extinction rates than would be observed in island ecosystems. The transport of propagules to green roof communities will be limited by the dispersal capabilities of individual taxa, as well as the distance a roof is from a source population of propagules. While it was historically thought that all microbes were everywhere (Baas-Becking 1934) we now know that dispersal limitation can occur for some microbial groups. For example, AM fungi are unlikely to be actively dispersed by animals across green roof habitats, unless they can be carried by birds and insects, as the dispersal of AM fungi is often accomplished by animals in non-engineered landscapes (Lekberg et al. 2011). Biogeographical structuring is also apparent for many microbial taxa, further supporting the notion that 'everything is not everywhere'. However, since dispersal is rarely assessed in microbial systems, the mechanisms of dispersal limitation versus environmental filtering cannot be disentangled without further studies that simultaneously address both processes.

7.5 Practical Applications and Future Research Priorities

Managing microbial inoculum to enhance green roof functionality can only be done with significantly more research on plant-microbial feedbacks in rooftop environments. However, additions of AM fungal inoculum have been standard practices in horticultural science, (Azcon-Aguilar and Barea 1997) and may also prove to be a useful management strategy for green roofs to maximize plant nutrient uptake, growth, and survival. The particular assemblages of AM fungi will need to be plant community specific, as the degree of benefit will likely vary with different plant-fungal combinations. Also, the particular mycorrhizal fungal taxa will need to be able to withstand the urban and rooftop environments. There are less than 500 species of AM fungi currently described, although total estimates are upwards of 1300 species (Kivlin et al. 2011; Opik et al. 2014). Considering that there are more than 300,000 species of described plants, and more than 75 % of them form AM associations (Wang and Qiu 2006), there are clearly many species of plants that share the same AM fungi. While most AM fungi are considered to be host-generalists, different combinations of AM fungi can have differential effects on plant performance (Helgason et al. 2002).

In addition to the practical considerations, green roofs can also be utilized to study basic ecological processes such as microbial community assembly and population dynamics. For instance, a microbe that is beneficial to the plant in one interaction can be detrimental to another host genotype. Additionally, plant-microbe mutualisms can evolve into parasitism in certain environmental contexts. With this in mind, practical benefits can be gained to ensure plant-microbe compatibility when selecting green roof vegetation.

Urban centers experience greater rates of deposition of heavy metals and other organic pollutants compared to non-urban areas (Chillrud et al. 1999). These potentially toxic compounds pose as a threat to human health since they can leach into local water sources from vehicles and streets or be inhaled. A major research priority is to determine if the various green roof-associated microbes are able to actively degrade organic contaminants and bioaccumulate heavy metals. Upon identifying various microbial strains possessing pollutant and metal detoxification capabilities, they can be inoculated in green roofs, which would serve as an appealing bioremediation effort in urban spaces.

7.6 Summary

Although microbes are essential to the functioning of green roofs as ecosystems, there is still much to understand about the drivers of microbial diversity and their spatial distribution throughout urban centers. Microbial interactions and their relationships to aboveground plant communities are inherently complex (Bonfante and Anca 2009). First, we must identify which plants both persist best on green roofs and provide high levels of desired functions (e.g., cooling, transpiration, habitat, appearance, etc.). Upon selecting types of plants, long-term persistence of these species on rooftop environments is intricately linked to how microbial dynamics contribute to their survival or failure to thrive (Table 7.1). Plant-associated microbes that enhance survival may be inoculated in establishing green roofs to increase plant longevity in the harsh conditions. These beneficial microbes can prevent colonization by pathogens, mediate host immunity, and help plants distinguish between mutualists versus pathogens. By utilizing a combination of culture-based and molecular techniques, microbes should be identified and studied to understand their interaction with green roof vegetation. Further research priorities should include determining to what extent green roof microbial communities are shaped by abiotic versus biotic factors over time since establishment, how microbial taxa can disperse via air and establish in green roof environments, and how local microsite conditions modify the novel microbial communities planted on green roofs. Ultimately, this information will be invaluable to the design of optimal green roof communities and will enhance sustainability efforts in urban environments.

References

Akita M, Lehtonen MT, Koponen H, Marttinen EM, Valkonen JPT (2011) Infection of the Sunagoke moss panels with fungal pathogens hampers sustainable greening in urban environments. Sci Total Environ 409(17):3166–3173

Auge RM (2001) Water relations, drought and vesicular-arbuscular mycorrhizal symbiosis. Mycorrhiza 11(1):3–42

Azcon-Aguilar C, Barea JM (1997) Applying mycorrhiza biotechnology to horticulture: significance and potentials. Sci Hortic 68(1–4):1–24

Baas-Becking LGM (1934) Geobiologie of Inleiding Tot de Milieukunde. Van Stockkum & Zoon, The Hague

Bardgett RD, Shine A (1999) Linkages between plant litter diversity, soil microbial biomass and ecosystem function in temperate grasslands. Soil Biol Biochem 31(2):317–321

Bell T, Newman JA, Silverman BW, Turner SL, Lilley AK (2005) The contribution of species richness and composition to bacterial services. Nature 436(7054):1157–1160

Berendsen RL, Pieterse CMJ, Bakker P (2012) The rhizosphere microbiome and plant health. Trends Plant Sci 17(8):478–486

Berthrong ST, Yeager CM, Gallegos-Graves L, Steven B, Eichorst SA, Jackson RB, Kuske CR (2014) Nitrogen fertilization has a stronger effect on soil nitrogen-fixing bacterial communities than elevated atmospheric CO2. Appl Environ Microbiol 80(10):3103–3112

Bertrand H, Plassard C, Pinochet X, Touraine B, Normand P, Cleyet-Marel JC (2000) Stimulation of the ionic transport system in *Brassica napus* by a plant growth-promoting rhizobacterium (*Achromobacter* sp.). Can J Microbiol 46(3):229–236

Bonfante P, Anca IA (2009) Plants, mycorrhizal fungi, and bacteria: a network of interactions. Annu Rev Microbiol 63:363–383 (Annual Reviews, Palo Alto)

Booth MS, Stark JM, Rastetter E (2005) Controls on nitrogen cycling in terrestrial ecosystems: a synthetic analysis of literature data. Ecol Monogr 75(2):139–157

Bossio DA, Scow KM, Gunapala N, Graham KJ (1998) Determinants of soil microbial communities: effects of agricultural management, season, and soil type on phospholipid fatty acid profiles. Microb Ecol 36(1):1–12

Bru D, Ramette A, Saby NPA, Dequiedt S, Ranjard L, Jolivet C, Arrouays D, Philippot L (2011) Determinants of the distribution of nitrogen-cycling microbial communities at the landscape scale. ISME J 5(3):532–542

Canfield DE, Glazer AN, Falkowski PG (2010) The evolution and future of Earth's nitrogen cycle. Science 330(6001):192–196

Chen CF (2013) Performance evaluation and development strategies for green roofs in Taiwan: a review. Ecol Eng 52:51–58

Chillrud SN, Bopp RF, Simpson HJ, Ross JM, Shuster EL, Chaky DA, Walsh DC, Choy CC, Tolley LR, Yarme A (1999) Twentieth century atmospheric metal fluxes into Central Park Lake, New York City. Environ Sci Technol 33(5):657–662

Clark SE, Steele KA, Spicher J, Siu CYS, Lalor MM, Pitt R, Kirby JT (2008) Roofing materials' contributions to storm-water runoff pollution. J Irrig Drain Eng-ASCE 134(5):638–645

Clay K, Schardl C (2002) Evolutionary origins and ecological consequences of endophyte symbiosis with grasses. Am Nat 160:S99–S127

Conrad R (1996) Soil microorganisms as controllers of atmospheric trace gases (H-2, CO, CH4, OCS, N2O, and NO). Microbiol Rev 60(4):609–640

Cookson WR, Marschner P, Clark IM, Milton N, Smirk MN, Murphy DV, Osman M, Stockdale EA, Hirsch PR (2006) The influence of season, agricultural management, and soil properties on gross nitrogen transformations and bacterial community structure. Aust J Soil Res 44(4):453–465

Curtis TP, Head IM, Lunn M, Woodcock S, Schloss PD, Sloan WT (2006) What is the extent of prokaryotic diversity? Philos Trans R Soc Lond B Biol Sci 361(1475):2023–2037

Davidson EA, Janssens IA (2006) Temperature sensitivity of soil carbon decomposition and feedbacks to climate change. Nature 440(7081):165–173

de Boer W, Folman LB, Summerbell RC, Boddy L (2005) Living in a fungal world: impact of fungi on soil bacterial niche development. FEMS Microbiol Rev 29(4):795–811

Degens BP, Harris JA (1997) Development of a physiological approach to measuring the catabolic diversity of soil microbial communities. Soil Biol Biochem 29(9–10):1309–1320

Dickie IA, Fukami T, Wilkie JP, Allen RB, Buchanan PK (2012) Do assembly history effects attenuate from species to ecosystem properties? A field test with wood-inhabiting fungi. Ecol Lett 15(2):133–141

Drew EA, Murray RS, Smith SE, Jakobsen I (2003) Beyond the rhizosphere: growth and function of arbuscular mycorrhizal external hyphae in sands of varying pore sizes. Plant Soil 251(1):105–114

Duchicela J, Vogelsang KM, Schultz PA, Kaonongbua W, Middleton EL, Bever JD (2012) Non-native plants and soil microbes: potential contributors to the consistent reduction in soil aggregate stability caused by the disturbance of North American grasslands. New Phytol 196(1):212–222

Fields S (2004) Global nitrogen—cycling out of control. Environ Health Perspect 112(10):A556–A563
Fierer N, Jackson RB (2006) The diversity and biogeography of soil bacterial communities. Proc Nat Acad Sci U S A 103(3):626–631
Fierer N, Grandy AS, Six J, Paul EA (2009) Searching for unifying principles in soil ecology. Soil Biol Biochem 41(11):2249–2256
Fravel DR (1988) Role of antibiosis in the biocontrol of plant diseases. Annu Rev Phytopathol 26:75–91
Friesen ML, Porter SS, Stark SC, von Wettberg EJ, Sachs JL, Martinez-Romero E (2011) Microbially mediated plant functional traits. In: Futuyma DJ, Shaffer HB, Simberloff D (eds) Annual Review of Ecology, Evolution, and Systematics. Annu Rev Ecol Evol Syst 42:23–46 (Annual Reviews, Palo Alto)
Gaffin SR, Khanbilvardi R, Rosenzweig C (2009) Development of a green roof environmental monitoring and meteorological network in New York City. Sensors 9(4):2647–2660
Garbaye J (1994) Helper bacteria—a new dimension to the mycorrhizal symbiosis. New Phytol 128(2):197–210
Gomez S, Ferrieri RA, Schueller M, Orians CM (2010) Methyl jasmonate elicits rapid changes in carbon and nitrogen dynamics in tomato. New Phytol 188(3):835–844
Gopal M, Gupta A, Thomas GV (2013) Bespoke microbiome therapy to manage plant diseases. Front Microbiol 4:355
Gregoire BG, Clausen JC (2011) Effect of a modular extensive green roof on stormwater runoff and water quality. Ecol Eng 37(6):963–969
Hanif MA, Bhatti HN, Ali MA, Asgher M, Bhatti IA (2010) Heavy metals tolerance and biosorption potential of white rot fungi. Asian J Chem 22(1):335–345
Hawksworth DL (2001) The magnitude of fungal diversity: the 1.5 million species estimate revisited. Mycol Res 105(12):1422–1432
Helgason T, Merryweather JW, Denison J, Wilson P, Young JPW, Fitter AH (2002) Selectivity and functional diversity in arbuscular mycorrhizas of co-occurring fungi and plants from a temperate deciduous woodland. J Ecol 90(2):371–384
Hobbs RJ, Arico S, Aronson J, Baron JS, Bridgewater P, Cramer VA, Epstein PR, Ewel JJ, Klink CA, Lugo AE, Norton D, Ojima D, Richardson DM, Sanderson EW, Valladares F, Vila M, Zamora R, Zobel M (2006) Novel ecosystems: theoretical and management aspects of the new ecological world order. Glob Ecol Biogeogr 15(1):1–7
Ji SH, Gururani MA, Chun SC (2014) Isolation and characterization of plant growth promoting endophytic diazotrophic bacteria from Korean rice cultivars. Microbiol Res 169(1):83–98
John J, Lundholm J, Kernaghan G (2014) Colonization of green roof plants by mycorrhizal and root endophytic fungi. Ecol Eng 71:651–659
Jumpponen A, Jones KL (2010) Seasonally dynamic fungal communities in the Quercus macrocarpa phyllosphere differ between urban and nonurban environments. New Phytol 186(2):496–513
Kiers ET, Palmer TM, Ives AR, Bruno JF, Bronstein JL (2010) Mutualisms in a changing world: an evolutionary perspective. Ecol Lett 13(12):1459–1474
Kivlin SN, Hawkes CV, Treseder KK (2011) Global diversity and distribution of arbuscular mycorrhizal fungi. Soil Biol Biochem 43(11):2294–2303
Klironomos JN (2002) Feedback with soil biota contributes to plant rarity and invasiveness in communities. Nature 417(6884):67–70
Kraiser T, Gras DE, Gutierrez AG, Gonzalez B, Gutierrez RA (2011) A holistic view of nitrogen acquisition in plants. J Exp Bot 62(4):1455–1466
Kremen C (2005) Managing ecosystem services: what do we need to know about their ecology? Ecol Lett 8(5):468–479
Kristin A, Miranda H (2013) The root microbiota-a fingerprint in the soil? Plant Soil 370(1–2):671–686
Lauber CL, Hamady M, Knight R, Fierer N (2009) Pyrosequencing-based assessment of soil pH as a predictor of soil bacterial community structure at the continental scale. Appl Environ Microbiol 75(15):5111–5120

Lekberg Y, Meadow J, Rohr JR, Redecker D, Zabinski CA (2011) Importance of dispersal and thermal environment for mycorrhizal communities: lessons from Yellowstone National Park. Ecology 92(6):1292–1302

Liu JW, Lovisolo C, Schubert A, Cardinale F (2013) Signaling role of Strigolactones at the interface between plants, (micro)organisms, and a changing environment. J of Plant Int 8(1):17–33

Lundholm J, MacIvor JS, MacDougall Z, Ranalli M (2010) Plant species and functional group combinations affect green roof ecosystem functions. PLoS One 5(3):e9677

Manzoni S, Schimel JP, Porporato A (2012) Responses of soil microbial communities to water stress: results from a meta-analysis. Ecology 93(4):930–938

Matsumura E, Fukuda K (2013) A comparison of fungal endophytic community diversity in tree leaves of rural and urban temperate forests of Kanto district, eastern Japan. Fungal Biol 117(3):191–201

McGuire KL, Payne SG, Palmer MI, Gillikin CM, Keefe D, Kim SJ, Gedallovich SM, Discenza J, Rangamannar R, Koshner JA, Massmann AL, Orazi G, Essene A, Leff JW, Fierer N (2013) Digging the New York City skyline: soil fungal communities in green roofs and city parks. PLoS One 8(3):e58020

Mishra PK, Bisht SC, Jeevanandan K, Kumar S, Bisht JK, Bhatt JC (2014) Synergistic effect of inoculating plant growth-promoting *Pseudomonas* spp. and *Rhizobium leguminosarum*-FB1 on growth and nutrient uptake of rajmash (*Phaseolus vulgaris* L.). Arch Agron Soil Sci 60(6):799–815

Nadeem SM, Ahmad M, Zahir ZA, Javaid A, Ashraf M (2014) The role of mycorrhizae and plant growth promoting rhizobacteria (PGPR) in improving crop productivity under stressful environments. Biotechnol Adv 32(2):429–448

Nikel PI, Silva-Rocha R, Benedetti I, de Lorenzo V (2013) The private life of environmental bacteria: pollutant biodegradation at the single cell level. Environ Microbiol 16(3):628–642

Oberndorfer E, Lundholm J, Bass B, Coffman RR, Doshi H, Dunnett N, Gaffin S, Kohler M, Liu KKY, Rowe B (2007) Green roofs as urban ecosystems: ecological structures, functions, and services. Bioscience 57(10):823–833

Opik M, Davison J, Moora M, Zobel M (2014) DNA-based detection and identification of *Glomeromycota*: the virtual taxonomy of environmental sequences. Bot-Bot 92(2):135–147

Pace NR (1997) A molecular view of microbial diversity and the biosphere. Science 276(5313):734–740

Philippot L, Raaijmakers JM, Lemanceau P, van der Putten WH (2013) Going back to the roots: the microbial ecology of the rhizosphere. Nat Rev Microbiol 11(11):789–799

Prabha C, Maheshwari DK, Bajpai VK (2013) Diverse role of fast growing rhizobia in growth promotion and enhancement of psoralen content in Psoralea corylifolia L. Pharmacogn Mag 9(36):57–65

Redecker D, Kodner R, Graham LE (2000) Glomalean fungi from the Ordovician. Science 289(5486):1920–1921

Rodriguez RJ, Henson J, Van Volkenburgh E, Hoy M, Wright L, Beckwith F, Kim YO, Redman RS (2008) Stress tolerance in plants via habitat-adapted symbiosis. ISME J 2(4):404–416

Rodriguez RJ, White JF, Arnold AE, Redman RS (2009) Fungal endophytes: diversity and functional roles. New Phytol 182(2):314–330

Rumble H, Gange AC (2013) Soil microarthropod community dynamics in extensive green roofs. Ecol Eng 57:197–204

Saikia SP, Goswami A, Mudoi KD (2014) Effect of 2, 4-D treatment and *Azospirillum* inoculation on growth of *Cymbopogon winterianus*. Afr J Microbiol Res 8:955–960

Sanchez-Pardo B, Zornoza P (2014) Mitigation of Cu stress by legume-Rhizobium symbiosis in white lupin and soybean plants. Ecotoxicol Environ Safe 102:1–5

Schnitzer SA, Klironomos JN, HilleRisLambers J, Kinkel LL, Reich PB, Xiao K, Rillig MC, Sikes BA, Callaway RM, Mangan SA, van Nes EH, Scheffer M (2011) Soil microbes drive the classic plant diversity-productivity pattern. Ecology 92(2):296–303

Schultz JC, Appel HM, Ferrieri AP, Arnold TM (2013) Flexible resource allocation during plant defense responses. Front Plant Sci 4:324

Schussler A, Schwarzott D, Walker C (2001) A new fungal phylum, the *Glomeromycota*: phylogeny and evolution. Mycol Res 105:1413–1421
Schwartz MW, Hoeksema JD (1998) Specialization and resource trade: biological markets as a model of mutualisms. Ecology 79(3):1029–1038
Shipton PJ (1977) Monoculture and soilborne plant pathogens. Annu Rev Phytopathol 15:387–407
Six J, Frey SD, Thiet RK, Batten KM (2006) Bacterial and fungal contributions to carbon sequestration in agroecosystems. Soil Sci Soc Am J 70(2):555–569
Smith JE, Read D (2008) Mycorrhizal symbiosis, 3rd edn. Academic Press, San Diego
Srogi K (2007) Monitoring of environmental exposure to polycyclic aromatic hydrocarbons: a review. Environ Chem Lett 5(4):169–195
Strong DT, Sale PWG, Helyar KR (1999) The influence of the soil matrix on nitrogen mineralisation and nitrification. IV. texture. Austr J Soil Res 37(2):329–344
Swift MJ, Heal OW, Anderson JM (1979) Decomposition in terrestrial ecosystems. Blackwell Scientific Publications, Oxford
Trivedi P, Anderson IC, Singh BK (2013) Microbial modulators of soil carbon storage: integrating genomic and metabolic knowledge for global prediction. Trends Microbiol 21(12):641–651
Turner TR, James EK, Poole PS (2013) The plant microbiome. Genome Biol 14(6):209
Van der Ent S, Van Hulten M, Pozo MJ, Czechowski T, Udvardi MK, Pieterse CMJ, Ton J (2009) Priming of plant innate immunity by rhizobacteria and beta-aminobutyric acid: differences and similarities in regulation. New Phytol 183(2):419–431
van der Heijden MGA, Klironomos JN, Ursic M, Moutoglis P, Streitwolf-Engel R, Boller T, Wiemken A, Sanders IR (1998) Mycorrhizal fungal diversity determines plant biodiversity, ecosystem variability and productivity. Nature 396(6706):69–72
van der Heijden MGA, Bardgett RD, van Straalen NM (2008) The unseen majority: soil microbes as drivers of plant diversity and productivity in terrestrial ecosystems. Ecol Lett 11(3):296–310
Vullo DL, Ceretti HM, Daniel MA, Ramirez SAM, Zalts A (2008) Cadmium, zinc and copper biosorption mediated by Pseudomonas veronii 2E. Bioresour Technol 99(13):5574–5581
Wakelin SA, Macdonald LM, Rogers SL, Gregg AL, Bolger TP, Baldock JA (2008) Habitat selective factors influencing the structural composition and functional capacity of microbial communities in agricultural soils. Soil Biol Biochem 40(3):803–813
Wang B, Qiu YL (2006) Phylogenetic distribution and evolution of mycorrhizas in land plants. Mycorrhiza 16(5):299–363
Wardle DA (1992) A comparative assessment of factors which influence microbial biomass carbon and nitrogen levels in soil. Biol Rev Camb Philos Soc 67(3):321–358
Wardle DA, Bardgett RD, Klironomos JN, Setala H (2004) Ecological linkages between aboveground and belowground biota. Science 304(5677):1629–1633
Waring BG, Averill C, Hawkes CV (2013) Differences in fungal and bacterial physiology alter soil carbon and nitrogen cycling: insights from meta-analysis and theoretical models. Ecol Lett 16(7):887–894
Weller DM, Raaijmakers JM, Gardener BBM, Thomashow LS (2002) Microbial populations responsible for specific soil suppressiveness to plant pathogens. Annu Rev Phytopathol 40:309–348
Xu HJ, Li S, Su JQ, Nie SA, Gibson V, Li H, Zhu YG (2014) Does urbanization shape bacterial community composition in urban park soils? A case study in 16 representative Chinese cities based on the pyrosequencing method. FEMS Microbiol Ecol 87(1):182–192
Yang J, Yu Q, Gong P (2008) Quantifying air pollution removal by green roofs in Chicago. Atmos Environ 42(31):7266–7273

Dr. Krista L. McGuire is an Assistant Professor of Biology at Barnard College, Columbia University and in the department of Ecology, Evolution and Environmental Biology at Columbia University. Her research focuses on feedbacks between soil microbes, plant communities, and global change.

Sara G. Payne received a B.A. in Biological Sciences from Barnard College. She is currently conducting blood stem cell research at Albert Einstein, and will later pursue a PhD at Duke University.

Giulia Orazi received a B.A. in Biology from Barnard College. She will pursue a PhD in Molecular and Cellular Biology at Dartmouth, with a focus in Microbiology and Immunology.

Dr. Matthew I. Palmer is a senior lecturer in the department of Ecology, Evolution, and Environmental Biology at Columbia University. His research includes work on community and ecosystem ecology in cities.

Chapter 8
Plant Biodiversity on Green Roofs

Susan C. Cook-Patton

Abstract Experimentation in grasslands and other ecosystems suggest that diverse plant communities grow more vigorously than simple communities, support a more robust animal community, and better resist stressors like disease, herbivory, and invasion. Despite the potential advantages of plant diversity on green roofs, many green roof communities consist of a few hardy species that are known to tolerate the harsh conditions on green roofs. Moreover, experimental tests of diversity on green roofs are infrequent. I therefore review the ecological literature in the context of green roof design to suggest ways to increase plant diversity on green roofs and hypothesize how increasing diversity might improve green roof function. Although it is unlikely that the complex, ecological dynamics of natural ecosystems will map directly onto the simplified, highly engineered ecosystem of a green roof, I argue that the lessons learned from decades of ecological experimentation can be adapted to green roof design to improve long-term plant performance and enhance the services provided by green roofs to urban communities. Ultimately, diversity experiments on green roofs will be required to prove whether similar ecological dynamics can exist in natural ecosystems and on rooftops, and whether or not the parallels I draw are justified. Therefore I end this chapter with a research agenda for the future, suggesting experiments that would greatly enhance our understanding of green roofs as ecosystems.

Keywords Genotypic richness · Species richness · Functional group diversity · Functional trait diversity · Phylogenetic diversity · Habitat · Ecosystem stability · Plant-animal interactions

S. C. Cook-Patton (✉)
Forest Ecology Lab, Smithsonian Environmental Research Center,
Edgewater, MD 21037, USA
e-mail: cookpatton@gmail.com

R. Sutton (ed.), *Green Roof Ecosystems,* Ecological Studies 223,
DOI 10.1007/978-3-319-14983-7_8

8.1 Introduction

Plants on green roofs must tolerate harsh environmental stressors such as temperature fluctuations, drought, and high winds (Boivin et al. 2001; Dunnett and Kingsbury 2004; Snodgrass and Snodgrass 2006). As a result, there has been a strong focus on finding individual plant species that can endure green roof conditions (Monterusso et al. 2005; Durhman et al. 2007; Oberndorfer et al. 2007) and the majority of green roofs have a limited palette of hardy, drought-tolerant *Sedum* species (Dunnett and Kingsbury 2004). However, decades of ecological research show that the performance of plants in diverse communities is often superior to their performance in isolation (Hooper et al. 2005; Cardinale et al. 2011). Diverse communities may also provide greater ecosystem services than simplified communities (Hooper et al. 2005; Cardinale et al. 2011). This has two important implications for green roof design. First, there may be species suitable for green roofs, which have been prematurely excluded because they were tested in isolation. Second, employing a diverse palette of species may enhance the functioning of a green roof ecosystem.

Although diversity is often cited as a valuable attribute for green roof design (e.g., Dunnett and Kingsbury 2004; Snodgrass and Snodgrass 2006; Dvorak and Volder 2010), empirical work backing these claims is almost entirely absent. By the date of this book's publication, I could locate only nine peer-reviewed studies in either English or German that have experimentally manipulated diversity on green roofs (Kolb and Schwarz 1986; Dunnett et al. 2008; Lundholm et al. 2010, 2014; Nagase and Dunnett 2010; Butler and Orians 2011; MacIvor et al. 2011; Nagase et al. 2013; Heim and Lundholm 2014). Several of the authors behind these key studies are contributors to this book (Chaps. 9 and 10), so I will not dwell extensively on the specific results of their experiments. Instead, the goal of this chapter is to review the ecological literature in the context of green roof design to suggest how plant diversity might increase the sustainability and functionality of green roofs.

8.2 What is Diversity?

There are multiple ways to characterize diversity within a community (Fig. 8.1). The most frequently employed metric for quantifying diversity is richness, which simply counts the number of entities present. ***Genotypic richness*** refers to the number of distinct genotypes of a single species present in the community. On green roofs, one might employ only a single *Sedum* species, for example, but increase genotypic richness by employing seed sources or cuttings from multiple, distinct populations.

A step above genotypic richness is ***species richness***, which refers to the number of different species present in a community. Species richness is the most commonly manipulated level of diversity in ecological diversity experiments (Hooper et al. 2005). However, one might have two communities that are equally species-rich, but have vastly different compositions, such a *Sedum*-only community versus a

Fig. 8.1 If we consider the flowering species depicted in this photo (*Rudbeckia hirta, Leucanthemum vulgare, Asclepias incarnata*, and *Penstemon digitalis*), we might note that they possess different inflorescence shapes (functional trait diversity), belong to three plant families (phylogenetic diversity), or are both native and non-native (plant origin). We might consider this community diverse because there are four species present (species richness) or simplistic because all species are herbaceous forbs (functional group diversity). (Photo Susan Cook-Patton)

mixture of grasses, forbs, and *Sedum* species. Because *Sedums* do not optimize all the functions one might desire on a green roof (Wolf and Lundholm 2008; Dunnett et al. 2008), one might expect the grass/forb/succulent community to better enhance multi-functionality on a green roof. Thus, richness may be an overly simplistic metric that ignores potentially important variation about how plants interact with their biotic and abiotic environments.

A more detailed diversity metric is ***functional group diversity***, which clusters species by broad morphological or physiological characteristics (Hooper et al. 2005). For example, does the plant have a C3, C4 or CAM photosynthetic pathway? Is the species a grass or an herbaceous forb? The few experiments examining diversity on green roofs have primarily manipulated functional group diversity, employing a mix of grasses, forbs, and succulents (Dunnett et al. 2008; Lundholm et al. 2010; Nagase and Dunnett 2010; Heim and Lundholm 2014). However, just as species richness treats all species within a community as functionally equivalent, functional group diversity treats all species within a group as equivalent when in reality there may be important distinctions among them. Indeed, at least one ecological study showed that functional group classifications explained the data no better than random clustering of species (Wright et al. 2006).

A finer-resolution metric is ***functional trait diversity***, which focuses on how species vary along key trait axes (Lavorel and Garnier 2002). For example, a green roof designer who wants to optimize floral display on a publicly visible green roof might select species that flower at different times to extend flowering duration on the rooftop. However, because green roof communities are often designed to serve multiple functions simultaneously (aesthetics, evapotranspirative cooling, rainwater retention, habitat formation, etc.), it quickly becomes difficult to select species with the right mixture of traits. In addition, it can be difficult to even know *a priori* which traits are important, as well as time-intensive to measure those traits in candidate species. One potential solution is to instead maximize ***phylogenetic diversity***, or the amount of evolutionary distance between species, with the assumption that more distantly related species will have more divergent traits (Cadotte 2013). More distantly related species are expected to compete less and use the total resource pool more completely (Cavender-Bares and Wilczek 2003; Burns and Strauss 2011), and a small but growing number of ecological experiments suggest that phylogenetic diversity is the best predictor of ecosystem function (Cadotte et al. 2008, 2009; Cadotte 2013). Although phylogenetic diversity has not been manipulated explicitly on green roofs, Lundholm et al. (2010) observed higher biomass and water capture, and lower roof temperatures in a phylogenetically diverse mixture of tall forbs, grasses, and succulents than in a more simplified community.

Beyond species and trait variation, another axis to consider is ***plant origin***, or whether the plants are native to the same geographic area as the green roof. Native plants may better support local wildlife, replace vegetation destroyed by development, and be less likely to become invasive (McKinney 2002; Tallamy 2007; Burghardt et al. 2010; Cook-Patton and Agrawal 2014). Moreover the few ecological experiments that compare native and non-native diversity suggest that, compared to non-natives, native species are more likely to grow synergistically with neighboring species (Wilsey et al. 2009; Cook-Patton and Agrawal 2014). However, an important caveat is that green roofs are modern, human constructions. Although natural analogs to green roof habitats do exist (Lundholm and Richardson 2010), the native species in a region may not be best suited to green roof conditions and there may be situations where non-native, non-invasive species are more appropriate (Hitchmough 2011). An experimental test of eighteen native species and nine non-native *Sedum* species, for example, found that only four native species maintained coverage after three years compared to all of the *Sedum* species (Monterusso et al. 2005). However, because all of the non-natives were *Sedum* species this experiment confounded phylogenetic relationships with plant origin. A more precise test of native versus non-native diversity should compare similar pools of native and non-native species, by employing congeneric pairs for example (e.g., Agrawal et al. 2005).

8.3 Linking Biodiversity and Green Roof Function

8.3.1 Community Growth and Vigor

In prairie and grassland ecosystems, manipulations of multiple types of diversity (genotypic richness, species richness, functional group diversity, functional trait diversity and phylogenetic diversity) have all shown that diverse communities produce more biomass on average than communities with a single species or genotype (Hooper et al. 2005; Cardinale et al. 2011; Cook-Patton et al. 2011; Cadotte 2013). In natural settings, more vigorously growing plant communities capture and retain more soil nutrients (Tilman et al. 1996), support more abundant and diverse animal communities (Srivastava and Lawton 1998; Cook-Patton et al. 2011), reduce soil temperature (Spehn et al. 2000), and limit the ability of invasive species to become established (Levine 2000). These may all be valuable functions on green roofs, if the green roof is intended to extract nutrient pollution from rainwater, serve as habitat for urban wildlife, or provide rooftop cooling, or if designers hope to limit the intrusion of weed species. In addition, increased productivity may enhance other green functions such as rainwater retention after a storm event, evapotranspirative cooling, or rooftop insulation (reviewed in Cook-Patton and Baüerle 2012).

There are multiple potential mechanisms by which diverse communities might grow more vigorously than communities composed of a single species. Competition, for resources and other facets of niche space, is expected to be stronger within species than among species (Darwin 1859; Macarthur and Levins 1967; Carroll et al. 2011). As diversity increases, the density of a given species declines. This allows species to interact with other species that are presumably less intense competitors. Diluting the density of individual species is also expected to mitigate other negative density-dependent interactions that suppress plant growth, such as disease transmission and herbivory (Janzen 1970; Connell 1971; Tahvanainen and Root 1972; Keesing et al. 2006). These latter mechanisms we discuss in more detail in Sect. 8.4.

In addition to loss of negative, conspecific interactions, the community may also gain positive/facilitative interactions from neighboring species as diversity increases. Diverse neighbors, for example, may reduce abiotic stress due to drought or cold, thereby increasing plant performance. For example, Callaway et al. (2002) found that in high-altitude, alpine communities the neighboring species ameliorated wind sheer, extreme temperatures, and soil instability. These are also common stressors on green roofs (Dunnett and Kingsbury 2004; Snodgrass and Snodgrass 2006). Similarly, Mulder et al. (2001) found that drought-stressed, bryophyte communities were more productive in diverse communities relative to monocultures and suggested that complementary moss architecture improved water retention.

Despite a substantial ecological literature linking plant diversity to biomass production and other ecosystem functions, there have been almost no equivalent experiments on green roofs. Heim and Lundholm (2014) measured the growth of *Solidago bicolor* in green roof modules with and without moss and lichen neighbors. Although *Solidago bicolor* had similar growth when grown alone or in mixture, mosses and lichens reduced substrate temperatures, and the authors suggest that this could improve plant performance through time. Kolb and Schwarz (1986) also observed that increasing green roof diversity decreased rooftop temperatures and suggested that this was because diverse communities had taller and more complex vegetation, which formed air pockets and enhanced insulation. Dunnett et al. (2008) similarly observed that increasing structural complexity (but not diversity) reduced rainwater runoff from green roofs. Clearly we need more experiments testing whether diversity impacts both plant growth and biomass-related services on green roofs to determine whether the same ecological dynamics operate in natural ecosystems and rooftops.

It is possible that the shallow soil substrates found on green roofs will constrain the impact of diversity on plant biomass relative to what is seen in natural communities. Indeed, at least one ecological experiment shows that the effect of diversity increases with soil depth (Dimitrakopoulos and Schmid 2004). However, overly vigorous plant growth on green roof could be problematic due to wind sheer, load-bearing capacity, and fire risk (Snodgrass and Snodgrass 2006), so moderate increases in biomass may be more desirable. A promising first step would thus be to identify species combinations that maximize structural complexity rather than biomass *per se,* while minimizing negative interactions such as competition, herbivory, and disease.

8.3.2 Green Roofs as Animal Habitat

Plant diversity not only affects the performance of plant communities (Hooper et al. 2005; Cardinale et al. 2011), but also has cascading effects through the food web (Siemann et al. 1998; Haddad et al. 2009). Although the fauna in urban environments is generally simpler than in natural communities (McKinney 2002), green roofs are not depauperate of species. A wide variety of animals have been documented on green roofs, including soil invertebrates, bees, other insects, birds, and even some endangered species (Hauth and Liptan 2003; Gedge and Kadas 2004; Schrader and Boening 2006; Brenneisen 2006; Colla et al. 2009; Dvorak and Volder 2010; Fernandez-Canero and Gonzalez-Redondo 2010; MacIvor and Lundholm 2011a). Green roofs may thus be able to support urban animal biodiversity and help replace habitat lost to development.

Diversity at higher trophic levels generally increases as plant diversity increases (Murdoch et al. 1972; Haddad et al. 2009; Cook-Patton et al. 2011). This may occur because plant biomass increases with diversity, creating more resources and habitat space to support more animals. With more animals comes an increased likelihood of sampling rare animal species. Animal diversity can thus increase with plant diversity via changes in plant biomass and animal abundance ("more individuals hypothesis", Srivastava and Lawton 1998). However, this mechanism may be the least likely to occur on green roofs because of limited biomass production on rooftops.

Even if biomass and animal abundance do not differ between single-species and diverse plant assemblages, faunal diversity may still increase with plant diversity if different plant species attract a unique assemblage of animals. This is especially true for specialist animals, that visit only one or a few plant species ("resource specialization hypothesis," Hutchinson 1959). However, specialist species are generally expected to decline with increasing urbanization (Sorace and Gustin 2009; Gagne and Fahrig 2011) and thus may be unlikely on green roofs.

Yet, even animals with more generalized diets may be more common in diverse plant communities. Some generalists, for example, have higher performance in diverse plant communities, because of the beneficial effects of mixing different plants into their diet (DeMott 1998). Generalist pollinators may also occur more frequently in diverse plant assemblages, where floral displays are more stable either because there are more flowers overall (Ebeling et al. 2008) or because those flowers are available more consistently throughout the season due to asynchronously flowering species (Moeller 2004). In addition, other insects, especially those at higher trophic levels such as predators, require higher levels of structural and resource complexity and are more likely to persist in diverse communities (Root 1973; Cook-Patton et al. 2011). Thus, a diverse green roof plant community may be better poised than a standard *Sedum* green roof to support a rich, multi-trophic community, even if this animal community is composed of more generalist species than specialists.

However, there has been little work linking plant and animal diversity on green roofs. A few studies hypothesized that plant structural diversity was an important determinant of faunal diversity (Brenneisen 2003, 2006; Gedge and Kadas 2004). Madre and colleagues (2013) compared green roofs that ranged in complexity from moss/sedum mixtures to moss/sedum/herbaceous plant/woody shrub communities, and found that animal abundance and richness generally increased with increasing plant diversity. They also emphasized the importance of plant structural complexity. In general both the ecological and green roof literatures suggest that suites of species with complementary architecture, phenologies, and nutritional resources may best maximize the habitat potential of green roofs for urban biodiversity.

Not only does plant diversity impact animal diversity, but conversely animal diversity can impact plant diversity. Ecological studies show that the structure of the food web can modulate the effect of plant diversity on the plant community (e.g., Duffy et al. 2007; Parker et al. 2010; Schnitzer et al. 2011; Cook-Patton et al. 2014). For example, in the presence of a vertebrate herbivore Common Evening Primrose (*Oenothera biennis*) produced 200 % more seeds in genotypically diverse plots compared to monocultures, but only 59 % more seeds in diverse plots compared to monocultures in the absence of the herbivore (Parker et al. 2010). Schnitzer et al. (2011) similarly found that diverse plant communities only produced more biomass than monocultures in the presence of a soil microbial community; otherwise plant monocultures and mixtures did not differ. Thus, to maximize the beneficial effects of plant diversity on green roofs, it may be necessary to deliberately foster a multi-trophic green roof community.

8.4 Diversity and Stability

The plants on a green roof will not be able to support a diverse animal fauna or provide other services if their growth and survival are negatively impacted over time. In natural communities, multiple pests diminish plant performance, including disease, invasive weeds, and herbivory. While these threats are not often assessed on green roofs, they have the potential to similarly harm green roof plant communities, especially because green roof plants are already stressed by abiotic factors that may limit their ability to respond defensively. However, diversity may help to mitigate the negative impacts of pests on green roofs. Plant diversity may also help to preserve green roof function over longer time frames via compensatory dynamics. If one species is attacked by a pest or is unable to withstand abiotic conditions, there may be other plant species within a diverse community that are less impacted, and can maintain coverage and green roof function.

8.4.1 Diversity and Disease

To my knowledge, there have been no investigations of disease prevalence on green roofs, nor examinations of how plant diversity might modulate those effects. Yet, plant diseases are a common feature in natural landscapes (Campbell and Madden 1990). It is likely that they also impact green roof plant communities. The typical relationship between disease and diversity is a negative one, though it is important to note that it is possible for disease prevalence to increase with diversity if community members amplify the disease (Elton 1958; Keesing et al. 2006).

There are several ways that plant diversity might modify the relationship between plants and disease organisms (reviewed in Keesing et al. 2006). First, non-host plant species might prevent disease organisms from encountering hosts, because non-hosts physically disrupt connections between hosts or because the host is less common in that community. This suggests that in addition to considering diversity *per se*, green roof designers may also want to consider the physical distribution of plant species and avoid spatial clumping. Second, non-host plant species can reduce disease on green roofs by improving the vigor of host species in that community. As mentioned previously, plant performance usually increases with plant diversity due to reductions in plant-plant competition or increases in facilitative interactions (Hooper et al. 2005; Cardinale et al. 2011). A vigorously growing host will likely have more resources to either prevent infection in the first place or recover from the disease if acquired. Finally, increasing genotypic diversity in a species can dampen the negative effects of an outbreak. If genotypes vary in their susceptibility to a given pathogen (Power 1991), then increasing genotypic diversity reduces the likely that an entire species will be eliminated from a green roof community. Genetic variation also increases the capacity of the population to evolve in response to future disease threats.

8.4.2 Diversity and Weed Invasion

While plant diversity will likely not prevent the establishment of all weedy species on green roofs, it may limit the abundance and vigor of invaders that do arrive (Levine et al. 2004), and reduce the need for costly weeding and maintenance. However, to my knowledge there are also no published studies investigating the relationship between plant diversity and weed invasion on green roofs.

In natural ecosystems, the relationship between plant diversity and resistance to invasion appears to depend on geographic scale (Levine 2000). At large geographic scales, areas rich in native species diversity are actually more likely to be invaded, presumably because the environmental factors that support high native richness also support high invader abundance (Lonsdale 1999). Even so phylogenetic diversity may still help to limit invasion at these scales, because the invasives that do become established tend to be more distantly related to the native community than expected by chance (Strauss et al. 2006; Davies et al. 2010). This implies that the resident vegetation fills the available niche space and that only phylogenetically distant species with different ways of interacting with the environment can enter.

At spatial scales more similar to a rooftop, invasibility can decline with native diversity because a diverse native community is expected to grow more vigorously and use the available resource space more fully than a less diverse community (Levine 2000; Levine et al. 2004). Again this presumably leaves less niche space free for undesirable species to enter the community and become established.

8.4.3 Diversity and Herbivory

Diverse green roofs systems may also be more stable because they are less susceptible to herbivore outbreaks. To my knowledge there have been no peer-reviewed investigations of herbivory on green roofs, except a note that *Sedum spurium* produced low cover in one experiment because of a severe aphid infestation (MacIvor and Lundholm 2011b). However, herbivore outbreaks could seriously impair green roof functioning if additional damage to already stressed plants leads to poor growth or significant mortality.

Much of the theory underlying patterns of herbivory in diverse communities focuses on specialist herbivores. For example, the resource concentration hypothesis (Root 1973) predicts that specialist herbivore outbreaks will be more frequent in monocultures than diverse mixtures because high concentrations of their preferred resource allow specialist herbivore populations to grow to epidemic levels. Although most herbivorous insects in natural ecosystems are specialists rather than generalists (Bernays and Graham 1988), in urban environments generalist fauna appear to be more common (Bonier et al. 2007; Gagne and Fahrig 2011; Bates et al. 2011). Although many of these studies focus on birds or non-herbivorous insects, the general pattern suggests that the herbivores on green roofs will be primarily generalists.

Studies have shown that damage by generalist insects can also decline with increasing plant richness (Unsicker et al. 2006; McArt and Thaler 2013). Generalist herbivores can have difficulty switching among plants with different chemistry and thus consume less in diverse plant communities (McArt and Thaler 2013). Moreover, predators are often more common in diverse environments, leading to stronger top-down control of herbivorous insects (Root 1973; Siemann et al. 1998; Unsicker et al. 2006; Cook-Patton et al. 2011). Thus there exists a strong potential for diversity to improve green roof performance by limiting herbivory, but these dynamics remain untested.

8.4.4 *Fluctuations of Green Roof Conditions Through Time*

Species not only vary in how susceptible they are to specific pathogens or herbivorous insects, but also in how well they tolerate abiotic factors like aridity versus soil saturation, or high temperatures versus freezing conditions. Because biotic and abiotic conditions can fluctuate from year to year, increasing the diversity of a green roof community increases the likelihood that at least one species will be able to grow and maintain coverage despite environmental change. In the ecological diversity literature, this is termed the *insurance effect* (Yachi and Loreau 1999).

This phenomenon of fluctuating cover was well-illustrated by Köhler (2006), who followed plant diversity on an extensive green roof in Berlin for a 20-year period and recorded 110 species in total, with 10–15 species consistently present. He observed that dieback was common, but that in wet years in particular, annual plants would fill in the gaps. Thus, the high diversity on this rooftop allowed a dynamically changing, but consistently present green roof community. Similarly, Lundholm et al. (2014) found that increasing functional group diversity enhanced the consistency of plant coverage through time. During a wet year, increased growth of *Solidago bicolor* compensated for decreased growth of *Sedum acre*.

Finally, even when growing conditions are ideal and pests are infrequent, the growth of a species will naturally peak and wane over the growing season. By selecting a diversity of species that vary in their phenology, it is possible to select a suite of species that maintain more uniform coverage throughout the year. Consistency in cover will lead to consistency of green roof services.

8.5 Diversity on Green Roofs: A Research Agenda

With only nine published studies manipulating green roof diversity (Kolb and Schwarz 1986; Dunnett et al. 2008; Lundholm et al. 2010, 2014; Nagase and Dunnett 2010; Butler and Orians 2011; MacIvor et al. 2011; Nagase et al. 2013; Heim and Lundholm 2014), many interesting questions remain about how diverse communities function on green roofs and how diversity might be used to increase the functionality and sustainability of green roofs. The experiments that have been conducted hint at parallel processes in diverse green roof and natural communities, but the unique abiotic and biotic conditions on green roofs likely limit similarities to some degree. It will take much additional research to determine when and why we might expect similar versus different patterns to emerge. In Sect. 8.5.2, I list four potential research questions that if answered would immensely improve the state of our knowledge.

Diversity manipulations on green roofs also have the potential to elucidate dynamics in natural ecosystem. For example, most ecological diversity manipulations have been conducted in aquatic or mesic terrestrial ecosystems (Hooper et al. 2005;

Cardinale et al. 2011), and not in temperature- or water-stressed ecosystems like roof tops where the relative strength of abiotic versus biotic stressors likely shifts to the abiotic. As a result, much of the ecological research has focused on how diversity alters biotic interactions (i.e., competition, herbivory). However, increasing diversity may also ameliorate abiotic stressors and improve plant performance by increasing water retention or reducing wind sheer, for example (Mulder et al. 2001; Callaway et al. 2002; Rixen and Mulder 2005). Because extreme climatic events are expected to become more common in the future (IPCC 2014), the observed dynamics on green roofs may help scientists predict how changes in biodiversity will interact with climate change to impact ecosystem function in natural communities.

In addition, there is a growing interest in the ecological literature about how changes in food web structure alter biodiversity-ecosystem function relationship, although experimental work is still limited (e.g., Duffy et al. 2007; Parker et al. 2010; Schnitzer et al. 2011; Cook-Patton et al. 2014). Green roofs may be ideal systems in which to study ecological trophic dynamics, because the relative simplicity of the food web makes them a more tractable system than natural ecosystems.

8.5.1 Designing a Diversity Manipulation on a Green Roof

The designs of early ecological manipulations of diversity were criticized for including "hidden treatments", which confounded a researcher's ability to attribute results to differences in plant diversity versus to some other factor (Huston 1997). As a result, a general experimental design has emerged, where one ideally selects a type of diversity upon which to focus, chooses a species pool larger than the highest diversity manipulation, and includes monoculture plots.

There are multiple ways to characterize the diversity of a community (Sect. 8.1). Although richness is a very commonly employed metric in ecological experiments (Hooper et al. 2005), selecting species randomly to increase diversity is probably impractical on green roofs given the severe abiotic constraints, and high costs of establishing and maintaining a green roof community. Many species may be inappropriate, as MacIvor and colleagues (2011) observed when they added wetland plants to green roof mixtures and found that they diluted the benefits of increasing species diversity. A more effective approach to random species draws may be to focus on plant functionality (Butler and Orians 2009; Lundholm et al. 2010), or even better plant functional traits. One could optimize variation along different trait axes, depending on what green roof functions are most valued. Several studies, for example highlighted the importance of choosing species that had different architectural structures (Kolb and Schwarz 1986; Dunnett et al. 2008; Madre et al. 2013). If pollinator recruitment is important, then one might want to select a floral community with different flowering windows (Moeller 2004), whereas variation in plant chemistry might help limit herbivory on green roofs (Barbosa et al. 2009; McArt and Thaler 2013). It is also important to remember that if one wants to optimize multiple green roof functions and/or if the traits of different species are unknown, it may be easier to simply increase phylogenetic diversity rather than guess at optimal trait combinations (Cadotte 2013).

The second experimental consideration is to have a larger species pool than the highest diversity manipulation. This ensures that the highest diversity treatment does not contain the exact same species composition. When this happens, the experiment becomes a test of species composition rather than diversity *per se* and it becomes impossible to say that diversity affected green roof performance.

Lastly, it is very important to incorporate monocultures into the design to untangle the mechanisms underlying the observed patterns (Huston 1997; Loreau and Hector 2001; Hooper et al. 2005). Higher performance in diverse mixtures may be due to the increased probability of including a highly productive species (the *sampling effect*, Huston 1997) and if this occurs, then it would be better to plant that highly productive species in a monoculture rather than in a mixture. In contrast, higher performance in mixture may be due to niche partitioning or facilitation among species, in which case it is better to plant a diverse community than a monoculture. Without monocultures as a baseline measurement, it is difficult to distinguish among these mechanisms.

8.5.2 Research Questions for the Future

What type of diversity is important on green roofs? Should green roof designers plant natives or exotics, phylogenetically-diverse mixtures, and/or species with specific combinations of traits? If increasing functional trait diversity improves green roof function, then on which trait axes should green roof designers focus? Comparative studies of different types of diversity are relatively new additions to the ecological literature. They suggest that increasing genotypic diversity can have surprisingly large positive effects (Cook-Patton et al. 2011). Phylogenetic diversity may also predict function better than functional group or functional trait diversity, if the functional traits measured do not encompass the important variation (Cadotte et al. 2008, 2009). Also, increasing native diversity can improve ecosystem function more than non-native diversity (Cook-Patton et al. 2014). The ecological literature is too young to entirely inform green roof design, even if ecological dynamics were the same in these two very different habitats. However, early data suggest that the best way to optimize green roof performance is to employ diverse mixtures of phylogenetically-distinct native species, especially if those species were represented by different genotypes. Green roof experiments that simultaneously test assemblages that differ in the type of diversity would provide valuable information to both the green roof literature and the ecological literature.

Are there species that improve green roof function, but have been discarded for poor survivorship or poor function when they may actually be useful in a diverse green roof community? MacIvor and colleagues (2011) found that including the dryland grass *Danthonia spicata* maximized water capture on green roofs even though *D. spicata* captured very little water in monoculture. Butler and Orians (2011) found that *Sedum* species enhanced the growth of the other species in dry conditions, relative to how they performed when growing alone. Thus, the performance or contribution of individual species to green roof function when grown alone may undervalue their performance or contribution to a diverse green roof mixture. Ideally, candidate species should be tested alone and in combination to determine their suitability as green roof species.

Does animal diversity increase with plant diversity on green roofs, as it does in ecological communities? Is it possible to use native plant species to increase the habitat potential of a green roof? MacIvor and Lundholm (2011) compared insect richness and abundance on intensive green roofs to ground-level areas, and found a wide variety of insects in their surveys. They found no strong differences in richness and abundance. While this experimental approach asks whether green roof communities differ from other urban green spaces, a similar approach could be employed to examine green roof communities that differ in diversity. Ideally, plots that differ in plant diversity would be established randomly on a single rooftop. Comparing across rooftops increases the risk of some additional variable confounding the treatments. Plot sizes would also have to be large enough to reduce spillover effects from adjacent plots. Arthropods could then be sampled with established protocols like pitfall traps or sweep nets.

Do pests (herbivores or pathogens) negatively affect green roof performance and if so, can plant diversity help ameliorate these impacts? Given that little to nothing is known about how herbivores and pathogens impact plant performance on green roofs, a logical first experiment would thus be to apply pesticides (insecticides or fungicides, for example) to subplots on a green roof to determine whether disease agents are indeed reducing plant growth. This experiment could be crossed with a diversity manipulation to examine how interactions between diversity and disease impact the performance of green roof plants. Since pests have been ignored to date, its possible that they diminish plant vigor and that increasing diversity could improve overall green roof function.

8.5.3 Conclusions

Green roofs clearly differ from natural ecosystems in that they are highly manufactured, human constructs. The complex ecological dynamics found in natural ecosystems will likely not map directly onto green roofs. However, experimental manipulations of diversity in systems ranging from mosses to algae to forest trees generally find positive effects of increasing diversity (Mulder 2001; Cardinale et al. 2011; Cook-Patton et al. 2014), suggesting that green roofs too may share similar dynamics. Given the many potential benefits of green roofs for urban environments (Dunnett and Kingsbury 2004), it is well worth investing in research that could optimize the ability of green roofs to provide their numerous services.

References

Agrawal A, Kotanen P, Mitchell C et al (2005) Enemy release? An experiment with congeneric plant pairs and diverse above- and below ground enemies. Ecology 86:2979–2989
Barbosa P, Hines J, Kaplan I et al (2009) Associational resistance and associational susceptibility: having right and wrong neighbors. Annu Rev Ecol Evol Syst 40:1–20
Bates A, Sadler J, Fairbrass A et al (2011) Changing bee and hoverfly pollinator assemblages along an urban-rural gradient. PLoS One 6:e23459. doi:10.1371/journal.pone.0023459
Bernays E, Graham M (1988) On the evolution of host specificity in phytophagous arthropods. Ecology 69:886–892
Boivin M, Lamy M, Gosselin A et al (2001) Effect of artificial substrate depth on freezing injury of six herbaceous perennials grown in a green roof system. HortTechnol 11:409–412
Bonier F, Martin P, Wingfield J (2007) Urban birds have broader environmental tolerance. Biol Lett 3:670–673
Brenneisen S (2003) The benefits of biodiversity from green roofs—key design consequences. Greening Rooftops for Sustainable Communities Conference
Brenneisen S (2006) Space for urban wildlife: designing green roofs as habitats in Switzerland. Urban Habitats 4:27–36
Burghardt K, Tallamy D, Philips C et al (2010) Non-native plants reduce abundance richness and host specialization in lepidopteran communities. Ecosphere 1:1–22
Burns J, Strauss S (2011) More closely related species are more ecologically similar in an experimental test. Proc Natl Acad Sci U S A 108:5302–5307
Butler C, Orians C (2009) Sedum facilitates the growth of neighboring plants on a green roof under water limited conditions. Greening Rooftops for Sustainable Communities Conference
Butler C, Orians C (2011) Sedum cools soil and can improve neighboring plant performance during water deficit on a green roof. Ecol Eng 37:1796–1803
Cadotte M (2013) Experimental evidence that evolutionarily diverse assemblages result in higher productivity. Proc Natl Acad Sci U S A 110:8996–9000
Cadotte M, Cardinale B, Oakley T (2008) Evolutionary history and the effect of biodiversity on plant productivity. Proc Natl Acad Sci U S A 105:17012–17017
Cadotte M, Cavender-Bares J, Tilman D et al (2009) Using phylogenetic, functional, and trait diversity to understand patterns of plant community productivity. PLoS One 4:e5695. doi:10.1371/journal.pone.0005695
Callaway R, Brooker R, Choler P et al (2002) Positive interactions among alpine plants increase with stress. Nature 417:844–848
Campbell C, Madden L (1990) Introduction to plant disease epidemiology. Wiley, New York
Cardinale B, Matulich K, Hooper D et al (2011) The functional role of producer diversity in ecosystems. Am J Bot 98:572–592
Carroll I, Cardinale B, Nisbet R (2011) Niche and fitness differences relate the maintenance of diversity to ecosystem function. Ecology 92:1157–1165
Cavender-Bares J, Wilczek A (2003) Integrating micro- and macroevolutionary processes in community ecology. Ecology 84:592–597
Colla SE, Willis E, Packer L (2009) Can green roofs provide habitat for urban bees (Hymenoptera: Apidae)? Cities and the Environment 2:1–12
Connell J (1971) On the role of natural enemies in preventing competitive exclusion in some marine animals and in rain forest trees. In: Boer den PJ, Gradwell G (eds) Dynamics of populations. Center for Agricultural Publishing and Documentation Wageningen, Netherlands pp 298–312
Cook-Patton S, Agrawal A (2014) Exotic plants contribute positively to biodiversity functions but reduce native seed production and arthropod richness. Ecology 95:1642–1650
Cook-Patton S, Bauerle T (2012) Potential benefits of plant diversity on vegetated roofs: a literature review. J Environ Manage 106:85–92

Cook-Patton S, McArt S, Parachnowitsch A et al (2011) A direct comparison of the consequences of plant genotypic and species diversity on communities and ecosystem function. Ecology 92:915–923

Cook-Patton S, LaForgia M, Parker J (2014) Positive interactions between herbivores and plant diversity shape forest regeneration. Proc Roy Soc Lond B-Biol Sci 281:20140261. doi:10.1098/rspb.2014.0261

Darwin C (1859) On the origin of species. John Murray, London

Davies K, Cavender-Bares J, Deacon N (2010) Native communities determine the identity of exotic invaders even at scales at which communities are unsaturated. Divers Distrib 17:35–42

DeMott W (1998) Utilization of a cyanobacterium and a phosphorus-deficient green alga as complementary resources by daphnids. Ecology 79:2463–2481

Dimitrakopoulos P, Schmid B (2004) Biodiversity effects increase linearly with biotope space. Ecol Lett 7:574–583

Duffy J, Cardinale B, France K et al (2007) The functional role of biodiversity in ecosystems: incorporating trophic complexity. Ecol Lett 10:522–538

Dunnett N, Kingsbury N (2004) Planting green roofs and living walls. Timber Press, Portland

Dunnett N, Nagase A, Booth R et al (2008) Influence of vegetation composition on runoff in two simulated green roof experiments. Urban Ecosyst 11:385–398

Durhman A, Rowe D, Rugh C (2007) Effect of substrate depth on initial growth coverage and survival of 25 succulent green roof plant taxa. HortScience 42:588–595

Dvorak B, Volder A (2010) Green roof vegetation for North American ecoregions: a literature review. Landsc Urban Plan 96:197–213

Ebeling A, Klein A-M, Schumacher J et al (2008) How does plant richness affect pollinator richness and temporal stability of flower visits? Oikos 117:1808–1815

Elton C (1958) The ecology of invasions by animals and plants. University of Chicago Press, Chicago

Fernandez-Canero R, Gonzalez-Redondo P (2010) Green roofs as a habitat for birds: a review. J Anim Vet Adv 9:2041–2052

Gagne S, Fahrig L (2011) Do birds and beetles show similar responses to urbanization? Ecol Appl 21:2297–2312

Gedge D, Kadas G (2004) Bugs, bees, and spiders: green roof design for rare invertebrates. Greening Rooftops for Sustainable Communities

Haddad N, Crutsinger G, Gross K et al (2009) Plant species loss decreases arthropod diversity and shifts trophic structure. Ecol Lett 12:1029–1039

Hauth E, Liptan T (2003) Plant survival findings in the pacific northwest. Greening Rooftops for Sustainable Communities

Heim A, Lundholm J (2014). Species interactions in green roof vegetation suggest complementary planting mixtures. Landsc Urban Plan 130:125–133

Hitchmough J (2011) Exotic plants and plantings in the sustainable designed urban landscape. Landsc Urban Plan 100:380–382

Hooper D, Chapin F, Ewel J et al (2005) Effects of biodiversity on ecosystem functioning: a consensus of current knowledge. Ecol Monogr 75:3–35

Huston M (1997) Hidden treatments in ecological experiments: re-evaluating the ecosystem function of biodiversity. Oecologia 110:449–460

Hutchinson G (1959) Homage to Santa-Rosalia or why are there so many kinds of animals. Am Nat 93:145–159

IPCC (2014) Climate change 2014: impacts adaptation and vulnerability. Intergovernmental Panel on Climate Change

Janzen D (1970) Herbivores and the number of tree species in tropical forests. Am Nat 104:501–528

Keesing F, Holt R, Ostfeld R (2006) Effects of species diversity on disease risk. Ecol Lett 9:485–498

Köhler M (2006) Long-term vegetation research on two extensive green roofs in Berlin. Urban Habitats 4:3–25

Kolb W, Schwarz T (1986) Zum Klimatisierungseffekt von Pflanzbeständen auf Dächern Teil I. Zeitschrift für Vegetationstechnick 9. Republished in Veit- shoechheimer Berichte Heft 39 Dachbegrünung. (cf. Dunnett and Kingsbury 2004)
Lavorel S, Garnier E (2002) Predicting changes in community composition and ecosystem functioning from plant traits: revisiting the Holy Grail. Funct Ecol 16:545–556
Levine J (2000) Species diversity and biological invasions: Relating local process to community pattern. Science 288:852–854
Levine J, Adler P, Yelenik S (2004) A meta-analysis of biotic resistance to exotic plant invasions. Ecol Lett 7:975–989
Lonsdale W (1999) Global patterns of plant invasions and the concept of invasibility. Ecology 80:1522–1536
Loreau M, Hector A (2001) Partitioning selection and complementarity in biodiversity experiments. Nature 412:72–76
Lundholm J, Richardson P (2010) Habitat analogues for reconciliation ecology in urban and industrial environments. J Appl Ecol 47:966–975
Lundholm J, MacIvor J, MacDougall Z et al (2010) Plant species and functional group combinations affect green roof ecosystem functions. PLoS One 5:e9677. doi:10.1371/journal.pone.0009677
Lundholm J, Heim A, Tran S et al (2014) Leaf and life history traits predict plant growth in a green roof ecosystem. PLoS One 9:e101395. doi:10.1371/journal.pone.0101395
Macarthur R, Levins R (1967) The limiting similarity convergence and divergence of coexisting species. Am Nat 101:377–385
MacIvor J, Lundholm J (2011a) Insect species composition and diversity on intensive green roofs and adjacent level-ground habitats. Urban Ecosyst 14:225–241
MacIvor J, Lundholm J (2011b) Performance evaluation of native plants suited to extensive green roof conditions in a maritime climate. Ecol Eng 37:407–417
MacIvor J, Ranalli M, Lundholm J (2011) Performance of dryland and wetland plant species on extensive green roofs. Ann Bot-London 107:671–679
Madre F, Vergnes A, Machon N et al (2013) A comparison of 3 types of green roof as habitats for arthropods. Ecol Eng 57:109–117
McArt S, Thaler J (2013) Plant genotypic diversity reduces the rate of consumer resource utilization. Proc Roy Soc B-Biol Sci 280:20130639. doi:10.1098/rspb.2013.0639
McKinney M (2002) Urbanization biodiversity and conservation. BioScience 52:883–890
Moeller D (2004) Facilitative interactions among plants via shared pollinators. Ecology 85:3289–3301
Monterusso M, Rowe D, Rugh C (2005) Establishment and persistence of Sedum spp. and native taxa for green roof applications. HortScience 40:391–396
Mulder C, Uliassi D, Doak D (2001) Physical stress and diversity-productivity relationships: the role of positive interactions. Proc Natl Acad Sci U S A 98:6704–6708
Murdoch W, Evans F, Peterson C (1972) Diversity and pattern in plants and insects. Ecology 53:819–829
Nagase A, Dunnett N (2010) Drought tolerance in different vegetation types for extensive green roofs: effects of watering and diversity. Landsc Urban Plan 97:318–327
Nagase A, Dunnett N, Choi M-S (2013) Investigation of weed phenology in an establishing semi-extensive green roof. Ecol Eng 58:156–164
Oberndorfer E, Lundholm J, Bass B et al (2007) Green roofs as urban ecosystems: ecological structures functions and services. BioScience 57:823–833
Parker J, Salminen J-P, Agrawal A (2010) Herbivory enhances positive effects of plant genotypic diversity. Ecol Lett 13:553–563
Power A (1991) Virus spread and vector dynamics in genetically diverse plant populations. Ecology 72:232–241
Rixen C, Mulder C (2005) Improved water retention links high species richness with increased productivity in arctic tundra moss communities. Oecologia 146:287–299

Root R (1973) Organization of a plant-arthropod association in simple and diverse habitats: the fauna of collards (*Brassica oleracea*). Ecol Monogr 43:95–124

Schnitzer S, Klironomos J, HilleRisLambers J et al (2011) Soil microbes drive the classic plant diversity-productivity pattern. Ecology 92:296–303

Schrader S, Boening M (2006) Soil formation on green roofs and its contribution to urban biodiversity with emphasis on Collembolans. Pedobiologia 50:347–356

Siemann E, Tilman D, Haarstad J et al (1998) Experimental tests of the dependence of arthropod diversity on plant diversity. Am Nat 152:738–750

Snodgrass E, Snodgrass L (2006) Green roof plants: a resource and planting guide. Timber Press, Portland

Sorace A, Gustin M (2009) Distribution of generalist and specialist predators along urban gradients. Landsc Urban Plan 90:111–118

Spehn E, Joshi J, Schmid B et al (2000) Above-ground resource use increases with plant species richness in experimental grassland ecosystems. Funct Ecol 14:326–337

Srivastava D, Lawton J (1998) Why more productive sites have more species: an experimental test of theory using tree-hole communities. Am Nat 152:510–529

Strauss S, Webb C, Salamin N (2006) Exotic taxa less related to native species are more invasive. Proc Natl Acad Sci U S A 103:5841–5845

Tahvanainen J, Root R (1972) The influence of vegetational diversity on the population ecology of a specialized herbivore *Phyllotreta cruciferae* (*Coleoptera: Chrysomelidae*). Oecologia 10:321–346

Tallamy D (2007) Bringing nature home. Timber Press, Portland

Tilman D, Wedin D, Knops J (1996) Productivity and sustainability influenced by biodiversity in grassland ecosystems. Nature 379:718–720

Unsicker S, Baer N, Kahmen A et al (2006) Invertebrate herbivory along a gradient of plant species diversity in extensively managed grasslands. Oecologia 150:233–246

Wilsey B, Teaschner T, Daneshgar TP et al (2009) Biodiversity maintenance mechanisms differ between native and novel exotic-dominated communities. Ecol Lett 12:432–442

Wolf D, Lundholm J (2008) Water uptake in green roof microcosms: effects of plant species and water availability. Ecol Eng 33:179–186

Wright J, Naeem S, Hector A et al (2006). Conventional functional classification schemes underestimate the relationship with ecosystem functioning. Ecol Lett 9:111–120

Yachi S, Loreau M (1999) Biodiversity and ecosystem productivity in a fluctuating environment: the insurance hypothesis. Proc Natl Acad Sci U S A 96:1463–1468

Dr. Susan C. Cook-Patton studies biodiversity-ecosystem function relationships. She is interested in using ecological knowledge to foster healthy forest and urban environments. Formerly a Smithsonian Fellow, she is now an AAAS Science & Technology Policy Fellow at the USDA Forest Service.

Chapter 9
Effects of Vegetation on Green Roof Ecosystem Services

Jeremy T. Lundholm and Nicholas S. G. Williams

Abstract The ecosystem services green roofs provide are influenced by both the engineered and biotic components of green roof systems. This chapter focuses on how the functioning of green roofs is controlled by plant species and the synthetic vegetation communities created by them. Plant species can differ greatly in their ability to provide services such as roof cooling and stormwater retention. Newer work, emphasizing less-well-characterized benefits such as reduction of heat loss in winter, air pollution mitigation and carbon sequestration (Chap. 2), also shows significant effects of plant species. The species that best perform a particular service differ between services; other research shows performance advantages in combining species or functional groups of plants into communities. Optimizing green roof benefits thus requires close attention to plant properties, and even superficially similar plant groups (e.g. succulents) can show large performance differences among species. Characterizing green roof vegetation by plant traits, such as leaf area, leaf thickness and photosynthetic pathway, could be a useful way to select green roof species, allowing rapid screening of regional floras for potential species. Plant traits are often directly linked to ecosystem processes that provide economically and environmentally valuable services. Consequently a trait-based approach can help elucidate the relationships among the performance of individual species, the role of plant diversity and the ecosystem services provided by green roofs. This should allow the design of purpose-specific green roofs that provide higher levels of ecosystem services.

Keywords Ecosystem functioning · Plant communities · Stormwater capture · Urban heat island · Plant traits · Water quality · Environmental psychology · Pollution mitigation

J. T. Lundholm (✉)
Department of Biology, Saint Mary's University, 923 Robie St., B3H 3C3 Halifax, NS, Canada
e-mail: Jeremy.Lundholm@SMU.CA

N. S. G. Williams
School of Ecosystem and Forest Sciences, The University of Melbourne, 500 Yarra Blvd, 3121 Richmond, VIC, Australia

R. Sutton (ed.), *Green Roof Ecosystems*, Ecological Studies 223,
DOI 10.1007/978-3-319-14983-7_9

9.1 Introduction

Green roofs are constructed primarily because they contribute valuable ecosystem services to humans and the urban environment. These services derive from living and non-living components of the ecosystem and are well described in the literature (Getter and Rowe 2006; Oberndorfer et al. 2007). The living component of green roofs consists of plants and associated soil organisms and animals that colonize or use plants and substrates. Plants drive many of the ecosystem services provided by green roofs including aesthetic appeal, moderation of heat fluxes, surface temperature reductions, stormwater retention and provision of habitat.

Green roof vegetation research has emphasized plant selection, largely to ensure plant survival and growth in the artificial rooftop environment. Plant survival and coverage is important for ecosystem service provisioning as dead plants make little or no contributions to ecosystem services (Speak et al. 2013b). A number of studies have used plant functional traits to help select species. Plant functional traits are the morphological, anatomical, physiological or phenological features that reflect species' ecological strategies, determine how plants respond to their environment and disturbances and influence ecosystem processes and other trophic levels (Pérez-Harguindeguy et al. 2013). Because different traits represent fundamental ecological tradeoffs, facilitate a mechanistic understanding and allow generalization of ecological knowledge across species and floras, they are an increasingly popular methodology for ecological research.

Van Mechelen et al. (2014) used plant traits to identify potential green roof species from the Mediterranean vegetation communities of southern France. They identified traits that would confer drought tolerance and regeneration capacity to species, helping them persist on green roofs. These included Grime's CSR strategies of ruderality or stress tolerance, evergreen leaf phenology, needle-like leaf shape, annual or perennial plant longevity, succulence, facultative CAM photosynthetic pathway, shallow rooting depth, plant height less than one meter, chamaephyte, geophyte or therophyte lifeform. Traits that confer survival may vary with climatic region. For example high leaf succulence, measured as water content divided by leaf area, increases survival on green roofs in hot dry climates (Farrell et al. 2012b) but may decrease survival in areas subjected to freezing conditions because the high water content increases leaf susceptibility to freezing and tissue death.

Recent research investigating green roof vegetation has expanded to examine the differential provisioning of ecosystem services by different vegetation types or plant species. Differential effects of plant types on ecosystem processes derive from the physiological, anatomical and morphological traits of plants that affect the flow of energy and materials through the green roof system. These traits can be used to predict ecosystem service provisioning (Lavorel et al. 2011), and there is hope that green roof designers could use plant traits to select species to optimize certain functions without laboriously testing each species separately. While much green roof research uses plant life-form groupings such as succulents, grasses or shrubs to differentiate plant types, evidence exists for substantial functional differences within a life-form grouping (e.g. Reich et al. 2003; Lundholm et al. 2010), thus specific traits of plant species may be more valuable than broad morphological or life-history groupings in predicting their contribution to ecosystem services. In addition, since the species available for green roofing differ across regions and continents, a trait-based approach transcends species identity in that different species with similar traits may provide the same ecosystem services.

In this chapter, we summarize the empirical literature describing differential performance of green roof ecosystem services by different types of plants or vegetation. While little work has been conducted to evaluate plant traits as predictors of green roof ecosystem services, we use ecological theory and literature from other systems to suggest traits that are likely to be important in driving ecosystem processes and resulting services on green roofs.

9.2 Thermal Benefits

Green roofs are promoted as providing two kinds of thermal benefits: energy savings to the building supporting the green roof and reductions in the urban heat island (Oberndorfer et al. 2007). Green roofs reduce building energy consumption when outside temperatures exceed those inside the building (1) by evapotranspirative cooling (Del Barrio 1998; Bass and Baskaran 2003), (2) increased albedo compared with conventional roof surfaces (Eumorfopoulou and Aravantinos 1998) and (3) insulation provided by the growing medium (Sailor 2008). Reductions to a building's contribution to the urban heat island effect result from decreased surface temperatures (Bass and Baskaran 2003) via evapotranspiration and albedo. When outdoor temperatures are colder than those inside, green roofs can also provide energy savings (Niachou et al. 2001; Liu and Baskaran 2005; Getter et al. 2011) although the mechanisms for thermal benefits in cold climates are not as well investigated.

The engineered components of green roofs make large contributions to these thermal benefits. Sailor (2008) showed significant effects of different growing media on thermal conductivity, heat capacity and albedo, all of which could influence thermal performance in both hot and cold conditions. Increased substrate depth also decreases heat gain during hot conditions (Sailor 2008; Permpituck and Namprakai 2012).

For roof cooling, the role of vegetation in optimizing green roof thermal performance relies on maximizing transpiration from the plant canopy, providing shade or increasing the roof's albedo. All of these can be directly affected by vegetation types and plant species (Lundholm et al. 2010). Recent studies suggest that increasing the reflectivity (albedo) of a building surface makes a larger impact on net global cooling than evaporative or transpirative cooling (Sproul et al. 2014) so different mechanisms of cooling may differ in their overall environmental benefits.

Plants can also indirectly affect substrate thermal properties such as heat capacity, thermal conductivity and albedo by altering soil moisture (Sailor et al. 2008) but this has largely not been studied empirically. Key plant traits that will drive cooling benefits include (1) specific leaf area (SLA) as plants with high SLA generally have faster gas exchange rates and hence greater transpirative cooling potential; (2) overall leaf area, plant height and growth form, which will interact to determine shading;

and (3) leaf hairs and waxes which will determine leaf color and hence leaf albedo. Species successful in the sunny and dry conditions typical of green roofs are likely to have relatively low SLA (Ackerly et al. 2002) but may vary in the other traits that determine thermal performance. Models tend to use vegetation height and leaf area index (LAI) as indicators of shading (due to absorption and reflection of radiation by plant leaves) and transpiration (Theodosiou 2003) but overall coverage, direct measures of albedo and stomatal conductance are also incorporated in some models (Sailor 2008).

The empirical studies examining thermal performance in hot conditions generally show large effects of plant type, indicating substantial potential to optimize thermal performance (Table 9.1; Chap. 3). Studies examined various indicators of thermal performance, with substrate surface temperatures relevant to building energy savings and urban heat island mitigation. Leaf surface temperatures integrate albedo and transpiration rates, and lower leaf surface temperatures lead to reduced heat transfer into the building and also lower air temperatures (Liu et al. 2012; Blanusa et al. 2013; Table 9.1). Even succulent species that are often considered functionally equivalent showed a 9.7 % difference between best and worst performing species in substrate surface temperatures relative to control roofs (Dvorak and Volder 2013; Table 9.1). Studies examining a greater range of vegetation types (Lundholm et al. 2010; MacIvor and Lundholm 2011; MacIvor et al. 2011) found 14–24 % differences in substrate temperature between vegetation types, with some evidence that this differentiation increased over time as vegetation cover increased. Direct heat flux measurements through green roofs with different vegetation are more directly relevant to calculating energy savings for buildings, and results show up to a 325 % increase from the lowest to highest performing vegetation type (Spolek 2008). There is a need to empirically compare heat flux and other indicators of thermal performance on green roofs planted with different species, but current studies show modest variation between vegetation types that should have an impact on building energy and urban heat island mitigation. Additionally, given that modeling studies routinely make assumptions about key vegetation parameters such as LAI or stomatal conductance, empirical characterization of the differences in plant species used on green roofs needs to extend to these variables.

In cold conditions, increased insulation due to the growing medium is the main mechanism proposed for energy savings with green roofs (Niachou et al. 2001; Sailor 2008), but plants can affect winter performance as well via LAI. In contrast to summer conditions, greater LAI or transpiration (if plants are physiologically active in the cold season) reduces solar heat gain by the building resulting in lower energy savings (Sailor et al. 2012). Snow coverage could also be affected by plants on green roofs, leading to lower temperature fluctuations (Teemusk and Mander 2007). Only one study to date has examined differential effects on winter performance of plant species (Lundholm et al. 2014a) and it found substantial differences in maximum and minimum substrate temperatures and in snow accumulation, all of which can affect building energy savings (Table 9.1). Vegetation with greater dead biomass in winter tended to trap more snow and lead to more moderate temperatures (lower maxima and higher minima).

Table 9.1 Empirical studies demonstrating differential performance of ecosystem functions and services by different plant types

Region/Climate	Substrate depth	Plant types tested (comparison)	Ecosys. functions	Effect size min-max (% diff.)	Reference
SW NA/humid subtropical	71 mm	Succulents (species)	Leachate Nitrate (mg/L) Ammonium (mg/L) Dissolved Organic C (mg/L) Dissolved Organic N (mg/L) Orthophosphate P (mg/L) Bicarbonate (mg/L) Magnesium (mg/L) Calcium (mg/L) Potassium (mg/L) Sodium (mg/L)	0.5–6.6 (1120%) ns (0%) ns (0%) ns (0%) ns (0%) ns (0%) 4.5–7.5 (66.6%) ns (0%) 2.5–8 (220%) ns (0%)	Aitken-Peterson et al. 2011
W NA/maritime temperate	127 mm	Mosses succulents graminoids forbs (mosses vs. vascular)	Saturated retention (% H_2O) 3 days dry retention (% H_2O) Natural rain event retention (% H_2O)	36.4–46.3 (27.2%) 71.7–83.9 (17.0%) 37.1–60.3 (62.5%)	Anderson et al. 2010
E NA/continental temperate grnhse.	89 mm	Succulents (species)	Cumulative ET (mL)	10–17 (70%)	Berghage et al. 2007
UK/maritime temperate	200 mm	Succulents forbs (species)	Leaf surface temperature (°C) Air temp. 300 mm above substrate (°C)	27–35 (29.6%) 32.2–34.4 (6.8%)	Blanusa et al. 2013
UK/maritime temperate	150 mm (outdoor expt.) 57.2 mm (indoor expt.)	Graminoids forbs (species, mixtures)	Total runoff vol. over 4 yrs (outdoor expt.) (L) Mean heavy rain runoff (mL) Mean light rain runoff (mL)	415–360 (13.2%) 1435–1099 (23.4%) 649–365 (43.7%)	Dunnett et al. 2008a
SW NA/humid subtropical	89 mm	Succulents (species)	Difference between substrate surface temperature and control (°C) (day delta max.)	−31 to −34 (9.7%)	Dvorak and Volder 2013

Table 9.1 (continued)

Region/Climate	Substrate depth	Plant types tested (comparison)	Ecosys. functions	Effect size min-max (% diff.)	Reference
Australia/ Mediterranean (greenhouse)	160 mm	Succulents (species)	Cumulative ET dry treatment (gH_2O) Cumulative ET wet treatment (gH_2O) Cumulative transp. dry (gH_2O) Cumulative transp. wet (gH_2O)	1527–1631 (6.8 %) 6112–11,964 (95.7 %) 9.5–32.9 (246.3 %) − 103.4–227.4 (319.9 %)	Farrell et al. 2012a (substrate with highest ET only shown)
Australia/ Mediterranean (greenhouse)	190 mm	Succulents Forbs Monocots Shrubs (species)	Cumulative transpiration per pot (wet)(gH_2O/day) Cumulative transpiration per pot (dry)(gH_2O/day) Cumulative transpiration per leaf area (wet)(gH_2O/m^2/day) Cumulative transpire./leaf area (dry)(gH_2O/m^2/day)	4.3–125.7 (2823.2 %) 5.2–28.5 (448.1 %) 38–2272 (5878.9 %) 90–2276 (2428.9 %)	Farrell et al. 2013b
Virtual (digital image preference analysis)		Succulents graminoids shrubs trees (veg.type)	Preference of green roof vegetation type (Likert scale: 1–5)	1.62–3.66 (125.9 %)	Fernandez-Cañero et al. 2013
Central NA/continental temperate	60 mm	Succulents (species)	Carbon in aboveground biomass (g C/m^2) Carbon in belowground biomass (g C/m^2)	64–239 (273.4 %) 37–187 (405.4 %)	Getter et al. 2009
Virtual (digital image preference analysis)		Succulents graminoids mixed (vegetation type)	Lifeform: shrubby succulent<grass Foliage colour: red<grey<green Height: lower<taller Flowers: absent<present	6.12–6.76 (10.4 %) 5.95–7.07 (18.8 %) 6.22–6.66 (7.1 %) 7.68–8.54 (11.2 %)	Lee et al. 2014

Table 9.1 (continued)

Region/Climate	Substrate depth	Plant types tested (comparison)	Ecosys. functions	Effect size min-max (% diff.)	Reference
Taiwan/humid subtropical	100 mm	Succulents graminoids forbs shrubs	Maximum lower leaf surface temperature (°C) Maximum difference between lower leaf surface and substrate temperature (°C)	40.6–30.3 (34.0 %) 10.2–17.5 (71.6 %)	Liu et al. 2012
E NA/maritime cold	65 mm	Succulents graminoids forbs shrubs (species, mixtures)	ET (g/hr/m^2) Water retention (%H_2O) Albedo (% reflection) Substrate temperature (°C)	105.1–143.5 (36.5 %) 52.3–69.2 (32.3 %) 16–21 (31.2 %) 26.8–21.77 (23.3 %)	Lundholm et al. 2010
E NA/maritime cold	65 mm	Graminoids forbs shrubs (species)	ET (g/hr/m^2) Water retention (%H_2O) Albedo (% reflection) Substrate temperature (summer) (°C) Maximum winter substrate temperature (°C) Minimum winter substrate temperature (°C) Snow accumulation (mm)	1121.2–136.4 (11.6 %) 67.7–75.4 (11.4 %) 17–22 (29.4 %) 24.6–21.5 (14.4 %) 15–28 (86.6 %) − 12 to − 6 (100 %) 50–80 (60 %)	MacIvor and Lundholm 2011 Lundholm et al. 2014a
E NA/maritime cold	65 mm	Graminoids shrubs (species, mixtures)	ET (g/hr/m^2) Water retention (%H_2O) Albedo (% reflection) Substrate temperature (year 1) (°C) Substrate temperature (year 2) (°C)	75.52–85.5 (13.2 %) 54.5–73.5 (34.9 %) 15.8–19.2 (21.4 %) 34.4–31.5 (8.9 %) 30–24.3 (23.5 %)	MacIvor et al. 2011
UK/maritime temperate (greenhouse)	50 mm	Succulents graminoids forbs (species, mixtures)	Runoff (mL)	650–355 (45.4 %)	Nagase and Dunnett 2012

Table 9.1 (continued)

Region/Climate	Substrate depth	Plant types tested (comparison)	Ecosys. functions	Effect size min-max (% diff.)	Reference
Korea/humid continental	hydroponic (wetland)	Wetland forbs (species)	ET (June) (L/m^2)	250–575 (130%)	Song et al. 2013
UK/maritime temperate	?	Graminoids forbs succulents (species)	PM_{10} capture by leaves ($g/m^2/yr$)	0.42–3.21 (664.3%)	Speak et al. 2012
Controlled environment chamber	65 mm	Succulents forbs graminoids shrubs (species)	R value (low temperature) (°F*hr*ft²/Btu) R value (high temperature) Water retention (low intensity rain)(%) Water retention (high intensity rain)(%) Lag to runoff (low intensity) (s) Lag to runoff (high intensity) (s)	0.8–3.4 (325%) 1.8–4.6 (155.5%) 29–54 (86.2%) 3–34 (325%) 31–52 (67.7%) 11–15 (36.4%)	Spolek et al. 2008
SW NA/humid subtropical	89 mm	Succulents (species)	Water retention	(7.5%)	Volder and Dvorak 2014
New Zealand (greenhouse)	70 mm	Succulents (species)	Cumulative ET over 29 days (mm)	20–30 (50%)	Voyde et al. 2010
Virtual (digital image preference study)		Turf sedum (flowering) tall flowering meadow brown roof	Preference (7 point scale) Beauty rating (7 point scale) Affective quality rating (7 point scale) Restoration rating (7 point scale) Preference (5 point scale)	3.54–4.05 (14.4%) 3.25–3.8 (16.9%) 4.34–4.82 (11.0%) 3.75–4.1 (9.3%) 2.75–3.3 (20.4%)	White and Gatersleven 2011

Two studies directly compared the effects of species mixtures relative to monocultures on thermal performance, but the results varied depending on the mixtures tested: Lundholm et al. (2010) showed the best monoculture substrate temperature to be 1.4% lower than the best mixture treatment, while in a different experiment (MacIvor et al. 2011) the best mixtures had temperatures 2% (year 1) and 12% (year 2) lower than the best monocultures. While these are modest differences, plant diversity warrants further investigation in improving the thermal functioning of green roofs.

Studies comparing combined evapotranspiration (ET) or transpiration are relevant to both thermal performance, via the dissipation of latent heat from the vegetation canopy and to stormwater retention as water loss from the growing medium increases its capacity to hold more water in subsequent precipitation events. Again, succulents showed substantial variation in transpiration (Farrell et al. 2012a) and a greater range when more plant types were included (Farrell et al. 2013b) but in general there was great variation in transpiration or total ET rates attributable to species selection, from 6.8% variation in dry conditions with succulents-only to over 2800% in wet conditions with succulents, dicot forbs, monocots and shrubs (Farrell et al. 2012a; Farrell et al. 2013b; Table 9.1). These differences can play a major role in determining the overall cooling potential of green roofs in hot conditions depending on water availability.

9.3 Stormwater Retention

Green roofs retain more stormwater than conventional roof surfaces due to storage in the substrate, and also reduce peak flows and increase lag times until runoff (Stovin et al. 2013). As well as plant choice, green roof design parameters such as substrate depth, slope and growth medium properties can all affect the provision of hydrological ecosystem services (VanWoert et al. 2005) with substrate depth being the major driver (Mentens et al. 2006). Roof slope can also have large effects with greater slopes leading to lower overall retention (Getter et al. 2007). The configuration of green roof water retention layers can affect plant water uptake (Savi et al. 2013), which, in turn, affects stormwater retention capability (Berndtsson et al. 2009; Stovin et al. 2013).

Plants can differentially affect the amount of stormwater retained on green roofs via active uptake of water from the soil (transpiration) (VanWoert et al. 2005), prevention of evaporation from the soil surface due to dense canopies (Lundholm et al. 2010) and interception of rain by the vegetation canopy (Dunnett et al. 2008b). Transpiration rate is the main driver of soil moisture in green roof systems that is directly affected by the vegetation and is highly variable across species used on green roofs (Table 9.1). Plant traits that are good predictors of ET rates are SLA and stomatal conductance. Nagase and Dunnett (2012) also found that plant height was a strong positive predictor of retention, likely due to increased interception (but this could also result from greater leaf area for transpiration); high root biomass was also a positive predictor of retention.

The volume of stormwater retention due to green roof vegetation ranges from 7.5 % in a study with three succulent species in Texas (Volder and Dvorak 2014) to 325 % in a controlled chamber containing succulents, graminoids, forbs and shrubs (Spolek et al. 2008). While we have little ability to generalize due to the paucity of empirical studies directly comparing vegetation types, graminoids are the standout vascular plants in providing stormwater capture services (Dunnett et al. 2008b; Lundholm et al. 2010; Nagase and Dunnett 2012) and mosses can significantly outperform vascular species (Anderson et al. 2010). However, more empirical studies are necessary to confirm the generality of these findings. Overall performance will depend heavily on the intensity, volume and frequency of precipitation and antecedent substrate moisture levels (Villarreal and Bengtsson 2005) but many of the studies in Table 9.1 examine cumulative retention rates that integrate a range of storm sizes, intensities, durations and substrate moisture conditions.

Four studies examined plant species mixtures and monocultures with respect to stormwater retention (Table 9.1). Dunnett et al. (2008a) found that the best mixtures retained less stormwater than the best monocultures in both greenhouse and field studies (1.4 % less outdoors, although this was not statistically different from the best mixture; 11.6 % and 8.6 % less in heavy and light greenhouse rain events, respectively). MacIvor et al. (2011) found a 20 % increase in retention in the best mixture treatment compared with the best monoculture; Lundholm et al. (2010), found an 8.4 % increase, in a study with more species. These results likely reflect the effect of previous evapotranspiration only (linked to plant diversity in that study), as water was added directly to the substrate surface (Lundholm et al. 2010), as opposed to the Dunnett et al. (2008a), Nagase and Dunnett (2012) studies, which added water above the vegetation canopies. Additionally, the latter studies used more realistic rainfall durations, thus it is difficult to directly compare the two sets of experiments.

9.4 Water Quality

Rowe (2011) summarizes the basis for expecting pollution abatement services from green roofs, including water quality. One of the main effects of green roofs on water quality of runoff is the overall reduction of runoff quantity, thereby reducing the quantity of pollutants reaching urban waterways or sewer infrastructure (Rowe 2011). However, green roofs can also contribute to water pollution by leaching nutrients (Montrerusso et al. 2004; Berndtsson et al. 2006) and metals from the growing medium/substrate (Alsup et al. 2010; Vijayaraghavan et al. 2012; Chap. 5). Green roofs can also buffer acid rain but this effect is thought to be temporary and may require soil amendments (liming) to perpetuate (Berghage et al. 2007). Other design parameters of green roofs affecting the quality of water runoff include substrate depth (Seidl et al. 2013) but the effects of increased depth can be positive or negative depending on the pollutant examined.

Plants can alter pollutant levels in soil by direct uptake, volatilization and stabilization (leading to lower solubility in runoff) (Chaney et al. 1997). Plant effects on water quality in green roof systems include increased leaching of some metals such as cadmium, possibly due to rhizosphere influence on pH, leading to greater solubility of some chemical species (Alsup et al. 2010). The same study showed reductions in lead concentrations in leachate from planted systems (Alsup et al. 2010). Several studies (e.g. Berndtsson et al. 2009) show reduced nitrate runoff in green compared to conventional roof systems, likely due to uptake by the vegetation. Plant species differ in their uptake rates for nutrients (Chapin 1980) and thus the potential to leach nutrients into the runoff.

Studies of ground-level, vegetated biofiltration systems suggest that pollution reduction depends greatly on the plants used (Payne et al. 2014). Read et al. (2010) found that plant traits had little effect on metal concentrations of stormwater effluent and that most metals were locked in the biofilter substrates. However some plant traits had substantial effects on N and P removal. Rooting depth, longest root, total root length and root mass were most influential, perhaps because they mediate contact between the plant and soil microbes (Payne et al. 2014), but relative growth rate (RGR) was also important (Read et al. 2010). A subsequent review of the plant traits most likely to influence nutrient uptake in biofilters recommended focusing on the following: (1) early successional species with reasonably high relative growth rates (because they often utilize nitrate and ammonium), (2) roots that have a high specific root length (long, dense roots), tolerate waterlogging and have mycorrhizal associations; (3) low litter decomposition rates; and (4) high evapotranspiration rates (Payne et al. 2014). However, some of these traits may be contradictory to fundamental plant strategy trade-offs, for example plants with low litter decomposition rates are often more slow growing than others (Cornelissen and Thompson 1997). Other traits, such as long, dense roots may be incompatible with the environment on a green roof.

Only two studies examined differential water quality attributable to green roof plant species. Aitkenhead-Peterson et al. (2011) compared succulent species and found an 1120% difference in leachate nitrate between the best and worst performing species, 67% for magnesium and 220% for potassium but no differences for the other pollutants examined (Table 9.1). An earlier study (Monterusso et al. 2004) found nitrate concentrations in runoff from prairie vegetation to be more than 100x lower (0.22 ppm) than runoff from sedum species (22.7 ppm), but this comparison is between green roof systems that also differed in substrate depth and the engineered components, making it difficult to attribute these differences to vegetation type alone.

9.5 Air Pollution Benefits

Plants can reduce air pollution by two main mechanisms: physical trapping of particulate matter (PM) or other pollutants on plant surfaces (Yang et al. 2008), and uptake of pollutants into plant tissues (Clark et al. 2008; Currie and Bass 2008). Absorbing pollutants, capturing PM and tolerating stress are important traits for the selection of vegetation for this ecosystem service.

Pollutant uptake rates vary greatly across species (Morikawa et al. 1998), suggesting that plant type can determine the extent of air pollution reduction benefits.

There is evidence that plants with high SLA values, higher stomatal conductance and faster growth rates are likely to be negatively affected by air pollution (Power and Ashmore 2002). This may be because high SLA is associated with faster growth and greater stomatal conductance that results in greater pollutant uptake at a given ambient concentration (Bassin et al. 2007). Similarly, plants possessing traits that confer drought stress resistance may also be resistant to air pollution by limiting gas exchange (Wilson 1995). Leaf hair densities, leaf wax quantities and plant height have been found to have a positive relationship with PM accumulation (Sæbø et al. 2012; Weber et al. 2014).

While trees are considered the most effective type of vegetation in reducing urban air pollution due to their large leaf surface area, shrubby and herbaceous vegetation may provide important supplementary benefits. Often the latter is more easily incorporated into dense urban landscapes particularly close to roads that are a major pollutant source (Pugh et al. 2012; Weber et al. 2014). Consequently, green roofs have been viewed as potential new locations for air pollution mitigation.

A number of studies have modeled the amount of pollutants that may be removed by green roof plants (Currie and Bass 2008; Yang et al. 2008; Pugh et al. 2012). However, these models were not parameterized for the type of vegetation typically grown on green roofs. Only one study has empirically quantified the effect of green roof plants on air pollution abatement (Table 9.1). Speak et al. (2012) found a 664 % difference between the species that trapped the most particulates and the least. They attribute the difference to leaf characteristics including microfeatures such as leaf hairs and ridges, suggesting an additional suite of traits that are important in determining the function of green roof vegetation.

9.6 Carbon Sequestration

Capture and storage of atmospheric carbon is an important role of vegetation, in the face of increasing burning of fossil fuels. Plants and soil build up carbon stores over time leading to the potential of green roofs as carbon sinks. While plant productivity is an important driver of overall carbon storage, plant type should make a large difference especially when comparing woody with herbaceous vegetation types (e.g. Jobbágy and Jackson 2000). Given that green roofs are planted with a large variety of vegetation types, studies of differential contributions to carbon sequestration are warranted. Only one study has done so thus far. Getter et al. (2009) compared sedum species for the accumulation of carbon (Table 9.1). They detected large differences among species in both above- (273 %) and below-ground carbon (405 %) storage, thus plant selection is very important for this function even within a single plant genus. Incorporation of other life forms in future studies should yield even larger differences in effect size. It should be mentioned that the overall carbon storage potential in extensive green roofs is low due to low biomass and shallow substrate layers. Intensive green roofs planted with trees could make more important contributions toward carbon capture and storage.

9.7 Psychological Benefits

Green spaces in urban areas are known to provide psychological benefits to their human inhabitants (Hansmann et al. 2007). Many studies have compared un-vegetated views to views of different vegetation types and examined preferences or actual psychological effects such as attention restoration and stress recovery (Hartig et al. 2003). Green roofs have been hypothesized to provide similar psychological benefits if they are viewed by building occupants (Lee et al. 2014). More generally, preferences for green roof vegetation are important in increasing public acceptability of green roofs (White and Gatersleben 2011), thus differential effects of vegetation on aesthetic evaluation and other aspects of human perception constitute important ecosystem services (Sutton 2014).

Landscape preference studies show that people can often have strong feelings about different vegetation types (Orians 1986; Kaplan 2001), thus we can anticipate that plant characteristics will influence the aesthetic appeal of green roofs. Foliage color, vegetation height, leaf width, presence of flowers and plant density are all plant traits found to influence people's preferences for herbs and shrubs at ground level (Kendal et al. 2012). In general green foliage and flowers are highly preferred, perhaps for evolutionary reasons as they are thought to indicate a productive environment (van den Berg et al. 2003). More recent work indicates that people tend to prefer more species-rich vegetation (Lindemann-Matthies et al. 2010), and there are measurable psychological benefits associated with higher plant diversity in urban ecosystems (Fuller et al. 2007). However, human preferences are complex and individuals respond very differently to plant traits (Kendal et al. 2012). Green roofs are also very different from ground landscapes, and preferred traits may not be applicable on a green roof due to constraints on the types of plants than can be grown.

Because no studies have investigated mental health benefits from the active use of green roofs, we focus on the effects of viewing vegetation. Four studies have examined human preferences for different green roof vegetation, using either site visits or digitally altered images (Table 9.1). White and Gatersleven (2011) found significant differences (9.3–20.4 %) in preferences among vegetation types including turf, sedum, a "tall flowering meadow" and a brown roof (image features a diversity of plant species, imitating a UK "biodiverse roof") (e.g. Kadas 2006). The meadow had the highest ratings for most variables, and the brown roof had the lowest. Fernandez-Cañero et al. (2013) found that people preferred a sedum extensive roof the least and an intensive roof with trees and shrubs the most (76 % difference). Conversely, Jungels et al. (2013) found that green roofs planted with stoloniferous grasses were least preferred compared to those planted with sedums or mixed perennials probably because the grasses were perceived as messy.

Lee et al. (2014) systematically varied 40 green roof images manipulating plant life-form, foliage color, flower presence, vegetation diversity and plant height. Office workers were then asked to rank their preference for the images and how restorative they thought each roof was. Preferences varied with plant traits. The most preferred and restorative living roof had taller, green, grassy and flowering vegetation, while lower-growing red succulent vegetation was least preferred. The results of this study were then used to demonstrate that people viewing the most preferred green roof image for brief periods had better attention control and mental arousal than those viewing a concrete roof (Lee 2014).

While green roofs are often designed using aesthetic criteria, these studies suggest that the traits of the plants chosen to grow on them can significantly influence people's responses to green roofs and the mental health benefits they provide.

9.8 Conclusions

9.8.1 *Research Questions for the Future*

Which plant traits optimize multiple ecosystem services? How can species diversity be used to design green roofs that overcome tradeoffs in plant performance? The traits required to maximize one ecosystem service may not be optimal for another or the desired trait combinations may violate fundamental plant strategy tradeoffs or reduce survivability specifically within a rooftop environment. For example while leaf succulence (or CAM photosynthetic pathway) generally increases plant survival on green roofs, it limits the ability of the green roof to absorb rainfall and retain stormwater runoff (Farrell et al. 2013b). Farrell et al. (2013b) have identified species that not only optimize the stormwater retention of green roofs by using high amounts of water when it is available to reduce runoff, but also persist and survive periods of drought. This combination of physiological traits was not expected as it seems to break a fundamental plant trade-off. Detailed research of this type can help green roof designers optimize service provisioning: green roofs could be built with a particular ecosystem service in mind and plants selected specifically for this purpose. Other research has identified species that may be complementary or act as facilitators, representing the potential to overcome tradeoffs by combining species that differ in traits (see Chap. 8 for a summary).

How do the effects of plant traits on green roof function vary across climates? While shallow substrates present a common set of challenges to plant growth and survival, the relative effect of traits in predicting function may vary across climatic zones. Trait-based research needs to be carried out in different climates to determine the impact of temperature, moisture and wind regimes on optimal plant choice for ecosystem service provisioning. Maximizing ecosystem provisioning from green roofs will require a shift in research priorities toward a more detailed consideration of plant traits and their effects on ecological function.

9.8.2 *Summary*

Understanding of green roofs as ecosystems requires attention to the relationships between the living and non-living components and their combined effects on ecosystem service provisioning. The engineered components of green roofs—including the growing media or substrate, membranes and water retention layers—can also influence the provision of these services. We have only addressed the vegetation components of green roof ecosystems. Other biological components (including the rhizosphere; Chap. 6, Chap. 7) have received little research but could have substantial impacts on nutrient leaching, carbon and pollutant sequestration, hydrology, carbon and indirect impacts on functioning via effects on the plants.

Most of the studies we have cited in this review directly consider how different vegetation types affect the ecosystem services provided by green roofs. Many other studies of plant growth, size and condition on green roofs are indirectly relevant to thermal and stormwater functions, as plant size (Sailor 2008) and vegetation damage influence ecosystem services (Speak et al. 2013a). Another important component of the green roof literature documents plant responses to substrate depth (e.g. Dunnett et al. 2008a; Getter and Rowe 2008), which is also important to overall ecosystem function as depth influences plant growth and survival. Similarly, the effects of irrigation regimes (Nagase and Dunnett 2013; Rowe et al. 2014) and soil amendments that aim to improve plant nutrient (Clark and Zheng 2013) and water status (Farrell et al. 2013a) on growth and survival have also been investigated. Both will have direct effects on ecosystem functioning.

The most frequently investigated green roof performance measures are related to thermal and hydrological functions (Table 9.1). Theoretical and empirical studies evaluating summer cooling and stormwater retention benefits have been conducted in many regions due to their incorporation into cost-benefit analyses (e.g. Carter and Keeler 2008) and incentive programs. However, there is a need for more empirical work for all the benefits considered here.

Most of the functions considered showed substantial variation attributable to plant species or general vegetation type. This suggests close attention to plant selection is required in order to optimize green roof functioning. While the green roof industry has focused on plant survival and growth as key criteria for optimizing green roofs, plants that do well in green roof conditions do not provide key services equally. Other ecosystem services such as the attenuation of noise pollution have not been addressed by plant evaluation studies, although plant cover has been shown to make a difference (Van Renterghem et al. 2013).

To maximize ecosystem services, the ideal green roof plant species would have the following traits (1) relatively large leaves, (2) a dense canopy and (3) low-growing, mat-forming structure to (4) maximize shade and minimize moisture loss from the substrate while allowing transpiration when water is available thus increasing stormwater retention of the green roof. It would also have (5) light-colored foliage due to leaf hairs or waxes to reflect heat and trap particulate pollutants but also be (6) green (7) with conspicuous flowers to maximize aesthetic and social benefits (Table 9.2). However, finding a single species with all these characteristics will be difficult, thus species diversity may be harnessed to optimize service provision (Patton-Cook and Bauerle 2012).

Acknowledgements This chapter has benefited from discussion with Claire Farrell and Kate Lee. NSGW was partially funded by Australian Research Council grants LP0990704 and LP130100731.

Table 9.2 Plant traits likely to influence the ecosystem services provided by green roofs

Plant or vegetation property	Definition[a]	Ecosystem services	Comments
Specific leaf area (SLA)	Leaf area per unit leaf mass	Cooling, stormwater retention	Positively related to photosynthetic rate and transpiration
Leaf area	Area of fully expanded leaf	Cooling, stormwater retention	Species with larger area of individual leaves are associated with higher productivity and transpiration, but plants with small individual leaves could also have high coverage overall, leading to shading; leaf area index is a property of vegetation canopies
Stomatal conductance	mmol $H_2O/m^2/s$	Cooling, stormwater retention	Stomatal conductance is directly related to transpiration rate; stomatal conductance of a species depends greatly on environmental conditions (e.g. humidity, soil moisture); species with high diurnal conductance exhibit greatest water uptake from substrate and greatest stormwater retention
Height	Height of mature plant in natural environment	Cooling, temperature moderation in winter, snow trapping, particulate trapping, aesthetics	Taller plants create more shade, reflect more incoming radiation, and generally have higher photosynthetic rates; tall plants could have positive or negative effects on stormwater retention depending on diurnal stomatal conductance; plants with greater necromass left in canopies in winter trap more snow and regulate temperatures but also reduce heat gain during sunny periods; taller plants with greater overall leaf area likely retain more water in the canopy during rain events, leading to overall greater retention; people tend to prefer taller vegetation in green roof images
Relative growth rate (RGR)	Dry weight at time two/initial weight; determined under controlled (optimal) conditions	Carbon sequestration, nutrient uptake	Growth rates in green roof systems (e.g. Lundholm et al. 2014b) will be more predictive of services than generalized RGR; RGR is likely positively correlated with other services related to overall productivity: nutrient uptake, transpiration, cooling, and stormwater retention

Table 9.2 (continued)

Plant or vegetation property	Definition[a]	Ecosystem services	Comments
Leaf reflectivity	% incoming radiation reflected by leaf	Cooling	Depends on presence of hairs on leaf surface and pigments within leaf cells; may have negative effects on winter thermal performance as reflective leaves will reduce heat gain during sunny periods
Root length, rooting depth, root mass	Various	Nutrient removal, carbon sequestration	All are general indicators of root activity and the area of contact between roots and soil microbes
Leaf hair density	# hairs per unit leaf area	Particulate trapping	Leaf hairs may increase leaf reflectivity; increased particulate loads on leaf surfaces will likely reduce reflectivity
Flower color, size and duration	Various	Aesthetic, psychological	Might also affect pollinator visitation, and thus overall plant production and stability of vegetation

[a] For traits used in the plant ecology literature, we provided the most common units used

References

Ackerly DD, Knight CA, Weiss SB, Barton K, Starmer KP (2002) Leaf size, specific leaf area and microhabitat distribution of chaparral woody plants: contrasting patterns in species level and community level analyses. Oecologia 130(3):449–457

Aitkenhead-Peterson JA, Dvorak BD, Volder A, Stanley NC (2011) Chemistry of growth medium and leachate from green roof systems in south-central Texas. Urban Ecosyst 14(1):17–33

Alsup S, Ebbs S, Retzlaff W (2010) The exchangeability and leachability of metals from select green roof growth substrates. Urban Ecosyst 13(1):91–111

Anderson M, Lambrinos J, Schroll E (2010) The potential value of mosses for stormwater management in urban environments. Urban Ecosyst 13(3):319–332

Bass B, Baskaran B (2003) Evaluating rooftop and vertical gardens as an adaptation strategy for urban areas. Institute for Research and Construction, National Research Council, Canada, Ottawa

Bassin S, Volk M, Fuhrer J (2007) Factors affecting the ozone sensitivity of temperate European grasslands: an overview. Environ Pollut 146(3):678–691

Berghage R, Jarrett A, Beattie D, Kelley K, Husain S, Rezai F, Long B, Nagase A, Cameron R, Hunt W (2007) Quantifying evaporation and transpirational water losses from green roofs and green roof media capacity for neutralizing acid rain. National Decentralized Water Resources Capacity Development Project

Berndtsson JC, Emilsson T, Bengtsson L (2006) The influence of extensive vegetated roofs on runoff water quality. SciTotal Environ 355(1–3):48–63

Berndtsson JC, Bengtsson L, Jinno K (2009) Runoff water quality from intensive and extensive vegetated roofs. Ecol Eng 35(3):369–380

Blanusa T, Vaz Monteiro MM, Fantozzi F, Vysini E, Li Y, Cameron RWF (2013) Alternatives to sedum on green roofs: can broad leaf perennial plants offer better 'cooling service'? Build Environ 59:99–106

Carter T, Keeler A (2008) Life-cycle cost-benefit analysis of extensive vegetated roof systems. J Environ Manage 87(3):350–363. doi:10.1016/j.jenvman.2007.01.024

Chaney RL, Malik M, Li YM, Brown SL, Brewer EP, Angle JS, Baker AJ (1997) Phytoremediation of soil metals. Curr Opin Biotechnol 8(3):279–284

Chapin FS (1980) The mineral nutrition of wild plants. Annu Rev Ecol Syst 11:233–260

Clark MJ, Zheng Y (2013) Plant nutrition requirements for an installed sedum-vegetated green roof module system: effects of fertilizer rate and type on plant growth and leachate nutrient content. HortScience 48(9):1173–1180

Clark C, Adriaens P, Talbot FB (2008) Green roof valuation: a probabilistic economic analysis of environmental benefits. Environ Sci Technol 42(6):2155–2161. doi:10.1021/es0706652

Cook-Patton SC, Bauerle TL (2012) Potential benefits of plant diversity on vegetated roofs: a literature review. J Environ Manage 106:85–92

Cornelissen JHC, Thompson K (1997) Functional leaf attributes predict litter decomposition rate in herbaceous plants. New Phytol 135(1):109–114

Currie BA, Bass B (2008) Estimates of air pollution mitigation with green plants and green roofs using the UFORE model. Urban Ecosyst 11:409–422

Del Barrio EP (1998) Analysis of the green roofs cooling potential in buildings. Energy Build 27(2):179–193

Dunnett N, Nagase A, Booth R, Grime P (2008a) Influence of vegetation composition on runoff in two simulated green roof experiments. Urban Ecosyst 11(4):385–398

Dunnett N, Nagase A, Hallam A (2008b) The dynamics of planted and colonising species on a green roof over six growing seasons 2001–2006: influence of substrate depth. Urban Ecosyst 11(4):373–384

Dvorak B, Volder A (2013) Rooftop temperature reduction from unirrigated modular green roofs in south-central Texas. Urban For Urban Green 12(1):28–35

Eumorfopoulou E, Aravantinos D (1998) The contribution of a planted roof to the thermal protection of buildings in Greece. Energ Build 27(1):29–36

Farrell C, Mitchell RE, Szota C, Rayner JP, Williams NSG (2012a) Green roofs for hot and dry climates: interacting effects of plant water use, succulence and substrate. Ecol Eng 49:270–276

Farrell C, Szota C, Rayner JR, Williams NSG (2012b) Hot, high, dry and green?—Research supporting green roof plant selection for arid environments. Paper presented at the Cities Alive: 10th Annual Green Roof and Wall Conference, Chicago, October 17–20, 2012

Farrell C, Ang XQ, Rayner JP (2013a) Water-retention additives increase plant available water in green roof substrates. Ecol Eng 52:112–118

Farrell C, Szota C, Williams NSG, Arndt SK (2013b) High water users can be drought tolerant: using physiological traits for green roof plant selection. Plant Soil 372:177–193

Fernandez-Cañero R, Emilsson T, Fernandez-Barba C, Herrera Machuca MT (2013) Green roof systems: a study of public attitudes and preferences in southern Spain. J Environ Manage 128:106–115

Fuller RA, Irvine KN, Devine-Wright P, Warren PH, Gaston KJ (2007) Psychological benefits of greenspace increase with biodiversity. Biol Letters 3(4):390–394

Getter KL, Rowe DB (2006) The role of extensive green roofs in sustainable development. HortScience 41(5):1276–1285

Getter KL, Rowe DB (2008) Media depth influences sedum green roof establishment. Urban Ecosyst 11:361–372

Getter KL, Rowe DB, Andresen JA (2007) Quantifying the effect of slope on extensive green roof stormwater retention. Ecol Eng 31(4):225–231

Getter KL, Rowe DB, Robertson GP, Cregg BM, Andresen JA (2009) Carbon sequestration potential of extensive green roofs. Environ Sci Technol 43:7564–7570

Getter KL, Rowe DB, Andresen JA, Wichman IS (2011) Seasonal heat flux properties of an extensive green roof in a Midwestern U.S. climate. Energ Build 43(12):3548–3557

Hansmann R, Hug S-M, Seeland K (2007) Restoration and stress relief through physical activities in forests and parks. Urban For Urban Green 6:213–225

Hartig T, Evans GW, Jamner LD, Davis DS, Gärling T (2003) Tracking restoration in natural and urban field settings. J Environ Psychol 23:109–123

Jobbágy EG, Jackson RB (2000) The vertical distribution of soil organic carbon and its relation to climate and vegetation. Ecol Appl 10(2):423–436

Jungels J, Rakow DA, Allred SB, Skelly SM (2013) Attitudes and aesthetic reactions toward green roofs in the Northeastern United States. Landsc Urban Plan 117:13–21

Kadas G (2006) Rare invertebrates colonizing green roofs in London. Urban Habitats 4(1):66–86

Kaplan R (2001) The nature of the view from home psychological benefits. Environ Behav 33(4):507–542

Kendal D, Williams KJH, Williams NSG (2012) Plant traits link people's plant preferences to the composition of their gardens. Landsc Urban Plan 105(1–2):34–42

Lavorel S, Grigulis K, Lamarque P, Colace M, Garden D, Girel J, Pellet G, Douzet R (2011) Using plant functional traits to understand the landscape distribution of multiple ecosystem services. J Ecol 99(1):135–147

Lee KE (2014) The role of green roof views in attention restoration and work performance. PhD Thesis. The University of Melbourne

Lee KE, Williams KJH, Sargent LD, Farrell C, Williams NS (2014) Living roof preference is influenced by plant characteristics and diversity. Landsc Urban Plan 122:152–159

Lindemann-Matthies P, Junge X, Matthies D (2010) The influence of plant diversity on people's perception and aesthetic appreciation of grassland vegetation. Biol Conserv 143(1):195–202

Liu K, Baskaran B (2005) Thermal performance of extensive green roofs in cold climates. National Research Council (Canada), Construction Technology Update No. 65, NRCC-48202. Ottawa, Ontario

Liu T-C, Shyu G-S, Fang W-T, Liu S-Y, Cheng B-Y (2012) Drought tolerance and thermal effect measurements for plants suitable for extensive green roof planting in humid subtropical climates. Energy Build 47:180–188

Lundholm J, MacIvor J, MacDougall Z, Ranalli M (2010) Plant species and functional group combinations affect green roof ecosystem functions. PLoS One 5(3):art. no. e9677

Lundholm J, Weddle B, MacIvor J (2014a). Snow depth and vegetation type affect green roof thermal performance in winter. Energy Build 84:299–307

Lundholm J, Heim A, Tran S, Smith T (2014b). Leaf and life history traits predict plant growth in a green roof ecosystem. PLoS One 9(6):e101395. doi:10.1371/journal.pone.0101395

MacIvor JS, Lundholm J (2011) Performance evaluation of native plants suited to extensive green roof conditions in a maritime climate. Ecol Eng 37(3):407–417

MacIvor JS, Ranalli MA, Lundholm JT (2011) Performance of dryland and wetland plant species on extensive green roofs. Ann Bot-London 107(4):671–679

Mentens J, Raes D, Hermy M (2006) Green roofs as a tool for solving the rainwater runoff problem in the urbanized 21st century? Landsc Urban Plan 77(3):217–226

Monterusso MA, Rowe DB, Rugh CL, Russell DK (2004) Runoff water quantity and quality from green roof systems. Acta Hortic 639:369–376

Morikawa H, Higaki A, Nohno M, Takahashi M, Kamada M, Nakata M, Toyo-hara G, Okamura Y, Matsui K, Kitani S (1998) More than a 600-fold variation in nitrogen dioxide assimilation among 217 plant taxa. Plant Cell Environ 21(2):180–190

Nagase A, Dunnett N (2012) Amount of water runoff from different vegetation types on extensive green roofs: effects of plant species, diversity and plant structure. Landsc Urban Plan 104(3–4):356–363

Nagase A, Dunnett N (2013) Establishment of an annual meadow on extensive green roofs in the UK. Landsc Urban Plan 112(1):50–62

Niachou A, Papakonstantinou K, Santamouris M, Tsangrassoulis A, Mihala-kakou G (2001) Analysis of the green roof thermal properties and investigation of its energy performance. Energ Build 33(7):719–729

Oberndorfer E, Lundholm J, Bass B, Coffman RR, Doshi H, Dunnett N, Gaffin S, Kohler M, Liu KKY, Rowe B (2007) Green roofs as urban ecosystems: ecological structures, functions, and services. Bioscience 57(10):823–833

Orians GH (1986) An ecological and evolutionary approach to landscape aesthetics. In: Penning-Rowsell EC, Lowenthal D (eds) Landscape meanings and values. Allen and Unwin, London, pp 3–25

Payne EG, Fletcher TD, Cook PL, Deletic A, Hatt BE (2014) Processes and drivers of nitrogen removal in stormwater biofiltration. Crit Rev Env Sci Technol 44:796–846

Pérez-Harguindeguy N, Díaz S, Garnier E, Lavorel S, Poorter H, Jaureguiberry P, Bret-Harte MS, Cornwell WK, Craine JM, Gurvich DE, Urcelay C, Veneklaas EJ, Reich PB, Poorter L, Wright IJ, Ray P, Enrico L, Pausas JG, De Vos AC, Buchmann N, Funes G, Quétier F, Hodgson JG, Thompson K, Morgan HD, Ter Steege H, Van Der Heijden MGA, Sack L, Blonder B, Poschlod P, Vaieretti MV, Conti G, Staver AC, Aquino S, Cornelissen JHC (2013) New handbook for standardised measurement of plant functional traits worldwide. Aust J Bot 61(3):167–234

Permpituck S, Namprakai P (2012) The energy consumption performance of roof lawn gardens in Thailand. Renew Energy 40(1):98–103

Power SA, Ashmore MR (2002) Responses of fen and fen-meadow communities to ozone. New Phytol 156(3):399–408

Pugh TAM, MacKenzie AR, Whyatt JD, Hewitt CN (2012) Effectiveness of green infrastructure for improvement of air quality in urban street canyons. Environ Sci Tech 46(14):7692–7699

Read J, Fletcher TD, Wevill T, Deletic A (2010) Plant traits that enhance pollutant removal from stormwater in biofiltration systems. Int J Phytoremediation 12(1):34–53

Reich PB, Buschena C, Tjoelker MG, Wrage K, Knops J, Tilman D, Machado JL (2003) Variation in growth rate and ecophysiology among 34 grassland and savanna species under contrasting N supply: a test of functional group differences. New Phytol 157(3):617–631

Rowe DB (2011) Green roofs as a means of pollution abatement. Environ Pollut 159(8–9):2100–2110

Rowe DB, Kolp MR, Greer SE, Getter KL (2014) Comparison of irrigation efficiency and plant health of overhead, drip, and sub-irrigation for extensive green roofs. Ecol Eng 64:306–313

Sæbø A, Popek R, Nawrot B, Hanslin HM, Gawronska H, Gawronski SW (2012) Plant species differences in particulate matter accumulation on leaf surfaces. Sci Total Environ 427–428:347–354

Sailor DJ (2008) A green roof model for building energy simulation programs. Energy Build 40(8):1466–1478

Sailor DJ, Hutchinson D, Bokovoy L (2008) Thermal property measurements for ecoroof soils common in the western U.S. Energy Build 40(7):1246–1251

Sailor DJ, Elley TB, Gibson M (2012) Exploring the building energy impacts of green roof design decisions-a modeling study of buildings in four distinct climates. J Build Phys 35(4):372–391

Savi T, Andri S, Nardini A (2013) Impact of different green roof layering on plant water status and drought survival. Ecol Eng 57:188–196

Seidl M, Gromaire MC, Saad M, De Gouvello B (2013) Effect of substrate depth and rain-event history on the pollutant abatement of green roofs. Environ Poll (Barking, Essex: 1987) 183:195–203

Song U, Kim E, Bang JH, Son DJ, Waldman B, Lee EJ (2013) Wetlands are an effective green roof system. Build Environ 66:141–147

Speak AF, Rothwell JJ, Lindley SJ, Smith CL (2012) Urban particulate pollution reduction by four species of green roof vegetation in a UK city. Atmos Environ 61:283–293

Speak AF, Rothwell JJ, Lindley SJ, Smith CL (2013a) Rainwater runoff retention on an aged intensive green roof. SciTotal Environ 461–462:28–38

Speak AF, Rothwell JJ, Lindley SJ, Smith CL (2013b) Reduction of the urban cooling effects of an intensive green roof due to vegetation damage. Urban Climate 3:40–55

Spolek G (2008) Performance monitoring of three ecoroofs in Portland, Oregon. Urban Ecosyst 11:349–359

Spolek G, Sailor D, Ervin D (2008) Optimization of green roof design–preliminary results. In: Proceedings of the 6th Annual Greening Rooftops for Sustainable Communities Conference, Baltimore, MD, April 30-May 2008

Sproul J, Wan MP, Mandel BH, Rosenfeld AH (2014) Economic comparison of white, green, and black flat roofs in the United States. Energy Build 71:20–27

Stovin V, Poë S, Berretta C (2013) A modelling study of long term green roof retention performance. J Environ Manage 131:206–215

Sutton R (2014) Aesthetics for green roofs and green walls. J Living Archit 1(2):1–20. http://greenroofs.org/resources/JOLA2014Volume1(Issue2)Sutton

Teemusk A, Mander U (2007) Rainwater runoff quantity and quality performance from a green-roof: the effects of short-term events. Ecol Eng 30(3):271–277. doi:10.1016/j.ecoleng.2007.01.009|ISSN0925-8574

Theodosiou TG (2003) Summer period analysis of the performance of a planted roof as a passive cooling technique. Energy Build 35(9):909–917

van den Berg AE, Koole SL, van der Wulp NY (2003) Environmental preference and restoration: (how) are they related? J Environ Psychol 23(2):135–146

Van Mechelen C, Dutoit T, Hermy M (2014) Mediterranean open habitat vegetation offers great potential for extensive green roof design. Landsc Urban Plan 121:81–91

Van Renterghem T, Hornikx M, Forssen J, Botteldooren D (2013) The potential of building envelope greening to achieve quietness. Build Environ 61:34–44

VanWoert ND, Rowe DB, Andresen JA, Rugh CL, Fernandez RT, Xiao L (2005) Green roof stormwater retention: effects of roof surface, slope, and media depth. J Environ Qual 34(3):1036–1044

Vijayaraghavan K, Joshi UM, Balasubramanian R (2012) A field study to evaluate runoff quality from green roofs. Water Res 46(4):1337–1345

Villarreal EL, Bengtsson L (2005) Response of a sedum green-roof to individual rain events. Ecol Eng 25(1):1–7. doi:10.1016/j.ecoleng.2004.11.008

Volder A, Dvorak B (2014) Event size, substrate water content and vegetation affect storm water retention efficiency of an un-irrigated extensive green roof system in Central Texas. Sustain Cities Soc 10:59–64

Voyde E, Fassman E, Simcock R (2010) Hydrology of an extensive living roof under sub-tropical climate conditions. J Hydrol 394:384–395
Weber F, Kowarik I, Säumel I (2014) Herbaceous plants as filters: immobilization of particulates along urban street corridors. Environ Pollut 186:234–240
White EV, Gatersleben B (2011) Greenery on residential buildings: does it affect preferences and perceptions of beauty? J Environ Psychol 31(1):89–98
Wilson SA (1995) Are Mediterranean plant species likely to have a distinctive response to SO_2 pollution? Agric Ecosyst Environ 55(2):71–93
Yang J, Yu Q, Gong P (2008) Quantifying air pollution removal by green roofs in Chicago. Atmos Environ 42(31):7266–7273

Chapter 10
Ruderal Green Roofs

Nigel Dunnett

Abstract An awareness of the ecological theory relating to the colonization, early successional stages and persistence of ruderal communities and their role within a matrix of other plant communities and plant types on green roofs provides an important basis for increased understanding of the long-term resilience of dynamic green roof vegetation assemblages. This chapter discusses the concept of the ruderal green roof, with its highly dynamic nature and inclusion of colonization, succession and change as core functioning elements. The theoretical background of a trait-based or functional type approach to working with green roof vegetation will be explored, and the wider role of 'ruderal' or disturbance-tolerant plant species in creating resilient and climate-adapted green roofs will be reviewed. Dynamic colonization processes have wide applications across typical extensive, semi-intensive and intensive green roof types where designers and users desire greater biodiversity, a more sustainable approach to long-term management, increased local distinctiveness, climate adaptation, and greater aesthetic and visual interest.

Keywords Bio-diverse roofs · Community dynamics · Functional ecology · Stress · Disturbance · Competition · Colonization

10.1 Introduction

As living systems, green roofs are subject to the same environmental pressures, processes and limits that are part of all ecosystems. While the popular view of a green roof is of a static layer of vegetation, the reality is that all but the most highly maintained are dynamic systems subject to ecological processes and change on the short, medium and long timescales. An understanding of these ecological processes, such as colonization and succession, that operate within green roof systems, widens the scope of possibilities for green roof vegetation and thereby its associated fauna. This understanding becomes critical for the development of diverse and ecologi-

N. Dunnett (✉)
Department of Landscape, University of Sheffield, Arts Tower,
Western Bank, Sheffield S10 2TN, UK
e-mail: n.dunnett@sheffield.ac.uk

R. Sutton (ed.), *Green Roof Ecosystems*, Ecological Studies 223,
DOI 10.1007/978-3-319-14983-7_10

cally attuned green roof systems that exhibit persistence, resilience, and adaptability in the face of environmental challenges and change.

The use of plants with traits that allow them to colonize and recolonize space fosters a self-repairing green roof system that responds to severe environmental change and perturbation. Combining these plants with others that are adapted to resist and survive severe environmental stresses, results in green roof vegetation that retains its integrity over time, but in a way that is constantly changing. This dynamic view of vegetation provides a central principle of a whole category of green roofs that, at least in part, rely on natural regeneration (Fig. 10.1) for success. For example, in the United Kingdom, promotion of urban biodiversity has become a leading driver for green roof installation (Chap. 15). 'Biodiverse' roofs (formerly known as brown roofs) have become well established as the primary means by which this is achieved. Creation of these roofs attempts to mimic the ecological conditions of urban brownfield or post-industrial sites on the ground, which are often biodiversity hot spots in cities. The application of a range of substrate or surface materials (often 'urban', recycled or secondary in nature, or local soils and materials), the use of different depths of substrate, and the inclusion of designed habitat structures and features lead to a heterogeneous environment that maximizes ecological diversity (Fig. 10.2). Moreover, they emphasize local plant communities and vegetation sources by promoting the natural colonization of green roofs with wind-blown or bird borne propagules, or from the seed and vegetation bank within natural soils that might be incorporated with the green roof substrate (Chap. 6). These techniques give rise to a distinctive 'ruderal' flora, but also have many other ecological differences, compared with conventional green roof types, that are linked to variations in substrate depth, aspect and microclimate, and soil moisture availability. Likewise,

Fig. 10.1 Green roof vegetation created through spontaneous colonization, with log pile *(Nigel Dunnett)* (Sharrow School Green roof, Sheffield)

Fig. 10.2 Mixed growing substrate, a variety of textures and aggregate sizes, and varying growing media depths, spread over a green roof prior to encouraging colonization by ruderal species. *(Nigel Dunnett)*

standard approaches of seeding and planting can be used to augment the spontaneous ruderal flora.

Dynamic colonization processes widely apply across typical extensive, semi-intensive and intensive green roof types where greater biodiversity, a more sustainable approach to long-term management, increased local distinctiveness, climate adaptation, and greater aesthetic and visual interest are desired objectives. However, when assembled in an urban context, challenging questions arise about the true nature of plant communities and ecosystems: the resultant vegetation may bear no relation to any reference communities in the wild and instead represent a form of new or 'novel' ecosystem (Chap. 1) of effective urban ruderals or colonizers. In this chapter, the concept of the biodiverse green roof, with its highly dynamic nature and inclusion of colonization, succession and change as core elements of its functioning, will be discussed; the theoretical background of a trait-based or functional type (Chaps. 8 and 9) approach to working with green roof vegetation will be explored; and the wider role of 'ruderal' or disturbance-tolerant plant species in creating resilient and climate-adapted green roofs will be reviewed.

10.2 The Dynamic Green Roof

In order to fully understand the dynamics of plant communities (Chap. 12) it is necessary to move beyond the usual taxonomic approach for describing vegetation to one that is based upon functional types (Chap. 9). In other words, while it is usual to think of plant communities as being composed of a list of species that form repeatable associations under the same environmental conditions, it is also important to consider the traits and characteristics of that vegetation that enables it to be fully fitted to the environmental conditions pertaining to that system. For example, most designed landscape systems (including green roofs) use a pre-determined set of species as their vegetation components, and the system is managed to keep those species in place. But even from a more ecological viewpoint, restoration ecology or habitat template approaches to green roof creation usually rely on a tick-box inclusion of the suitable and typical representative list of species of a reference plant community (Chaps. 6 and 11). Thus, the inclusion of individual species becomes the main focus for the vegetation or plant component of the system.

However, this represents a relatively static snapshot of how vegetation works. The idea that plant communities in the wild are dependable entities, with little alteration in their composition or appearance from year to year is of course a misconception: change is fundamental to the processes that operate within natural or semi-natural plant communities. And while dynamic change in time and space is an inherent component of ecological systems, ironically, the objective of most landscape management operations is to halt, arrest, prevent or reverse that change. Furthermore, in the context of urban green roofs, the notion that a reference semi-natural plant community can be recreated is also open to question. A more ecologically-informed approach to creating biologically-diverse green roof systems is to consider how the *process* of change can be harnessed from the very beginning and throughout the life of the roof to foster a green roof ecosystem that is in tune with its wider environment.

By using this approach, it becomes more important that suitable types of plants initiate the establishment of a self-sustaining or integral vegetation into the long-term, i.e. the plants possess the best set of traits and adaptations for the given environmental conditions, rather than that there is a pre-determined set of species from the onset. In other words, do the components of the vegetation possess the best set of functional adaptations to those environmental conditions? Achieving this outcome when an inherent unpredictability about the precise species composition of the resulting vegetation exists challenges more traditional landscape planting approaches.

10.3 The Origins of Biodiverse Green Roofs

Largely as a response to the static nature of the ubiquitous thin and lightweight roofs which vary little in their design as well as the lack of any local and regional identity with such technologies, a revolution has been sweeping European green roofs to place biodiversity as one of the main drivers for green roof installation, and to develop green roof types that maximize the delivery of biodiversity objectives (Chaps. 8 and 15).

This movement originated in Switzerland in the early 2000s with the work of Dr. Stephan Brenneisen, largely centered on green roofs in the Swiss city of Basel. Brenneisen considered development initiatives for the potential of habitat creation on roof surfaces to compensate for the loss of habitats destroyed or damaged on the ground as a result of buildings. The key concept was to use local soils and substrate materials, with the recommendation that the top 15 cm (6 inches) of material from a construction site be removed and carefully stored so that some existing vegetation, seed bank and soil organisms can be preserved (Brenneisen 2006) (Chap. 6) in order to support locally appropriate plant and animal communities. Secondly, seed mixtures of vegetation types typical of the area are used, or roofs are simply allowed to colonize spontaneously with vegetation (both with propagules already present in that soil, or from wind or animal borne external sources). Brenneisen's original work focused specifically on the riverbanks and floodplain of the Rhine River which is particularly important for bird species that favor gravely river terraces or open meadows. Design features that have been found to be advantageous for birds in this region include dead wood (i.e. branches and tree trunks) that provide bird perches (and offer invertebrate habitat in their own right) and open areas of spread crushed roof tiles, pebbles or sparsely vegetated areas to provide foraging or nesting space. Study of bird usage of such roofs revealed that the principal reason for birds visiting the green roofs was foraging for food. The most frequently recorded species were black redstarts (*Phoenicurus ochruros*), wagtails (*Motacilla* sps), rock doves (*Columba livia*), and house sparrows (*Passer domesticus*), species naturally occurring in open landscapes such as higher mountain areas, on river banks, or in steppes with grasslands and bare stony ground and patchy vegetated areas (Baumann 2006).

However, the value of this approach to the invertebrate diversity of the Swiss green roofs soon became apparent. For example, seventeen green roofs in the city of Basel were monitored, including turf roofs, *Sedum* spp. roofs, and the specially designed roofs with landscaped surfaces created using local waste material substrates and rubble which were either left to colonize spontaneously or capped with thin layers of regionally distinctive topsoil (Brenneisen 2006). Two groups of invertebrates that are good indicators of vegetation structure were monitored: ground beetles and spiders. In the first 3-year period of the study, 78 spider and 254 beetle species were found. Fourteen (18 %) of the spider species and 27 (11 %) of the beetle species were classified as rare or endangered. Older green roofs tended to support more species than younger ones. Importantly, this and similar studies found thin layers of substrate dry out very quickly and develop less diverse ecologies than those with thicker substrate depths. A further key principle for design of roofs to support biodiversity includes variations in substrate depth across the roof. Deeper areas hold more moisture and also provide more opportunities for soil-dwelling organisms. Shallow, dry areas provide opportunities for specialist species. Where appropriate, drainage can be impeded or reduced to allow areas that remain moist during wet weather and dry out in dry weather in a way that mimics seasonal habitats in the wild, enabling a completely different assortment of plants to be used compared with the usual dryland species (MacIvor et al. 2011).

While these approaches to creating locally-distinctive green roofs, employing ruderal and spontaneous vegetation and locally-derived growing media, are seen as contemporary developments, they actually hark back to the earliest examples of green or sod roofs, such as those in Scandinavia, where soils from the immediate vicinity of the building were used, and the vegetation developed of its own accord (Dunnett and Kingsbury 2008). Never the less, the current concept of the biodiverse roof has changed profoundly perceptions of what green roofs are and can be, and has resulted in policies in several cities across Europe and elsewhere (Chap. 15) to promote such green roofs (Brenneisen 2006).

10.4 Redefining Urban Biodiverse Roofs

This value for urban birds of the Swiss biodiverse roofs heavily influenced a new generation of green roofs in London because they help conserve the habitat of the black redstart, a rare and protected bird restricted to industrial and post-industrial sites in several British cities. Mainly native to continental Europe on gravelly river bank habitats, the black redstart colonized bombed sites in the UK after the Second World War and post-industrial derelict urban sites in the 1960s (Grant 2006), where the warmer temperatures of the inner city and the stony, rubble surfaces provided an analog to its original habitat. Urban regeneration initiatives tend to clean up and eliminate these derelict brownfield sites, which can be urban ecological hotspots. Post-industrial brownfield sites, such as demolition sites, vacant sites awaiting development, abandoned railway sidings are characterized by free draining stony, low fertility surfaces that can be ecologically rich, particularly for their specialist invertebrate fauna and interesting spontaneous plant communities. If black redstarts are found to be breeding on a site then compensatory measures must be put into place if development is to occur. In response, the Black Redstart Action Plan has initiated a green-roof program in London to provide this endangered bird with increased urban breeding habitat (Wieditz 2003). New developments must include measures to protect against loss of its habitat. This is being achieved by designing biodiverse roofs with the original footprint of the building prior to development and which aim to recreate urban brownfield or post-industrial conditions on the rooftop.

Originally, these roofs were known as *brown roofs*: a term to describe roofs that use 'urban substrates' such as brick rubble, crushed concrete, sands, gravels and subsoils, often derived from the development site of the new building (Gedge 2003). Such roofs aim to recreate the conditions found in typical urban 'wastelands' or brownfield sites and are promoted for their potential value to rare invertebrates and ground-nesting birds. In the United Kingdom, where no central or local support is given for green-roof implementation, biodiversity and the mitigation for loss of brownfield land is one of the few levers that has been used to promote their wider use.

Currently the term 'Brown Roofs' is used less frequently, because of its negative aesthetic connotations, and the term 'Biodiversity Roof' or 'Biodiverse Roof' is used more commonly. While the same general principles apply, there is now greater rigor and control over substrate selection. The use of raw urban substrates can be problematic because of toxicity and texture, and many early examples failed to support vegetation because of substrate problems. Instead, known and defined materials or substrate formulations are used.

10.5 Functional Ecology, Diversity, and the Ruderal Strategy

Biodiverse roofs provide ideal conditions for so-called 'ruderal' plants: plants that have high dispersal and colonizing capacity, and which show adaptations to withstanding severe disturbance events such as droughts (Grime 2001). An awareness of the ecological theory relating to the colonization, early successional stages and persistence of ruderal communities and their role within a matrix of other plant communities and plant strategy types on biodiverse roofs provides an important basis for increased understanding about the long-term resilience of dynamic green roof vegetation assemblages. Applying ecological theory to green roof design and management extends the discussion around the biodiversity value of green roofs beyond whether or not they support species or groups of species to a much wider consideration of their functioning as ecosystems, the central topic of this book. As discussed later in this chapter, this does require an acceptance that urban systems, and artificial, created systems, have equal validity as models for studying and applying ecological theory as semi-natural or natural systems (Collins et al. 2000).

The most useful means to understanding the dynamic processes and ecological interactions on biodiverse roofs, and the plant communities that develop there, comes from Plant Strategy Theory, more recently known as Universal Adaptive Theory (Grime 1977; Grime and Pierce 2012), which categorizes plant species into competitor, stress-tolerant and ruderal (CSR) functional types. The great value of this theory comes from providing an overarching model for how plant diversity can be maintained over the long-term on green roofs, and helps us understand how dynamic ecological processes can be managed. The CSR model has proved to be a remarkably powerful tool for predicting how plants and other organisms react to changes within their environment (Dickinson and Murphy 1998).

Promoting diversity in vegetation in any situation relies primarily on damping the vigor of potential dominant species (i.e. those species that if left unmanaged will eliminate most or all other species in a plant community). It is far too simplistic to assume that the way to achieve vegetation diversity is by including a large number of plant species in the original planting scheme for a green roof because that greater diversity of species has to be resistant to competition and elimination from aggressive species. Dominant species are those that, in the absence of constraining factors, tend to eliminate other species through *competition*, resulting in low diversity or mono-specific stands of vegetation. It is easy to think of plants as being essentially passive organisms, unlike animals that actively hunt and compete with each other for food resources. However, where resources are abundant, plants can compete equally, fighting for the same unit of water, nutrient or light, and often in an aggressive manner, moving both roots, shoots and foliage to capture those resources. In this way, and in the absence of constraining factors, the best competitor for those resources will tend to be the winner in terms of space, eventually excluding less competitive species. This pattern holds for fertile, productive 'high energy' environments, but the importance of aggressive competition is reduced when certain constraining factors occur or are introduced to a habitat or ecosystem. It is, therefore, of great importance to understand what constitutes constraining factors that increase the diversity of plant communities (through reducing the vigor of aggressive species), and equally, to understand how to specify the conditions that will enhance that diversity.

Applying basic CSR theory starts with two fundamental sets of environmental constraints that limit the growth and survival of aggressive, potentially dominant species: (1) those that hinder the functioning of the plant, and thereby its growth rate and production of biomass, and (2) those that physically damage or destroy plant tissues or biomass already present. The first set of constraints is termed **stress** factors, involving abiotic constraints that affect the physiological processes of the plant. Such factors include extreme low or high temperatures, heavy shade, drought or low nutrient availability. The second set of constraints is termed **disturbance** factors and these biotic factors include herbivory, trampling, burning and cultivation. The relative combinations and proportions of stress and disturbance factors that operate within it can define every habitat on the Earth's surface, including the various types of green roofs. Over the course of evolutionary time, natural selection has resulted in plants that grow in environments subject to such pressures exhibiting adaptations that aid their survival and regeneration in those environments (Chaps. 6 and 11). Remarkably, unrelated species growing in geographically separated parts of the world show very similar responses to the same environmental pressures or constraints. Grime (1977) identified three basic responses or 'strategies' for survival in environments that are subject to the various combinations of high and low stress or disturbance (Table 10.1).

Table 10.1 Combinations of environmental stress and disturbance resulting in the three basic plant response strategies (Grime 1977)

		Intensity of stress	
		Low	**High**
	Low	Competitors	Stress- tolerators
		(C-strategists)	(S-strategists)
Intensity of disturbance	**High**	Disturbance	Uninhabitable
		Tolerators or Ruderals	
		(R-strategists)	

10.5.1 Competitors: The C Strategy

The combination of low environmental stress and disturbance is characteristic of typical 'productive' conditions (i.e. where nutrients and water are not in limited supply and regular physical damage is rare) that encourage vigorous plant growth and the dominance of aggressive species. Such conditions may be found, for example, on abandoned fertile agricultural fields, old unworked allotments or gardens, or unmanaged productive grasslands. Species well adapted to these environments tend to be tall herbaceous perennials, have spreading clonal growth and rapid summer growth rates. They are extremely effective **competitors** and tend to dominate vegetation, crowding out less vigorous species and resulting in low diversity stands. In effect, the competitive strategy is to maximize the capture of resources (light, water, nutrients) and to invest these in further growth to capture still more resources.

In terms of green roofs, the typical intensive roof or roof garden, where irrigation is used, substrate depths are generous, and fertilizers may be applied, fulfills these conditions. However, this type of green roof requires high-input maintenance to prevent dominance by competitors.

10.5.2 Stress-Tolerators: The S Strategy

Environmental stress and disturbance tend to limit the ability of competitive species to dominate. Restricted availability of resources (stress) prevents rapid growth (both in height and spread) thereby allowing species better adapted to growth under harsh conditions. Where resources occur in very limited supply (i.e. in stressed environments) plants evolved very different strategies to those of the competitors. Rather than exhibiting rapid rates of growth, **stress-tolerant** species tend to grow slowly, produce persistent foliage often with modified protective tissues, and utilize specialized physiologies (Chap. 11). In general, stress tolerant vegetation tends to be unproductive, relatively sparse and with low biomass. In such 'low energy systems' (Dickinson and Murphy 1998) plants tend to reproduce primarily through vegetative growth rather than through seed. In effect the stress-tolerant strategy is one of thrift: to make the most of captured resources by sitting tight rather than investing in rapid growth to capture more resources. Examples of relatively stressed habitats include low fertility, acidic or calcareous, and drought stressed grasslands (Chap. 6) and the shady, understory habitat of woodlands.

In terms of green roofs, the typical extensive green roof, with thin layers of low-fertility substrate, and minimal irrigation fulfills these conditions. The low-growing, evergreen, highly drought-tolerant *Sedum* spp. layers that most typically cover these types of green roofs possess many of the typical features of stress-tolerant vegetation. *Sedum* spp. use specialist CAM photosynthesis for adaptation to arid conditions whereby plants close their stomata during the day to reduce evapotranspiration, and open them at night to absorb carbon dioxide thus conserving water. Also some Sedum species can switch between C_3 and CAM photosynthesis.

10.5.3 The R Strategy: Ruderals

Environments where disturbance or destruction of vegetation is a regular occurrence have given rise to plant strategies that either avoid or enable rapid recovery from that disturbance. Although naturally disturbed environments include screes and landslides, shingle beaches and sand dunes, the majority of disturbed environments are human-influenced (e.g. cultivated fields and agricultural grasslands). Plants adapted to such environments tend to show rapid growth rates and a reliance on reproduction through seed as well as vegetative expansion to allow rapid colonization of bare and disturbed areas. For example, annuals are adapted to regular severe disturbance: their rapid growth rate and copious seed production enable them to take quick advantage of bare ground following a disturbance event, and to ensure their survival into future generations before another disturbance. Biennials and short-lived perennials are similarly adapted to disturbances on a longer time cycle. In effect the **disturbance tolerant** strategy or **ruderal** strategy (named after the roadside habitats from which the disturbance-tolerant life-history was first described) provides an insurance policy: investing resources in mechanisms that ensure a rapid response to predictable disturbance.

In terms of green roofs, ruderal species rely on the dynamic processes of colonization and regeneration to sustain their integrity into the long-term. Appropriate substrate depth is key to maximizing ruderality. Too shallow or unfertile substrates make it uninhabitable thus impossible for ruderal species to develop sufficient biomass to establish and reproduce effectively and therefore to maintain vegetation cover. Too deep and productive and aggressive, competitive species will dominate. Ruderals are generally short-term occupants of space, unless regular disturbance opens gaps in the vegetation for their regeneration.

The three main strategies listed above are extremes. In reality most species exhibit combinations of traits from the different strategies and possess intermediate strategies, depending upon the exact environmental conditions to which they are adapted (Fig. 10.3). The crucial point is that, in terms of the maintenance of diversity in vegetation, low stress combined with low disturbance is not good, favoring the aggressive competitor species. Equally, combinations involving high intensities of stress and/or disturbance produce hostile conditions for plant growth, restricting vegetation to limited numbers of highly adapted species. In general, greatest species diversity is promoted at moderate intensities of environmental stress and/or disturbance. This is easily illustrated with reference to various grassland types. The more species-rich semi-natural grassland types tend to occur on relatively low fertility, free draining acid or calcareous soils (moderately stressed) or, in the case of traditional hay meadows, on relatively fertile sites subject to moderate disturbance (hay cutting and after-grazing) (Smith et al. 1996). Addition of fertilizers (reducing stress) or removal of maintenance (reducing disturbance) will result in these grasslands becoming dominated by aggressive competitive grasses, with associated loss of diversity.

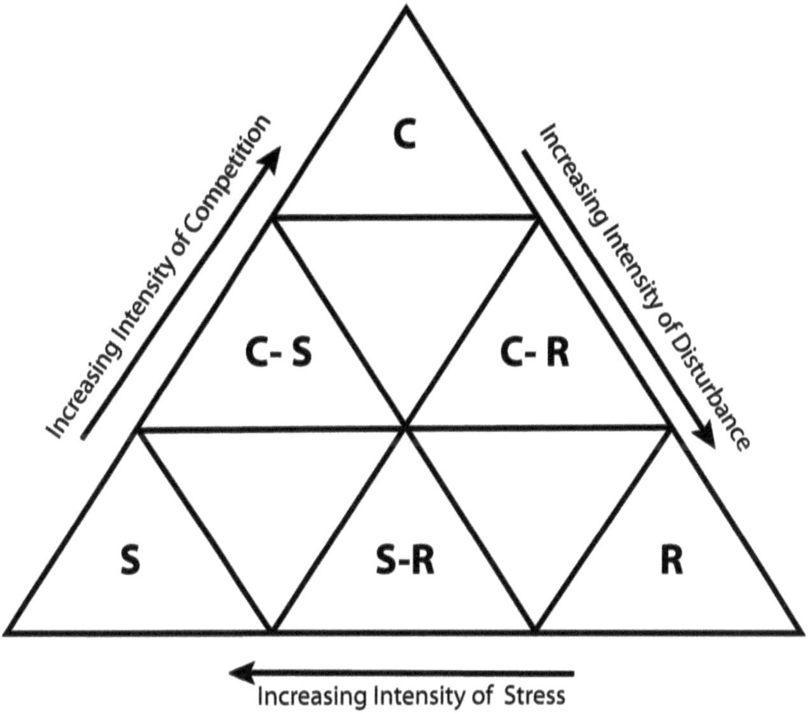

Fig. 10.3 The three primary plant strategies, (S, C and R) and intermediate strategies. (Adapted from Grime 1977)

10.5.4 Disturbance and Diversity

Ecologists have recognized that periodic or 'intermediate' disturbance (as opposed to severe or very limited disturbance) can be the key to the maintenance of high diversity in ecosystems (Connell 1978). This mitigates against the still often held view of 'climax' or permanent vegetation being at stable equilibrium with its environment and having some form of long-term steady state. Indeed, static, non-disturbed plant communities can drift towards a lower diversity state. Ruderal species that are able to respond to these disturbance events contribute to the greater diversity of an ecosystem, and its greater resilience to that periodic disturbance. This resilience and adaptability is potentially of great importance in maintaining the integrity of ecosystems and plant communities into the future. Indeed, while it is generally assumed that ecosystems respond to environmental change slowly and gradually, the reality is that loss of resilience in ecosystems can result in rapid and dramatic change and responses to environmental change, and potentially a complete shift to an alternative, less diverse state (Scheffer et al. 2001). Ecosystems therefore have more than one potential stable state (Schröder et al. 2005), some of which may be more desirable than others. In theoretical terms it is vital to consider the inclusion of 'stabilizing forces' in ecosystems to mitigate against the destructive effects of external 'destabilizing forces' (Holling 1996).

10.6 A Closer Look at Spontaneous Urban Ruderals, and Novel Ecosystems

The biodiverse green roof is just one example of the application of ruderal plant communities in an urban context. Interest in these urban communities originated in the former West Berlin, where the whole modern science of urban ecology as a separate field of study was realized (Sukopp et al. 1979; Sukopp 2003). Interestingly, the same group of urban ecologists who recognized the value of urban ruderal communities also recognized the special nature of the *Sedum*-based vegetation and grasslands that developed on flat rooftops, and that gave rise to the modern German green roof industry (Chap. 1).

Ruderal (literally meaning 'rubble') plants had previously been so-named because of the their tendency to be found in the human-disturbed edges of roads, tracks and railway lines, as well as in agricultural fields. Botanists had recognized from the nineteenth century that alien plants were commonly found in such situations. However, it was in Berlin (and many other European cities of course) that massive areas were transformed into derelict and abandoned rubble fields following the Second World War. Interestingly, because of the special political situation in Berlin, vast areas were left undeveloped and a unique urban flora became established (Lachmund 2013). This resource was studied intensively and detailed classifications of the different urban plant communities were constructed (e.g., Sukopp et al. 1979). A striking observation was the large number of 'neophytes' or non-native species that made up these urban plant communities, and that most of these originated from further south in Germany; such species were responding to the warmer conditions in the city compared to the surrounding hinterlands. German green roof plants were also the focus of studies at this time (e.g., Kreh 1945, (in German), cited by Kohler 2006). The studies indicated that early constructed green roofs in Germany (created from the end of the nineteenth century onwards) with growing media of sand and gravel at depths of up to 200 mm (8 inches) developed grassy or sedum-moss vegetation but also supported many ephemeral annual plant species.

Exactly the same urban rubble conditions are created as part of the general cyclical processes of urban development and economic boom and bust. The associated urban ruderal and neophyte communities are a common feature of most cities, and their ecological value has come to be widely recognized. As discussed in Sect. 10.4, these brownfield sites, which most people think of as demolition sites, wasteland or derelict land, are often very rich in wildlife, with a great diversity of ecological conditions, with very free-draining substrates ideal for the development of diverse vegetation, but also in places areas of poor drainage give rise to mini wetlands. Because they are seen as derelict, few people visit them, so wildlife remains undisturbed (Gedge 2011).

One of the first people to draw wider attention to urban brownfields was the ecologist Oliver Gilbert who in his book 'The Ecology of Urban Habitats' (Gilbert 1991) coined the phrase 'urban common' to denote the vegetation that developed on derelict urban sites, and considered the successional changes that ensue from open bare rubble, through ruderal phases, to the long-term development of urban woodland. In terms of biological diversity, Gilbert identified the early successional stages as by far the most valuable. He also took a pragmatic view of their temporality, noting their relationship to the economic cycle and that brownfield and development sites tend to get built upon eventually. However, he pointed out that in cities there tends to be a patchwork or mosaic of such sites being continuously produced and liberated, and that the ruderal nature of the typical plants of such sites means that there tends to be a continuous source of propagules and sites, despite the lack of long-term futures for many individual sites.

Outside of Europe, the positive viewpoint of the value of urban ruderal communities has had a slower response, although increasing interest occurs from ecologists in such plant community dynamics (Gallagher et al. 2011). In North America, Peter Del Tredici (2010) argues that spontaneous urban ruderal species that are fully adapted to urban conditions are the true urban plants of the future, even though most of them may not be native. He argues that traditional restoration ecology principles of looking backward to native plant communities of the region, and removing alien species, is an approach that is no longer suited to cities subjected to a changing climate. Instead, he advocates a whole-hearted embrace of ruderal communities and a celebration of 'cosmopolitan urban vegetation'.

This attitude reflects much current European thinking on the integration of ruderal plant communities more formally into the network of urban green infrastructure. Two approaches are emerging: (a) because of significant concerns about public responses to the aesthetics of spontaneous urban ruderal communities, additional decorative or ornamental species may be added (e.g. Koppler et al. 2014); or (b) native species of nature conservation value may be added into the background matrix of the urban ruderals (e.g. Fischer et al. 2013).

This formal adoption of spontaneous urban ruderal communities, and the human manipulation of them again raise the issue of novel plant communities and ecosystems. The concept of a novel ecosystem, bearing no ecological or geographical affinity to a recognized semi-natural community was defined by Hobbs et al. (2006), and the place of urban wasteland ruderal communities of native and non native species within that concept highlighted by Kowarik (2011). But the human manipulation of urban ruderal vegetation (itself already highly responsive to human interference) in a designed or planned human landscapes (such as on green roofs) by adding additional species to it, takes the concept of novel ecosystems a stage further, to that of designed novel ecosystems, whereby species without geographical affinity, but which are ecologically fitted are used to maximize ecological functioning.

These are all exciting and radical concepts, many of which turn perceived wisdom and accepted ecological practice on its head. These concepts, however, become fully justified in ecological terms if one accepts the premise outlined at the beginning of this chapter. The premise shifts our attitude from seeing plant communities as a pre-determined list of names, to one where plant communities are seen as an assemblage of plant species that contain particular functional types fitted to a situation. This attitude shift opens up new possibilities for planned urban landscapes including green roofs.

10.6.1 Ruderal Green Roofs

The preceding account touches many possibilities and permutations for working with ruderal vegetation on green roofs. Because one of the main defining traits of plants with a ruderal strategy is that they produce large amounts of easily dispersed seed, that germinates readily in suitable conditions, the actual colonization or vegetation establishment aspects can be relatively straight-forward. However, because ruderals also tend to be short-lived, questions arise about their persistence. Furthermore, ruderal vegetation does not necessarily abide by the rules of more conventional vegetation aesthetic. Therefore issues relating to public perception need to be considered.

One commonly used approach to heterogeneity is to vary both the depth and type of growing medium within a green roof. Varied depth of the growing medium will allow for a range of microhabitats and growing conditions, thus providing both dry (xeric) conditions with sparsely vegetated xeric areas and more lush areas of taller vegetation (Chaps. 4, 5, 6 and 11). The effect of the varied substrate depth will also provide slopes for a range of burrowing invertebrates to find opportunities to nest (Chap. 14). Mounding also creates differences in aspect and sun and shade—translating into different abiotic conditions. Varying the type of substrate will provide different opportunities for invertebrates. Rubble and stony areas will provide shelter and nesting opportunities for certain species. Sandy areas and commercial brick based substrates/media will allow burrowing bees and other invertebrates to have nesting opportunities.

As well as maximizing opportunities for faunal diversity, these techniques also maximize floral diversity. Returning to Plant Strategy Theory, and the environmental conditions that favor ruderals, a mosaic or variety of environmental conditions on the roof will favor the persistence of the ruderal vegetation. Ruderal species require disturbance for their regeneration, creating patches of open substrate for seed regeneration. Where growing medium depths are so thin as to create severe moisture or nutrient stress then only stress-tolerant species will persist (or no vegetation at all). Where depths are sufficiently deep to foster productive, competitive vegetation then there will be little opportunity for re-colonization. Therefore, intermediate depths will support biomass production whilst either producing sufficiently sparse vegetation to enable regeneration, or fostering vegetation that is susceptible to damage by disturbances such as drought. Typical biodiverse roofs that encompass ruderals usually employ other elements to further increase biodiversity: log piles and stones/boulders can add another dimension to a small-scale roof. Not only do they provide habitat and opportunity to create some interesting designs, they also provide additional ecological niches for plants through provision of some shade and protection from the wind (Gedge 2011). By creating the range of conditions across a single roof, the interplay between productivity, stress and disturbance will encourage a greater diversity of species across the range of conditions on that roof.

10.6.2 Spontaneous Colonization

Using spontaneous colonization of bare substrates is the most 'ecological' method for generating ruderal vegetation: only those plants that are locally available and totally suited to the roof environment will establish and those that survive are clearly fitted to the prevailing conditions. It is also the least expensive vegetation establishment method. But there are some disadvantages. Firstly, only those species that are able to reach the roof will have a chance of establishing. A source of propagules may be lacking in the vicinity, reducing the potential diversity of the roof. Secondly, by definition, many ruderal species that are likely to make it to the roof are by their very nature 'weedy' species. Far from creating a species-diverse roof, which in turn will support a wider faunal diversity, spontaneously colonized roofs can be characterized by a very low diversity flora dominated by a small number of rampant, aggressive weedy species. For example, in a survey of the medium-term dynamics of a spontaneously colonizing green roof in Sheffield, UK, just three species made up 50 % of all the individual plants on the roof (Dunnett et al. 2008). There is also danger in relying on spontaneous colonization on dry and skeletal substrates that may result in low or failed vegetation cover (Grant 2006).

10.7 Beyond Spontaneous Colonization: Biodiverse Green Roofs and Aesthetics

The few studies that have been undertaken into public aesthetic preferences relating to green roofs indicate a general unease and dislike for the more wild types of green roof in cities. For example, in a study of the attitudes of office workers in Toronto and Chicago, Loder (2014) found that 'prairie' type green roofs were generally not well liked because of their perceived untidiness and lack of suitability to the urban context, although they did foster feelings of interest and intrigue. Certainly, aesthetic criteria have generally not been as high of a priority in the creation of biodiverse green roofs as ecological considerations (Dunnett 2006). While it is argued that widening the concept of aesthetics for green roofs to encompass notions of 'ecological beauty' (Sutton 2014) will enable people to better understand the ideas behind biodiverse roofs, long-established research (e.g., Nassauer 1995) indicates that consideration has to be given to enhancing purely wild vegetation with designed inputs such as greater color, structure or order if they are truly to become mainstream, accepted and 'loved' as the default green roof condition in cities.

Combining opportunities for natural colonization, with deliberate introduction of additional species by seeding and/or plug or container-grown planting will introduce a greater floral diversity, and will provide the means for establishing known and desirable plants (those that have particular ecological or aesthetic value). Adding additional species is a useful strategy for overcoming public concerns over the aesthetic impact of visible or accessible biodiverse green roofs.

10.8 Case Study: Sharrow School, Sheffield, UK

Started in 2007, this ambitious project created a ruderal biodiverse roof on top of a new school building. There was limited space on the ground, so playgrounds, outdoor teaching spaces and gardens were installed at different levels of the building itself. The whole of the upper level of the school is a biodiverse roof, supporting a range of habitats and vegetation types including dry perennial meadow, annual self-seeding meadow, limestone grassland, and urban brownfield vegetation types. Webcams allow children to view the bird and insect life on the roof and a weather station provides continuous information on weather conditions (Dunnett and Kingsbury 2008).

The roof was initially spread with a standard layer of 100 mm (4 inches) of a crushed-brick-based commercial green roof substrate. Additional materials were placed above this base layer to create a varied, mounded topography, with typical mounds of up to 350 mm (14 inches) in height, and one reaching 450 mm (18 inches) where the underlying building structural support was suitable. Substrate materials included the commercial free draining substrate, crushed limestone, sands and gravels. The substrate was installed and the roof was left for 1 year to enable spontaneous colonization. All plants reaching the roof were either wind-blown or reached the roof through some other non-deliberate means (for example via birds or bird droppings) because the substrates used contained no seed bank or viable vegetative fragments.

After 1 year, the main plant species that had colonized were typical urban ruderals, resulting in a cosmopolitan mix of native and non-native species (Nigel Dunnett, unpublished data). The most abundant were mullein (*Verbascum thapsus*), red valerian (*Centranthus ruber*) (Fig. 10.4), purple toadflax (*Linaria purpurea*) (Fig. 10.5), common wormwood (*Artemisia vulgaris*), butterflybush (*Buddliea davidii*) and bull thistle (*Cirsium vulgare*). However, 50% of the surface of the roof remained uncolonized. In the following 2 years, aggressive colonizers such as *A. vulgaris* and *B. davidii* became dominant, resulting in a low diversity, and very weedy-looking vegetation in the areas of deeper substrate. Active management resulted in the removal of the majority of individuals of these two species, and additional species, both native and non-native but typical of dry meadow habitats were introduced through seed mixes. Importantly, these included many floriferous species with colorful flowers, such as common viper's bugloss (*Echium vulgare*). However, at the same time, maintenance was minimal, allowing ongoing spontaneous colonization, as well as regeneration of the introduced species. The addition of colorful flowering species was an essential component of the success of this green roof because of its high usage and visibility to school children and teachers. The green roof is now a highly diverse assemblage of species, with a strongly urban character. Its wider value for biodiversity has been recognized by the UK government's nature conservation agency, Natural England, and in 2010, the Sharrow School was given an official government designation of a 'Local Nature Reserve'. To date, the Sharrow School is the only such purposely-designed green roof in the world to have received this official government nature conservation designation, and it is of note that this is a 'novel ecosystem' in the sense described earlier in this chapter.

Fig. 10.4 Colonizing ruderal species on the roof of Sharrow School, Sheffield, including red valerian (*Centranthus ruber*) (*Nigel Dunnett*)

Fig. 10.5 Purple toadflax (*Linaria purpurea*), a common ruderal plant of urban brownfield sites in the UK, colonizing the green roof on Sharrow School (*Nigel Dunnett*)

10.9 The Wider Role of Ruderals on Green Roofs

Plant species with a ruderal character have a wider role in promoting greater diversity in green roof vegetation, and in developing new types of climate-adapted green roofs that are resistant to, and resilient in the face of the challenges of an unpredictable and changing climate. Whilst ruderals in natural or semi-natural contexts are colonists, invaders and opportunists, and arrive of their own accord or via other non-deliberate vectors, in designed or managed landscapes such as green roofs, they can also be introduced deliberately as components of the intended vegetation. In accordance with Plant Strategy Theory, plants with a ruderal character are not just annuals and biennials, but also perennial species that have a ready capacity for dispersal and regeneration.

Some annuals are suited to green-roof cultivation. These are generally desert annuals adapted to surviving in stressful dry, hot conditions, avoiding the most harsh time of the year as dormant seed and germinating, growing, and flowering during more benign periods. The most successful annuals for green roofs are those that will self-seed from year to year once established. In the UK, species such as *Linaria maroccana, Gypsophila muralis, Silene armeria* and *Alyssum maritima* will persist under typical extensive green roof conditions (Fig. 10.6). While many such species will respond to positively to increasing depth of growing medium, in more diverse mixes there can be differences in performance. For example, the low-growing, drought-tolerant *Alyssum maritima* will persist in more shallow growing media

Fig. 10.6 Introduced annuals and biennials and self-seeding perennials add a dynamic element to green roof vegetation, regenerating from year to year *Silene armeria* on the green roof at Sharrow School, Sheffield (*Nigel Dunnett*)

depths 50–100 mm (2–4 inches) where other species are unable to grow successfully, but is out-competed by taller species at growing media depths greater than 100 mm (4 inches). Under green-roof conditions it is most likely that self-seeded annuals will germinate in autumn, overwinter, and flower in spring. Nagase and Dunnett (2013) found that annual mixes for green roofs can be established successfully and will result in very long flowering periods, depending on the species included. They also found that, in the UK climate, irrigation was not necessary for successful establishment, however, higher sowing densities were required to compensate for lack of irrigation. Annuals can persist indefinitely within green roof systems and can be important components in filling fluctuating gaps from more permanent vegetation (Kircher 2004). Because they are very floriferous, often with brightly colored flowers, annuals can be important in raising the aesthetic impact of ruderal roofs, but can also be a significant resource for pollinating insects.

Biennials and short-lived perennials perform similar roles but clearly work on longer timescales. However, the possession of ruderal characteristics may be very important in the persistence of diverse assemblages of species with a more perennial character. For example, in a long-term study of the vegetation on the green roof at Moorgate Crofts, Sheffield, UK (Fig. 10.7), the change in abundance of all included species was monitored over a 7-year period. The green roof was designed to have minimal input of resources (i.e. no fertilizer, minimal or no irrigation) and maximum visual impact. The aim was to use naturalistic species mixtures to achieve the sort of visual qualities normally associated with intensive roof gardens. Substrate depth varies from 10 to 20 cm (4–8 inches). Forty-two species, typical of dry grassland

Fig. 10.7 Vegetation on Moorgate Crofts green roof in the fall, with abundant seed heads (*Nigel Dunnett*)

Fig. 10.8 Vegetation on Moorgate Crofts green roof in the spring, with forbs such as *Pulsatilla vulgaris* and *Primula veris*, which have increased dramatically across the roof through self-seeding (*Nigel Dunnett*)

habitats in Western and Southern Europe were planted. The exact numbers of each species were known at time of planting because all the species, including those with ruderal character, were introduced as pot-grown plants, with the exact quantities of each known. A full survey of the roof was undertaken in summer 2012, and the exact numbers of individuals of all species counted. Those species that had increased in abundance (13 out of the original 42), such as *Primula vulgaris* and *Pulsatilla vulgaris* (Fig. 10.8), shared ruderal characteristics of effective seed or vegetative regeneration. The more stress-tolerant species included in the original plantings had been eliminated through competition with the more ruderal species, while the more competitive species also had not survived into the long-term because of the inherent low productivity of the site (Dunnett, N. unpublished data). This indicates that the most important factors for the long-term integrity of dynamic or biodiverse green roofs (at least in temperate climates) may be as much related to the regenerative capacity of the component species, as well as to the more conservative strategies of long term vegetative persistence under conditions of severe stress.

10.10 Conclusion

This overview of the place of ruderals in green roof design and function ultimately presents, to some extent, the picture of a dynamic, unpredictable, system. It is one where chance factors dictate the precise species composition of the vegetation and dramatic may change the composition from year to year. And yet ironically, ruderals have a role to play in the maintenance of a resilient and stable long-term green roof ecosystem. This apparent contradiction becomes less of a paradox if we take a functional view of green roof vegetation dynamics, rather than a rigid, taxonomic view of the vegetation composition. By applying plant strategy theory it is possible to interpret the composition of a green roof through the equilibrium or balance between species. Those able to regenerate effectively and respond dynamically to environmental change and disturbance contrast with those able to sit tight and ride out any severe external stresses and disturbances. Typical extensive green roof vegetation mostly contains species that fall into the second category. In this chapter, it is argued that not only does inclusion of a more dynamic ruderal element increase the visual interest and wider biodiversity value of a green roof, but that it becomes an essential component if we are to truly apply ecological theory and processes to the creation of ecologically-functioning green roof systems.

10.11 Research Questions

Many research questions remain unanswered regarding the application of this dynamic approach that applies ecological theory and process to green roof ecosystems. Those questions span both theory and application.

- Is it possible to set off a predicable, long-term succession from the outset of green roof creation?
- How does inclusion of a compatible mix of shorter-term ruderals combined with longer term and more persistent species alter the later composition?
- Are regeneration characteristics of component species more important, or as important, as those that encourage persistence under stress?
- Are periodic management interventions necessary or desirable to maintain the ruderal element in a green roof?
- Does the type and timing of the disturbance/intervention affect later species composition?
- How can the encouragement of a dynamic and spontaneous approach to green roof vegetation be reconciled with human aesthetic aspirations?

References

Baumann A (2006) Ground-nesting birds on green roofs in switzerland: preliminary observations. Urb Habitats 4:37–50

Brenneisen S (2006) Space for wildlife: designing green roofs for habitat in Switzerland. Urb Habitats 4:27–36

Collins JP, Kinzig A, Grimm NB, Fagan WF, Hope D, Wu JG, Borer ET (2000) A new urban ecology. Am Sci 88:416–425

Connell JH (1978) Diversity in tropical rain forests and coral reefs. Science 199(4335):1302–1310
Del Tredici P (2010) Wild urban plants of the northeast: a field guide. Cornell University Press, NY
Dickinson G, Murphy K (1998) Ecosystems. Routledge, London
Dunnett N (2006) Green roofs for biodiversity: reconciling aesthetics with ecology. In Proceedings, Greening Rooftops for Sustainable Cities Boston 2006. Green Roofs for Healthy Cities, Toronto
Dunnett N, Kingsbury N (2008) Planting green roofs and living walls. Timber Press, Portland
Dunnett N, Nagase A, Hallam A (2008) The dynamics of planted and colonising species on a green roof over six growing seasons (2001–2006): the influence of substrate depth. Urb Ecosyst 11:373–384
Fischer L, von der Lippe M, Rillig M, Kowarik I (2013) Creating novel urban grasslands by reintroducing native species in urban grassland. Biol Conserv 159:119–126
Gallagher F, Pechmann I, Holzapfel C, Grabosky J (2011) Altered vegetative assemblage trajectories within an urban brownfield. Environ Pollut 159(5):1159–66
Gedge D (2003) From rubble to redstarts. Greening rooftops for sustainable communities, Proceedings of the First North American Green Roofs Conference, Chicago, May 2003. Toronto: The Cardinal Group
Gedge D (2011) Green roofs for biodiversity. In: Dunnett NGD, Little J, Snodgrass E (eds) Small green roofs. Timber Press, Portland
Gilbert O (1991) The ecology of urban habitats. Chapman and Hall, London
Grant G (2006) Extensive green roofs in London. Urb Habitats 4:51–65
Grime J (1977) Evidence for the existence of three primary strategies in plants and its relevance to ecological and evolutionary theory. Am Nat 111:1169–1194
Grime J (2001) Plant strategies, vegetation processes and ecosystem properties. Wiley, Chichester
Grime J, Pierce S (2012) The evolutionary strategies that shape ecosystems. Wiley-Blackwell, Chichester
Hobbs R, Arico S, Aronson J, Baron J, Bridgewater P, Cramer V, Epstein P, Ewel J, Klink C, Lugo A, Norton D, Richardson D, Sanderson E, Vallandare F, Vila M, Zamora R, Zobel M (2006) Novel ecosystems: theoretical and management aspects of the new ecological world order. Glob Ecol Biogeogr 15:1–7
Holling CS (1996) Engineering resilience versus ecological resilience. In: Schulze P (ed) Engineering within ecological constraints. National Academies Press, Washington, pp 31–44
Kircher W (2004) Annuals and sedum cuttings in seed mixtures for extensive roof gardens. Int Conf Urb Hortic: Acta Hortic 643:301–303
Kohler M (2006) Long-term vegetation research on two extensive green roofs in Berlin. Urb Habitats 4:3–25
Koppler M, Kowarik I, Kuhn N, von der Lippe M (2014) Enhancing wasteland by adding ornamentals: opportunities and constraints for adding steppe and prairie species on urban demolition sites. Landsc Urban Plan 126:1–9
Kowarik I (2011) Novel urban ecosystems, biodiversity and conservation. Environ Pollut 159:1974–83
Kreh W (1945) Die Pflanzenwelt unserer Kiesdächer. Ver. f. vaterlaendische Naturkunde in Wuerttemberg 97–101:(1941–1945: 199–219)
Lachmund J (2013) Greening Berlin. The co-production of science, politics, and urban nature. MIT Press, Boston/Mass
Loder A (2014) 'There's a meadow outside my workplace': a phenomenological exploration of aesthetics and green roofs in Chicago and Toronto. Landsc Urban Plan 126:94–106
MacIvor J, Ranalli M, Lundholm J (2011) Performance of dryland and wetland plant species on extensive green roofs. Ann Bot 107(4):671–679
Molineux CJ, Fentima CH, Gange AC (2009) Characterising alternative recycled waste materials for use as green roof growing media in the UK. Ecol Eng 35:1507–1513
Nagase A, Dunnett N (2011) Establishment of an annual meadow on extensive green roofs in the UK. Landsc Urban Plan 112:50–62

Nagase A, Dunnett N (2013) Establishment of an annual meadow on extensive green roofs in the UK. Landsc Urban Plan 112:50–62

Nassauer J (1995) Messy ecosystems, orderly frames. Landsc J 14:161–170

Scheffer M, Carpenter S, Foley JA, Folke C, Walker B (2001) Catastrophic shifts in ecosystems. Nature 413(6856):591–596

Schröder A, Persson L, De Roos AM (2005) Direct experimental evidence for alternative stable states: a review. Oikos 110(1):3–19

Smith RS, Buckingham H, Bullard MJ, Shiel R, Younger A (1996). The conservation management of mesotrophic (meadow) grassland in northern England. 1. Effects of grazing, cutting date and fertilizer on the vegetation of a traditionally managed sward. Grass Forage Sci 51:278–291

Sukopp H (2003) Flora and vegetation reflecting the urban history of Berlin. http://www.die-erde.de/DIE_ERDE_2003-3_SukoppX.pdf. Accessed 6 May 2015

Sukopp H, Blume H, Kunick W (1979) The soil, flora and vegetation of Berlin's wastelands. In: Laurie I (ed) Nature in cities. John Wiley, Chichester, pp 115–132

Sutton R (2014) Aesthetics for green roofs and green walls. J Living Archit March 2014 1(2):22

Wieditz I (2003) Urban biodiversity: an oxymoron? Green Roofs Infrastruct Monit 5(1):8–9

Dr. Nigel Dunnett is Reader in Urban Horticulture at the University of Sheffield. His interdisciplinary research focuses on the interface between ecology, design and horticulture in urban green space and built development.

Chapter 11
Assembling Prairie Biome Plants for Minnesota Green Roofs

L. Peter MacDonagh and Nathalie Shanstrom

Abstract Green roofs are currently seen as an ecologically sustainable practice, but in fact they can be unstable and vulnerable. Low-diversity green roofs can fail, because they are not resilient. Within prairies many specialized plant communities occur in hot, dry, windy places with thin, poorly developed soils. These diverse but stressed communities offer suitable local templates for assembling sustainable and ecologically robust extensive green roof plants, since growing conditions in these communities closely mimic those found on green roofs. As the nascent green roof movement of North America establishes, the opportunity arises to embed these new landscapes with ecological robustness.

Keywords Prairie · Bedrock bluff prairie · Prairie green roof · Twin Cities: Minneapolis · Habitat analog · Habitat template · Plant assembly rules

11.1 Introduction

For green roof plants, understanding where they come from (plant geography of its related biome) plus its individual plant characteristics prove useful when identifying, assembling, growing, and managing potential plants. This chapter presents a longitudinal case study for prairie biome green roofs (MacDonagh et al. 2006). In it we look at the general characteristics of the North American prairie biome then a subset of prairie communities with shallow soil profiles and stressful growing conditions in Minnesota (Anderson 2007; Curtis 1959, 1971; Gruchow and Burkey-Harris 1999; MN DNR 2005b; Wovcha et al. 1995) as a template (Lundholm 2006; Sutton et al. 2012a) to find, select, combine, use, manage, and monitor perennial herbaceous forb and graminoid analogs on two Minneapolis green roofs (Fig. 11.1). Finally, we describe and discuss a decade of experience designing, planting and growing green roofs based on prairie community templates in the Minneapolis area. The ideas and process presented here could be applied worldwide to many types of prairie or meadow plantings for extensive semi-intensive green roofs.

L. P. MacDonagh (✉) · N. Shanstrom
The Kestrel Group, 7109 Ohms Lane, Minneapolis MN 55439, USA
e-mail: peter@tkdg.net

R. Sutton (ed.), *Green Roof Ecosystems,* Ecological Studies 223,
DOI 10.1007/978-3-319-14983-7_11

Fig. 11.1 Analogy of green roof as bedrock bluff prairie: both have thin soils, high wind and sun exposure, and are places of prospect over the surrounding landscape (Kestrel Design Group)

11.2 Plant Assembly Rules for Green Roofs

Restorationists use rules of species assembly to detail the context and requirements for selecting plants to fit environments they are considering to restore (Franklin 1987; Temperton et al. 2004) (Chap. 1). These can be applied in a general way for creating analogs of prairies on green roofs (Fig. 11.2) and that process structures

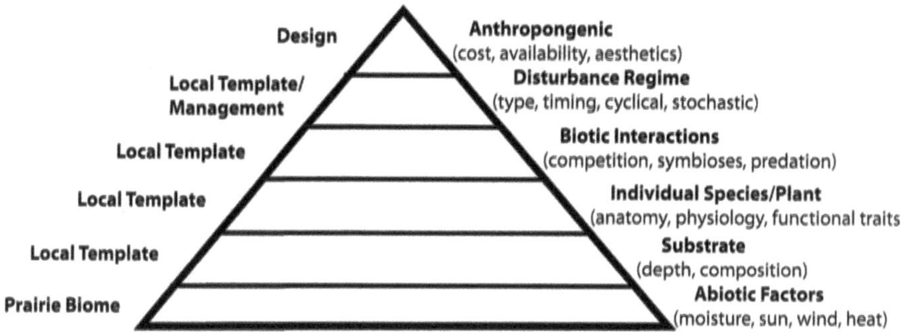

Fig. 11.2 Hierarchy of prairie green roof plant assembly factors and their applicable levels

this chapter. Firstly and most importantly we examine the abiotic factors of climate and microclimate (Chap. 3) including seasonal high and low temperatures, drought, sun/shade relationships, wind, humidity, and potential evapotranspiration (PET). Closely related are local edaphic characteristics like soil texture, drainage and nutrient availability (Chaps. 5 and 6), though they vary widely within the prairie biome (Sims and Risser 1999). Next come influences on individual species and plants such as phenology, plant and root architecture, leaf and stem characteristics, flowering, seed set and vegetative propagation (Weaver 1968). Even within a species broad ecotypes range across different latitudes (Cornelius 1947; Sutton 1990). Plant selection can tap plant functional groups tied to life-form, photosynthesis, respiration, and carbohydrate synthesis; functional group similarities may or may not embed within genetic similarities (Chaps. 8 and 9). Higher order interactions such as intra and inter-species competition, symbioses, mutualisms, and interactions between plants and animals may also be considered and tie to the local prairie template (Greenberg 2002; Wovcha et al. 1995). Disturbance regimes may include abiotic local influences such as drought, flooding, shading/insolation, fire, freezing, abrasion, wind dislodgement, and nutrient deficiencies or biotic influences such as herbivory, diseases, and maladapted genetics. Human constraints on green roof planting involve costs, availability, and our aesthetic affinities (Chaps. 10 and 16). In this process, all or some of the factors listed above may become filters to guide initial plant selection.

Individual plant candidates for extensive green roofs can be understood as following a hierarchy of factor importance that aids the sequence of species assembly (Fig. 11.2). The most fundamental is a species' tolerance of a green roof's ubiquitous drought, wind and heat stress such as found at the prairie biome level (MacDonagh 2006). Next, with the local template, comes the ability to grow in restricted root space with substrates of low to minimal nutrient richness. Third, the plant must possess the anatomical characteristics and ability to withstand cyclic and stochastic disturbance regimes and be able to reproduce and grow back; this again depends on those individuals and species in a local template and possibly proposed management regimes. Ideal green roof plants should neither be invasive nor require intensive management; these considerations are best understood at the local

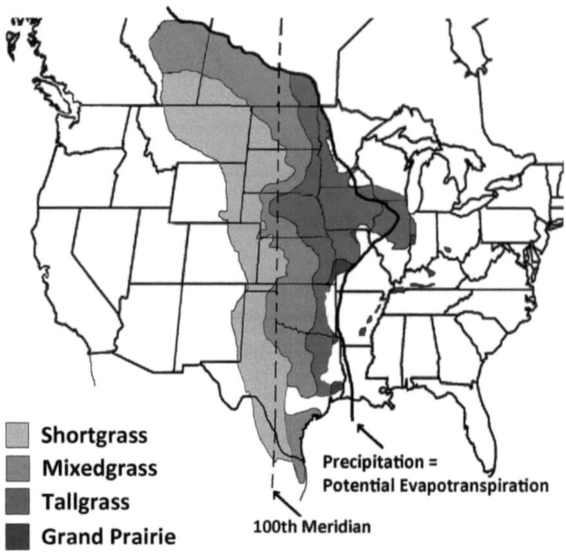

Fig. 11.3 Prairie types of North America related to latitude and moisture balance. (Modified from Sutton et al. 2012a)

template. A plant species' continuity with and linkages to local flora and fauna help secure it to any pre-existing site community and local template. Finally, because planted green roof ecosystems arise from human activities, the plants must be acceptable to the users and stakeholders of the green roof, located in suitable ecotypes readily available as seed or propagules, and inexpensive to plant and maintain.

We will use the North American prairie biome as a general starting point leading to a local template, by explaining how the abiotic conditions of prairies meet the basic condition for many green roof analogs. Next we will explore the selection of suitable templates for the Minneapolis-St Paul, Minnesota metropolitan area that set the context for creating several green roof analogs. This process can be applied generally to most any of the world's grassland biomes.

11.3 North American Prairies: Abiotic Characteristics

Prairies are North American grasslands and that word is derived from the Latin *pratum* for meadow (Merriam-Websters 2003, 2014). Over time, and with use, its meaning has evolved into meaning a more horizontal, spacious context of dry, treeless plant communities. Worldwide, other large grasslands exist that could be also used as a template for that particular region's biome: South American: Pampas (Spanish), South African: Veldt (Dutch), Eurasian: Steppe (Russian), Australian: Bush or Outback (Anderson et al. 2007; Lauenroth and Burke 2008; Paruelo and Lauenroth 1996; Sims 1999; Risser et al. 1981; Weaver and Albertson 1956).

Not monolithic, the prairies of North America change dramatically in stature from east (moist) to west (dry). Numerous landscape ecological researchers, as early as Henry Cowles (1899), offered three simple descriptors for this change in stature: Shortgrass, Mixedgrass and Tallgrass prairie (Fig. 11.3). But recent researchers (Greenberg 2002; Packard and Mutel 1997; Wilhelm 1998) have now divided the Prairie into four major types from east to west where height is largely based on precipitation:

- Grand Prairie (annual precipitation >40-inches (1016 mm))
- Tallgrass Prairie (annual precipitation ranging from 20 to 39-inches (508–991 mm))
- Mixedgrass Prairie (annual precipitation <19–12 inches (483–305 mm))
- Shortgrass Prairie (annual precipitation <12-inches (305 mm))

Prairies share similarities to green roofs. Precipitation defines growing conditions in dry prairies whether Shortgrass to the west or Tallgrass to the east (Chap. 6). Rainfall for over 80% of North America, and the entire prairie region arrives as Type II events (fast, high intensity, and with long return intervals between storms) (Sala and Lauenroth 1982; Winkler 1988). This rainfall gradient occurring over any prairie with shallow and well-drained soil, whether that of glades in moist Tennessee or shortgrass prairie of the sandhills of arid northeastern Colorado, means little moisture gets stored for plant growth (Banner 2005; Schiemann 2005; Shaw and On 1981). Thus dry prairies within the Mixedgrass and Tallgrass prairie regions possess relatively high species affinities with those in the Shortgrass prairie region (Greenberg 2002; Swink and Wilhelm 1979). And dry prairies are generally suitable templates for green roof design.

A second influence, potential evapotranspiration, also called moisture balance, tightly interconnects precipitation, insolation, and wind. High insolation and wind are features of most green roofs. Where annual precipitation equals PET, forest transitions to prairie. Precipitation decreases and PET generally increases as one moves west across North America.

In the prairie biome both gradual change and stochastic disturbance support species redundancy plasticity and resilience. A good example of this plant community plasticity in the landscape occurred with the movement northwards of a suite of southwestern plants during an inter-glacial warming period known as the Altithermal (5600 BP+/−). This warming coincided with dropping annual moisture levels presenting more arid conditions (Wright 1980). It led to several desert genera, like some Cacti (Gruchow and Burkey-Harris 1999; Djupstrom et al. 1995), joining the grasses and forbs of the prairie in moving among and into the Shortgrass, Mixedgrass, and Tallgrass prairies. With its affinities to the abiotic conditions present on most green roofs, prairies present a potential place to find plants meeting those conditions.

11.4 North American Prairie: Plant Characteristics and Species Composition

In addition to its broad abiotic features, looking at the North American prairie vegetation in general helps understand the special characteristics of local dry prairie templates occuring within this grassland matrix.

11.4.1 Grasses

The predominant cover of all Prairie types consists of graminoids, typically in a 60:40 ratio to forbs (Weaver 1968; Wilhelm 1998; Greenberg 2002), though some investigators have found original cover on prairie with grass/forb ratios as high as 75:25 (Weaver 1968) and up to 95:5 (Minnesota DNR 2005b). In relation to the size of the land area covered, by prairies—perhaps as much as 1,000,000 mi^2 (258 million ha) existed prior to plowing—a relatively small number of genera and species of grasses and sedges comprised it (Curtis 1959; Leopold 1949). Even combining the dominant and sub-dominant grasses of Grand, Tallgrass, Mixedgrass and Shortgrass prairies, under any accounting method and even with the tendency of taxonomists to finely split species, it would be difficult to list more than 40–50 grass species.

The two major functional categories of Prairie grasses are warm season (C_4) and cool season (C_3), which refer to both the season that the grass grows the most biomass and photosynthetic pathways for the manufacture of food or simple sugars (Greenberg 2002). The warm-season, C_4 photosynthesis pathway adapts plants to warm and dry environments; C_4 grasses most readily exploit the warmest part of the growing season, typically July and August, for prairies further south (Texas, Oklahoma) June and October are also warm season growing months. Cool season (C_3) grasses (e.g. *Koeleria macrantha*) do not have the photosynthetic pathway allowing them to take full advantage of the entire growing season. Thus C_3 cool-season grasses compete best during the cooler portion of the temperature gradient and comprise these months of the year: April, May, September, and October with less moist stress. The months of June and September are the most variable of months for Grand, Tall, Mixed and Shortgrass prairies, depending on the declining east to west moisture gradient. In some years June and September are characterized by cool season temperatures and in other years are characterized by warm season temperatures. All prairie plant communities (Grand, Tallgrass, Mixedgrass and Shortgrass) have a complement of both warm and cool season grasses, adapted to fully exploit all parts of an often-variable growing season and soil-defined niches to the fullest extent possible.

Grass anatomy has adapted to dry prairie conditions. Little bluestem (*Schizachryium scoparium*) has hairy leaves, green needlegrass very narrow leaves. Others respond to grazing pressure by growing serrated leaf edges and accumulating chloride in their leaves so when it cuts flesh, it causes pain. As well, high silica content in grass leaves acts like sandpaper wearing down herbivores' teeth. Some paleoecologists (Owen and Wiegert 1981) note that the growth of grasses from intercalary meristem may have co-evolved with the increased lengthening of grazing animal's teeth over evolutionary time. Some caespitose (low-growing) grasses spread from densely clustered clumps to further reduce herbivory.

11.4.2 Forbs

Forbs on the prairie richly vary in speciation, structure and form. While grass species number in the tens, forbs species number in the hundreds. The *Asteraceae* family is the best represented of the prairie forbs and its diversity is stunning in leaf and stem structure as well as root morphology. Generally forb leaves can be: clustered and compact (pussytoes, *Antenaria* spp.) or large and fanlike (compass-plant, *Silphium* spp). Some forbs form stem-like, thorny structures (prickly-pear cactus, *Opuntia* spp), or can be hirsute with light colored hairs (silky aster, *Symphyotrichum pratense*), be covered by a thickened or high gloss waxy cuticle (spiderwort, *Tradescantia* spp), grow diminutively or become narrow (narrow-leaved coneflower, *Echinacea angustifolia*). Other adaptations include light gray or white color for high reflectivity and cooling (wild indigo, *Baptisia*), thorns that pierce flesh (roses (*Rosa* spp.) and structures to hold onto water harvested from dew, rain, snow, and fog which include cupped, densely hirsute leaves, rosette whorls, and ridged petioles all directing water droplets to plant crowns (thistle, *Cirsium* spp). Geophytic features like fleshy corms, tubers and bulbs store both carbohydrates and water to take plants through lean times (e.g., blazing-star, *Liatris* spp., wild onion, *Allium* spp., and prairieclover, *Dalea* spp.). Chloride accumulation in leaves sting and serrations cut herbivore's flesh; higher lignin content discourages herbivory; and bioaccumulation of selenium poisons herbivores (Weaver 1968).

11.4.3 Prairie Roots: Biomass and Diversity of Structure

11.4.3.1 Grasses

By far the largest majority of roots in both C_4 warm and C_3 cool season grasses are very fine (<0.04 inches (1 mm) in diameter), (Weaver 1954; 1968) and occur in great numbers, orders of magnitude greater than the number of roots found in their associated forbs in the same location. An example of these multitudes of roots can be found in little bluestem. When mature, each individual will typically have in excess of a half a mile (0.80 km) of active living roots (Jackson 2002). The primary exception to these numerous fine roots of grasses are the modified underground stems or rhizomes and stolons often mistaken as true roots (Michael 2012) found at or just below the surface which may be as large as 0.20 inches (5 mm) in diameter (Weaver 1968) In addition to root elongation, trees, shrubs and many of the forbs have the ability to grow their roots radially, in other words to increase their girth in addition to their length, prairie grasses do not have this capacity (Weaver 1968). Another feature of these grass roots is their ability to replace large portions of their root network on an annual basis. Again, little bluestem the most common grass for most dry prairie types, will typically shed and regenerate approximately a fifth of its root mass on an annual basis (Weaver 1968) equalling one tenth of a mile or approximately >500 ft. (152 m) (Jackson 2002).

This constant addition and then abandonment of roots, structures the soil and creates narrow channels at an extremely fine grained level after death, these grass roots decay into networks of tiny tubes that distribute water and organic matter throughout the soil profile (Chap. 5). These soils quickly absorb extremely large quantities of precipitation, (even with Type II or high-intensity storms) (Weaver 1968). An analogy might be the capacity of sponges to absorb much greater volumes of water than their apparent solid dimensions would indicate (Weaver 1968; Greenberg 2002). The water absorbing capacity of these root curtains allows them to keep the bulk of their root systems close to the top of the soil surface, usually never deeper than 8 ft. (2.44 m) and more typically closer to between 2 and 4 ft. (0.61 and 1.22 m) deep.

On upper, hot, dry slopes or flatter dry prairies, the primary grasses are little bluestem (*Schizachryium scoparium*), side-oats grama (*Bouteloua curtipendula*), green needlegrass (*Nassella viridula*), junegrass (*Koeleria macrantha*), and prairie dropseed (*Sporobolus heterolepis*). On very dry sites needlegrass and prairie dropseed predominate, in drier places it is side-oats grama that dominates, (Weaver 1968) with the last grass standing, so to speak, blue grama (*Bouteloua gracilis*) (Weaver 1968). Root decay in blue grama is slower than the other dryland grasses. Little bluestem, the warm season bunchgrass, is the second most dominant plant of all prairie grasses covering the upland prairies of the East and moving westward, into mixed-grass prairies. In short prairie little bluestem is replaced by green needlegrass to the north (Weaver 1968). Needlegrass, a cool season bunch grass, often has roots descending 5 ft. (1.52 m) deep, but where it dominates in extremely dry upland sites, needlegrass roots also adapts by spreading widely through the upper soil to capture surface moisture (Weaver 1968). Its roots decay very rapidly, as quickly as junegrass (Weaver 1968). Prairie drop-seed an upland, warm season bunchgrass, produces many widely spread roots in the surface soil, with numerous short woody rhizomes with some roots extending vertically 5 ft. (1.52 m), similar to its other warm season cohort little bluestem (Weaver 1968). Side-oats grama is an extremely drought-resistant warm season grass in the Mixed and Shortgrass prairies. Some roots spread laterally but most grow down to 6 ft. (1.83 m) deep (Weaver 1968). Some of its drought strategy may be related to its reduced levels of evapotranspiration. Root decay in side-oats grama is almost as slow as blue grama (Weaver 1968). Junegrass the shallow rooted, short-lived, cool season grass (C3) had its roots systems or structures examined in four completely different ecosystems and was found to have the same pattern of development in all (Weaver 1968). Junegrass prolifically produces seed, which germinates readily; this strategy of constant replacement, allows it to reduce energy inputs into its root system so that they can remain shallow (Weaver 1968). Also root decay in junegrass is extremely rapid—more rapid than all other prairie grasses (Weaver 1968). This differential root decay between grass species has implications for addition of organic matter to green roof substrate and the fluxes and cycling of nutrients (Chap. 5).

The Shortgrass prairie covers the extreme western portion of the prairie biome south from Alberta and Wyoming to Texas (Banner 2005; Schiemann 2005; Shaw and On 1981). The dominant grass is blue grama that mixes with and may be co-dominant with buffalograss (*Bouteloua dactyloides*). During the growing season, blue grama, a C_4 grass also has the unusual ability to quickly go in and out of dormancy.

As little as 2 inches (12.7 mm) of precipitation can bring it back to active functioning in 12 hours (Sala and Lauenroth 1982). Coincidental with moisture activation of soil organic matter is activation of a suite of nutrient recycling microbes; pulsating in and out of dormancy, another strategy to avoid the impacts of dry spells.

11.4.3.2 Forbs

Many fine forb roots spread horizontally (e.g., prickly-pear, *Opuntia* spp.) and analysis of their large root volumes relative to crown size (Weaver 1968), shows as much as 80 % biomass below ground in root structures (Young 1994). Deceptively massive and deeply rooted, many forbs have excellent capacity for hydraulic redistribution (Weaver 1968; Natura 2003). Diverse mycorrhizal associations (Weaver 1968) also interact with both prairie forbs and grasses (Chaps. 5 and 7) and extend individual plant's ability to take up water and nutrients (Greenberg 2002).

Weaver's (1954) examples of deep forb root structure, literally uncovered by his graduate student's excavations to as deep as 20 ft. (6.3 m) include: prairie rose (*Rosa suffulta*) and rushlike lygodesmia (*Lygodesmia juncea*); Weaver found examples of leadplant (*Amorpha canescens*) rooting to 15 ft deep (4.57 m)—however it is this forb's other common name 'Devils Shoestrings' which hint at its potential root system depth; bush morning-glory (*Ipomoea leptophylla*), and dotted blazing-star (*Liatris punctata*) are also known to extend their roots 15 ft (4.57 m) or more into the ground. In the Nebraska Sandhills a sample of 45 Forbs, Weaver found that 4 or only about 10 % of the sample could be designated as shallow rooted (Weaver 1968).

11.4.4 Prairie Forb and Grasses Suitability for Green Roofs

The competitive advantage of the root systems of grasses over forbs, shrubs and trees, means that these other non-grasses need to invest in large tuberous roots or extremely deep roots to access water from below these grass root curtains. Most forbs' rather coarse, deep and elongated roots face the unusual difficulty of having to anchor their stems so that their roots do not push the plant's shoots, stems and leaves out of the soil, as their roots seek to exploit this available ecological niche. Many investigators have shown that root biomass is typically 80–90 % of mature prairie biomass and many prairie average grass root systems extend down twice the height of their aboveground portion (Curtis 1959, 1971; Greenberg 2002; Schulenberg 1997; Young 1994; Wilhelm 1998; Weaver 1968).

Weaver created a root classification system from a population of 80 different prairie forbs (Weaver 1968, pp. 251–252), consisting of four forb types:

1. Dotted Blazing-star (*Liatris punctata*) Type (DBT) 25 %: Taproots with low absorbing, infrequent, widely spreading branches in the first 3 ft (1 m) of soil, and then penetrating another 10 foot (3.05 m) or greater into the soil;

2. Broom Snakeweed (*Gutierrezia sarothrae*) Type (BST) 25 %: Shallow (1′–2′ (0.33–0.66 m)) lateral taproots, extensive lateral branching, absorptive surface roots;
3. Pale Purple Coneflower *Echinacea pallida*) Type (PPCT) (20 %: Taproots or several main routes with few or no branches but deeply penetrating > 10 ft (3.05 m);
4. Many-flowered Aster (*Aster ericoides*) (MAT) Type 30 %: Numerous main roots with rhizomes, root offshoots, and corms of equal size.

Adaptability to green roof environments: based on Weaver's forb root classification system would suggest that the root type with the greatest possibility of suitability is the broom snakeweed type (BST). However, anecdotal observations indicate that many prairie plant root types can adapt to shallow green roof soil profiles, as many deep rooted native prairie species in Minnesota, Michigan, Chicago, IL and Lincoln, NE, have been found to grow their roots horizontally and thrive on green roofs (Kestrel Design Group 2013; Sutton 2011; Grese 2008).

At the bottom of the moisture spectrum, short grasses, forbs and cushion and mat plants, adapted to arid conditions dominate dry prairies. Blue grama (*Bouteloua gracilis*), buffalograss (*Bouteloua dactyloides*), little bluestem, (*Schizachryium scoparium*), side-oats grama (*Bouteloua curtipendula*), silky aster (*Aster sericeus*), hairy grama (*Bouteloua hirsuta*), green needlegrass (*Nassella viridula*), western wheatgrass (*Pascopyron smithii*), and porcupinegrass (*Hesperostipa spartea)* are some of the typical, dominant C_3 and C_4 grasses; dotted (*Liatris punctata*) and Rocky Mountain blazing-star (*Liatris ligulistylis*), pasqueflower (*Pulsatilla patens*), pussytoes (*Antenaria* spp), birds-foot violet (*Viola pedafida*), beardtongues (*Penstemon* spp, spiderworts (*Tradescantia* spp), evening-primroses (*Oenothera* spp.), and puccoons (*Lithospermum* spp) are examples of characteristic forbs, as are cacti such as prickly-pear (*Opuntia* spp), and tree cholla, (*Cholla imbricata*).

11.5 Local Minnesota Templates

Unsurprisingly the high mineral, low organic, rapidly draining, thin substrates of green roofs mimic many dry prairie soils. Green roofs exhibit edaphic analogs and soils of dry prairies display the characteristics of both Lithosols, and in extremely dry conditions, Aridisols. They display little or no soil horizon or ped development or depth, with high sand or gravel content, lack organic matter, nutrients, cohesive parent material, display extremely shallow depths and low moisture availability (Chap. 5), contain few soil invertebrates (Chap. 14), have poorly developed fungal communities (Chap. 7), and drain rapidly, (Chaps. 5 and 6) (Greenberg 2002).

Rapidly draining sandy, gravelly and rocky soils in these places compound this pattern of scarcity so that a larger storm (e.g. one with infrequent storm return > 1.1 inches (28 mm)/24 h), will be very similar in amount, whether in Wyoming (95 % storm interval) or Arkansas (85 % storm interval) (Sala and Lauenroth 1982; Winkler 1988). Thus functional annual rainfall of southwestern Minnesota at < 25-inches (635 mm) with the geophytic pasque flower (*Anemone patens*) and common yucca (*Yucca filamentosa*) dominating reflects the same co-dominants as the >60-inches (1524 mm) Tennessee limestone glade. Likewise the Front Range bedrock bluff prairie at 5000 ft (1524 m) elevation in Montana with <15-inches (385 mm) annual precipitation has a plant community assemblage reminiscent of the <10-inches (253 mm) annual precipitation seen in the deserts of southwestern New Mexico (Banner 2005; Shaw and On 1979; Schiemann 2005). In functional terms however, precipitation amounts and patterns of infrequency means that whether in winter or summer, xeric. prairie patches may go months without significant moisture (<0.1 (2.5 mm) in 24 h).

The dry prairies of the Twin Cities region of Minnesota are divided into four subtypes (Minnesota DNR 2005b):

Dry Barrens Prairie occurs on deep deposits of sand left primarily by glacial meltwater rivers. Winds often reworked these deposits into dunes during subsequent periods of severe drought.

Dry Sand-Gravel Prairie occurs on nearly level to steeply sloping gravel-rich deposits left by melting glaciers or deposited along the shores or beaches of large glacial lakes and Kame glacial deposits.

Dry Hill Prairie is richest in species and lies on the steep slopes of loamy glacial till.

Dry Bedrock Bluff Prairie occurs along steep Mississippi River and Minnesota River bluffs in southeastern Minnesota, northwestern Iowa, southern Wisconsin, and northwest Illinois, often referred to as goat prairie; plants sprout from a thin layer of soil over bedrock (Minnesota DNR 2005a; Wovcha et al. 1995).

11.5.1 Substrate of Bedrock Bluff Prairies

In all four cases, frequent fires, shallow and poorly developed soils, plus low moisture levels help prevent the encroachment of trees and agriculture on to these bedrock prairies. Thus some of this landscapes character and a complement of arid species have been preserved (Greenberg 2002). The dry bedrock bluff prairie subtype has growing conditions most similar to those found on extensive green roofs (Fig. 11.1). These prairies occur on thin soils over bedrock on steep, often south or west facing slopes on bluffs along the Mississippi River and its tributaries in southeastern Minnesota, as well as occasionally along the St. Croix River (Minnesota DNR Natural Heritage Program 1993). Soil depth typically ranges from 0 to 4 ft (0–1.23 m). Bedrock outcrops as well as cobble to boulder-sized rubble are common, creating a variety of microhabitats (Minnesota DNR Natural Heritage Program 1993). Soils are excessively well drained to well drained (Minnesota DNR Natural Heritage Program 1993) and vary in composition depending on whether they are derived from sandstone or limestone bedrock and the amount of unconsolidated material, such as till, alluvium, or loess, that mantles the bedrock slopes; sandy loams to silty loams (Wovcha et al. 1995) are common. Bedrock bluff prairies are therefore similar to many extensive green roofs in that they have shallow, well drained, soil profiles and are exposed to considerable sun, drought, and wind.

11.5.2 Plants of Bedrock Bluff Prairies

The primary native plant community used to inform the case studies in Minneapolis, Minnesota, described below, emphasizes Minnesota's bedrock bluff prairie, a subtype of dry prairie (Minnesota DNR Natural Heritage Program 1993; Minnesota DNR 2005b). Minnesota's Eastern Broadleaf Province, Minnesota DNR (2005a) summarizes the vegetation of these southern Minnesota dry prairies:

Graminoid cover:

- Patchy to continuous with 50–100 % vegetation cover
- Overall species composition varies depending on soils and topography, but,
- Mid-height and shortgrass species predominate, with little *bluestem (Schizachryium scoparium)* generally the dominant grass, and sideoats grama (Bouteloua curtipendula) and prairie dropseed (*Sporobolus heterolepis*) next in frequency
- Tallgrass species can be seen occassionally with big bluestem most frequent

Forb cover:

- Sparse to patchy, with 5–50 % cover
- Forb species composition more variable than that of graminoids

Bedrock bluff prairies are characterized by a high diversity of native prairie plants: grasses and forbs (more than 25 species per 30 × 30 ft (9.14 m.), mostly 2–30 inches (5–76 cm) tall (Wovcha et al. 1995). Based on data from nine vegetation plots in east central Minnesota, Minnesota DNR and Great River Greening (2004) found the grasses with the highest frequency were: little bluestem *(Schizachryium scoparium*), sideoats grama (*Bouteloua curtipendula*), prairie dropseed (*Sporobolus heterolepis*), hairy grama (*Bouteloua hirsuta*), and big bluestem (*Andropogon gerradii*). The forbs with the highest frequency were: silky aster (*Symphyotrichum sericeum)*, pussytoes (*Arennaria* sps.), harebell (*Campanula* sps.), purple prairie-clover (*Dalea purpurea*), and whorled milkweed (*Asclepias verticillata*).

11.6 Using Dry Prairies as a Template for Designing Green Roof Analog in Twin Cities Region

Study of local plant community publication guides, site visits to local template dry prairie communities, (Minnesota Scientific and Natural Areas (SNA's)) and desktop analysis of local plant community data, informed the design of the two case study extensive green roofs. Since growing conditions on bedrock bluff prairies are most similar to green roofs, analysis focused most heavily on these communities. Because growing conditions on other dry prairie communities are also similar to green roofs in many ways, and to increase our sample size, other dry prairie communities were also investigated.

Local plant community resources analyzed included:

- Minnesota's St. Croix River Valley and Anoka Sandplain: A Guide to Native Habitats (Wovcha et al. 1995)
- Minnesota's Native Vegetation: A Key to Natural Communities Version 1.5 (Minnesota DNR Natural Heritage Program 1993)
- Minnesota DNR's County Biological Survey's plant lists for Minnesota's bedrock bluff prairies (unpublished data from the Minnesota DNR County Biological Survey)
- Field Guide to the Native Plant Communities of Minnesota (Minn. DNR Guide to Minnesota Scientific and Natural Areas (1995) 2005a, b)
- Our field and desktop analysis of template plant communities examined:
- Species frequency (number of plots in which species occurs, divided by total number of plots)
- Species abundance (average percent within the community)
- Where species are located within the template community (microhabitat, especially soil depth)
- Apparent drought adaptations

To maximize resilience, we selected a diversity of species, as well as include the major functional groups, with early and as late successional species when possible (Paruelo 1996) (Chap. 8).

11.7 Ten Years' Experience Applying Native Templates to Design Green Roofs in Minnesota

Several green roofs have been designed using a dry prairie template in Minnesota's Twin Cities Region. To provide a snapshot of 10 years of experience using native templates to design analog green roofs in Minnesota, two are described below in more detail:

1. Municipal Building Commission Green Roof: Planted in 2008
2. Phillips Eco-Enterprise Center (PEEC) Green Roof: planted in 2004

The following climate and context information is applicable to both case studies:

- USDA Hardiness Zone 4
- Elevation above sea level: 871 ft. (265.5 m).
- Normal yearly precipitation: 29.41 inches (747 mm) (based on 30-year normal for Minneapolis, MN, 1971–2000)
- ET/P = 1:1.13
- Mean daily maximum temperature in the hottest month of the year (July): 83.4 °F (28.56 °C) (based on 30 year normal for Minneapolis, MN, 1971–2000)
- Mean daily minimum in the coldest month of the year (January) 4.4 °F (− 15.33 °C) (based on 30 year normal for Minneapolis, MN, 1971–2000)

11.7.1 Municipal Building Commission Green Roof, Minneapolis, MN (a.k.a. Minneapolis City Hall)

We include the Municipal Building Commission Green Roof as a case study because it features several species that have been grown on very few, if any, other green roofs (Table 11.1). Located in a courtyard that has both sunny and shaded microhabitats, this green roof provided a unique opportunity to include a high diversity of species. A total of 44 native species were planted in 2008 consisting of 38 herbaceous species; 3 vines, and 3 shrubs. Species were chosen based on frequency and abundance in template plant communities as observed in field observations, published literature, and informed by performance on previous local green roof projects. Template communities included both sunny and shaded river bluffs of the Mississippi and Minnesota Rivers with shallow soils. Species growing in the thinnest soil profiles and cliff surfaces were especially noted during field visits to template plant communities. In 2009, 1 year after planting, 43 of the 44 species planted were still viable (Fig. 11.4) on the roof! Thirty-four of these were still viable in late summer of 2012. Some of the ten species that were not observed in late summer of 2012 may still have been present but not observed because of the season. Regardless, high species diversity was still observed 4 years after planting. Species with the greatest cover in 2012 include *Aquilegia canadensis, Aster macrophyllus, Cassia fasciculata, Fragaria virginiana, Geranium maculatum, Polemonium reptans, Solidago flexicaulis, Parthenocissus quinquefolia.* In 2014, the green roof vegetation was still thriving and the species composition largely unchanged since 2012.

Table 11.1 Minneapolis city hall green roof details

Year planted	2008
Size	4,200 ft^2 (390.2 m^2)
Height	Roof at floor level of 2nd story courtyard, surrounded by walls
Sun exposure	Mix of full sun, part sun, continuous shade
Context	Urban, courtyard in downtown Minneapolis
Substrate composition and depth	Construction documents called for 4–9 inches but as-built conditions are primarily 4 inches in depth 101.6–228.6 mm deep growing substrate that conforms to FLL requirements for multi-course extensive green roofs, with a maximum saturated weight no more than 85 lbs/foot^3 1285.33 kg/m^3 Depth varies by roof location
Layers/Materials	Drainboard and filter fabric
Composition methods	44 species of native plugs planted 12 (304.8 mm) OC
Irrigation system	Drip irrigation system that uses rain collected from taller surrounding roofs in a cistern and switches to potable water when not enough water is available from the cistern. Soil moisture sensors indicate when the green roof needs to be watered
Irrigation & maintenance regime	Weeded and irrigated as needed, approximately 3 times annually

Fig. 11.4 Minneapolis city hall green roof (**a**) August 7, 2009, 1-year after planting: *Allium stellatum, Tradescantia ohiensis, Fragaria virginiana*, various *Carex* spp., *Cassia fasciculata* (**b**) August 7, 2013, 5-years after planting: *Cassia fasciculata, Fragaria virginiana, Campanula rotundifolia* (**c**) May 29, 2009, 1-year after planting: *Aquilegia canadensis, Heuchera americana, Phlox divaricata, Polemonium reptans, Fragaria virginiana, Geranium maculatum* (**d**) May 29, 2009, 1-year after planting: *Aquilegia canadensis, Heuchera americana, Penstemon grandiflorus*. (Kestrel Design Group)

11.7.2 PEEC Green Roof, Minneapolis, MN

We include PEEC Green Roof in Minneapolis as a case study because it shows the response of native plants on a green roof to drought and lack of irrigation (Table 11.2). The roof was not irrigated at all in some years and had below average rainfall from May through July in 5 out of the 10 years since it was planted. While it possesses less than 75 % native plant cover in some years, in 2010 it recovered to 100 % native cover with normal rainfall with no irrigation. In 2012 the roof had 100 % vegetation cover, with 20 % invasive cover. The native species with the highest cover included the following, listed in order from highest to lowest cover: *Allium stellatum, Schizachyrium scoparium, Bouteloua gracilis, Sporobolus heterolepis, Bouteloua curtipendula*, and *Geum triflorum*.

Plant species for the PEEC Green Roof were selected based on the authors' field studies of bedrock bluff and other dry prairie communities, as well as published dry prairie descriptions and species lists, and unpublished data from the Minnesota DNR. Especially targeted were plants with drought adaptations, (e.g. fuzzy or hairy leaf surface, sticky surfaces, waxy leaves, succulent plants, plants with water storage organisms, thicker leaves, CAM plants, shiny, reflective leaves, small leaves, extensive surface roots, spines instead of leaves) as discussed earlier. Ten years after installation, in early summer of 2014, the dominant native species were *Koeleria macrantha, Bouteloua gracilis*, and *Allium stellatum*. Most of the origin al species that were planted still remained, with 16 out of 18 native species still present albeit with variable coverage. Based on our anecdotal observations, in the first 10 years of the PEEC green roof growth, species composition has changed considerably year to

Table 11.2 PEEC Green roof details

Year planted	2004
Size	4000 ft^2 (371.61 m^2)
Height	On top of 2nd floor
Deck slope	1 %
Sun exposure	Full sun
Context	Integrated into industrial/commercial neighborhood and residential neighborhood
Substrate composition and depth	Intensive green roof mix; depth varies 2–6 inches (51mm - 15.3 cm)
Layers/materials	Moisture retention layer, drainboard, filter fabric
Composition methods	Native plugs at 2/foot^2 (0.05 m^2) on 90 % of roof (13 forb species and 5 grass species); 7 species of sedums at 1 foot^2 (0.10 m^2) on 10 % of roof
Irrigation system	Manual overhead sprinkler and hose
Irrigation and maintenance regime	2004–2006: irrigated and weeded as needed 2007: no irrigation or weeding 2008–2009: irrigated and weeded as needed starting mid-June 2008 2010: no irrigation or weeding 2012: no irrigation or weeding 2014: no irrigation or weeding

Fig. 11.5 PEEC Green Roof (**a**) May 31, 2005, 1-year after planting: *Tradescantia ohiensis, Geum triflorum* (**b**) May 31, 2005, 1-year after planting: *Geum triflorum* closeup (**c**) September 23, 2005, 1 year after planting: *Solidago nemoralis, Schizachyrium scoparium, Aster sericeus* (**d**) July 9, 2014, 10 years after planting: native still dominant where planted (Sedums were planted inside axes) (Kestrel Design Group)

year, yet the prairie green roof plant assemblies appear to be resilient, and in general, return to full coverage in moist years (Fig. 11.5). Because all of the green roofs we have designed with prairie species in Minnesota, except for one, receive regular irrigation, we have limited experience with species survival without irrigation on extensive green roofs in Minnesota. Table 11.3 lists non-Minnesota references to more widely gauge the relative need for irrigation.

Table 11.3 Native species successfully grown on extensive green roofs in Minnesota, Kestrel Design Group 2004–2014

Scientific name	Common name	Regular irrigation	Minimal irrigation
Bouteloua curtipendula	Side-oats grama		X[ab]
Bouteloua graciis	Blue grama		X[abcf]
Campanula rotundifolia	Harebell		
Carex pensylvanica	Pennsylvania sedge		
Carex vulpinoidea	Brown fox sedge		
Chamaecrista fasciculata	Partridge pea		
Coreopsis palmata	Bird's foot coreopsis		
Dalea purpurea	Purple prairie clover	X[e]	X[1]
Fragaria vesca	Wild strawberry		
Fragaria virginiana	Wild strawberry	X[e]	X[ab]
Geranium maculatum	Wild geranium		
Geum triflorum	Prairie smoke		X[ab]
Heuchera richardsonii	Alumroot		
Koeleria macrantha	Junegrass	X[e]	X[adf]
Liatris aspera	Rough blazing star	X[e]	
Liatris cylindracea	Cylindric blazing star		
Penstemon grandiflorus	Large-flower beardtongue		
Phlox divaricata	Woodland phlox		
Polemonium reptans	Jacob's ladder		
Ruellia humilis	Wild petunia		
Schizachyrium scoparium	Little bluestem	X[e]	
Solidago nemoralis	Gray goldenrod		
Solidago ptarmicoides	Upland white aster		
Sporobolus heterolepis	Prairie dropseed	X[e]	X[b]
Thalictrum dioicum	Early meadow-rue		
Tradescantia bracteata	Bracted spiderwort		
Tradescantia occidentalis	Western spiderwort		
Tradescantia ohiensis	Ohio spiderwort	X[de]	
Viola pedatifida	Birdfoot violet		
Penstemon grandiflorus	Large-flower beardtongue		
Phlox divaricata	Woodland phlox		
Polemonium reptans	Jacob's ladder		
Ruellia humilis	Wild petunia		
Schizachyrium scoparium	Little bluestem	X[e]	

Table 11.3 (continued)

Scientific name	Common name	Regular irrigation	Minimal irrigation
Solidago nemoralis	Gray goldenrod		
Solidago ptarmicoides	Upland white aster		
Sporobolus heterolepis	Prairie dropseed	X[e]	X[b]
Thalictrum dioicum	Early meadow-rue		
Tradescantia bracteata	Bracted spiderwort		
Tradescantia occidentalis	Western spiderwort		
Tradescantia ohiensis	Ohio spiderwort	X[de]	
Viola pedatifida	Birdfoot violet		

[a] Goes dormant or turns brown with little or no irrigation but rebounds with water
[b] Based on trials at Chicago Botanical Garden, Richard Hawke, Personal Communication 2013
[c] Based on Kevin Carroll, personal communication 2013
[d] Based on research at Michigan State University, (Rowe in Sutton et al. 2012a)
[e] Based on research at Mich. State Univ, (Monterusso et al. 2005). Their plants were irrigated the first growing season then abruptly stopped July 10 of the 2nd growing season, during an unusually warm and dry summer; plants un-irrigated during 3rd growing season
[f] Observations at PEEC green roof, The Kestrel Design Group 2014

11.7.3 Case Conclusions

The use of prairie plants on green roofs is becoming more and more common throughout the world, and many others have collected anecdotal observations on the performance of prairie plants on extensive green roofs in the Midwest of North America. Many other case studies using prairies to inform green roof design have been previously published, for example, in Dvorak and Volder (2010) and Sutton et al. (2012a). The project database at greenroofs.com also includes several green roofs designed using native plant community templates.

Although data on the above green roof case studies is observational only, without replication, it is nonetheless clear that prairie plants from a local prairie community template in Minnesota can thrive with minimal irrigation when carefully assembled and designed. To date, these roofs appear to be dynamic communities (Chap. 12) that adapt to seasonal and yearly weather variations and irrigation regime. Because prairie green roof plant assemblies appear so resilient they make a promising option for green roofs.

11.8 Prairie Green Roof Management

Bluff bedrock, scree beds and other lean substrate prairie plant communities within larger prairie patches were typically visited by fire on a 1–5-year cycle but fuel loads then and now in those naturally occurring environments were so scarce it led to small and less intense burns. In these naturally occurring xeric prairie plant communities it was the aridity, exposed bedrock, rapid drainage and evaporation, extreme insolation, and thin, organically, poor soils that maintained an open canopy of forbs and grasses, with inadequate fuel to carry a fire.

11.8.1 Influence of Irrigation

Six out of the seven prairie green roofs we have designed in the Minneapolis area include an automatic irrigation system. Most include soil moisture sensors to make irrigation as efficient as possible. We are finding that most prairie species can thrive on green roofs even in very thin soil profiles if they are irrigated (also see Sutton 2013). On the Target Arena Green Roof, (Fig. 11.6) at least 21 prairie species are thriving in only 2.75-inches (70 mm) of growing substrate (3.75 inches (95 mm) on a small portion of the roof) with irrigation as of the summer of 2014, 5 years after installation. The irrigation systems on the green roofs we have designed are used only as needed. Most include soil moisture sensors. In 2014 the irrigation system at the Target green roof was not turned on until mid summer because there was adequate rain fall until then. At the PEEC green roof, 16 prairie species are thriving without an automatic irrigation system 10 years after installation. More long-term research is needed, however, to confirm which prairie species can thrive without irrigation or with very minimal irrigation, in addition to minimum soil depths without irrigation.

11.8.2 Thatch Build-Up

Thatch is regularly removed on most of the green roofs we have designed, mostly in the spring. Soil tests on the Target Center Green Roof are showing that soil organic matter content is increasing even with thatch removal. This is perhaps not surprising considering how much organic matter the roots of native prairie plants produce every year (Chaps. 5 and 13).

11.8.3 Seeding

The Target Center Green Roof vegetation consisted of a pre-grown mat, supplemented with native plugs and seed. Seed has proved to be an effective tool for increasing species diversity. Species that are thriving on the Target Green roof that were installed as seed only include *Asclepias verticillata, Cassia fasciculata*, and *Coreopsis palmata*. Many species, including several *Liatris* species, several *Aster* species, *Coreopsis lanceolata*, and *Penstemon grandiflorus* are self-seeding prolifically on the Target Arena Green roof. Because the plugs were planted in large

Fig. 11.6 Target center arena green roof (**a**) July 10, 2012, 3 years after planting: diverse mix of native species, and Sedums including Allium cernuum and Liatris cylindracea (**b**) July 10, 2012, 3 years after planting: diverse mix of native species, and Sedums including Dalea purpurea (**c**) June 4, 2014, 5 years after planting: diverse mix of native species, and Sedums including Tradescantia occidentalis (**d**) August 26, 2014, 5 years after planting: diverse mix of native species, and Sedums including Liatris aspera (Kestrel Design Group)

groups on the Target Green roof, it is particularly evident when self-seeding occurs on this roof, as the species spread outside of the original planted groupings. Most of those native forb species also self-seed on other green roofs we have designed. The ability to recruit new plants by seed becomes important in buffering plant cover dynamic and is an important aspect of resilience.

11.8.4 Cuttings

Sedum cuttings were installed on the Minneapolis Central Library green roof to supplement plugs installed 1foot (305 mm) O.C. At first (Seasons 1 and 2) these Sedum cuttings filled in very vigorously between the 1 foot (305 mm) O.C. plugs. Sedum cuttings are also used on the Target Center green roof as needed to fill in sparse areas. Both harvested and purchased cuttings have been successfully used. Over time the Sedums have remained on both roofs, but the density of *Sedum* sps. cover has diminished dramatically, as natives seed and grow vigorously among and above them.

11.8.5 Pre-Grown Mats

Pre-vegetated *Sedum* spp. mats (with non-biodegradable Erosion Control Mat ECM and expanded shale) were installed on the Target Center Arena. Various types of pre-vegetated mats are available from several suppliers as proprietary systems. What these pre-grown mats have in common is: a base 3 dimensional erosion control blanket or mat (either biodegradable or persistent) about 1 inch (25 mm) in thickness; mineral rich green roof substrate (mainly comprised of various lightweight material) to fill the voids in the ECM mat, with *Sedum* sps. cuttings spread on top. These pre-grown vegetated mats are grown in intensive nursery situations with waterproof membranes below and irrigation above, as was the case with the Target Center Arena's pre-grown mats. A typical start to delivery timetable for these vegetated mats is at least one full growing season to fill in so that less than 15% of the base 3 dimensional ECM remains exposed. Pre-grown mats were used on the Target Center Arena Green Roof to allow plants to gain strength before facing the harsh conditions on the roof; as well as to maximize plant cover and wind resistance. Wind resistance is important on this green roof because of the greater than 165 feet (50.3 m) height of the roof, its size (115,000 ft^2 > 2.5 acres (> 1 ha), geographic location (northern end of "Tornado Alley"), and location in the metro at the southwest approach track of summer storms toward central Minneapolis.

The Target Center Arena pre-vegetated mats are ending their 5th growing season. Throughout that time they have received vegetation inspections twice a year for percent total cover, percent acceptable plant species cover, percent native plant species cover and percent weed cover. This extensive green roof has a 20-year vegetation warranty and maintenance contract as recommended by the authors. To date, this green roof has met and exceeded performance requirements.

11.8.6 Native Plant Plugs

Seven of the constructed extensive green roofs in the MSP metro region designed by the authors employ prairie plant plugs. These plugs are always small and planted on 6" to 1-ft. centers (152.4 mm–305 mm). The soil/root portion of the plug is

1.5 inches (38 mm) in diameter and 4 inches (100 mm) in height. These plugs have been reliable performers on all of the referenced extensive green roofs.

Those prairie plant plugs that have not done well over time have been most related to the unsuitability of that particular species, not a result of plugs as a poor method of vegetating roofs or poor plant growing production. Another advantage of using live plant plugs to vegetate extensive green roofs is that design intent can be executed much more flexibly and reliably than cuttings, seed or pre-grown mats.

A suite of native plants (forbs) was installed as plugs to supplement the vegetation cover of the pre-grown mats on the Target Center Arena. These forb plugs were planted in large single species groups to increase visual impact for the residents of the surrounding buildings. Two more unusual benefits of these supplemental plantings on the Target Center Arena occur on game days when blimps are used in game day TV coverage requiring long shots of the stadiums for these two professional basketball and baseball franchises. Lastly, the Target Center Arena is on one of the major incoming flight paths for the main metro airport MSP so these large bands of flower color combined with the complimentary curves of the 10-ft. (2.8 m) wide firebreak paths on the roof become visually arresting.

11.8.7 Manufactured Fertilizer

Unlike most other kinds of horticultural projects, extensive prairie green roofs need not be the targets of additional N-P-K nutrient fertilization. Too much fertilizer on an extensive prairie green roofs can result in a number of potential problems: nutrient leaching into downstream receiving waters (Chaps. 4 and 5), promotion of annual agricultural weed growth, and eventual dominance of weeds because of their efficiency at converting surges of freely available nutrients into plant growth. Additionally the selected native bedrock plant species fare better in stressed ecological circumstances and simply do not require fertility. The Minneapolis City Hall and the PEEC Green Roof are not fertilized; their suite of native plants shows no signs of nutrient deficiency.

11.9 Application and Further Study

Below we recommend long-term quantitative studies to understand these changes in more depth (Chap. 13). The same plant assembly rules used for the Minnesota case studies (see Fig. 11.2) can be applied in other grassland regions with suitable template communities, that is, plant communities with growing conditions similar to green roofs. The steps outlined at the beginning of the chapter and noted in Fig. 11.2 detail the steps taken for this case study, but we note in several places that those steps can be applied in other parts of the world. Any designer or ecologist should be knowledgeable enough about his or her own local plant communities and be able to find wind resistant and moisture stress resistant biomes with communities on shallow soils. (Chap. 6 is a good example of this approach.) Select plants from those locales that support other organisms, are not invasive, can easily be propagated and produced. Test the plants and make detailed observations of the growing conditions and the plant growth and survival.

Because moisture availability is perhaps the most limiting factor for green roof vegetation, more long-term research is needed to confirm which prairie species can thrive without irrigation or with very minimal irrigation.

- For a locale's given PET what strategies help plant establishment?
- What are the ET rates for prairie forbs and grasses?

Prairie plants are already known as parsimonious users of nutrients but…

- Will long-term deficiencies appear?
- How do complementary C_3 and C_4 plants cycle, and change nutrient availability?
- How does management (e.g. mowing with removal) affect nutrient availability?
- What can be learned from turfgrass science and range management about nutrient availability?

Long-term research is also needed to determine successional patterns of prairie plants on green roof environments to answer questions such as the following:

- Does initial species establishment determine composition or will species composition change over time?
- How does type of initial propagule affect long-term composition?
- Which species will be early, mid, and late successional species?
- What do comparisons with restored, at grade prairies' successional patterns, timeframe, or species composition reveal?
- How will weed species and pressure change over time?
- How will prairie species versus *Sedum* spp. change substrate over time?
- Does the use of prairie plants on green roofs translate to increased animal diversity?
- Does proposed diversity contribute equally as habitat when compared to at grade prairie plantings?
- Is burning feasible and desirable for prairie-based green roofs?

References

Anderson RC, Fralish JS, Baskin JM (eds) (2007) Savannas, barrens, and rock outcrop plant communities of North America. Cambridge University Press, New York

Banner R, Pratt M, Bowns J (2005) Grasses and grasslike plants of Utah: a field guide. Utah State University Extension Press, Salt Lake City

Carroll K (2013) Personal communication, Chicago City Hall Green Roof Field Observations. Chicago IL

Cornelius DR (1947) The effect of source of little bluestem grass seed on growth, adaptation, and use in revegetation seedings. J Agric Res 74:133–143

Cowles HD (1899) Ecological relations of the vegetation of the Sand Dunes of Lake Michigan. Vol 27 University of Chicago
Curtis JT (1959) The vegetation of Wisconsin: an ordination of plant communities, 1st edn. The University of Wisconsin Press, Madison WI
Curtis JT (1971) The vegetation of Wisconsin: an ordination of plant communities, 2nd edn. The University of Wisconsin Press, Madison
Darwin C (1859) The origin of species. John Murray Publishing. London
Djupstrom R (1995) Guide to Minnesota's Scientific and Natural Areas. MN Department of Natural Resources, St. Paul MN
Dvorak B, Volder A (2010) Green roof vegetation for North American Ecoregions: a literature review. Landscape Urban Plan 96:197–213
Franklin C (1987) Personal Communication, Plant Community Analogs Dow Headquarters Campus, New Jersey at Chicago Botanic Gardens, Chicago, IL
Getter K, Rowe DB (2008) Selecting plants for extensive green roofs in the United States. Michigan State Univ Ext Bull E-3047. Lansing, MI. http://www.hrt.msu.edu/greenroof/PDF/08%20GetterRoweExtensionBulletin.pdf. Accessed 1 May 2013
Greenberg J (2002) A natural history of the Chicago Region. University of Chicago Press, Chicago
Grese R (2008) Personal communication, Field Observations of Green Roofs, Nichols Arboretum, Univeristy of Michigan Ann Arbor
Gruchow P, Burkey-Harris R (1999) Worlds within a world: reflections on Visits to Minnesota Scientific and Natural Area Preserves; Minnesota's Bookstore Conservation Volunteer, Minnesota Department of Natural Resources; Minnesota Department of Administration Communications Media Division, St. Paul MN
Hawke R (2013) Personal communication, Chicago Botanic Gardens Green Roof Trials at 2013 Green Roofs for Healthy Cities Conference San Francisco CA
Jackson W (2002) Natural systems agriculture: a truly radical alternative. Agr Ecosyst Environ 88(2):111–117
Kestrel Design Group (2013) Field Observations of Green Roofs Twin Cities MN
Lauenroth WK, Burke IC (eds) (2008) Ecology of the shortgrass steppe: a long-term perspective (Long-Term Ecological Research Network Series). Oxford University Press, NewYork
Leopold A (1949) A sand county Almanac and sketches here and there. Oxford University Press, New York
Lundholm JT (2006) Green roofs and facades: a habitat template approach. Urban Habitats 4(1):87–101
MacDonagh P, Hallyn N, Rolph S (2006) Midwestern USA plant communities + Design = Bedrock Prairie Greenroofs. In: Proceeding of 4th North American Green Roofs Conference: greening rooftops for sustainable communities, Boston May 11–12, 2006. Toronto: The Cardinal Group
Michaels T (2012) Personal Communications University Minnesota Horticultural Department. St. Paul
Minnesota Department of Natural Resources (1993) Minnesota county biological survey map series: No. 5
Minnesota Department of Natural Resources (2005a) Field guide to the Native Plant Communities of Minnesota: The Eastern Broadleaf Forest Province. Ecological Land Classification Program, Minnesota County Biological Survey, and Natural Heritage and Non-game Research Program. MNDNR St. Paul
Minnesota Department of Natural Resources (2005b) Field guide to the Native Plant Communities of Minnesota: The Prairie Parkland and Tallgrass Aspen Parklands Provinces. Ecological Land Classification Program, Minnesota County Biological Survey, and Natural Heritage and Non-game Research Program. MNDNR St. Paul
Monterusso MA, Rowe DB, Rugh CL (2005) Establishment and persistence of Sedum spp. and native taxa for green roof applications. HortScience 40:391–396.
Natura H (2003) Conservation design resource manual. Northeastern Illinois Planning Commission. http://chicagowilderness.org/files/1413/3087/0449/Conservation_Design_Resource_Manual.pdf. Accessed 3 Sept 2014

Owen DF, Wiegert RG (1981) Mutualism between grasses and grazers: an evolutionary hypothesis. Oikos 36:376–378
Packard S, Mutel CF (1997) The tallgrass restoration handbook: for Prairies, Savannas, and Woodlands. Island Press, Washington
Paruelo JM, Lauenroth WK (1996) Relative abundance of plant functional types in grasslands and shrublands of North America. Ecol Appl 6:1212–1224. ESA Publications, Ithaca NY
Merriam-Webster's (2003) Collegiate dictionary, 11th edn. Encyclopedia Britanica
Merriam-Webster's (2014) In Merriam-Webster.com. http://www.merriam-webster.com/dictionary/hacker. Accessed 8 May 2011
Risser P, Birney EC, et al (1981) The true prairie ecosystem. Hutchinson Ross Publishing Company, Stroudsburg
Rowe DB, Rugh CL, Durhman AK (2006) Assessment of substrate depth and composition on green roof plant performance. Proc. of 4th North American Green Roof Conference: Greening Rooftops for Sustainable Communities, Boston, Mass, 10–12 May (2006) The Cardinal Group, Toronto
Rowe DB, Getter KL, Durhman AK (2012a) Effect of green roof media depth on Crassulacean plant succession over seven years. Landsc Urban Plan 104: 310–319
Sala OE, Lauenroth W (1982) Small rainfall events: an ecological role in semiarid regions. Oecologia 53(3):301–304
Schiemann DA (2005) Wildflowers of Montana. Mountain Press Publishing, Missoula
Schulenberg R (1997) Personal communication, Field Observations of Schulenberg Prairie Morton Arboretum, Lisle IL
Shaw R, On D (1981) Plants of Waterton-Glacier National Parks and The Northern Rockies. Mountain Press Publishing Company, Missoula MT
Sims PL, Risser P (in MG Barbour and WD Billings (Editor)) (1999) North American terrestrial vegetation; Chaper 9 Grasslands, 2nd edn. Cambridge University Press, New York USA
Sutton, RK (1990) Seed sources. Landscape Architect 80(5):68–69
Sutton R (2011) Personal communication
Sutton R (2013) Seeding Green Roofs with Native Grasses. J Living Architect 1(1):26. http://greenroofs.org/resources/JOLA[2]Sutton.pdf
Sutton RK, Harrington JA, Skabelund L, MacDonagh LP, Coffman RR, Koch G (2012a) Prairie-based green roofs: literature, templates, and analogs. J Green Build 7(1):143–172
Sutton R, Rowe DB, Acomb G, Lambrinos J, MacDonagh LP, Hawke R (2012b) New plant performance for 21st century green roof ecosystems. Cities Alive Conference, Chicago
Swink F, Wilhelm G (1979) Plants of the Chicago region; The Morton Arboretum, Lisle
Temperton VK, Hobbs RJ, Nuttle T, Halle S (2004) Assembly rules and restoration ecology: bridging the gap between theory and practice. Island Press. Washington
Weaver JE (1954) North American Prairie University of Nebraska Press, Lincoln
Weaver JE (1968) Prairie plants and their environment: a fifty-year study in the midwest. University of Nebraska Press, Lincoln
Weaver JE, Albertson FW (1956) Grasslands of the great plains: Their Nature and Use University of Nebraska Press, Lincoln
Whittinghill LJ, Rowe DB (2011) Salt tolerance of common green roof and green wall plants. Urban Ecosyst 14(4):783–794
Whittinghill LJ, Rowe DB (2012) The role of green roof technology in urban agriculture. Renew Agri Food Syst 27(4):314–322
Whittinghill LJ, Rowe DB, Cregg BM (2013) Evaluation of vegetable production on extensive green roofs. Agroecol Sust Food Syst 37(4):465–484
Wilhelm G (1998) Personal communication, Field Observations, Dolomite Prairie Restoration Peck Farm Park, Geneva IL
Winkler JA (1988) Climatological characteristics of summertime extreme rainstorms in Minnesota. Ann Assoc Am Geogr 78(1):57–73
Wovcha DS, Delaney BC, Nordquist GE (1995) Minnesota's St. Croix River Valley and Anoka Sandplain, a guide to native habitats; Minnesota County Biological Survey, Minnesota Department of Natural Resources; U of Minnesota Press

Wright HE (Fall 1980) Lecture on glacial geology at University Minnesota Department of Geology. Minneapolis

Young D (1994) Kane county wild plants and natural areas, 2nd edn. Kane County Forest Preserve District

Peter MacDonagh FASLA, PLA is Adjunct Faculty, University of Minnesota College of Design, registered landscape architect, and Director of Design + Science, Kestrel Design Group. He specializes in landscape restoration, urban forestry, stormwater planning, and green roof technologies since 1985.

Nathalie Shanstrom PLA is a landscape architect with Kestrel Design Group, in Minneapolis, MN, since 1999, specializing in sustainable landscape architecture.

Chapter 12
Green Roof Plant Assemblage and Dynamics

Stephanie Carlisle and Max Piana

Abstract The benefits of green roofs derive from their existence as functional, living ecosystems. Despite an emerging understanding of green roofs as dynamic ecosystems, most green roof vegetative studies treat plant communities as static assemblages. As designed ecosystems, green roofs display complex growth dynamics stemming from a combination of initial site conditions (shading, thermal exposure, wind, moisture), roof design (vegetation, growing substrate, drainage), inter- and intra-species interactions, and disturbances (extreme climate conditions, weeding, disease, emergent species, fertilization). Ultimately, ecologically grounded research directed at increasing our fundamental understanding of green roof plant dynamics and plant assemblages has the potential to improve design and maintenance practices, supporting high-performance and resilient green roof systems.

Keywords Vegetation dynamics · Vegetation assemblage · Vegetation ecology · Ruderal plants · Plant biodiversity · Floristic composition · Long-term monitoring · Vegetation census

12.1 Introduction

While seemingly natural in appearance, a green roof is a wholly constructed ecosystem—disconnected from the ground, composed of prepared materials assembled on site, exposed to human mediation, and carefully engineered to function as an infrastructural element. A green roof is both a landscape system and a facet of the built environment, recognized for its unique capacity to provide valuable ecosystem services (Chap. 9) and positioned as an alternative to traditional infrastructure mechanisms. No matter how well a green roof is initially designed and specified, all living systems grow and change over time. Over the life cycle of a building, plants installed on a green roof become established, mature, die, and regenerate as the roof is exposed to disturbances (Fig. 12.1). Vegetation dynamics reflect ecological

S. Carlisle (✉) · M. Piana
KieranTimberlake, 841 North American St., Philadelphia, PA 19123, USA
e-mail: scarlisle@kierantimberlake.com

R. Sutton (ed.), *Green Roof Ecosystems,* Ecological Studies 223,
DOI 10.1007/978-3-319-14983-7_12

Fig. 12.1 The green roof at Yale School of Art Sculpture Gallery was left undisturbed for 6 years after establishment. Vegetation at time of planting (*left*) and after 6 years of growth (*right*). (Left Image by Peter Aaron, Right Image by KieranTimberlake)

processes intertwined with engineered building systems and human intervention. They provide us with a means of better understanding the functioning of green roofs as novel ecosystems (Hobbs et al. 2006; Kowarik 2011, Chaps. 1, 10 and 15) and create a lens through which we may evaluate a roof's performance.

Vegetation is not the only aspect of green roof systems that change over time. Environmental context and conditions on green roofs are themselves spatially and temporally variable. Solar radiation is rarely evenly distributed over a site, and while sun paths remain fixed, shading dynamics can change dramatically as surroundings become developed and landscape elements grow. Climate varies from year to year. Even growing substrate, and as a result roof hydrology, is not stable over the life of a green roof installation (Schrader and Böning 2006; Getter et al. 2009; Thurning and Dunnett 2014; Berghage et al 2007). Organic content (C_{org}) has been found to fluctuate but ultimately increase over time (Getter et al. 2009; Thuring and Dunnett 2014), while pH and substrate depths have been found to decrease (Schrader and Böning 2006; Thuring and Dunnett 2014). Such changes ultimately impact roof floristics (Thuring and Dunnett 2014) and biodiversity as well as thermal and hydrologic properties/performance (Schrader and Böning 2006; Berghage et al. 2007; Thuring and Dunnett 2014).

If such changes in green roof context and composition are inevitable, what is their effect on performance? How do installed green roofs change over time, and what expectations do designers, engineers, and managers have regarding the long-term trajectories and goals of green roof planting?

In this chapter, we address such questions by examining ecological theory and existing research, identifying avenues for future inquiry in both design and research concerning green roof vegetation. This chapter proposes that applying an ecological perspective to the assessment of green roof systems, specifically the observation and study of plant dynamics and plant assemblages, has the potential to improve design and maintenance practices, supporting high-performance and resilient green roof systems. To illustrate this point, this chapter introduces a case study of two mature green roofs whose plant dynamics have been and continue to be observed. Observations from these two roofs are a source of provocations regarding the behavior and performance of green roofs over time. This and similar research can improve our approach to the design, management, and study of green roofs.

12.2 Vegetation Dynamics and Green Roof Performance

A green roof introduces vegetation and landscape elements to the surface of a building or structure, transforming otherwise unused roof surface into a living medium. As discussed in Chap. 9, a green roof system improves building performance and positively impacts local climate, hydrology, energy consumption, and human comfort and well-being (Del Barrio 1998; Bass et al. 2003; Theodosiou 2003; Villarreal and Bengtsson 2005; Lazzarin et al. 2005; Rowe 2011; Susca et al. 2011; Stovin et al. 2013). While the engineered components of a green roof greatly contribute to thermal benefits and hydrologic performance of a building (e.g. Takakura et al. 2000; Mentens et al. 2005), research has determined that these benefits along with other ecosystem services can be at least partially attributed to vegetation and the plant communities established on a green roof (Lundholm et al. 2010, 2014).

Early horticultural research on green roofs emphasized the attributes and fitness of individual species, focusing on plant establishment and performance. Increasingly, researchers have begun to study co-habiting species and diverse plant assemblages. Terrestrial vegetation research has found a positive correlation between ecosystem function and both species diversity (Naeem and Tjossem 1999; Hooper et al. 2005) and richness (Naeem et al. 1994; Tilman and Downing 1996; Aarssen 1997; Freitas 1999; Spehn et al 2000). Similarly, green roof studies have found diverse plantings to increase productivity when compared to monoculture systems (Lundholm et al. 2010; MacIvor and Lundholm 2011). With respect to broader habitat function, diverse floristic assemblages on green roofs have also been shown to support broader habitat function (Baumann 2006; Kadas 2006).

Green roof vegetation research has also found that the biological function and physical properties of specific species groupings impact green roof function and performance (Lundholm et al. 2010; 2014). The functional diversity and structural complexity of plant assemblages have been correlated with a range of green infrastructure attributes, such as decreasing hydrological and thermal loads (Kolb and Schwarz 1993; Lundholm et al. 2010, 2014), while complimentary planting may improve plant growth (Heim and Lundholm 2014). Several mechanisms support such behavior. Carefully selected plant assemblages may benefit roof performance through complimentary resource use over space and time. Species may selectively uptake different nutrient or maximize productivity at different times of the year. Complimentary resource use may also minimize inter-species competition, allowing different species to cohabit the same area. In other instances, green roof plant species may facilitate the growth of neighboring species (Butler and Orians 2011).

The research above emphasizes the relationships between plant functional traits as well as the usefulness in considering species by functional traits in order to describe collective behavior and life form strategies. Plant functional types (PFTs) or plant functional groups are defined as species that exhibit similar responses to the environment and impact on ecosystem function (Diaz and Cabido 1997). Plant strategy theories and species groupings, such as Grime's Competitor-Stressor-Ruderal (CSR) triangle (Grime 1977) (Chap. 10), examine the relationship between habitat characteristics and plant life strategies. Such models provide a framework for understanding plant community dynamics such as competition and facilitation, and for determining how species dynamics relate to site conditions, stressors, and disturbances.

In addition to life form strategies, broader conceptual understanding of vegetation dynamics may be applied to green roofs. A concept that extends from succession theory, vegetation dynamics consider changes in structure and composition, "the three dimensional architecture" of a plant community through space and time (Pickett and Cadenasso 2013, p. 107; Glenn-Lewin and van der Maarel 1992). Changes in vegetation may occur at the individual plant-plant level, between plant communities, and most broadly, at the landscape scale (Dunnett and Hitchmough 2007). The drivers of vegetation dynamics are numerous and in the framework shown in Fig. 12.2 are grouped by three mechanisms: differential site availability, differential species availability, and differential species performance (Picket and Cadenasso 2013). Within each of these processes are multiple potential interactions, constraints, and resource conditions that govern and direct plant dynamics including climate, abiotic environment, biotic interactions, and disturbance history.

While developed for "natural" plant communities, the vegetation dynamics framework depicted in Fig. 12.2 is just as applicable to a constructed or managed urban ecosystem such as a green roof. Like at-grade landscapes, green roof vegetation is initially governed by abiotic factors (e.g. precipitation, temperature, drainage and substrate) (Chaps. 1 and 11). Vegetation dynamics, however, extend beyond planting and early establishment. As illustrated in Fig. 12.2, environmental stressors and disturbances complicate them—both natural (e.g. drought, emergent vegetation and competition, shade regimes, and disease) and human imposed (e.g. weeding, mowing, fertilization).

Green roof vegetation research suggests that functionally diverse green roofs may perform better than single functional trait plantings. Current vegetation ecology research has begun to examine how species dynamics, specifically inter-species and inter-functional grouping interactions, can affect roof performance. How green roof vegetation dynamics extend over time as plant assemblages respond to natural and human imposed stressors, however, is less well understood. Ecological study of other landscape types indicates that the structure and composition of a plant community will contribute to the productivity, health, and resilience of landscape systems to future disturbances Tilman et al. 2001; Cardinale et al. 2007), but further research is needed to establish how these principles apply to the dynamics of green roof systems over time.

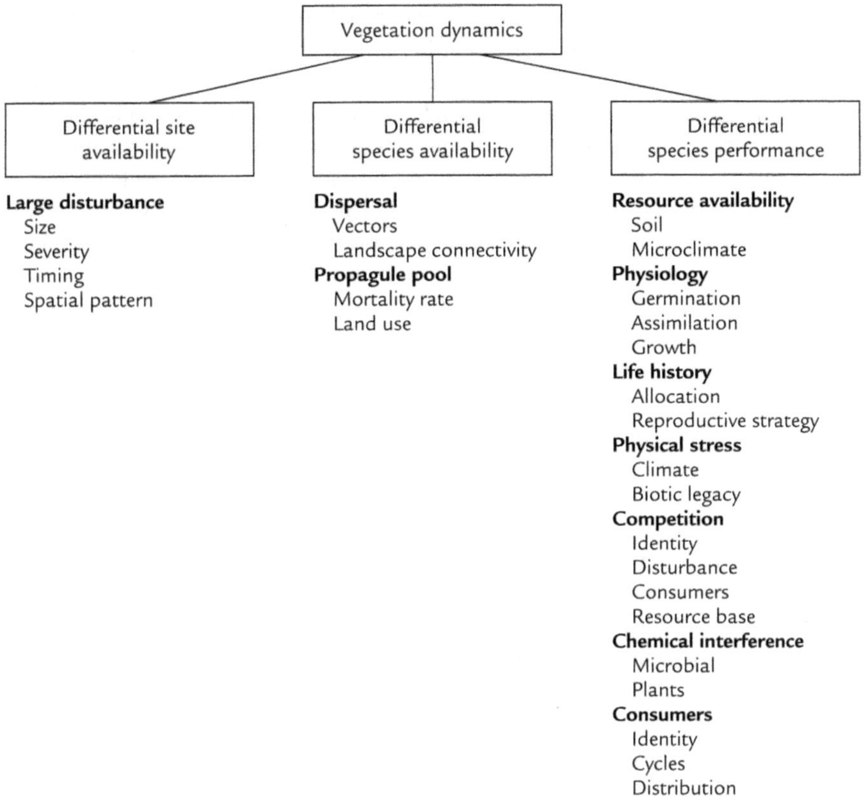

Fig. 12.2 The vegetation dynamics framework. Three general causes drive changes in plant community composition and structure: differential site availability, differential species availability, and differential species performance. (Pickett and Cadenasso 2013)

12.3 Applying Ecological Theory to Green Roof Research

Nearly a decade ago, citing the relationship between vegetation and green roof performance, Oberndorfer et al. (2007) declared the need for ecologically driven research on green roofs, including the study of plant assemblages. In response, a growing body of research activity has begun to employ ecological theory to explore spatial and temporal relationships between vegetation and roof environments through the study of floristic relationships to resource distribution and micro-heterogeneity in roof systems (Brenneisen 2003; Kohler 2006; Martin 2007; Dunnett et al. 2008a; Kohler and Poll 2010; Nagase and Dunnett 2010, 2011; Olly et al. 2011; Nardini et al. 2012).

Even as an ecological perspective on green roofs has gained traction within the research community, our understanding of the complex dynamics of green roof vegetation and resultant system function and performance over the life of buildings remains limited (Cook-Patton and Bauerle 2012; Oberndorfer et al. 2007). As green roofs (and other forms of landscape on structure) grapple with increasing complexity in design and goal setting, there is utility in looking to both the long-term study of natural systems and to existing literature on human-made and managed landscape systems such as engineered wetlands, where long-term performance has been studied for decades. Efforts by environmental engineers and landscape architects to accurately approximate the performance of existing green roofs and apply these lessons to the design of future green roofs are presently challenged by a lack of long-term data on real buildings with diverse plant communities (Cook-Patton and Bauerle 2012) that can be compared to design proposals.

In order to understand the performance of green roof systems over their expected life spans, there is a need for increased study of the long-term performance of plant communities (Thuring and Dunnett 2014) (Chap. 13). Presently, due to the constraints of experimental design, funding and academic practice, the majority of green roof vegetation studies are conducted for less than 2 years (Dvorak and Volder 2010). While short-term empirical studies may yield valuable insight into initial plant establishment, they have limited usefulness in informing design decisions and expectations of long-term behavior (Dunnett et al. 2008; Rowe et al. 2012; Thuring and Dunnett 2014).

To date, only five commonly cited long-term studies on green roofs ranging in age from 6 to 100 years appear in peer-reviewed literature (Kohler 2006; Dunnett et al. 2008; Kohler and Poll 2010; Rowe et al. 2012; Thuring and Dunnett 2014). These studies all seek to explore variables that affect survival and resilience of green roof vegetation in realistic roof settings.

These five long-term studies have compelled researchers to assert that initial findings on plant suitability and community dynamics may not accurately reflect long-term behavior. For example, Rowe et al. (2012) tracked the growth trajectories of 25 succulents planted in three different substrate depths (2.5, 5.0, and 7.5 cm), noting changes in absolute cover, survival and dominance over 7 years. By the third growing season, 17 initially planted species had completely died including species which had appeared successful during initial establishment, leading to misleading initial results being published 2 years into the study (Durhman et al. 2007; Rowe et al. 2012). Similarly, Dunnett et al. (2008) explored the ecological characteristics of 15 perennial grass and forb species over 6 years across varying substrate depths (10 and 20 cm). The study observed fluctuations in abundance over time, as species that initially thrived were eventually replaced by species that were initially less productive. Dunnett et al. notes that earlier results from the same experimental plots were unable to capture long-term population dynamics of slow growing taxa, including both planted and colonizing species.

While green roofs have increased in popularity and prominence in North America in the last 10 years, Germany has a long-standing tradition of green roof construction and experimentation that stretches back to the 1830s (Köhler and Poll 2010).

Three studies in Germany have expanded the time scale of green roof research by providing data on mature green roofs ranging in age from 20 to 100 years. In the longest running published study of green roof vegetation available, Köhler (2006) studied the plant communities on two building complexes through bi-annual surveys over a period of 20 years. Over the full study period, plant populations continued to shift in relation to changes in climate, resource availability, and human interference in the form of maintenance activities.

Thuring and Dunnett (2014) recently conducted a study on a similar generation of German extensive green roofs, surveying nine green roofs built in the 1980s in order to better understand the correlations of roof age with vegetation and substrate properties. The study found significant decline in soil pH and biomass, less substrate depth, and an increase in soil organic content since time of installation. The paper proposes that changes in substrate characteristics may affect vegetation cover and species diversity on such roofs as they transition from a pure sedum mat to a more diverse mixed succulent/meadow community.

Expanding the time horizon for green roofs yet further, Köhler and Poll (2010) revisited green roofs initially studied in the 1960s and 1980s in order to compare tar-and-paper green roofs built in the 1930s with modern extensive green roofs built in the 1980s. While this study also found changes in substrate texture, composition, and organic content over time, the authors also observed that initial planting had a persistent effect on roof biodiversity over the full building timespans.

These studies represent model efforts contributing valuable observations and encouraging further long-term observation. However, they also reflect other elements of conventional green roof research, namely a focus on minimally diverse extensive roof systems as objects of study (Köhler 2006; Köhler and Poll 2010; Rowe et al. 2012; Thuring and Dunnett 2014), and a tendency to remove emergent vegetation over the course of studies (Thuring and Dunnett 2014; Rowe et al. 2012).

In recent years, green roof designs have begun to include an increasingly diverse range of grasses, herbaceous forbs, shrubs, and trees in addition to sedum and other hardy succulents (Sutton et al. 2012). Despite the growing complexity of green roof designs, published research has emphasized methods that utilize easily replicable shallow-tray, sedum-dominated systems with minimal diversity that are examined through the use of highly mediated small-plot studies (Dvorak and Volder 2010).

Additionally, despite an emerging understanding of the complex behavior of green roofs as dynamic ecosystems, the majority of green roof vegetative studies continue to treat green roof plant communities as static assemblages. Conventional experimental design often encourages the maintenance of original planting treatments through rigorous removal of all emergent vegetation (i.e. weeding) and the periodic cutting and weighing of biomass, a degree of human disturbance that may alter plant development and life growth strategies. More importantly, these practices are neither common nor feasible on many installed green roofs.

Such studies approach the optimization of green roof systems through an engineering or horticultural perspective in which individual system components (vegetation, substrates, drainage layers, etc.) may be tested analytically through the limiting of context and complex variables (Chap. 10) (Dunnett and Kingsbury 2004; Oberndorfer et al. 2007). For such research, the primary goal is the understanding of plant performance and function under controlled conditions (Blank et al. 2013) in which success is primarily measured by survival rates of initial vegetation or in average percent cover for highly controlled and species-limited scenarios (Dvorak and Volder 2010). While such studies are valuable, they are unable to fully capture realistic long-term dynamics or represent the spatial heterogeneity of green roofs systems.

Such methodologies and industry trends communicate a deeply held assumption that the success of a green roof is tied to maintaining a static expression of initial designed conditions (Beck 2013; Sutton 2014). This model of stasis and stability fits with a desire to achieve an idealized condition—a state in which performance is understood and quantifiable—through a combination of careful detailing, specification, and ongoing, rigorous maintenance activities.

12.4 Ecologically Informed Green Roof Design

The design, construction, and maintenance of green infrastructure elements such as green roofs present the daunting requirement of approximating and predicting the future performance of ecological systems that are not entirely within our control. In landscape design, great care is made to select planting patterns that express an aesthetic vision while also responding to site conditions and performance expectations. Over time, vegetative communities, both planted and emergent, also have the capacity to respond to their context and to transform their environment. Changes in growing substrate structure, the creation of microclimates within multi-strata planting, and shifting thermal loads on building surfaces are all examples of emergent properties that tie vegetative communities to their context.

In constructed landscapes, natural processes and human intervention are particularly intertwined, and the line between a "disturbance" and a natural "perturbation" is blurred, particularly as a landscape moves farther away from its initial designed state. Landscape maintenance activities (e.g. weeding, fertilization, watering) introduce disturbance and shape ecological communities by influencing population dynamics and succession (Ranalli and Lundholm 2008). Maintenance activities aimed at freezing vegetative communities as static forms require a high degree of control, which can often translate into material and labor costs to the building owner (Sutton 2013; Beck 2013). Perhaps more importantly, landscape maintenance with aesthetic goals alone rarely enhances ecological function (Nassauer 1995). As vegetation can be tied to many attributes of green roof performance, such disturbance regimes can be understood to influence the full range of ecosystem services, from climate regulation to aesthetic and human health benefits.

Review of the literature and experience in the field of architecture and engineering indicate that practitioners often approach the question of maintenance from the perspective of controlling liability and risk (Weiler and Scholz-Barth 2009). Furthermore, green roof management plans are often created by designers and engineers to cover only the establishment of a roof system during its warranty period. The industry trend of moving towards single-source suppliers for the entirety of a green roof installation (from waterproofing to substrate to plants) supports the view of a green roof as a holistic system. But it may cause us to question the conflict between a preference for stability and predictability connected to engineered elements and a desire to allow for increasing complexity and adaptation over time with regard to biological elements of the green roof ecosystem.

If green roofs are to be taken seriously as infrastructural elements, then designers and managers must increase their ability to estimate future performance. Landscape design and management techniques that take time, disturbance, and resilience into account require informed decision-making. Understanding the dynamics of plant assemblies through empirical research or observation increases the ability for designers to escape the static expression of the planting plan and perhaps more readily engage with the relationships between landscape pattern and process.

At times, ecological function and other systems attributes may be at odds with aesthetic visions and expectations (Gobster et al. 2007), reminding us that the "design problem" of green roofs is not just one of maintaining vegetative cover or supporting biodiversity. It is in this very distance between ecological function and the appearance of nature that the design problem lies (Nassauer 1995). For designers, increased understanding of the functional role of vegetative communities and legibility of ecological processes may lead to a reconsideration of the role of initial planting and roof detailing—re-casting both the biotic and abiotic components of the roof, not as final product but as critical infrastructure, enabling the roof system to grow and develop over time (Beck 2013).

12.5 Case Study: Long-Term Observations of Two Green Roofs

In the following section, we present a case study of two green roofs, each with nearly 10 years of undisturbed growth. Surveyed in 2012 and 2013, these two roofs present a unique opportunity to examine changes in designed plant assemblages in two distinct green roof systems, and to consider the relationship between plant dynamics, green roof design, and performance over time.

This research is part of a larger green roof monitoring agenda initiated by the project architect KieranTimberlake. From this monitoring, a methodology based on full plant census and spatial mapping has been established and applied to six roofs, ranging from 2 to 10 years in age (Carlisle and Piana 2014). Given that the data is spatially explicit, census results may be assessed visually, in graphic form, as well as quantitatively, thereby introducing a novel means of data communication that seeks to increase the legibility of complex spatial-temporal data. This case study allows for examination and discussion of (1) the plant assemblages and changes in two green roofs following initial establishment (2) the relationship between assemblages, their change over time, and green roof performance, and (3) the development of green roof maintenance plans and owner expectations that consider vegetation dynamics and the presence/role of emergent species.

12.5.1 Study Site

The two green roofs examined in this study are located on the Cornell University campus in Ithaca, New York, atop separate dormitory buildings: the Alice H. Cook House (House 1) and the Carl L. Becker House (House 2) (Fig. 12.3). The two roofs were constructed 1 year apart, in 2004 and 2005 respectively. Census of green roof vegetation was completed in August 2012 and 2013.

Ithaca is characterized by a moderate continental climate, with extended winters and an average growing season of 141 days (NOAA 2014). During July of 2012 the region experienced higher than average temperatures of 29 °C (84.2 °F) and below average precipitation events of 4 cm (1.59 in) (NRCC 2013). The effects of this drought on green roof vegetation were apparent during the 2012 site visit (Fig. 12.3).

The House 1 roof is an intensive green roof system covering 329 m^2 (4617 ft^2, 47 % of total roof area) divided by elevated skylights running east to west, which effectively establish four separate bays of vegetation. Originally planted with 16 species, the roof was designed to include a warm-season meadow mix of grasses and herbaceous forbs (See Appendix for full list of planted species). The green roof has an approximate depth of 24 cm (9.5 in) (including drainage layer) with approximately 20 cm (8 in) of growing substrate, above a combination of Fabrene fabric, PVC membrane, and tapered rigid insulation. Two four-story dormitories to the north and south of House 1 effectively "canyonize" the roof, contributing uneven shading of roof surface.

The House 2 roof is physically similar to House 1, with vegetation representing 418 m^2 (4500 ft^2, 50 % of total roof area) divided by elevated skylights into four separate bays. Unlike House 1, this roof is an extensive green roof, originally designed to include only five species (*Sedum sexangulare, Sedum spurium, Sporobolus*

Fig. 12.3 The green roofs examined in this case study are the Alice H. Cook House (House 1, *left*), an intensive roof, and the Carl L. Becker House (House 2, *right*), an extensive roof. (Bing Maps, Image modified by KieranTimberlake)

heterolepis, Allium cernuum, Sempervivum spp.). The green roof features 12.70 cm (5 in) of growing substrate on top of a sheet water retention layer and filter fabric. Adjacent buildings shade the roof, while a knee wall on the eastern edge acts as a significant wind block. Both the House 1 and House 2 green roofs were irrigated during the first year of establishment, after which neither roof experienced any irrigation, supplemental planting, fertilization, mowing, or weeding.

12.5.2 Methods

To assess green roof plant dynamics and vegetative performance, this study utilized a spatially explicit survey methodology of plant species and roof conditions (Carlisle and Piana 2014). The study method is based on the comparison of annual surveys to the original green roof installation, creating snapshots of the vegetative composition and associated vegetation performance metrics (percent cover, species richness, species diversity).

The field methodology utilized in this case study is based on the Relevé Method (Poore 1955) and Braun-Blanquet cover/abundance scale (Braun-Blanquet 1932). In the field, each roof was segmented into 2 m^2 (21.5 ft^2) quadrants for full census of green roof vegetation. Percent cover was recorded by calculating relative area occupied by the vertical projection of all aerial parts of plants, expressed as a percentage of the surface area of the sample plot. Vegetative cover, species richness, and species diversity were computed across the entire roof as well as for each plot, allowing for a compiled plot-based view of the roof. Analysis of census results focused on a species-specific level but also considered plant type (herbaceous forb, grass, woody, and succulent) and plant family to understand functional trait diversity on each green roof. Species diversity was calculated using the Shannon Index (Shannon 1948) and converted to true diversity ($TD = e^{\text{Shannon Index}}$) (Hill 1973).

Microclimatic conditions on the roof related to solar access were assessed through a three-dimensional context model of the building site created in Rhinoceros3D and analyzed using a custom Grasshopper plug-in to calculate daylight hours across peak growing season (May–October). The model utilizes model geometry (including adjacent buildings) and NOAA's solar angle calculator to determine solar angles appropriate for the site (NOAA 2014, USDE 2013). Solar analysis was run on 30 cm^2 (4.65 in^2) grids across each roof. Model results were averaged within each census plot and linearly regressed with plot-level vegetative cover and biodiversity scores using R software (Team R.C. 2008). Architects and engineers often utilize modeling and simulation to quantify environmental loads on building designs. Direct measurements gathered by sensors deployed across a green roof can also be used for this purpose, but they pose trade-offs in cost and data resolution. One benefit of modeling versus monitoring is the ability to utilize large data sets representing historic conditions for a set period of time or adjusted "typical meteorological year" (TMY).

Data interpretation was achieved through traditional quantitative analysis and visually based diagramming of vegetation composition and spatial relationships, at a species-specific or plot level. The combination of qualitative, quantitative, and graphic processing of survey results allows observation of plant community relationships over time and focus on specific zones or growth patterns that may not be apparent from numerical outputs alone. This methodology allows for additional and increasingly complex study of green roof ecosystem dynamics impacting vegetation while allowing for comparison between multiple roofs varying in design intent, site context, or geographic location.

12.5.3 Results

In the 9 years since initial planting, House 1 transitioned from a mixed meadow roof to a single-species dominated roof system, in which the designed planting zones have been obscured by the colonization of *Schizachyrium scoparium* (Little bluestem) and interspersed emergent species. In 2012, after 8 years of undisturbed growth, census results found the roof to include 39 distinct species, including 14 of the original 16 species (Table 12.1, Fig. 12.4). However, *S. scoparium* represented more than 55% of total vegetative coverage on the roof. In 2013, while still the

Table 12.1 House 1 and 2 survey results (Kieran Timberlake)

Building	Year	% coverage	Species richness	Plant families	Diversity (True Diversity)
House 1	2005	NA	16	7	8.41
	2012	94	38	16	6.98
	2013	99	52	21	12.57
House 2	2006	NA	5	3	3.67
	2012	94	64	29	14.89
	2013	99	73	29	15.87

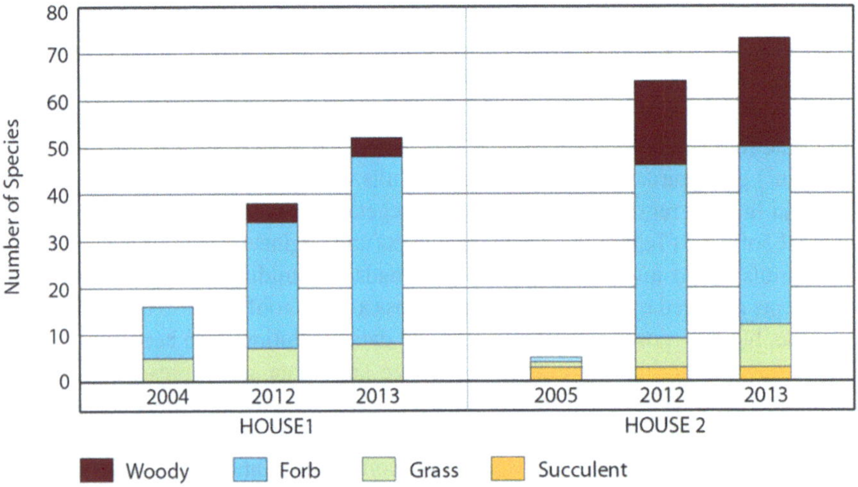

Fig. 12. 4 Plant distribution by life form groups. (Image by KieranTimberlake)

most dominant species on House 1 and covering nearly 60 % of the roof area (same as 2012), *S. scoparium* represented just 38 % of the total vegetative cover on the roof. Of the other originally planted species, none were found to contribute more than 5 % to total roof vegetation in either 2012 or 2013.

Collectively, ruderal species represented 31.25 % of all vegetative cover in 2012 and 49.22 % in 2013. Fluctuations in the population of the most abundant ruderals were observed between 2012 and 2013, most commonly with biannual species such as *Melilotus officinalis*, replaced in 2013 by *Lotus corniculatus* as the most abundant emergent species. While the total number of species more than doubled in 2012, the green roof was found to be less diverse in 2012 (True Diversity (TD) = 6.74) than at time of planting in 2005 (TD = 8.67). In 2013, however, increases were observed in species richness (52 species), as well as diversity (TD = 12.57), nearly doubling observations from the year before Table 12.1. While such data merely represents a snapshot of continuously changing conditions, it hints at the variability of community attributes over time and the limitations of considering any particular year of data as representative of the success or limitations of a landscape system.

The dynamics of the House 2 green roof vegetation over the past 8 years are defined by the resilience and persistence of designed planting zones, which have effectively supported increased diversity and ecological complexity across the roof. Originally planted with only five species (listed earlier), the roof increased in richness to include 65 distinct species in 2012 and 73 species in 2013.

The roof census reveals that sedum plantings still dominate the system and have maintained coverage in designated planting zones, but they have allowed for the integration of a variety of ruderal species, including forbs, shrubs, and small trees, throughout the roof. Despite drought conditions, the roof has maintained nearly full coverage, with multi-strata communities appearing as succulents that occupy area beneath emergent forbs and trees. Additionally, House 2 species diversity has shown continued increase from original planting (TD = 3.67) to 2012 (TD = 14.80) and 2013 (TD = 15.87).

12.5.4 Shifts in Vegetative Assemblages

With minimal human intervention, dynamic changes in the floristic and structural composition of the vegetation of the House 1 and House 2 green roofs can be observed (Fig. 12.5). Some planted species have established and successfully colonized the roof, while other species populations have been out-competed. All along, new species have emerged with varying presence within the context of the larger system. While this study is too short in duration to yield definitive results by comparing annual census results over several years, it may be possible to differentiate between short-term fluctuations and longer-term shifts in the composition and structure of plant communities through observations of dominant and emergent plant species (Figs. 12.6 and 12.7).

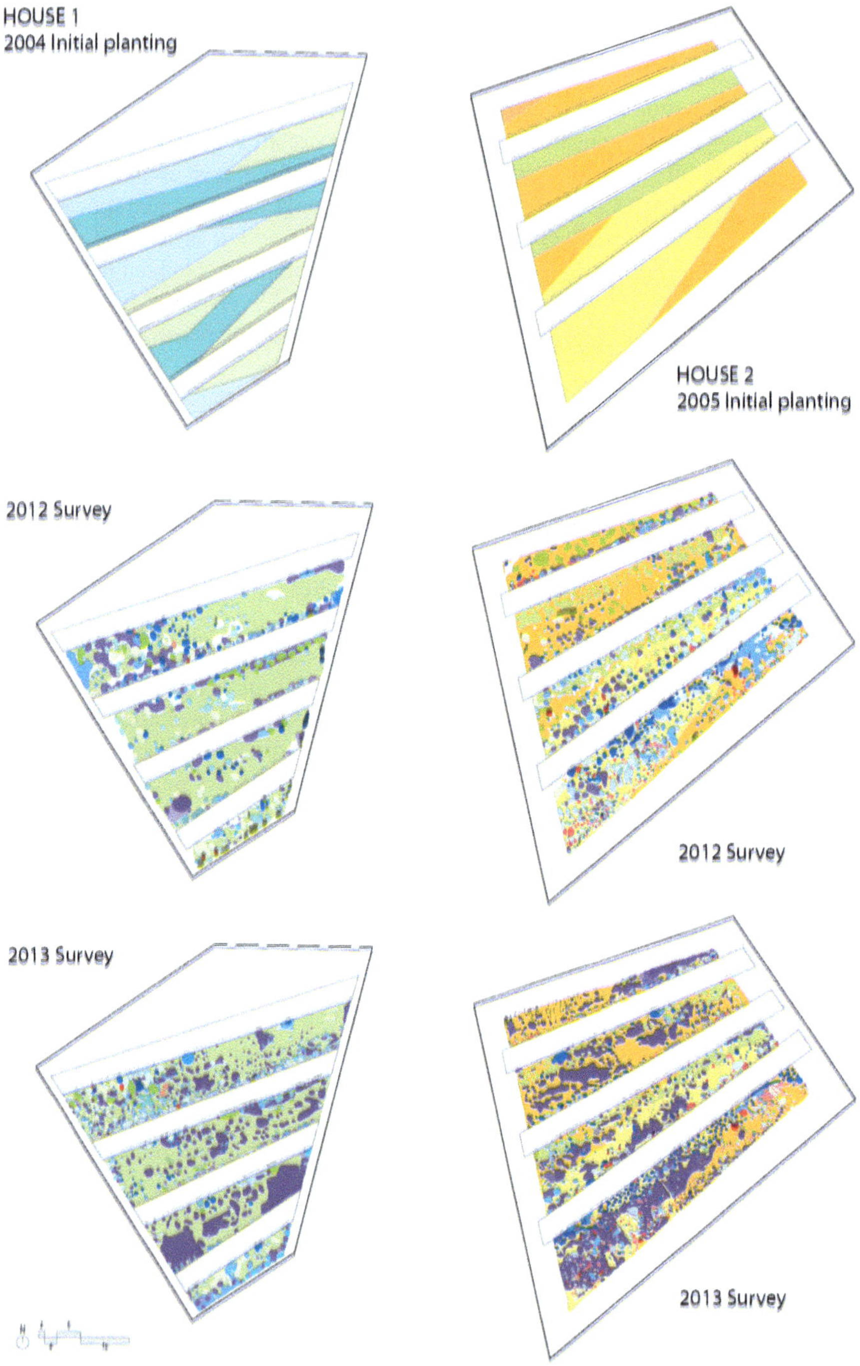

Fig. 12.5 House 1 and 2 vegetation survey maps. Each plant species is represented by a single color. Plant groups are represented by shade of color: grasses (*greens*), herbaceous forbs (*blue-purple*), succulents (*oranges*), and woody plants (*reds*). (Image by KieranTimberlake)

Fig. 12.6 House 1 changes between 2012 (*left*) and 2013 (*right*). (Image by KieranTimberlake)

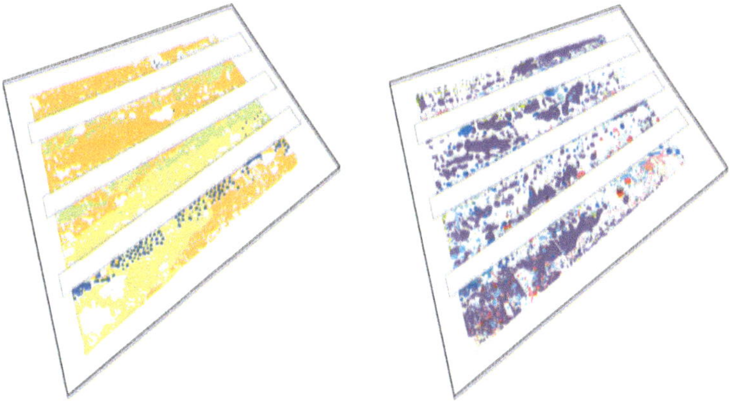

Fig. 12.7 Comparison of initially planted species (*left*) and emergent vegetation (*right*) from the 2013 House 2 survey. (Image by KieranTimberlake)

On each roof, a few select species have established themselves as dominants—those species that contribute significantly to total biomass but that represent less than 25 % of total species richness (Grime 1998; Schwartz et al. 2000). From an ecological perspective, dominant plant species, such as *S. scoparium* on House 1 and *Sedum sexangulare* and *Sedum spurium* on House 2, are indicators of ecosystem health function and productivity. These species' impacts on vegetation dynamics and the observed composition and structure of each roof are very different.

On House 1, *S. scoparium*, the lone dominant species, has successfully out-competed other plants, obfuscating the original planting design. If one were to consider 2012 alone, the changes in plant composition and structure on House 1 may be interpreted as a system moving towards a state of reduced diversity and floristic complexity. Such a decline in diversity may also be thought to leave the green roof more vulnerable to environmental change or disturbance events, as suggested by ecological studies on prairie grass systems (Tilman 1996; Tilman and Downing 1996), while also potentially reducing building benefits (increased thermal performance) achieved by green roofs (Kolb and Schwarz 1993). However, it is possible that the establishment of *S. scoparium*, like other pioneering grasses and cryptogams (a plant that bears no true flowers), may actually support the future emergence and colonization of stable, "higher-level" plant species and communities (Sutton et al. 2012). Such a hypothesis is supported by 2013 findings, where despite *S. scoparium*'s remaining the most dominant species both in terms of roof area and total number of individuals, species richness has continued to increase, along with species diversity. It is possible that *S. scoparium* may in fact have facilitated growth in the midst of resource stress during the 2012 drought.

Conversely, during the same climate conditions, the original planting design of House 2 has remained legible, while the composition and structure of vegetation has changed dramatically from a minimally diverse, sedum-dominated system to a rich multi-strata assembly. The increase in diversity and system complexity on House 2 is likely in part the result of interspecies facilitation by the sedum species and microclimate conditions defined by site context. As stated in Sect. 12.2, research has found that sedum, like some desert nurse species, may create localized microclimates where soil is cooled and moisture increased (Franco and Nobel 1989; Turner et al. 1966), facilitating growth and increasing survival rates of neighboring plant species during periods of environmental stress (Butler and Orians 2011). Recent research supports this perspective, relating not only to plant diversity but also to specific combinations of plant species achieving greater performance and ecosystem function (Lundholm et al. 2010). So while *S. sexangulare* and *S. spurinum* may have provided a means for establishment and persistence in times of severe climate conditions, the complexity of this rooftop system may be changing to further enable growth and in turn, perhaps, system productivity.

While the few dominant species of House 1 and 2 maintain high levels of coverage and are spatially defined by large, low diversity patches, much of the roofs' vegetation consists of scattered clusters of ruderal species characterized by fluctuating populations. Opportunistic by nature, ruderal species populations fluctuate based on their life history strategies and response to stressors, disturbances, and competition specific to the roof system (Grime 1977) (Chap. 10).

While potentially disruptive to the integrity of the original planting design, ruderals may also be viewed as vegetation that represents an increase in climate-adapted, resilient species that perform important service functions (Del Tredici 2010a, b). In other instances, population fluctuations are a response to the autoecology and phenology of neighboring plants. For instance, *Melilotus officinalis*, a large biannual, was found to represent 16.28 % (House 1) and 6.26 % (House 2) of all vegetation present in 2012, only to be replaced by *Lotus corniculatus*, which increased from 2.47 to 17.15 % on House 1 and 0.24 to 31.41 % on House 2 in 2013 as *M. officinalis* lay dormant. Many of the ruderal species fluctuate based on opportunity and neighboring plant life histories and phenology; a high-functioning and productive system is maintained through staggered emergence and maturity (Chap. 10).

12.5.5 Spatial Dynamics and Landscape Heterogeneity

In addition to the plant strategies and species interactions described above, the vegetation dynamics of House 1 and House 2 are also governed by abiotic factors and constraints. From an ecological perspective, it is understood that microhabitat heterogeneity related to the uneven distribution of resources (light, nutrients, and moisture) and the contribution of context features (adjacent buildings, skylights, and site walls) impacts the structure, composition, and therefore performance of green roof vegetation.

On both House 1 and House 2, the neighboring buildings and raised skylights have created microhabitats across each roof, with unique solar access, moisture regimes, and exposure profiles. Solar modeling allows for the quantification and mapping of direct and incident solar radiation loads on the roof surface (Fig. 12.8). The map in Fig. 12.8 shows that direct solar access, commonly expressed as "daylight hours," is variable across the roof surface with House 1 (mean = 10.18 h) receiving more hours of direct sunlight than House 2 (mean = 7.48 h), $F(1, 209) = 90.22$, $p = 0.000$ in a typical year.

Plant diversity is also variable for each roof and can be defined as diverse and heterogeneous. Qualitatively, from field observations, it was apparent that in areas of greatest solar exposure (those with more than 10 h per day), vegetative coverage and species diversity decreased on both House 1 and House 2. Greatest species diversity was observed in the shaded areas along the skylights of each roof. Regression analyses (See Sect. 12.5.2) confirmed these field observations, and found that increased solar exposure had a significantly negative impact on House 2 species diversity ($R^2 = 20.79$, $p < 0.000$) and vegetative cover ($R^2 = 26.30$, $p < 0.000$) at the plot

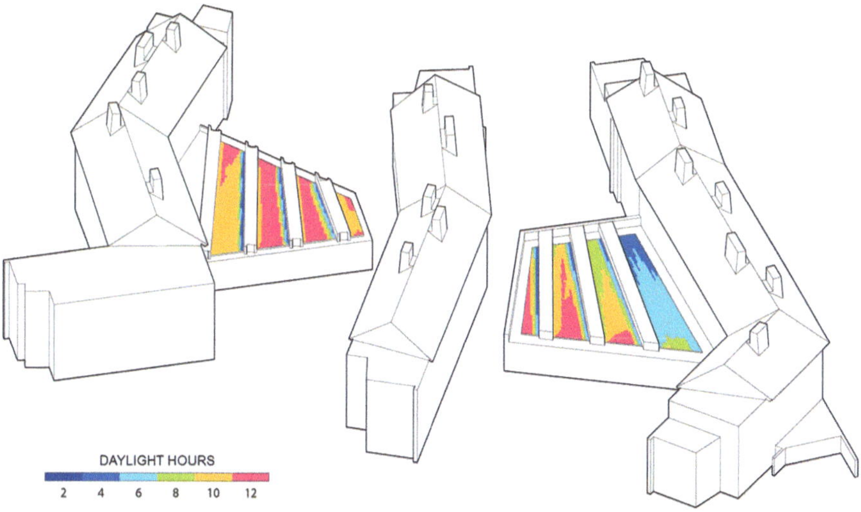

Fig. 12.8 Solar access analysis for House 1 (*left*) and House 2 (*right*). (Image by KieranTimberlake)

scale. House 1 results did not prove to be statistically significant, a finding that is likely the result of modeling resolution (30 cm^2 (4.65 in^2) grid), which did not capture smaller scale solar patterns associated with conditions along House 1 skylights.

Given that previous studies have related moisture stress and solar exposure to plant performance (Dunnett and Kingsbury 2004; Getter et al. 2009; Martin 2007), it is not surprising that areas with greater solar access exhibited less plant diversity and lower vegetative coverage. It is important to note however that in the modeling exercise above, the impact of solar access variability across each of these roofs cannot be isolated from other abiotic factors impacting vegetation. Solar access may impact plant productivity through changes in soil moisture content, but it may also impact vegetation productivity directly through decreased photosynthetic potential. It is also possible that physical attributes of the roof, such as the skylights, may allow for increased ponding, thereby directly altering moisture regimes.

Given the time scales of these interactions and the complexity of multiple variables, a complete understanding of how macro and microclimate factors govern plant dynamics is challenging. Additional complexity may be found in relating these vegetation dynamics to performance dynamics (Fig. 12.9)—the primary rea-

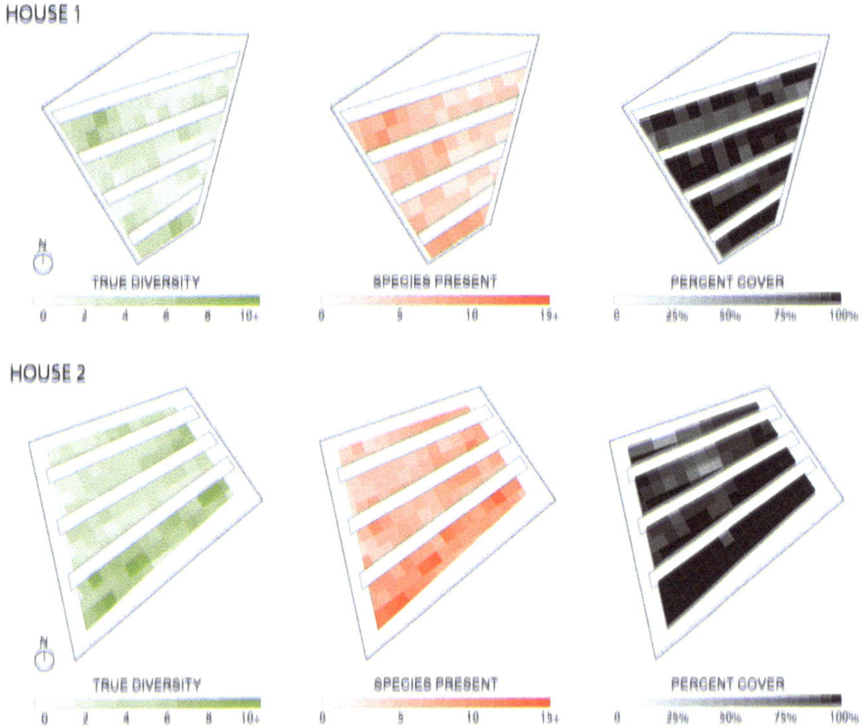

Fig. 12.9 Vegetative performance measures expressed across sample plots, 2012 survey. Areas of highest biodiversity, species richness, and percent cover occur in bays closest to an adjacent building face (to the north and south of House 1, and to the south and west of House 2). (Image by KieranTimberlake)

son these building systems are installed. Further study may allow for designers to make connections between microclimatic conditions created by design and detailing decisions, and goals for roof plantings, such as coverage, biodiversity, or thermal performance. Additionally, a deeper understanding of the variables driving green roof ecosystems may aid landscape managers who are seeking to minimize maintenance inputs while achieving multiple performance goals.

12.6 Future Considerations: Ecological Green Roof Design

Green roofs are designed landscapes, installed as part of constructed and managed ecosystems. From the time of installation, plant communities on green roofs respond to the heterogeneity of the constructed landscape—collectively becoming tuned to the stable microclimates of the site (such as solar access and wind patterns) while responding to irregular disturbances such as drought, disease, and competing emergent species.

This chapter explores concepts of vegetation dynamics within the context of green roofs, while examining empirical research and case study observations of plant assemblages and dynamics over long time periods. There is need for continued empirical study of green roofs, with an emphasis on researching plant communities in realistic roof conditions—roofs without artificial stressors associated with small plots, raised trays, rigorous weeding, or annual harvesting of biomass. An ecological approach, emphasizing vegetation dynamics, should avoid the casual dismissal of spontaneous ruderal vegetation as simply "weeds." Case studies such as the one presented in this chapter support the examination of the functional value of both intentionally planted, nursery-grown meadow species and spontaneous urban meadow species and seek to better understand the relationships within such dynamic communities. A shift to a performance-based and community ecology perspective allows for exploration of the possible benefits of establishing a diverse pallet of spontaneous climate- and site-adapted plants over time (Del Tredici 2010a, b).

Conventional landscape design and management tend to focus on assuring successful initial establishment and maintaining initial design intent through short-term warranty periods. As our focus and timescale of consideration shifts—toward the long-term resilience of a green roof system and of life cycle building performance—can rapid adaptation to unpredictable changes and resistance to disturbance in fact be seen as overlooked benefits of emergent species? Or do the presence of such species represent a trend towards a less ecologically rich, single-species dominated system, in which system function is diminished? Is the presence of emergent species on these roofs the sign of benign neglect or a symbol of increased biodiversity and community complexity?

Deepening our understanding of vegetative dynamics and growth trajectories in novel ecosystems such as green roofs allows the managers of these spaces to critically evaluate conventional management goals and practices. Integrating vegetation

Fig. 12.10 Pattern or process: For what are we designing and managing? Static designs, such as those above, may rely on continuous inputs of labor and planting. Ultimately, the design of green roof planting can work against community dynamics through a reliance on fixed pattern-making or it can embrace change and growth over time. (Image by KieranTimberlake)

dynamics theory with observation on installed roofs reminds us that all landscapes are characterized by change (Pickett and Cadanesso 2013) and that resilient landscapes make use of this capacity for flux to buffer against stressors and disturbance. Landscape management can be motivated by a desire to project an impression of stasis (Fig. 12.10). Or, it can strive to make use of ecosystem dynamics and controlled species interaction (Rosenberg and Freedman 1984; Luken 1990; Pickett et al. 2009), steering plant communities towards increased biodiversity, performance, and resilience as they age.

An understanding of ecological principles is important for designers and engineers, who are particularly inclined to portray engineered systems as stable and optimized—built as specified to realize a particular vision (Beck 2013) and performance expectation, but also to endure a changing environment. For designers and managers, increased understanding of the functional role of vegetative communities and legibility of ecological processes may lead to a reconsideration of the role of initial planting and roof detailing—re-casting both the biotic and abiotic components of the roof, not as final product but as evolving, critical infrastructure, enabling the roof system to grow and develop over time.

Ultimately, exploring the ecological, social, and economic forces that drive landscape dynamics, both during the design process and over a green roof's lifetime, requires collaboration between all stakeholders: designers, engineers, ecologists, managers, building owners and users. By questioning the vegetation dynamics of green roofs and integrating such questions into the initial design and management of green roofs, current projects can serve as designed experiments, creating data in realistic contexts and fostering the conditions for collaboration and exploration (Felson and Pickett 2005, 2013). In the end, this conversation deepens our understanding of the design product and creates new opportunities and techniques for designers concerned with creating high-performing, sustainable green roofs.

Appendix

Table 12.2 House 1 summary of initially planted species and all species identified in the 2012 and 2013 surveys that were present on more than 1 % of total green roof area. Species are ranked in order of dominance (percent cover) at time of 2013 census. Species in bold are from original planting in 2005. (Note: All originally planted species may not have been found to demonstrate greater than 1 % cover in 2012 or 2013.)

Latin name	Initial cover	2012 % cover	2012 sociability	2013 % cover	2013 sociability
Schizachyrium scoparium	**23.23**	**65.00**	**66.71**	**61.44**	**63.90**
Lotus corniculatus	–	2.47	21.39	17.15	21.23
Erigeron annuus	–	3.59	11.67	9.23	19.46
Medicago lupulina	–	–	–	8.33	13.27
Vicia spp.	–	2.28	44.38	8.04	44.82
Festuca rubra	**5.00**	**4.04**	**13.70**	**7.47**	**22.40**
Aster pilosus	–	–	–	6.96	18.08
Panicum virgatum	**21.44**	**4.94**	**20.26**	**5.35**	**16.70**
Conyza canadensis	–	–	–	4.55	12.24
Oxalis stricta	–	0.87	5.63	4.13	14.66
Echinacea purpurea	**2.42**	**1.19**	**11.56**	**2.98**	**12.24**
Lactuca serriola		1.57	6.13	2.92	8.75
Rumex obtusifolius	–	0.13	2.50	2.63	22.78
Trifolium pratense	–	1.19	7.71	2.24	8.75
Ambrosia artemisiifolia	–	0.03	2.50	1.67	16.25
Asclepias tuberosa	**1.73**	**1.06**	**10.31**	**1.35**	**15.00**
Geranium maculatum	**2.20**	**0.03**	**2.50**	**0.77**	**15.00**
Hypericum perforatum	–	1.99	7.05	0.67	8.75
Helianthus mollis	**1.25**	**1.06**	**20.63**	**.58**	**5.63**
Liatris aspera	**1.73**	**3.11**	**7.35**	**0.48**	**7.50**
Melilotus officinalis	–	16.28	28.22	0.48	7.50
Heliopsis helianthoides	1.25	0.29	5.63	0.45	8.75
Festuca ovina	**5.00**	**0.22**	**8.75**	**0.22**	**8.75**
Aster divaricatus	–	–	–	0.19	15.00
Eupatorim hyssopifolium	**1.25**	**0.19**	**15.00**	**0.19**	**15.00**
Phlox spp.	–	1.38	26.88	–	–
Arctostaphylos uva-ursi	**22.75**	**0.22**	**8.75**	–	–
Elymis hystrix	**6.25**	–	–	–	–
Lupinus perennis	**2.20**	–	–	–	–

Table 12.3 House 2 summary of initially planted species and all species identified in the 2012 and 2013 surveys that were present on more than 1 % of total green roof area. Species are ranked in order of dominance (percent cover) at time of 2013 census. Species in bold are from original planting in 2006. (Note: All originally planted species may not have been found to demonstrate greater than 1 % cover in 2012 or 2013.)

Latin name	Initial cover	2012 % cover	2012 sociability	2013 % cover	2013 sociability
Sedum sexangulare	**40.00**	**41.84**	**54.56**	**52.69**	**63.70**
Sedum spurium 'Fuldaglut'	**32.00**	**32.80**	**55.22**	**34.00**	**61.11**
Lotus corniculatus	–	0.24	10.83	31.41	35.11
Sporobolus heterolepis	**20.00**	**7.80**	**37.05**	**10.32**	**49.02**
Medicago lupulina	–	2.50	11.08	8.57	15.20
Allium cernuum	**6.00**	**4.02**	**38.21**	**6.26**	**37.84**
Hypericum perforatum	–	3.21	9.94	6.13	15.98
Aster pilosus	–	6.45	16.49	5.90	16.02
Melilotus officinalis	–	6.26	13.88	5.49	33.18
Daucus carota	–	4.06	16.36	4.44	16.86
Solidago canadensis	–	3.61	13.71	4.42	16.32
Erigeron annuus	–	6.22	17.24	4.25	16.62
Prunus virginiana	–	0.71	11.88	3.33	11.96
Parthenocissus quinquefolia	–	1.47	10.26	2.93	16.25
Leontodon autumnalis	–	1.82	7.35	2.89	11.32
Trifolium pratense	–	2.61	15.11	2.65	16.02
Festuca rubra	–	2.58	10.38	2.37	14.32
Setaria faberi	–	–	–	2.33	10.69
Populus spp.	–	1.80	10.00	2.16	15.97
Acer rubrum	–	0.11	15.00	1.39	13.21
Setaria viridis	–	2.78	16.09	1.33	22.19
Oenothera biennis	–	–	–	1.13	13.64
Taraxacum officinale		2.16	11.50	0.94	6.25
Solidago graminifolia		1.56	10.92	0.51	22.50
Sempervivum spp.	**2.00**	**0.06**	**2.50**	**0.15**	**6.67**

References

Aarssen L (1997) High productivity in grassland ecosystems: effected by species diversity or productive species? Oikos 80:183–184

Bass B, Liu KKY, Baskaran BA (2003) Evaluating rooftop and vertical gardens as an adaptation strategy for urban areas. Ottawa (Canada) National Research Council Canada, Institute for Research in Construction Report no. NRCC-46737

Baumann N (2006) Ground-nesting birds on green roofs in Switzerland: preliminary observations. Urban Habitats 4(1):37–50

Beck T (2013) Principles of ecological landscape design. Island Press, Washington DC

Berghage RD, Jarrett A, Beattie DJ, Kelley K, Husain S, Rezai F, Long B, Nagassi A, Cameron R, Hunt W (2007) Green roofs and green roof media capacity for neutralizing acid rain. A report on the National Decentralized Water Resources Capacity Development Project. PennState University

Blank L, Vasl A, Levy S, Grant G, Kadas G, Dafni A, Blaustein L (2013). Directions in green roof research: a bibliometric study. Build Environ 66:23–28

Braun-Blanquet J (1932) Plant sociology: the study of plant communities. McGraw Hill, New York

Brenneisen S (2003) The benefits of biodiversity from green roofs: key design consequences. In: Cities alive, Proceedings of the 1st North American Green Roof Conference. Chicago, 29–30 May 2003

Butler C, Orians C (2011) Sedum cools soil and can improve neighboring plant performance during water deficit on a green roof. Ecol Eng 37(11):1796–1803

Cardinale BJ, Wright JP, Cadotte MW, Carroll IT, Hector A, Srivastava DS, Weis JJ (2007). Impacts of plant diversity on biomass production increase through time because of species complementarity. Proc Nat Acad Sci U S A 104(46):18123–18128

Carlisle S, Piana M (2014) Green Roofs over time: using a spatially explicit methodology to study long-term green roof vegetation performance and dynamics. Cities and the Environment (CATE)

Cook-Patton S, Bauerle T (2012) Potential benefits of plant diversity on vegetated roofs: a literature review. J Environ Manage 106:85–92

Del Barrio EP (1998) Analysis of the green roofs cooling potential in buildings. Energy Build 27:179–193. doi:10.1016/s0378-7788(97)00029-7

Del Tredici P (2010a) Spontaneous urban vegetation: reflections of change in globalized world. Nat Cult 5(3):299–315

Del Tredici P (2010b) Wild urban plants of the northeast. Comstock Publishing Ithaca, New York

Díaz S, Cabido M (1997) Plant functional types and ecosystem function In relation to global change. J Veg Sci 8(4):463–474

Dunnett N, Hitchmough J (eds) (2007). The dynamic landscape: design ecology and management of naturalistic urban planting. Taylor & Francis, London

Dunnett N, Kingsbury N (2004) Planting green roofs and living walls. Timber Press Portland, Oregon

Dunnett N, Nagase A, Hallam A (2008) The dynamics of planted and colonising species on a green roof over six growing seasons 2001–2006: influence of substrate depth. Urban Ecosyst 11(2008):373–384

Durhman AK, Rowe DB, Rugh CL (2007) Effect of substrate depth on initial growth, coverage, and survival of 25 succulent green roof plant taxa. HortScience 42(3):588–595

Dvorak B, Volder A (2010) Green roof vegetation for North American ecoregions: a literature review. Landsc Urban Plan 96(4):97–213

Felson A, Pickett S (2005) Designed experiments: new approaches to studying urban ecosystems. Front Ecol Environ 3(10):549–556

Felson A, Oldfield E, Bradford M (2013). Involving ecologists in shaping large-scale green infrastructure projects. BioScience 63(11):882–890

Franco A, Nobel P (1989) Effect of nurse plants on the microhabitat and growth of cacti. J Ecol 77(3):870–886

Freitas H (1999) Biological diversity and functioning of ecosystems. In: Pugnaire F, Valladares F (eds) Handbook of functional plant ecology. CRC Press, New York, pp 719–733

Getter K, Rowe DB, Cregg M (2009) Solar radiation intensity influences extensive green roof plant communities. Urban For Urban Green 8(4):269–281

Glenn-Lewin D, van der Maarel E (1992) Patterns and processes of vegetation dynamics. In: Glenn-Lewin D, Peet R, Veblen T (eds) Plant succession: theory and prediction. Chapman and Hall, New York, pp. 11–59

Gobster P, Nassauer J, Daniel T, Fry G (2007) The shared landscape: what does aesthetics have to do with ecology? Landsc Ecol 22:959–972

Grime J (1977) Evidence for the existence of three primary strategies in plants and its relevance to ecological and evolutionary theory. Am Nat 1169–1194
Grime J (1998) Benefits of plant diversity to ecosystems: immediate, filter and founder effects. J Ecol 86(6):902–910
Heim A, Lundholm J (2014) Species interactions in green roof vegetation suggest complementary planting mixtures. Landsc Urban Plan 130:125–133
Hill M (1973) Diversity and evenness: a unifying notation and its consequences. Ecology 54:427–432
Hobbs RJ, Arico S, Aronson J, Baron JS, Bridgewater P, Cramer VA, Epstein PR (2006). Novel ecosystems: theoretical and management aspects of the new ecological world order. Global Ecol Biogeogr 15(1):1–7
Hooper D, Chapin F III, Ewel J, Hector A, Inchausti P, Lavorel S, Lawton J, Lodge D, Loreau M, Naeem S, Schmid B, Setälä H, Symstad J Vandermeer J, Wardle D (2005) Effects of biodiversity on ecosystem functioning: a consensus of current knowledge. Ecol Monogr 75(1):3–35
Kadas G (2006) Rare invertebrates colonizing green roofs in London. Urban Habitat 4:66–86
Köhler M (2006). Long-term vegetation research on two extensive green roofs in Berlin. Urban Habitat 4(1):3–25
Köhler M, Poll P (2010) Long-term performance of selected old Berlin greenroofs in comparison to younger extensive greenroofs in Berlin. Ecol Eng 36(5):722–729
Kolb W, T Schwarz (1993) Zum Klimatisierungseffekt von Pflanzenbeständen auf Dächern. Veitshöchheimer Berichte 4:28–36
Kowarik I (2011) Novel urban ecosystems, biodiversity, and conservation. Environ Pollut 159(8):1974–1983
Lazzarin R, Castellotti F, Busato F (2005) Experimental measurements and numerical modeling of a green roof. Energy Build 37:1260–1267
Luken J (1990) Directing ecological succession. Chapman and Hall, New York
Lundholm J, MacIvor J, MacDougall Z, Ranalli M (2010) Plant species and functional group combinations affect green roof ecosystem functions. PloS ONE 5(3):9677
Lundholm J, Heim A, Tran S, Smith T (2014) Leaf and life history traits predict plant growth in a green roof ecosystem. PloS ONE 9(6):e101395
MacIvor JS, Lundholm JT (2011) Performance evaluation of native plants suited to extensive green roof conditions in a maritime climate. Ecol Eng 37:407–417. doi:10.1016/j.ecoleng.2010.10.004
Martin M (2007) Native Plant Performance on a Seattle Green Roof. Master's Thesis. University of Washington
Mentens J, Raes D, Hermy M (2005) Green roofs as a tool for solving the rainwater runoff problem in the urbanized 21st century? Landsc Urban Plan 77:21–226
Naeem S, Tjossem S (1999) Plant neighborhood diversity and production. Ecoscience 6(3):355–365
Naeem S, Thompson L, Lawler S, Lawton J (1994). Declining biodiversity can alter the performance of ecosystems. Nature 368:21
Nagase A, Dunnett N (2010) Drought tolerance in different vegetation types for extensive green roofs: effects of watering and diversity. Landsc Urban Plan 97(4):318–327
Nagase A, Dunnett N (2011). The relationship between percentage of organic matter in substrate and plant growth in extensive green roofs. Landsc Urban Plan 103(2):230–236
Nardini A, Andri S, Crasso M (2012). Influence of substrate depth and vegetation type on temperature and water runoff mitigation by extensive green roofs: shrubs versus herbaceous plants. Urban Ecosyst 15(3):697–708
Nassauer J (1995) Messy ecosystems orderly frames. Landsc J 14(2):161–170
National Oceanic and Atmospheric Administration (NOAA) (2014) NowData—NOAA online weather data. http://www.nws.noaa.gov/climate/xmacis. Accessed 10 Feb 2014
Northeast Regional Climate Center (NRCC) Cornell University (2013) Northeast regional climate center. http://www.nrcc.cornell.edu/. Accessed 28 May 2013

Oberndorfer E, Lundholm J, Bass B, Coffman R, Doshi H, Dunnett N, Gaffin S, Kohler M, Liu K, Rowe DB (2007) Green roofs as urban ecosystems: ecological structures, functions, and services. BioScience 57(10):823–833

Olly L, Bates A, Sadler J, Mackay R (2011) An initial experimental assessment of the influence of substrate depth on floral assemblage for extensive green roofs. Urban For Urban Green 10(4):311–316

Pickett S, Cadenasso M (2013) Vegetation dynamics. Vegetation ecology. Wiley-Blackwell, Hoboken, pp 172–198

Pickett S, Cadenasso M, Meiners S (2009) Ever since clements: from succession to vegetation dynamics and understanding to intervention. Appl Veg Sci 12(1):9–21

Poore MED (1955) The use of phytosociological methods in ecological investigations: I. The Braun-Blanquet system. J Ecol 43(1):226–244

Ranalli MA, Lundholm JT (2008) Biodiversity and ecosystem function in constructed ecosystems. CAB Reviews: PAVSNNR 3: #034. doi:10.1079/PAVSNNR20083034

Rosenberg DB, Freedman SM (1984) Application of a model of ecological succession to conservation and land-use management. Environ Conserv 11(04):323–330

Rowe DB (2011) Green roofs as a means of pollution abatement. Environ Pollut 159(8–9):2100–2110

Rowe DB, Getter K, Durhman A (2012) Effect of green roof media depth on Crassulacean plant succession over seven years. Landsc Urban Plan 104(3–4):310–319

Schrader S, Böning M (2006) Soil formation on green roofs and its contribution to urban biodiversity with emphasis on Collembolans. Pedobiologia 50(4):347–356

Schwartz M, Brigham C, Hoeksema J, Lyons K, Mills M, Van Mantgem P (2000) Linking biodiversity to ecosystem function: implications for conservation ecology. Oecologia 122(3):297–305

Shannon CE (1948) A mathematical theory of communication. Bell Syst Tech J 27:379–423, 623–656

Spehn E, Joshi J, Schmid B, Diemer M, Körner C (2000) Above-ground resource use increases with plant species richness in experimental grassland ecosystems. Funct Ecol 14(3):326–337

Stovin V, Poë S, Berretta C (2013) A modelling study of long term green roof retention performance. J Environ Manage 131:206–215

Susca T, Gaffin SR, Dell'Osso GR (2011) Positive effects of vegetation: urban heat island and green roofs. Environ Pollut 159(8):2119–2126

Sutton R (2013) Rethinking extensive green roofs to lessen emphasis on biomass. J Living Archit. November (10):2. http://greenroofs.org/resources/JOLA[8]Sutton.pdf

Sutton R (2014) Aesthetics for green roofs and green walls. J Living Archit Aesthet Green Roofs Green Walls 1(2):19

Sutton R, Harrington J, Skabelund L, MacDonagh P, Coffman R, Koch G (2012) Prairie-based green roofs: literature templates and analogs. J Green Build 7(1):143–172

Takakura T, Kitade S, Goto E (2000) Cooling effect of greenery cover over a building. Energy Build 31:1–6

Team RC (2008) R: a language and environment for statistical computing. R Foundation for Statistical Computing, Vienna ISBN 3-900051-07-0

Theodosiou T (2003) Summer period analysis of the performance of a planted roof as a passive cooling technique. Energy Build 35(9):909–917

Thuring C, Dunnett N (2014) Vegetation composition of old extensive green roofs (from 1980s Germany). Ecol Process 3(1):1–11

Tilman D (1996) Biodiversity: population versus ecosystem stability. Ecology 77(2):350–363

Tilman D, Downing JA (1996). Biodiversity and stability in grasslands. In ecosystem management. Springer, New York, pp. 3–7

Tilman D, Reich PB, Knops J, Wedin D, Mielke T, Lehman C (2001) Diversity and productivity in a long-term grassland experiment. Science 294(5543):843–845

Turner R, Alcorn SM, Olin G, Booth JA (1966) The influence of shade, soil, and water on saguaro seedling establishment. Bot Gazette 127:95–102

United States Department of Energy (USDE) (2013) Climate data based on EPW/STAT files from Elmira Regional Airport, NY located 35 miles to the Southwest. http://apps1.eere.energy.gov/buildings/energyplus/cfm/weather_data.cfm. Accessed 30 May 2013

Villarreal E, Bengtsson L (2005) Response of a Sedum green-roof to individual rain events. Ecol Eng 25(1):1–7

Weiler S, Scholz-Barth K (2009) Green roof systems: a guide to the planning, design, and construction of landscapes over structure. Wiley, New York

Stephanie Carlisle is a lecturer in Urban Ecology at the University of Pennsylvania School of Design and an environmental researcher at KieranTimberlake, where she investigates interactions between the constructed and natural environment, bringing data-driven analysis and complex systems theory to architectural problems at the building, landscape and urban scale.

Max Piana is an urban ecologist at Rutgers University and the Center for Urban Restoration Ecology, where his research focuses on urban vegetation dynamics and the integration of ecology, design, and land management. He is a former environmental researcher at KieranTimberlake.

Chapter 13
Long-term Rooftop Plant Communities

Bradley Rowe

Abstract Most plant studies of green roof taxa have only been conducted for a duration of 1 or 2 years. The problem with this scenario is that it can result in premature conclusions and misleading recommendations because green roofs are dynamic systems. Plants that initially survive may eventually experience reduced coverage or disappear completely due to competition, variability in climate, and other factors. Setting up long-term studies similar to the National Science Foundation (NSF) Long Term Ecological Research (LTER) model would provide the opportunity to follow changes to green roof habitats over time and also examine impacts and ecosystem service outputs on similarly designed roofs across geographic locations. Without consciously considering the effects and changes over time mistakes are not only made, but also repeated. We review several important longitudinal studies and discuss factors that impact long-term plant communities such as substrate composition and fertility, substrate depth, substrate moisture, microclimates, roof slope, orientation, and irradiance levels; as well as initial plant choices, functional diversity and complexity, and maintenance practices. In addition, we discuss the potential of applying the LTER model to green roofs and close with future research needs and questions.

Keywords Long-term ecological research · Plant performance · Plant selection · Plant succession · Substrate composition · Substrate depth · Substrate moisture

13.1 Introduction

The long-term plant communities that exist on green roofs can have a major impact on the ecological services provided (Oberndorfer et al. 2007; Rowe and Getter 2010). If green roofs are to deliver these benefits over time, as well as to meet

---

B. Rowe (✉)
Department of Horticulture, Michigan State University, 1066 Bogue Street,
A212 Plant and Soil Sciences Bldg, 48824 East Lansing, MI, USA
e-mail: rowed@msu.edu

R. Sutton (ed.), *Green Roof Ecosystems*, Ecological Studies 223,
DOI 10.1007/978-3-319-14983-7_13

long-term client expectations, then plant selection and long-term plant performance are extremely important. To date, most green roof research has been conducted to measure performance of engineering traits such as stormwater retention and heat flux through roofing membranes or has focused on whether a particular species will survive. Unfortunately, there has been much less research involving green roof ecological principles (Cook-Patton and Bauerle 2012). These facts raise questions regarding how time will influence changes in plant communities, how these changes influence the ecosystem services that a roof provides, and what can be done to address these issues.

13.2 Review of Several Important Longitudinal Studies

One problem with the green roof literature dealing with plant evaluations is the lack of long-term studies published in peer-reviewed journals. Here we review several of the longest green roof studies of record that have been published in English. Studies where initial plantings were recorded which provide baseline information include the Paul-Lincke-Ufer project, Berlin (20 years) (Köhler 2006; Köhler and Poll 2010); Communication Arts Building, Michigan State University (9 years) (Getter et al. 2009a); Horticulture Teaching and Research Center, Michigan State University (7 years) (Durhman et al. 2007; Rowe et al. 2012); a commercial building in Sheffield, U.K. (6 years) (Dunnett and Nolan 2004; Dunnett et al. (2008); and Seaton Hall, Kansas State University (5 years) (Skabelund et al. 2014). The study on the Church of Jesus Christ of Latter-day Saints Conference Center in Salt Lake City (Dewey et al. 2004) was only conducted for 2 years, but the roof still exists and provides an excellent opportunity to go back and survey the roof after 14 years. The Ufa-Fabrik Cultural Center, Berlin (13 years) (Köhler 2006) and the Thuring and Dunnett (2014) paper that examined numerous old green roofs in Germany are included even though the original plantings are unknown.

13.2.1 Paul-Lincke-Ufer Project, Berlin (20-years)

In the longest plant evaluation study of record, Köhler (2006) evaluated long-term vegetation succession on the Paul-Lincke-Ufer (PLU) project located in Berlin. The study area consisted of ten sub-roofs each with a substrate depth of 10 cm (4 in) with varying orientation and slope. The green roofs were installed in 1985 as pre-vegetated mats seeded with ten species : wild chives (*Allium schoenoprasum*), cheatgrass (*Bromus tectorum*), orchardgrass (*Dactylis glomerata*), sheep fescue (*Festuca ovina*), red fescue (*Festuca rubra*), *junegrass* (*Koeleria macrantha*), perennial ryegrass (*Lolium perenne*), Canada bluegrass (*Poa compressa*), Kentucky bluegrass (*Poa pratensis*), and yellow stonecrop (*Sedum acre*). Data were recorded almost every year from 1985 until 2005 and measurements included the number of plants, coverage for each species, plant heights, and percentage of "standing dead" (living plants with dead leaves and stems). Over the 20-year period, 110 species were observed, but only about 10–15 were present in large numbers. The average number of plant species present at any given time was 15. Of the original ten species only five were present every year (*A. schoenoprasum, B. tectorum, F. ovina, P. compressa*, and *S. acre*). *Dactylis glomerata* no longer existed after the first year, and *K. macrantha, P. pratensis, L. perenne*, and *F. rubra* disappeared after 3, 5, 7, and 8 years, respectively. By 2005, *A. schoenoprasum* became by far the dominant species covering 56% of the area followed by *F. ovina, P. compressa*, and *B. tectorum*. Initially, numerous weeds sprouted from the seed bank present in the growing substrate. However, these disappeared after a few years as the roofs were not irrigated. Wet summers tended to encourage spontaneous species and enrich plant diversity due primarily to colonization. Some colonizing species such as bulbous bluegrass (*Poa bulbosa*) persisted, likely because it forms a bulb, which allows it to survive during dry periods. Also, the lichen, *Cladonia coniocrea*, colonized and persisted because it can withstand dry periods.

Köhler attributed weather related factors such as temperature and rainfall to be more important than roof size, slope, or age in regards to species richness. After the initial decrease in plant diversity, roof age had limited impact on species richness (defined as the number of different species present in a given habitat). It varied from year to year due to weather conditions, but had more or less reached equilibrium. Some species such as loose silky-bent (*Apera spica-venti*) were more apparent during wet summers compared to dry ones.

13.2.2 Ufa-Fabrik Cultural Center, Berlin (13-years)

The Ufa-Fabrik Cultural Center located in suburban Berlin was installed in 1986, but the first data were not collected until 1992 and then data collection continued until 2005 (Köhler 2006). The roof was planted with seed of wildflower meadow species collected from the Alps, but it is not known exactly what species were originally sown. This roof was irrigated the first 11 years and exhibited higher species richness during this time period. When irrigation was discontinued in 1997, herbaceous plant species started to decline and *Sedum* species began to dominate. In fact, in 1998, the common green roof plants, *S. acre* and Caucasian stonecrop (*Phedimus spurius*) appeared for the first time. It is difficult to make major conclusions regarding this roof since the exact original species are unknown and there wasn't any data collected until 8 years after installation.

13.2.3 Communication Arts Building, Michigan State University, East Lansing (9-years)

A study is being conducted on the third-story rooftop of the Communications Arts and Sciences Building on the campus of Michigan State University to quantify the effect of solar radiation (full sun vs. full shade) on several U.S. native and non-native species (Getter et al. 2009a) (Fig. 13.1). Plugs of six native and three non-native species were planted in May 2005 on substrates of two different depths [8.0 cm (3.1 in) and 12.0 cm (4.7 in)] both in sun and shade. Species tested included wild nodding onion (*Allium cernuum*), heath sedge (*Carex flacca*), cascade stonecrop (*Sedum divergens*), narrow-petaled stonecrop (*Sedum stenopetalum*), largeflower fameflower (*Talinum calycinum*) (currently known as *Phemeranthus calycinus* by taxonomists), and sunbright *(Talinum parviflorum*)(currently known as *Phemeranthus parviflorus*), as well as three non-natives (*Sedum acre* (biting stonecrop), *Sedum album* (white stonecrop), and *Sedum urvillei* (stonecrop). Plots were irrigated during the first year of establishment, but relied on natural rainfall thereafter.

At the end of the first growing season, *C. flacca* was one of the most abundant species for both substrate depths in the shade. However, in subsequent years, it decreased in abundance during the driest portions of the summer that likely impacted overall regeneration. By the end of the 4 years, this species exhibited zero or near-zero absolute cover (AC). Absolute cover is defined as the total number of contacts recorded for each species divided by the number of data collection points.

Fig. 13.1 Replicated research plots located in the shade on the Communication Arts and Sciences Building at Michigan State University. (Photo DB Rowe)

In contrast, at the end of the second growing season, *S. acre* had established itself as the most abundant species for both substrate depths in the shade and exceeded an AC of 0.6 by the third growing season. For both substrate depths in the shade, *A. cernuum* was the next most abundant species by the fourth growing season, followed by *S. album* and *T. calycinum*. In the sun, by the second growing season both substrate depths were dominated by *S. album*, followed by *T. calycinum* and *S. acre*. At 12 cm (4.75 in), *A. cernuum* closely followed as the fourth most abundant, but this species was not nearly as abundant as it was in the shade at the same depth. By week 174 (23 Sept 2008), most species exhibited different AC within a depth between sun and shade. However, when all species were combined, overall AC did not differ between sun and shade within a depth. This indicated that while species make-up was changing among solar radiation levels, that overall coverage was not significantly different between sun and shade. For all substrate depths and solar levels, the most abundant species were *S. acre, A. cernuum, S. album*, and *T. calycinum*. With the exception of *T. calycinum*, native species were less abundant than non-native species. The native *Talinum* species (*T. calycinum* and *T. parviflorum*) were outside of their hardiness zone, but are prolific seeders. However, they need bare soil in order to germinate. By the end of 9 years, the only species that still existed were *A. cernuum, S. acre*, and *S. album* and represents a significant decrease in species richness over time. Plots were weeded the first 4 years, but have received no maintenance over the past 5 years.

13.2.4 Horticulture Teaching and Research Center, Michigan State University, East Lansing (7-years)

This study followed the succession of 25 succulents grown at three substrate depths over the course of 7 years (Durhman et al. 2007; Rowe et al. 2012). Absolute cover was determined using a point-frame transect every two weeks during the first three growing seasons and monthly during years four through seven to measure community composition and change (Fig. 13.2). At the 7.5 cm (3 in) depth, 22 species were present at the end of the first growing season, but these numbers were reduced to 13, 8, and 7 after 2, 3, and 5 years, respectively. Similar results occurred at the shallower depths except that the number of species was reduced at a faster pace. For the most part, the species present did not change after 4 years, but the relative abundance for each species continued to change. At 5.0 cm (2 in) and 7.5 cm (3 in), both Caucasian stonecrop (*Phedimus spurius*) and Chinese mountain sedum (*Sedum middendorffianum*) continued to expand through year 7 at the expense of the other remaining species. At 2.5 cm (1 in), *S. acre* and *S. album* were the dominant species.

Results show that the length of the study can have a dramatic effect on conclusions and plant recommendations. The initial paper published from the first two seasons of data from this study (Durhman et al. 2007) recommended *P. spurius, S. acre, S. album, S. middendorffianum*, Jenny's stonecrop (*S. reflexum*), pale stonecrop (*S. sediforme*), and *P. spurius* for extensive green roofs ranging from 2.5 cm (1 in) to

Fig. 13.2 Use of a point frame to measure absolute cover on the Horticulture Teaching and Research Center green roof at Michigan State University. (Photo DB Rowe)

7.5 cm (3 in) in depth. Additional recommendations for subsidiary species that were present at specific substrate depths, but may not exhibit an ability to cover large areas included Burnatti sedum (*S. dasyphyllum* 'Burnatii'), lilacmound sedum (*S. dasyphyllum* 'Lilac Mound'), diffuse sedum (*S. diffusum*), Spanish sedum (*S. hispanicum*), and orange stonecrop (*S. kamtschaticum* syn. *Phedimus kamtschaticus*). As can be seen from the results following 7 years, recommendations were misleading as *S. sediforme, S. dasyphyllum* 'Burnatii', *S. dasyphyllum* 'Lilac Mound', *S. diffusum*, and *S. hispanicum* no longer existed at any depth.

13.2.5 A Commercial Building in Sheffield, U.K. (6 years)

Dunnett and Nolan (2004) and Dunnett et al. (2008) conducted a study on top of a three story commercial building in Sheffield, U.K., over a period of 6 years from 2001–2006. The objectives of their study were to evaluate potential plant taxa for use on green roofs that experience a maritime U.K. climate and to test how substrate depth [10 cm (4 in) and 20 cm (8 in)] influenced plant establishment and survival, as well as visual and aesthetic criteria. Species included white thrift seapink (*Armeria maritima* 'Alba'), lesser calamint (*Calamintha nepeta*), maiden pink (*Dianthus deltoides*) dwarf blue fescue (*Festuca ovina glauca*), bearskin fescue (*Festuca scoparia*), Lindheimer's beeblossom (*Gaura lindheimeri*), white creeping babysbreath (*Gypsophila repens* 'Alba'), Border Ballet red hot poker (*Kniphofia* X'Border Ballet'), sea-lavender (*Limonium platyphyllum*), Fassen's catnip (*Nepeta Xfassenii*), Herrenhuasen oregano (*Origanum laevigatum* 'Herrenhausen'), roseroot (*Rhodiola rosea*), yellow stonecrop (*Sedum acre*), lambsear (*Stachys byzantina*), and dwarf spiked speedwell (*Veronica spicata* 'Nana').

Species tested were all native to dry and nutrient-stressed habitats, but differed widely in their heights, flowering times, life spans, growth forms, and locations where they are considered native. In addition to annually measuring plant height, spread, flowering performance and percent vegetation cover for the 15 planted species, they also recorded the numbers and percent cover for colonizing species.

The greatest survival, diversity, size, and flowering performance of planted species occurred at a substrate depth of 20 cm (8 in) relative to 4 in (10 cm) and the herbaceous species developed 85 and 58 % coverage at the end of two growing seasons at depths of 20 cm (8 in) and 10 cm (4 in), respectively. By the end of 5 years, all species survived at both depths, however, 14 of 15 species maintained at least 50 % of their original numbers at 20 cm (8 in) whereas, only eight did so at 10 cm (4 in). Bare ground and moss cover was greatest at 10 cm (4 in) as was diversity of colonizing species, presumably due to the presence of open space for invading seeds to germinate.

Species-richness (mean number of taxa per subplot) decreased over time at both substrate depths, but the rate of decline was greater at 10 cm (4 in). As mentioned, some species performed better at the 10 cm (4 in) or 20 cm (8 in) depth. The low-growing species such as sedum that are typical of shallow extensive roofs were not as competitive at 20 cm (8 in). Likewise, the perennial plants that normally possess greater biomass could not survive as well at 10 cm (4 in). Even when drought tolerant plant species are selected, the limiting factor for plant survival is often substrate moisture, which is often a function of substrate depth (Chaps. 4, 5). The authors emphasized the importance of long-term monitoring of green roofs because of the changes that occurred in plant communities from the first to the sixth year of their experiment.

13.2.6 Seaton Hall, Kansas State University, Manhattan, KS (5-years)

The first green roof project installed at Kansas State University was planted on Seaton Hall in May 2009 (Skabelund et al. 2014). The main goal of the project was to see if a semi-intensive green roof consisting of native grasses and forbs growing in a substrate profile ranging from 10 cm (4 in) to 18 cm (7.1 in) was feasible in this relatively dry climate with minimal maintenance and irrigation. The roof is south facing and receives reflected light off windows and limestone especially during spring and fall. The 28.3 m^2 (305 ft^2) roof was planted with plugs of five species of grasses, ten forbs, and one forb-like shrub. Grasses included side-oats grama (*Bouteloua curtipendula*), blue grama (*Bouteloua gracilis*), little bluestem (*Schizachyrium scoparium*), prairie dropseed (*Sporobolus heterolepis*), and Indian-grass (*Sorghastrum nutans*). Forbs consisted of smooth aster (*Aster laevis*), purple poppy-mallow (*Callirhoe involucrata*), purple prairieclover (*Dalea pupurea*), tall gayfeather (*Liatris aspera*), dotted gayfeather (*Liatris punctata*), prairie coneflower (*Ratibida columnifera*), gray-headed prairie coneflower (*Ratibida pinnata*), wild blue sage (*Salvia azurea*), rigid goldenrod (*Solidago rigida*), and common spiderwort (*Tradescantia ohiensis*). The forb-like shrub was New Jersey tea (*Ceanothus americanus*). The study is still ongoing and has yet to be published other than in a proceedings from a meeting (Skabelund et al. 2014).

Along with plant survival and dynamics, a range of climatic variables was monitored. A subset of selected grasses was evaluated for height, basal diameter, and number of flowering stalks at the end of each growing season between 2009 and 2013. In 2009 and 2010, supplemental irrigation was provided on an as-needed basis and growing conditions were favorable, resulting in nearly 100 % plant survival. Most grasses exhibited flowering stalks and increased basal diameter. The west side of the green roof was not irrigated in 2011, the entire roof was irrigated in 2012, and then supplemental irrigation ceased during mid-August 2012. Between 2010 and 2011 the original plantings decreased from 130 to 98 for individual grasses and from 98 to 39 for forbs. At the end of 2012 grasses exhibiting visibly-green above ground biomass remained at 98 while forbs increased to 54. By November 2013 original grasses numbered 68 and forbs 21. After the first year many new native grasses and forbs established themselves from germinating seeds produced by the original plantings. This was particularly pronounced in 2010 and 2012. Notably, plants of *B. gracilis* were taller in deeper substrates, with 12–18 cm (4.75–7.1 in) depths producing plants approximately 10.5 cm (4.1 in) taller than 7.5–9 cm (3.0–3.5 in) depths. Between 2009 and 2012, 15–18 cm (5.9–7.1 in) substrate depths produced *B. gracilis* 10.8 cm (4.2 in) taller than those at 10 cm (4 in) depths.

13.2.7 Church of Latter-Day Saints Convention Center, Salt Lake City, Utah

An example of a short-term study is the Church of Jesus Christ of Latter-day Saints Conference Center in Salt Lake City, Utah (Dewey et al. 2004) (Fig. 13.3). However, because original substrate conditions and some information on plantings were recorded, the opportunity exists to monitor this roof into the future.

The objective of the original study was to observe the relative competitiveness of native grass and wildflower species growing in a range of different radiation/temperature environments. For research purposes, the roof was partitioned into seven radiation zones: (1) maximum sunlight, maximum reflection/radiation, (2) maximum sunlight, moderate refection/radiation, (3) maximum sunlight only, (4)

Fig. 13.3 Meadow consisting of native plants on the Church of Latter-day Saints Convention Center in Salt Lake City, Utah. (Photo DB Rowe)

minimal shading, (6) moderate shading, and (7) maximum shading. Zone 5 was eliminated from the study as it was considered similar to zone 4. The main component of the substrate was heat-expanded shale and it was placed at a depth of 1 m (3.3 ft). The roof was planted with plugs during the summer of 2000, overseeded in April 2001 with some of the same species in addition to others. Weeds were pulled as needed and the roof was irrigated twice a week.

During fall 2001, the roof was evaluated by counting the number of plants present for each species in a given sample area. Twenty one species were identified that should at least be considered for future grass/wildflower green roofs. However, Canada bluegrass (*Poa compressa*) and white sage (*Artemisia ludoviciana*) were too aggressive when planted in this grass and wildflower mixture. In contrast, the alpine bluegrass (*Poa alpina*), big bluegrass (*Poa secunda*), mutton bluegrass (*Poa fendleriana*), blue bellflower (*Campanula rotundifolia*), columbine (*Aquilegia* spp.), purple meadowrue (*Thalictrum purpurea*), and tickseed. (*Coreopsis* spp.) may not be competitive enough. Since there was no experimental design to the original planting, no replication, and only an estimate of the number of plugs planted in each zone, the study is only observational. Still, if monitored in to the future it would provide valuable information as to the long term succession of a grass and wildflower meadow on a green roof.

13.2.8 Old Green Roofs in Germany

It would be a travesty to discuss long-term plant communities on green roofs without acknowledging the long tradition of over 100-years of green roofs in Germany. Unfortunately, much of the original information on these roofs was never recorded, has been lost, or was anecdotal; studies were observational in nature without replication and thus not scientifically sound by today's standards; were not published in peer-reviewed journals; or are not easily accessible to the scientific world as they were not written in English. However, in addition to the Paul-Lincke-Ufer project and the Ufa-Fabrik Cultural Center in Berlin (Köhler 2006) described above, two recent papers published in scientific journals have gone back and looked at some of these older German roofs (Köhler and Poll 2010; Thuring and Dunnett 2014).

The purpose of the Köhler and Poll (2010) study was to compare vegetation and substrate characteristics between the old Tar-Paper-Green roofs (TPG-roofs) that were installed between 1880 and 1930 to the first Modern Extensive Green roofs (MEG-roofs) that were established in the 1980's. These roofs, subjects of previously published studies written in German from 1960, 1982, 1986, 1987, 1990, and 1995 were surveyed in 2008. While the Paul-Lincke-Ufer project (Köhler 2006) discussed earlier focused on ecological succession, this study concentrated on growing substrate, vegetative quality, and species richness.

According to the specified criteria set by Köhler and Poll (2010), they concluded that the performance of the MEG-roofs with engineered substrates composed of heat expanded clay, etc. was higher than the older TPG-roofs that originally utilized sandy soils. Even so, both roof types were still functional after many years and exhibited an increase in pedogenesis, a trend toward higher organic carbon, and a neutral pH. The old TPG-roofs were significantly richer in humus (mean organic C content of 4 %) than the MEG-roofs. Initial mean organic carbon content on the MEG-roofs was 2.5 % and then declined to 1.9 % due to microbial oxidation. After the roof stabilized after about 10 years, their organic carbon content increased steadily for the next 25 years up to the point that by 2008, the organic C content of both roof types were not significantly different. Total porosity of the MEG-roof substrates rose over a period of 10 years from 50 to 60 %. This change is likely due to processes such as the continuous formation and decay of plant roots, microbial activity, freezing and thawing.

Regarding plant species, 70 different species were recorded on the MEG-roofs, compared to 45 on TPG-roofs. Of course, this difference in species richness could be due to differences in substrate properties, as well as many other factors such as initial plantings. The most successful species were generally grasses such as cheatgrass (*Bromus tectorum*), poverty brome (*Bromus sterilis*), fescues *Festuca* spp., perennial ryegrass (*Lolium perenne*), annual bluegrass (*Poa annua*), and Canada bluegrass (*Poa compressa*) (most common).

The second study took place in southwestern Germany where Thuring and Dunnett (2014) surveyed vegetation and substrates on nine of the oldest extensive green roofs in the Stuttgart area during 2010 and 2011. Roof ages at the time of the survey ranged from 20 to 33-years-old. Unfortunately, there was little information on original substrate composition and depth or original species planted on these roofs so the results serve as a snapshot in time of present conditions. They could only speculate on how the substrates and plant communities changed over time. However, the roofs likely all adhered to early FLL standards and had a substrate depth less than 20 cm (8 in), a pH between 6.5 and 8.0, and organic content below 4.1 lbs/ft^3 (65 g/L) when constructed (FLL 2008).

Results suggested a decrease in substrate depth, substrate pH, and plant biomass over time while substrate organic content increased. This increase in organic matter agrees with the Köhler and Poll (2010) study discussed above and with the findings of Getter et al. (2007) who reported that organic matter nearly doubled from 2.33 to 4.25 % in just 5 years where the primary component of the substrate was heat-expanded slate. Similarly, Getter et al. (2009b) reported that the amount of carbon sequestered on shallow sedum based roofs increased with age and that 100 g C/cm^2 (57.8 oz/in^3) were sequestered during the first 2 years after installation of 6 cm (2.4 in) deep plots. The increase in organic carbon makes sense when one considers that the engineered substrates often used on extensive green roofs have limited initial organic matter because they are designed to hold moisture by manipulating particle size distributions (FLL 2008). Also, low substrate pH could result in an accumulation of substrate organic matter because some microbes are adversely affected by low pH, thus reducing decomposition (Berendse 1998).

Regarding plant cover, Thuring and Dunnett (2014) reported that succulents dominated these roofs either by themselves or as a consistent groundcover underneath other herbaceous perennials or grasses. Over time species diversity decreased which agrees with the work of Liesecke (1998) who reported that one or two succulents, a single herb, and one or two moss species often dominated older, extensive green roofs or two moss species.

13.3 Factors Impacting Long-Term Plant Communities

Numerous factors impact long-term plant communities on green roofs. Factors include substrate composition and fertility, substrate depth, substrate moisture, microclimates, roof slope, orientation, and irradiance levels; as well as initial plant choices, functional diversity and complexity, and maintenance practices.

13.3.1 Substrate Composition and Fertility

Substrate composition influences plant communities primarily through moisture retention and nutrient availability. Ideally they should be lightweight, permanent, and able to sustain plant health without leaching nutrients that may pollute receiving water bodies. For these reasons substrates often incorporate aggregate materials such as heat expanded slate, shale, or clay as their main component (FLL 2008). Water holding capacity can be altered by manipulating the particle size distribution of the aggregates and by adding organic matter. Although organic matter will retain moisture and provide nutrients, high levels are not recommended because it decomposes resulting in substrate shrinkage and can leach nutrients such as nitrogen (N) and phosphorus (P) in the runoff (Rowe 2011). The same runoff problems can occur when fertilizer is applied. A detailed discussion of nutrient cycling in green roof ecosystems can be found in Chap. 5.

In a study that looked at the effects of substrate composition and fertility, Rowe et al. (2006) found that sedum achieved 100% cover regardless of the percentage of heat-expanded slate in the substrate, but that the herbaceous perennials and grasses required greater percentages of organic matter or supplemental irrigation. They also reported that a greater number of smooth aster (*Aster laevis*), junegrass (*Koeleria macrantha*), and showy goldenrod (*Solidago speciosa*) survived when they were not fertilized. Presumably, these plants could survive drought conditions for a longer period of time since they had less biomass to maintain. In contrast, if the purpose of the green roof is urban agriculture then fertility levels must be relatively high to produce acceptable yields for fruits and vegetables (Whittinghill and Rowe 2012a; Whittinghill et al. 2013).

Most commercial green roofs are built within German FLL guidelines (FLL 2008) and are composed of manufactured plastic layers topped with engineered growing substrates. These standards help to assure consistency of materials and success of green roof projects. However, many are being built without these expensive components, especially in Switzerland (Brenneisen 2006; Kiers 2013). Stephan Brenneisen, from the University of Applied Sciences Wädenswil, has been a proponent for the construction of green roofs with the primary purpose of promoting biodiversity. For example, some roofs utilize gravel or layers of straw or grasses such as maidengrass (*Miscanthus sinensis*) as the drainage layer, use native soils blended with other components such as lava rock or gravel, and are planted with native wildflowers. A commercial installer may be hesitant to go outside FLL substrate specifications, but other systems do work (Chaps. 3, 6).

An excellent example of a green roof constructed with non-standard green roofing materials is the Moos Lake water filtration plant in Wollishofen, Zürich, Switzerland. Installed in 1914, the roof was built long before German FLL guidelines were written and adopted (Fig. 13.4). The original drainage layer consisted of gravel topped with 12.5 cm (5 in) of sand and 15–20 cm (6–8 in) of local topsoil. After 100 years those layers are no longer distinguishable, but there are neither problems with drainage nor any negative effects on the vegetation, and the original roofing membrane is still in place. The 30,000 m^2 (322,917 ft^2) roof is home to 175 plant species, several of which are now endangered or rare. The roof consists of nine species of orchids and approximately 6000 specimens of green-winged orchid (*Orchis morio*) a species that is now extinct in the landscape surrounding Zürich. The roof reflects species richness of the surrounding area from 100 years ago as well as today. The original vegetation developed from the seed bank that was part of the

Fig. 13.4 The Moos Lake water filtration plant in Willishofen, Zürich, Switzerland was built in 1914 and is home to 175 plant species, many of which are now endangered or rare. (Photo DB Rowe)

original topsoil. Today, plant composition consists of these original species as well as any new species that colonized from the surrounding landscape.

13.3.2 Substrate Depth

Substrate depth has a major impact on plant survival and long-term plant communities. Depending on climate and the availability of supplemental irrigation, most shallow extensive green roofs are limited to drought tolerant species such as succulents. This is primarily due to a lack of moisture (Dunnett and Nolan 2004; Durhman et al. 2006), but some taxa such as *Sedum* spp. are naturally found in these conditions. However, even among succulents, substrate depth will influence total coverage and coverage of individual species. In Pennsylvania, Thuring et al. (2010) reported that white stonecrop (*S. album*) and six-sided stonecrop (*S. sexangulare*) survived in 3 cm (1.2 in), but produced greater biomass at depths of 6 cm (2.4 in) and 12 cm (4.7 in). Similarly, Getter and Rowe (2009) reported that the majority of the 12 species of *Sedum* tested in Michigan exhibited greater growth and coverage at a depth of 7.0 cm (2.7 in) and 10.0 cm (4 in) compared to 4.0 cm (1.6 in). At 5.0 cm (2 in) and 7.5 cm (3 in), *Phedimus spurius* and *Sedum middendorfianum* were the dominant species, but at 2.5 cm (1 in), *S. acre* and *S. album* covered the most area (Durhman et al. 2007).

In Sweden, *S. acre* and *S. album* were dominant at a depth of 4 cm (1.6 in) while the other succulents in the study, *S. reflexum* (syn. *S. rupestre*), *S. sexangulare*, pink Mongolian stonecrop (*Hylotelephium ewersii*), Chinese sedum (*Phedimus floriferus*), hybrid stonecrop (*Phedimus hybridus*), *Phedimus kamtschaticus* (syn. *S. kamtschatium*), and Caucasian stonecrop (*Phedimus spurius*) grown in various combinations had minimal coverage by the end of 3 years, generally 15 % or less for all other species combined (Emilsson and Rolf 2005; Emilsson 2008). Development over time varied depending on the original species mix planted, as well as substrate composition.

As depth increases, the number of potential species expands to grasses, many annual or herbaceous perennials, and even woody plants. Deeper substrates are beneficial for both increased water holding capacity (Durhman et al. 2006; VanWoert et al. 2005a; VanWoert et al. 2005b) and as a buffer for overwintering survival, as shallow substrates are more subject to fluctuations in temperature (Boivin et al. 2001). As discussed above, Dunnett et al. (2008) reported the greatest survival, diversity, size, and flowering performance of grasses and herbaceous perennials occurred at a substrate depth of 20 cm (8 in) compared to a depth of 10 cm (4 in). By the end of 5 years, all species survived at both depths, however, 14 of 15 species maintained at least 50 % of their original numbers at 20 cm (8 in) whereas, only eight did so at 10 cm (4 in). Likewise, in Southern Tuscany, most of the 20 Mediterranean xerophytic species tested exhibited greater growth and cover at 20 cm (8 in) relative to those grown at 15 cm (6 in) (Benvenuti and Bacci 2010). In addition to greater moisture stress, temperatures in the shallower substrate [15 cm (6 in)] reached a maximum of 90 °F (50 °C) and were on average 9 °F (5 °C) higher than the 20 cm (8 in) deep substrate. This could be partially explained by the fact that shallower substrate depths often have less coverage which exposes more substrate to direct sun resulting in higher substrate temperatures (Getter et al. 2009a).

13.3.3 Substrate Moisture

Substrate moisture is a function of substrate composition and depth and is often the limiting factor for plant survival on green roofs (Dvorak and Volder 2010). In the Getter and Rowe (2009) study discussed above, mean volumetric moisture content at the three substrate depths were correlated with plant growth and coverage. Similarly, Thuring et al. (2010) reported that the herbaceous species tested were severely affected by drought when grown in shallower substrates. In addition, Monterusso et al. (2005) found that only four of 18 species of native herbaceous perennials and grasses still existed after 3 years when grown at a 10 cm (4 in) depth without irrigation. The majority of the plants tested were considered to be drought tolerant, but their survival in a native environment relies on deep tap roots to obtain moisture. Survival and persistence could have been improved by increasing substrate moisture through changes in substrate composition, depth, or by providing irrigation.

However, deeper substrate depths that hold more moisture are not beneficial to all plant species as long-term survival of stress tolerant species often depends on shallow soil depths with limited moisture. Otherwise, species with greater growth potential will outcompete them. This was even evident in the Rowe et al. (2012) study as *S. acre* and *S. album* were dominant at 2.5 cm (1 in) whereas *P. spurius* and *S. middendorfianum* were most prevalent at deeper depths. Similarly, Emilsson (2008) reported that *S. acre* decreased in area of coverage after 2 years. This may be because *S. acre* allocates a relatively small percentage of plant carbon to the root system (Getter et al. 2009b) and these roots also tend to be shallow and less able to compete for water. Increasing substrate depth is of no advantage to this species, as it must then compete against more aggressive plants with greater biomass (Getter and Rowe 2009). Likewise, in the Dunnett et al. (2008) study, *Armeria maritima* performed better at 10 cm (4 in) relative to 20 cm (8 in). *Armeria maritima* is a self-seeder and likely took advantage of the greater bare space at the shallower depth. Other species such as the succulents *T. calycinum* and *T. parviflorum* also depend on bare space for long-term survival in climates such as that found in Michigan. These species are perennials, but are killed by cold winter temperatures in Michigan and reappear each year by reseeding. However, as the roof obtains 100 % coverage, there is little open space for germination to continue from year to year and the species eventually disappears (Getter et al. 2009a).

Supplemental irrigation can alleviate substrate moisture problems, but the use of potable water on green roofs is often problematic. If irrigation is to be supplied, then it should be done so with the most efficient and sustainable method for the particular application (Rowe et al. 2014). Irrigation is critical when growing vegetables on roofs (Whittinghill and Rowe 2012a; Whittinghill et al. 2013).

13.3.4 Microclimates, Roof Slope, Orientation, and Irradiance Levels

Microclimates present on a roof will dramatically influence short and long-term plant communities (see Chap. 3). They can be caused by variations in substrate composition and depth as described above, or from variations in irradiance levels due to shaded areas, roof slope, and roof orientation.

In the irradiance level (full sun versus full shade) study on the MSU Communication Arts Building described above it was found that regardless of depth, species differed depending on sun exposure (Getter et al. 2009a). After four growing seasons, heath sedge (*Carex flacca*), was still present at 12 cm (4.7 in), but only in the shade. After 9 years it has completely disappeared. Even though species mix was changing among solar radiation levels, overall coverage was not significantly different between sun and shade. Roof slope and orientation also influence substrate moisture and thus plant communities. Getter et al. (2007) reported that water retention was reduced by 10 % when slope increased from 2 to 25 %. Orientation is also important as evapotranspiration increases with solar exposure. Köhler and Poll

(2010) reported that the greatest plant coverage was found on north-facing sections of the roof on the Paul-Lincke-Ufer Building in Berlin. The most dominant species was *Allium schoenoprasum* while on south facing slopes *Sedum* spp. dominated. Grasses were least competitive on west facing slopes.

One example of a roof designed to create various microclimates that in turn promote diversity is the California Academy of Sciences in San Francisco (Hauser 2013). The seven domes create different microclimates due to variations in slope and sun exposure and thus substrate moisture. This in turn influences the plant communities that find their niche among the various microclimates where they have a competitive advantage. The roof was originally planted in 15 cm (6 in) of substrate with four perennial and five annual species uniformly spaced over the entire roof. Today, there are approximately 70 native species thriving where the environmental conditions are best for each individual species. This increase in species richness is contrary to the decreases that occurred when only one substrate depth was employed on the other roofs described above.

Therefore, it seems logical that one way to increase plant diversity on green roofs is to create multiple microclimates. If roof slope and orientation are not options then variations in substrate depth and composition can be created. Because different species can compete best in a specific environment, each species will find its niche location where it has advantages over competing species. This will likely increase the biodiversity potential and improve the species richness of the long-term plant community as environmental conditions change over time and species increase and decrease in numbers.

13.3.5 Initial Plant Choices, Functional Diversity and Complexity, and Maintenance Practices

The plants present on a green roof at any given time also depend on what was initially planted, the functional diversity and complexity of these species, and the intensity or lack thereof of maintenance. Some plants may be originally chosen for factors such as aesthetics, but may be ill suited for the particular environmental conditions and destined to fail. Others may be too aggressive and will crowd out everything else (Getter and Rowe 2009; Rowe et al 2012). If the overly aggressive species was not planted to begin with, then the dynamic would be completely different.

Maintenance is also a major factor. If 'weed' species are removed on a regular basis then they clearly will not be able to colonize a roof. In this case, weeds are defined as any species that was not planted in the original design. Colonizing species will also be influenced by the proximity to local seed sources. The height of the roof and the surrounding landscape will influences seed sources (Chap. 15). Maintenance practices such as irrigation and fertility management are also major factors as discussed in Chaps. 4 and 5.

There are also complex interactions among plants (Chap. 8). Nagase and Dunnett (2010) studied how plant diversity on a green roof influenced survival by testing combinations of three major taxonomic and functional plant groups that are commonly used for extensive green roofs (forbs, sedums and grasses). They concluded that under drought conditions, combinations of species differing in functional diversity and complexity exhibited greater survival rates and visual qualities than monocultures. They attributed this result to the fact that plants of the same taxonomic group compete for the same resources when grown together.

In addition, Butler and Orians (2011) showed that the drought tolerant succulent, *S. album*, could have a positive or negative influence on neighboring plants depending on substrate moisture content. When ample substrate moisture was present, *S. album* had an adverse effect on growth of threadleaf giant hysop (*Agastache rupestris*) and whorled milkweed (*Asclepias verticillata*). In contrast, during drought conditions the presence of *S. album* as an understory cover facilitated growth of these more water dependent herbaceous plants. The favorable response during drought is likely due to *S. album* shading the surface, reducing evaporation from the substrate surface, and from a reduction in substrate temperatures (Butler and Orians 2011). One might expect the same result for other plant species although the use of *S. album* cover crop for green roof production of an assortment of vegetables had no effect on crop yields (Whittinghill and Rowe 2012b). Vegetables tested were tomatoes (*Lycopersicon esculentum*), bush beans (*Phaseolus vulgaris*), bush pickle hybrid cucumbers (*Cucumis sativus*), sweet peppers (*Capsicum annuum*), and large-leaf Italian basil (*Ocimum basilicum*). However, these plants were irrigated regularly so water deficit conditions were never an issue.

13.4 The Long Term Ecological Research (LTER) Model Applied to Green Roofs

Long-Term Ecological Research (LTER) is a National Science Foundation (NSF) funded program that was created in 1980 (Callahan 1984; Kratz et al. 2003). The research network of scientists currently includes 26 research sites studying ecology over extended temporal and spatial scales. Long-term studies are important because the natural world is dynamic and with climate change, patterns of natural variation are occurring even faster. Plant communities take time to accumulate biomass, respond to disturbances such as invasions of native and non-native species, weather extremes, or disease and insect pressures, and there may be time lags between the cause and effect of ecological changes. They can provide a baseline from which to determine if an ecological system has changed over time and define the range of natural variability, they allow us to assess relationships and interactions among various components of the system, they allow us to detect cause and effect relationships among slowly changing variables, and data gathered across multiple sites can lead to stronger conclusions than those from single sites (Kratz et al. 2003).

Although many would argue that placing plants on top of buildings in artificial substrates is not a natural system, the same LTER concepts apply to green roofs. One difference is that studies of natural landscapes could span decades, centuries, or even thousands of years. Buildings do not last that long. Green roofs are limited in time as most roofing membranes are replaced within 40–50 years. So what constitutes a long-term study on a green roof? Regardless, studies that span years are critical for making sound conclusions on long-term plant communities. The short 1 and 2 year studies that are common in the green roof literature do not really tell us anything about what species will be populating a green roof in the future. These short-term experiments are really just studies of plant establishment. However, the prevalence of 1 or 2 year studies at single sites is not surprising as research funding is rarely guaranteed for more than a few years. Also, many studies are graduate student projects, which cannot be dragged out for years and years. Even so, when studies have been conducted for three or more years, conclusions drawn are often dramatically different than what would have been concluded following just one or two seasons. Three to 5 years seems sufficient to predict long-term plant communities on shallow roofs consisting of sedum. However, on deeper roofs or roofs where species are allowed to colonize, then a much longer period of time is needed.

The few longer-term green roof studies where the original plantings were recorded in order to provide a baseline from which to work from all point to the importance of long-term studies. As outlined above from the 7 year study on the MSU Horticulture Teaching and Research Center, conclusions drawn at the end of 2 years were significantly different than what was present following 7 years (Durhman et al. 2007; Rowe et al. 2012). Similar results were drawn comparing 12 species of stonecrop in terms of absolute cover (Getter and Rowe 2008; Getter and Rowe 2009). Likewise, Dunnett et al. (2008) emphasized the importance of long-term monitoring of green roofs because of the changes that occurred in plant communities from the first to the fifth year of their experiment with 15 herbaceous perennials and grasses (Dunnett and Nolan 2004; Dunnett et al. 2008). In all of the above studies, changes in plant community development occurred faster at shallower substrate depths relative to deeper ones.

Setting up green roof research sites similar to the NSF LTER program would provide opportunities to follow changes to green roof habitats for longer periods of time and also look at similarly designed roofs across geographic distances. Because of the relatively short life spans of roofing membranes and modern buildings, it may be more feasible to conduct replicated studies over numerous geographic locations with varying climates, etc. As with most research, the primary roadblock to doing so is funding.

13.5 Future Research Needs and Questions

Since there have only been a handful of green roof plant studies that were carried out for more than 1 or 2 years, an obvious place to start is to initiate more of these studies. One example is a study that was initiated in 2011 to evaluate establishment,

Fig. 13.5 The Molecular Plant Sciences Building at Michigan State University is being used to follow the green roof plant community over time. (Photo DB Rowe)

survival, and changes in plant community over time on the Molecular Plant Sciences Building at Michigan State University (Fig. 13.5). Plugs of four grasses and 13 herbaceous perennials native to Michigan were installed at substrate depths of 10 cm (4 in) and 20 cm (8 in). Up to 45 plugs of each species were planted on 20 cm (8 in) centers. Survival rates were recorded during June 2012 and as expected most species experienced greater survival when grown in 20 cm (8 in) relative to those at 10 cm (4 in). The roof will continue to be sampled every year into the distant future to record the presence of individual species.

In long-term studies it is important to record baseline plantings when the roof was first installed in order to know what changes occur over time. However, older existing roofs should to be sampled also even if it is not exactly known what existed there in the beginning. Estimates can often be made based on the type of roof, location, and who installed the roof. For example, even though it is not known exactly what was planted on day one and the roof was overseeded and additional species added the year after installation, the roof on the Church of Latter-day Saints Convention Center described above should be reevaluated. The roof is now 14-years-old and valuable information could be gleaned and compared to the original study. In addition, changes in substrate composition should be looked at on this roof as well as others. Since long-term studies may not always be possible, the LTER model could still be followed by replicating studies over multiple geographic locations to determine the role of site-specific environments on plant community development. Plants species should be tested by themselves and in combination with multiple species. Other factors that should be investigated include interactions among plant species, the effects of roof maintenance (pulling weeds or allowing other species to colonize), and how different plant combinations influence ecosystem services such as stormwater management, heat flux, aesthetics, and the ability of the roof to provide habitat for wildlife. Common sense would suggest that increasing plant diversity would increase the ability of a green roof to provide these services and reduce the impact of environmental change (Cook-Patton and Bauerle 2012). Although this statement is true for the most part, adding plant species without considering their interactions may actually decrease services (Lundholm et al. 2010, MacIvor et al. 2011). Research is needed to determine which combinations of species and functional groups will complement each other and maximize services over time (Chap. 8). It is a challenge to balance relative competition among species so that more aggressive species do not dominate the community and reduce biodiversity.

Lastly, more roofs need to be installed where multiple microclimates are created and then these interactions among the various microclimates need to be studied to see how they influence long-term plant communities. Different microclimates were achieved on the California Academy of Sciences Building in San Francisco due to roof slope and sun exposure. Other options include varying substrate compositions and depths on the same roof. All of these practices should increase plant diversity, green roof function, and long-term success.

References

Benvenuti S, Bacci D (2010) Initial agronomic performances of mediterranean xerophytes in simulated dry green roofs. Urban Ecosyst 13:349–363

Berendse F (1998) Effects of dominant plant species on soils during succession in nutrient poor ecosystems. Biogeochemistry 42:73–88

Boivin M, Lamy M, Gosselin A, Dansereau B (2001) Effect of artificial substrate depth on freezing injury of six herbaceous perennials grown in green roof system. Hort Technology 11(3):409–412.

Brenneisen S (2006) Space for urban wildlife: designing green roofs as habitats in Switzerland. Urban Habitats 4:27–36

Butler C, Orians C (2011) Sedum cools soil and can improve neighboring plant performance during water deficit on a green roof. Ecol Eng 37(11):1796–1803

Callahan J (1984) Long-term ecological research. BioScience 34(6):363–367

Cook-Patton S, Bauerle T (2012) Potential benefits of plant diversity on vegetated roofs: a literature review. J Environ Manage 106:85–92

Dewey D, Johnson P, Kjelgren R (2004) Species composition changes in a rooftop grass and wildflower meadow. Native Plants J 5(1):56–65

Dunnett N, Nolan A (2004) The effect of substrate depth and supplementary watering on the growth of nine herbaceous perennials in a semi-extensive green roof. Acta Hort. 643:305–309

Dunnett N, Nagase A, Hallam A (2008) The dynamics of planted and colonizing species on a green roof over six growing seasons 2001–2006: influence of substrate depth. Urban Ecosys 11:373–384

Durhman A, Rowe DB, Rugh C (2006) Effect of watering regimen on chlorophyll fluorescence and growth of selected green roof plant taxa. HortScience 41(7):1623–1628

Durhman A, Rowe DB, Rugh C (2007) Effect of substrate depth on initial growth, coverage, and survival of 25 succulent green roof plant taxa. HortScience 42(3):588–595

Dvorak BD, Volder A (2010) Green roof vegetation for North American ecoregions: a literature review. Landsc Urban Plan 96:197–213

Emilsson T (2008) Vegetation development on extensive vegetated green roof: Influence of substrate composition, establishment method and species mix. Ecol Eng 33:265–277

Emilsson T, Rolf K (2005) Comparison of establishment methods for extensive green roots in southern sweden. Urban for Urban Green 3:103–111

Forschungsgesellschaft Landschaftsentwicklung Landschaftsbau (FLL) (Research Society for Landscape Development and Landscape Design) (2008) Introduction to the FLL: guidelines for the planning, construction and maintenance of green roofing. FLL, Bonn, Germany

Getter K, Rowe DB (2008) Media depth influences Sedum green roof establishment. Urban Ecosys 11:361–372

Getter K, Rowe DB (2009) Substrate depth influences sedum plant community on a green roof. HortScience 44(2):401–407

Getter K, Rowe DB, Andresen JA (2007) Quantifying the effect of slope on extensive green roof stormwater retention. Ecol Eng 31:225–231

Getter K, Rowe DB, Cregg B (2009a) Solar radiation intensity influences extensive green roof plant communities. Urban For Urban Green 8(4):269–281

Getter K, Rowe DB, Robertson G, Cregg B, Andresen J (2009b) Carbon sequestration potential of extensive green roofs. Environ Sci Technol 43(19):7564–7570

Hauser K (2013) An evolving landscape: the living roof at California Academy of Sciences. Pac Hortic 74(4):24–27

Kiers H (2013) Back to nature: an alternative approach to planting green roofs. Pac Hortic 74(4):28–30

Köhler M (2006) Long-term vegetation research on two extensive green roofs in Berlin. Urban Habitiats 4(1):3–26

Köhler M, Poll P (2010) Long-term performance of selected old Berlin greenroofs in comparison to younger extensive greenroofs in Berlin. Ecol Eng 36:722–729

Kratz T, Deegan L, Harmon M, Lauenroth W (2003) Ecological variability in space and time: insights gained from the US LTER program. BioScience 53(1):57–67

Liesecke H (1998) Langzeitentwicklung einer extensiven Dachbegrünung: Untersuchungenzum Substratverhalten und zur Vegetationsentwicklung eines 1985 ausgeführten Objektes (Long-term development of extensive roof vegetation: a study on substrate and vegetation development over 12 years). Stadt-und-Grün 47:428–436

Lundholm J, MacIvor J, MacDougall Z, Ranalli M (2010) Plant species and functional group combinations affect green roof ecosystem functions. PLoS ONE 5(3):11

MacIvor J, Ranalli M, Lundholm J (2011) Performance of dryland and wetland plant species on extensive green roofs. Ann Bot 107:671–679

Monterusso M, Rowe DB, Rugh C (2005) Establishment and persistence of *Sedum* spp. and native taxa for green roof applications. HortScience 40(2):391–396

Nagase A, Dunnett N (2010) Drought tolerance in different vegetation types for extensive green roofs: effects of watering and diversity. Landsc Urban Plan 97:318–327

Oberndorfer E, Lundholm J, Bass B, Connelly M, Coffman R, Doshi H, Dunnett N, Gaffin S, Köhler M, Lu K, Rowe DB (2007) Green roofs as urban ecosystems: ecological structures, functions, and services. BioScience 57(10):823–833

Rowe DB (2011) Green roofs as a means of pollution abatement. Environ Pollut 159(8–9):2100–2110

Rowe DB, Getter K (2010) Green roofs and roof gardens. In: Aitkenhead-Peterson J, Volder A (eds) Urban ecosystems ecology. Agron. Monogr. 55. American Society of Agronomy. Crop Science Society of America. Soil Science Society of America, Madison, WI, p 391–412

Rowe DB, Monterusso M, Rugh C (2006) Assessment of heat-expanded slate and fertility requirements in green roof substrates. HortTechnology 16(3):471–477

Rowe DB, Getter K, Durhman A (2012) Effect of green roof media depth on Crassulacean plant succession over seven years. Landsc Urban Plan 104(3–4):310–319
Rowe DB, Kolp M, Greer S, Getter K (2014) Comparison of irrigation efficiency and plant health of overhead, drip, and sub-irrigation for extensive green roofs. Ecol Eng 64:306–313
Skabelund L, Blocksome C, Hamehkasi M (2014) Semi-arid green roof research 2009–2014: resilience of native species. Proc. of 12th North American Green Roof Conference: Cities Alive, Nashville, TN
Thuring C, Dunnett N (2014) Vegetation composition of old extensive green roofs (from 1980s Germany). Ecol Process 3:4. doi:10.1186/2192-1709-3-4
Thuring C, Berghage R, Beattie D (2010) Green roof plant responses to different substrate types and depths under various drought conditions. HortTechnology 20(2):395–401
VanWoert N, Rowe DB, Andresen J, Rugh C, Fernandez R, Xiao L (2005a) Green roof stormwater retention: effects of roof surface, slope, and media depth. J Environ Qual 34(3):1036–1044
VanWoert N, Rowe DB, Andresen J, Rugh C, Xiao L (2005b) Watering regime and green roof substrate design affect *Sedum* plant growth. HortScience 40(3):659–664
Whittinghill L, Rowe DB (2012a) The role of green roof technology in urban agriculture. Renew Agric Food Sys 27(4):314–322
Whittinghill L, Rowe DB (2012b) Vegetable production on extensive green roofs. Proc. of 10th North American Green Roof Conference: Cities Alive, Chicago, IL. 17–20 October, 2012
Whittinghill L, Rowe DB, Cregg B (2013) Evaluation of vegetable production on extensive green roofs. Agroecol Sustain Food Syst 37(4):465–484

Dr. Bradley Rowe has conducted green roof research at MSU since 2000, was the founding chair of the Green Roofs for Healthy Cities research committee, and received the GRHC Excellence in Research Award in 2008.

Chapter 14
Invertebrates on Green Roofs

J. Scott MacIvor and Kelly Ksiazek

Abstract Insects and other invertebrates provide essential ecosystem functions in designed habitats including green roofs. Services offered by invertebrates in these novel environments include pollination for plant reproduction and yield in cultivated crops, pest control to reduce damage to green roof vegetation, decomposition to retain organic matter and cycle nutrients in the substrate, and contribution to food webs for species like birds that frequent green roofs.

Although we may assume that beneficial invertebrates are desirable on green roofs, it is not clear whether they adequately provide habitat or not. Green roof design can vary, as can their suitability as habitat, some supporting almost no species and others meeting both the foraging and nesting requirements of many. When designers include plant, substrate and other microhabitat conditions to support certain at-risk species or functionally important groups, green roofs may act as analog habitat where it is limited at ground level. Green roofs are uniquely isolated and exposed to sun and wind and their relative value will ultimately depend on the invertebrates in question. If green roofs are to support invertebrate communities, elucidating habitat requirements and monitoring wildlife design successes are as essential as public outreach that encourages urban biodiversity conservation.

Keywords Insect ecology · Pollination · Soil stabilization · Pest control · Plant-insect interactions · Arthropods · Ecosystem services · Urban ecology

14.1 Introduction

Invertebrates (including insects, spiders, and soil arthropods) are a common sight on green roofs. One can often observe bees visiting *Sedum* flowers when in bloom, or a jumping spider scurrying along the edge of a green roof module in search of

J. S. MacIvor (✉)
Biology Department, York University, Toronto, ON M3J 1P3, Canada
e-mail: jsmacivor@gmail.com

K. Ksiazek
Plant Biology Chicago Botanic Garden, 1000 Lake Cook Rd, Glencoe, IL 60022, USA

R. Sutton (ed.), *Green Roof Ecosystems*, Ecological Studies 223,
DOI 10.1007/978-3-319-14983-7_14

prey. One's first impression is usually "how did it get up here?" especially if the green roof is high above ground level. Although some invertebrates will colonize during installation as hitchhikers on plants and in growing substrate, others have mechanisms that enable them to reach green roofs on their own while in search of suitable habitat (Fig. 14.1). Many invertebrates on green roofs are already abundant in the surrounding environment and among the earliest colonizers of new or recently disturbed or constructed habitat (McIntyre et al. 2000). Taxa having flexible habitat requirements, such as the ability to substitute native resources for novel or exotic ones after landscape change, are the most successful colonizers in urban areas (Savard et al. 2000), and tend to be the most prevalent on green roofs.

Green roofs provide habitat for many invertebrates, although the community may not be representative of those at the ground level. Studies sampling invertebrate communities on a variety of green roof types have found many hundreds of species (Mecke and Grimm 1997; Mann 1998; Jones 2002; Gedge and Kadas 2005; Brenneisen 2006; Schindler et al. 2011; Ksiazek et al. 2014), including rare, and

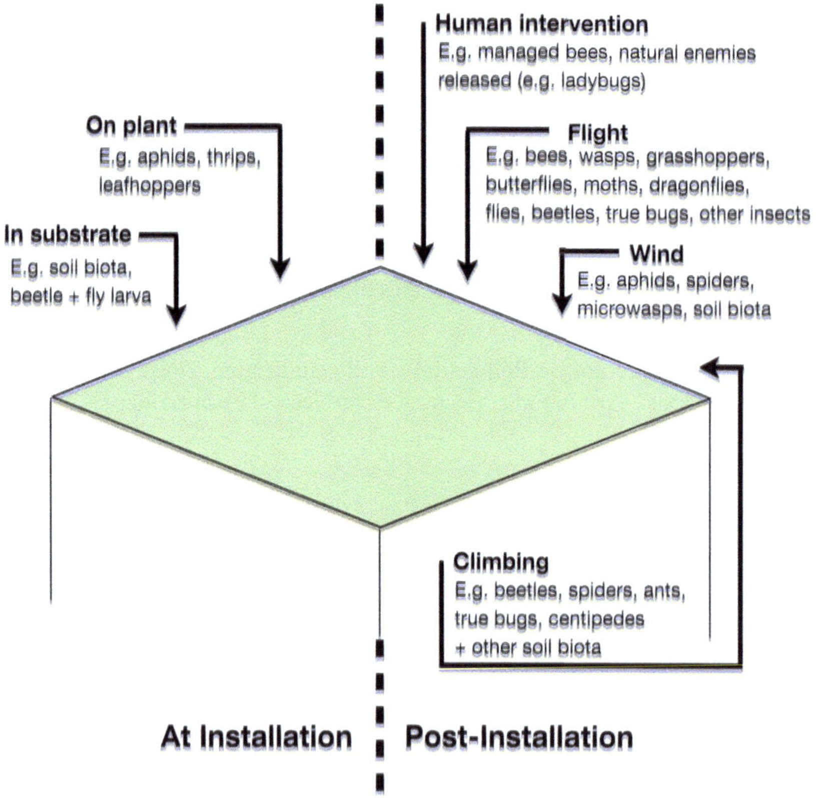

Fig. 14.1 Invertebrates colonize green roofs through various mechanisms both during and after installation: at installation, invertebrates may arrive on vegetation or in substrate; once installed, invertebrates will arrive by climbing or flying onto the roof, getting blown there by the wind or by human intervention

red listed species of conservation concern (Brenneisen and Hänggi, 2006; Kadas 2006; Kaupp et al. 2004); even a new record for a province in Canada (Majka and MacIvor 2009). These findings support the idea that green roofs can contribute to regional biodiversity by providing resources to a variety of species. Despite many studies conducted that compare invertebrate populations between different types of ground level urban green spaces (e.g. parks, community gardens, home gardens; see McIntyre 2000), few have compared diversity in these habitats to that of green roofs. In those that have used comparative studies, none recorded diversity on green roofs to be equivalent to or greater than sites at ground level (Colla et al. 2009; MacIvor and Lundholm 2011a; Tonietto et al. 2011). Thus, despite developers' and industry's touting of the benefits to bees and butterflies, green roofs do not compensate for habitat lost at ground level when a building is constructed (Chap. 15, Williams et al. 2014).

Because no two green roofs are alike, they cannot be expected to provide habitat or resources equally (Dunnett and Kingsbury 2004). The abundance and diversity of invertebrate species recorded on a green roof can vary greatly due to site-specific factors, such as plant and substrate composition (Brenneisen 2006; Madre et al. 2013), height (increase with proximity to ground level, (MacIvor 2013), or age (older roofs harboring different species but similar diversity; Schrader and Böning 2006). Landscape-scale factors also shape invertebrate communities on green roofs as they do in urban green spaces at ground level; for example, butterflies (Blair and Launer 1997), bees Cane et al. 2006, and beetles (Niemelä et al. 2002). Even green roofs not designed specifically to support invertebrate communities can unintentionally do so (MacIvor and Lundholm 2011a). Despite the design objectives, new niches develop on green roofs as they do in urban green spaces at ground level; adding substrate and plantings that interact with infrastructure creates unique microhabitats. However, features exclusive to green roofs including limited substrate depths and vertical isolation from ground level create wholly unique urban niches. The mobility and resource requirements of particular taxa will determine which species benefit from green roof habitat.

Invertebrates can be categorized according to how they use green roofs; either they permanently find refuge on a green roof for multiple generations over a single season or over multiple years, or they use a roof temporarily as part of a range of local habitats. Moreover, invertebrates can be categorized by their mechanisms of dispersal onto green roofs (Fig. 14.1). Species having low mobility, such as soil dwelling organisms that are common in ground level urban green spaces (McIntyre et al. 2000) can also be surprisingly abundant on green roofs (Schrader and Böning 2006). These organisms are not likely to move between roofs and the ground within a generation, and many are probable decedents of populations found in the substrate and planting material prior to installation. Other invertebrates use green roofs temporarily or in addition to other nearby habitats. Bees, for example, are central place foragers, nesting in one location and foraging within a range around the nest (Michener 2007). Many bee species could access resources on a green roof while nesting at ground level, or, conversely, could nest on a green roof while utilizing resources in the surrounding landscape.

To better understand the uniqueness of green roofs as habitat and their role in strategies that enhance the management of urban green space for biodiversity, researchers have begun to direct their studies towards ecological theory and empirically collected data. Green roofs, like other urban green spaces, are modified habitat, such that habitat structure and productivity rates are altered from that of natural areas (Shochat et al. 2006; Gaston 2010). Thus, we can apply relevant general trends in the ecology of urban green spaces to green roof design and maintenance in order to enhance urban wildlife.

In this chapter, we review the benefits provided by invertebrates on green roofs, as well as the contribution of green roofs as habitat for local species and for invertebrate biodiversity conservation strategies (Chap. 15, Goddard et al. 2010). Invertebrates play a central role in ecosystem functions such as nutrient cycling, pollination, and stabilization of complex food webs (Wilson 1987). Preliminary findings suggest that local and landscape-level variables important in determining invertebrate diversity and community structure in urban green spaces at ground level (e.g. floral diversity, structural heterogeneity, habitat isolation, Frankie and Ehler 1978; McIntyre 2000) also apply to green roofs. Given the rapidly expanding body of literature on trends in urban ecology and biodiversity (Forman 2014), these studies can shed light on how invertebrate communities assemble and improve the ecological and economic performance of green roofs.

14.2 Invertebrate Communities in Urban Areas

The composition and structure of invertebrate communities in urban areas typically differ from those in natural areas. Urban development reduces the amount of total green space and fragments natural habitats, thereby reducing connectivity of metapopulations of some species and subsequently decreasing their reproductive output (Hanski and Gilpin 1991). Although urban landscapes are referred to as heterogeneous (Grimm et al. 2008), in part due to diversity of ownership and partitioning of land by users (Colding and Barthel 2013), when aesthetics is a criteria in designed or restored urban landscapes, urban green spaces can become homogenized. For example, one study in Montreal found that garden design differed with increasing distance: neighbors tended to have similar vegetation, thereby creating pockets of homogenized habitat (Zmyslony and Gagnon 1998). Such homogenization in habitat may not provide sufficient resources to invertebrates. Many populations experience biotic homogenization in urban landscapes as a result of generalist species (often non-native) thriving and specialist species becoming locally extirpated (McKinney 2006; Hahs et al. 2009; Duncan et al. 2011). A loss of plant diversity in urban areas generally leads to a decrease in invertebrate diversity resulting in an uneven distribution of common and rare species (Burghardt et al. 2009; Crisp et al. 1998). For example, Shochat et al. (2004) found that urban 'productive' habitats like mesic yards had high spider abundance but low diversity. Deichsel et al. (2006) found only a few beetle species most common in urban habitats with urbanization a limiting factor for flightless species and forest specialists.

Altered environmental conditions typical of urban landscapes favor non-specialists that are flexible in their resource requirements in type, space and time. The vacancies created by the loss of native species due to urban pressures increases available habitat for non-specialist colonizers, such as cosmopolitan and exotic species that have physiological traits better adapted to both colonization and survival in disturbed habitats (Myers et al. 2000). Several traits have been linked to recruitment and tolerance in urban insect communities. For example, Banaszak-Cibicka and Żmihorski (2012) found that among urban bee assemblages 'winners' tended to be non-solitary, polylectic, smaller bodied bees whose foraging activity peaks in late summer. Bees (Banaszak-Cibicka 2014) and ants (Menke et al. 2011) that are thermophilic have also been found to be more suited to the warmer and drier conditions typical of urban landscapes due to the urban heat island effect (Oke 1973).

At the local level, many biotic and abiotic factors influence invertebrate community assembly. In natural habitats, the structure and diversity of invertebrate communities are often coupled with that of vegetation (Root 1973; Siemann et al. 1998; Siemann 1998; Smith et al. 2006). Different plant species and functional types provide the template upon which arthropod communities develop. The effect of vegetation can shape urban invertebrate communities at both local and landscape levels. Edge effects, fragment size and area, and dominant species all play a role in the diversity and abundance of invertebrates that are found in vegetated urban patches (Bolger et al. 2000). For example, local factors such as available sunlight (Matteson and Langellotto 2010) and habitat patch size (Williams and Winfree 2013) have been shown to affect pollinator diversity. Additionally, Pawelek et al. (2009) found that enhancing floral diversity in urban gardens resulted in increased bee diversity year after year.

Invertebrates are mostly mobile species, and many utilize several habitats within a landscape to obtain the resources needed to successfully mate and produce viable offspring. The need for several habitats to fulfill foraging, mating and nesting requirements make some species less suited to fragmented urban landscapes. For example, dragonflies that depend on both aquatic and terrestrial environments (for reproduction and predation, respectively) are particularly sensitive to human land use (Samways and Steytler 1996). Species with these types of diverse habitat requirements might not benefit from the proliferation of urban green spaces unless they are also near freshwater aquatic environments. At the landscape scale, habitat isolation, measured as the distance from proximal natural areas, results in declining pollinator diversity and visits to flowers in agroecosystems (Schüepp et al. 2014) and in urban areas (Hennig and Ghazoul 2011). More broadly, isolation from native habitat results in a decline in stability of insect pollinators (Kennedy et al. 2013).

14.2.1 Trends in Green Roof Invertebrate Communities

Green roofs are engineered habitats with designs constrained by factors such as industry standards, client desires, and weight restrictions. How these restrictions directly impact invertebrate communities remains unknown; however green roofs limit the types of invertebrate communities they support because of local and landscape factors common to all urban green spaces (Fig. 14.2). Local factors such as the diversity of vegetation is often linked to the assembly of the invertebrate community but studies have found few empirical correlations, possibly due to a lack of sampled sites. In Switzerland, Brenneisen (2006) found invertebrate populations were promoted on green roofs by increasing plant diversity. Additionally, Madre et al. (2013) found that increasing plant diversity and structure resulted in more diverse insect assemblages on green roofs throughout France. Although multi-decade plant community data has been collected from very few green roofs to date, one study indicates that green roofs experience seasonal and successional changes over time, just like at ground level (Köhler 2006). That is, initial colonization will include disturbance-tolerant species and those arriving during installation, followed by diversity that increasingly resembles the local urban community, with some adventive species whose occurrence is haphazard, all influenced by microclimate or other local or landscape factors. Other, theoretical concepts (Box 1) in ecology might also apply to the colonization, establishment, and persistence of invertebrate communities on green roofs in cities. Assembly of invertebrate communities on green roofs will also be a function of niches are created and not yet occupied. Green roofs often do not resemble the ground level habitat replaced by a building. Many local habitat types, such as wetlands and forests, could be difficult or cost prohibitive to mimic on a green roof. For example, installing a green roofs on top of a large

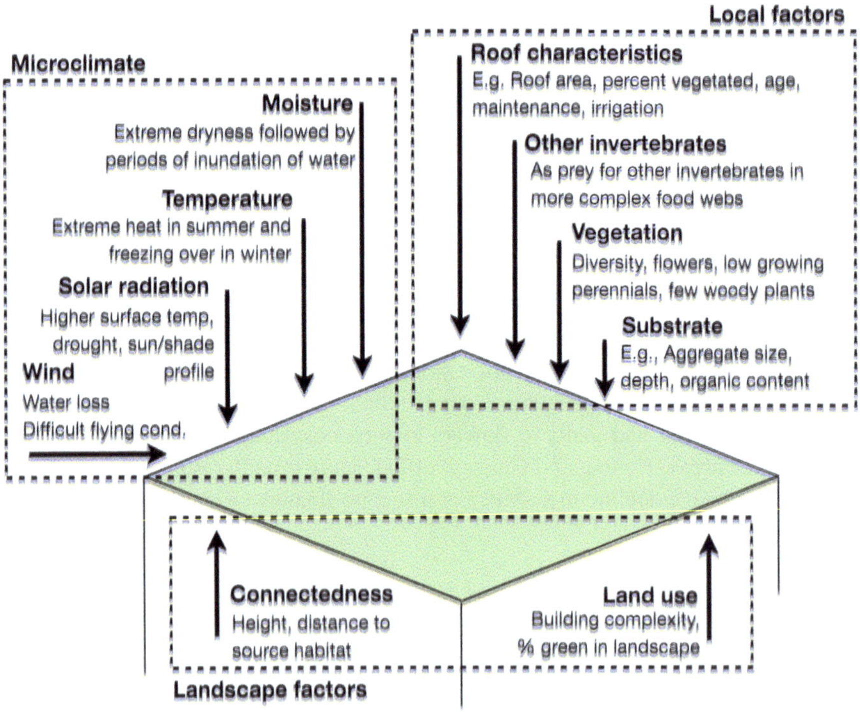

Fig. 14.2 Different biotic and abiotic factors affect invertebrates' colonization of green roofs

building constructed over a former wetland would likely not replicate all conditions of the original environment and the diversity of available aquatic niches would not represent that found on the site pre-construction. Despite this, many species, even native ones can find refuge among green roof vegetation, including exotics such as *Sedum* that dominate many extensive green roofs. Many macro-invertebrates like springtails, millipedes, and centipedes have been found colonizing the substrates of *Sedum*-based green roofs, likely because they provide shade and moisture on rooftops where these resources are limited (Schrader and Böning 2006). Some bee species, including honey bees, and native and non-native bumble bees, leaf cutting bees and sweat bees visit *Sedum* flowers (MacIvor et al. 2014). In their natural habitat, other insects will visit *Sedum* for pollen and nectar, including several species of ants, moths, butterflies, flesh flies (*Diptera: Sarcophagidae*) and bottle flies (*Diptera: Calliphoridae*) (Clausen 1975).

Box 1. Widely Accepted Concepts in Ecology Applicable to Green Roof Study

Island biogeography explains the effect of habitat size and isolation on species diversity (MacArthur and Wilson 1967), and has been used to describe the impact of habitat fragmentation in urban areas on the decay and dispersal of different assemblages of flora and fauna (McIntyre 2000). As an application of this theory, green roofs have been referred to as 'stepping stone habitat' (Kim 2004), connecting several fragmented patches throughout the landscape. In this scenario, green roofs replace unsuitable spaces (e.g. conventional rooftops) with habitat, thereby improving emigration and immigration between source habitat (e.g. large continuous urban parks or natural areas) and sink habitat (isolated urban green spaces, such as small gardens).

Invasion theory explains why sampled green roof populations tend to be dominated by a few mobile species and are more homogenous compared to ground level. Green roofs are new and novel habitat where habitat would not normally be found. Despite some green roofs designed using templates or characteristics of local xeric and exposed, rocky habitats (Lundholm 2006; FLL 2008; Sutton et al 2012), there are no true green roof analogs in nature. Thus, new green roofs could represent 'vacant niches' (Herbold and Moyle 1986), where species can establish and multiply. Many of these established introduced species will not impact species in the surrounding area (Simberloff 1981). However, where enemies are limited or absent (Liu and Stiling 2006) and/or resources are abundant, certain species could spread and impact the local ground level landscape (Elton 1958). Although invasion theory has been used to examine green roof design that resists invasion from plant colonizers (Nagase et al. 2013), studies have not yet examined invertebrate invaders on green roofs.

Microclimatic effects, caused by water, wind and substrate conditions exclusive to rooftop habitat in cities, will limit some invertebrate species. Such effects include xeric conditions prevalent at certain times of the year and winds that move across flat roofs at high speeds that can disturb substrate and roofing materials. Water can be designed to pool in certain areas on the green roof, but in general, excess water will run off the roof through drainage layers and roof drains or return to the atmosphere quickly through evapotranspiration. High winds may prevent some smaller species, such as small bees and flies, from flying up to the surface of a roof in search of resources or, once recruited, from remaining on a roof for more than a few seconds. Large bodied bumble bees and honey bees, both among the most capable of flying insects and able to forage more than a kilometer from the nesting site (Greenleaf et al. 2007), are often more common on green roofs than other species (Tonietto et al. 2011). Wind could also promote the colonization of other wind-dispersed species, such a web building spiders and aphids, from green roofs to new habitats.

Although all green roof substrates will be suitable for some species depending on their specific preferences, properties of substrates can limit the diversity of soil-dwelling invertebrates. The depth, texture of aggregates, and other biological features will also contribute to the success or failure of different colonizing invertebrates. For example, providing areas of bare substrate can promote perching and burrowing habitat, the latter consisting of more fine and non-compacted aggregates that could promote the nesting of solitary bees or pest controlling wasps. Shallow substrates in combination with exposure to sun and wind, can create very dry conditions hostile to many of the soft-bodied annelids and invertebrates typical of urban soils. In a survey of two extensive *Sedum*-based green roofs, Rumble and Gange (2013) noted dominance by a few and mostly xerophilic species. Periodic drought conditions, as well as the absence of deeper, warmer soil in the winter, likely limit survival of common urban soil organisms, such as many ant species and earthworms. Although exotic in many regions in North America, earthworms are the primary soil architects in terrestrial soil environments (Lee 1985). In contrast, they are rarely encountered on green roofs, in part because earthworms like other soil invertebrates have a limited ability to disperse to green roofs. Those that reach the roof may perish due to lack of moisture at peak summer periods or freezing temperatures in winter with restricted ability to move to deeper soil due to the shallow depth of extensive green roof substrates.

Including a diversity of substrate design on green roofs may allow for increased colonization and survival of invertebrates. Heterogeneity in soil properties has been linked to organismal diversity in ground level habitats (Ettema and Wardle 2002). To date, little published data support the conservation of these trends in green roof substrates (often designed to be uniform in depth and composition). However, it is possible that colonizing soil invertebrates might move between microhabitats created by roof features (e.g. machinery, parapets providing shade; drainage areas) to survive during extreme weather, thereby repopulating other, more exposed spots. Creating heterogeneity in substrate depths so that invertebrates can move to deeper areas during dry or cold conditions could enhance their populations in the future and therefore increase the degree to which they provide ecosystem services.

The surrounding landscape is also a primary driver affecting the assembly of invertebrates and, consequently, the services that such communities are able to provide (Isaacs et al. 2009). For example, in agricultural landscapes, the abundance of sheet web spiders (*Linyphiidae*), a highly mobile taxa, increases with the addition of native plantings in the region (Schmidt and Tscharntke 2005) and high bee abundance and species richness are correlated with a moderate amount of regional human disturbance (Winfree et al. 2007). On green roofs, the structure of the surrounding urban landscape is a more important predictor of invertebrate community than the size of the roof, especially for highly mobile species of bees and weevils (Braaker et al. 2013). Building height appears to limit pollinator-nesting activity (MacIvor 2013) however, bees are frequently observed foraging on green roofs (MacIvor et al. 2014). Therefore, in heavily urbanized areas, the retrofitting of green roofs onto existing buildings could have a positive impact on invertebrate populations at the landscape scale (Braaker et al. 2013), extending the urban habitat to include buildings that support urban biodiversity conservation goals.

14.3 Ecosystem Services of Invertebrates on Green Roofs

Invertebrates play a central role in ecosystem functions such as nutrient cycling, pollination and food web structuring (Wilson 1987). As green roof habitat increases the amount of vegetated spaces in cities compared to non-vegetated infrastructure, invertebrates can provide a variety of ecosystem services in urban environments. Research regarding the full potential of services provided by green roof invertebrates has not been given much attention to date but is becoming more widespread. Here, we highlight some of the ecosystem services provided by invertebrates that live on or use green roofs and explore possible contributions at both the site and landscape levels.

14.3.1 Substrate Stabilization

Invertebrates contribute to the maintenance and stabilization of soils. The physical and chemical breakdown of organic matter (plant, animal and microbial material) in soils is one of the principal factors determining the structure and function of ecosystems (Odum 1971). Invertebrates, in particular, play a critical role as agents of litter fragmentation and breakdown of coarse organic debris (Chapin et al. 2002). Fragmentation breaks down larger detritus to create increased surface area for microbial colonization and activity (Chaps. 6 and 7). Together, the activity of microflora (fungi), microfauna (e.g. bacteria), mesofauna (e.g. collembolans) and macrofauna (e.g. arthropods and earthworms) transform the composition of litter and form new substrate. This process increases the amount of nutrients that are then available for plants and other organisms (Singh and Gupta 1977). Invertebrate communities can be used as indicators of soil quality in urban areas (Hartley et al. 2008;

Santorufo et al. 2012), and their movement through soil (e.g. burrowing) creates channels that reduce compaction and improve pathways for water to flow. In natural systems, larvae from many invertebrates, for examples, flies (*Calliphoridae* and *Muscidae*) scarab beetles (*Scarabaeidae*) and carrion beetles (Silphidae) contribute to the breakdown of organic matter and soil respiration (Kim 1993). Soil invertebrates should be encouraged on green roofs to aid in substrate stabilization and maintenance.

Although substrates used on green roofs are often sterilized prior to installation to inhibit weeds, the colonization of green roofs by detrivorous organisms will eventually occur. Thus, studies on older roofs record high numbers of soil invertebrates (primarily collembolans) than younger roofs (Davies et al. 2010; Schindler et al. 2011; Rumble and Gange 2013). Other studies record more collembolans on older green roofs but no significant difference between older and younger roofs (Schrader and Böning 2006; MacIvor and Lundholm 2011a). Nevertheless, all green roofs, even *Sedum*-based extensive habitats contain soil biota involved in decomposition and nutrient cycling. The extent to which these communities need to be encouraged will depend on the level of maintenance. For example, green roofs containing cultivated crops could benefit from natural nutrient fixation (e.g. phosphorus, nitrogen, carbon) by soil organisms in the substrate, thereby reducing the need for manual fertilization. Low maintenance meadow-like or *Sedum*-based green roofs, often designed initially with a low-organic content to minimize weed proliferation, could benefit in perpetuity from decomposition and subsequent nutrient availability and soil formation provided by invertebrates in the substrate.

Biomass accumulation and decomposition contribute to ecological change inherent in natural habitats and including these functions in design and maintenance strategies on green roofs could promote diverse soil biota. The removal of excess accumulated biomass on green roofs, especially prior to and after the winter season, occurs primarily by human agency in response to aesthetic demands. Leaving some of this material for the invertebrate soil community benefits the plant community, thereby creating more niches on the roof and conferring benefits to a wider variety of species. The slow decomposition of this material will also enable other niches once decomposition is completed. Methods used to introduce mychorrizal fungi onto new green roofs to enhance soil productivity and plant growth (McGuire et al. 2013) could also be used to increase soil invertebrates that support decomposition and nutrient cycling in the substrate.

14.3.2 Pest Control

Predatory invertebrates that eat one another can reduce the threat of pests to plantings in urban environments. Numerous known pest-controlling invertebrate groups, such as spiders (Kadas 2006; Brenneisen 2006), assassin and damsel bugs, dragonflies, solitary wasps (MacIvor and Lundholm 2011a), and predatory ground beetles (Meierholfer 2013), among many others, have been documented on green roofs. These pest-controlling invertebrates are essential for the regulation of hyper-abundant and phytophagous insects (Kim 1993), yet there has been no study of their impact on control of pests affecting vegetation on green roofs. Pest control in agriculture and horticulture is well studied and management of local habitat in support of natural pest enemies appears to be the most effective strategy (Landis et al. 2000). Pests are most prevalent where their host is most abundant, for example in monoculture plant systems (Altieri and Nicholls 2004). Therefore, landscape designs that include homogenous plantings of one or a few similar species, such as extensive green roofs with *Sedum* mats, could be the most susceptible to pest damage.

Although little research to date has directly measured invertebrate pest control on green roofs, the presence of many organisms from representative guilds has been documented (Fig. 14.3a, b). On green roofs with low plant diversity, a pest outbreak could impact both aesthetics and performance. Aphids, as well as their ladybug predators, were common on green roofs in one study in London (Kadas 2006), and leafhoppers and aphids (both of which feed on the liquid contained in plant stems and leaves) were among the most collected species by Coffman and Waite (2011). Aphids, which are capable of moving great distances by wind, can be particularly damaging to many forbs including *Sedum* in natural environments (Clausen 1975), and are sometimes observed on *Sedum* on extensive green roofs, in

Fig. 14.3 Invertebrates provide many ecosystem services on green roofs **a** Ladybeetles, common as larva and adults on green roofs, control pest populations of aphids **b** Many spider species, such as this individual perched on some exposed root stabilization mat, predate on various phytophagous insects **c** Butterflies pollinate native plants, such as this coneflower **d** Pollination of *Sedum* species can improve population stability and resilience by increasing genetic diversity (JS MacIvor)

particular *Sedum spurium* and *Sedum kamtchaticum* (synonyms: *Phedimus spurius* and *Phedimus kamtschaticus*) (MacIvor, pers. observation). Pests can also include those that are a nuisance to human users of green roofs. Outbreaks are not described in the green roof literature to date, but for example, ants and their nests can occur in high numbers on green roofs (MacIvor and Lundholm 2011a; Madre et al. 2013). Although there is little study of pest insects on green roofs, the frequency at which they occur is expected to increase as green roofs increase in number.

14.3.3 Pollination

Pollinating insects are often encouraged in home and community gardens to augment cultivated crop yield, produce seeds and proliferate native plants (Fig. 14.3c, d). Bees are the most significant pollinators in terrestrial environments, but are in decline worldwide (Potts et al. 2010). In general, positive sentiments toward bees are growing among urban citizens. Often these translate into action in landscape design and enhancement strategies that reduce vulnerability to pollinator declines by considering alternative pollinators for management (e.g. not just honey bees). As this paradigm shift in viewing bees as essential and not just as stinging insects gains momentum, explicit designs for pollinating insects on green roofs is expected to increase.

In urban landscapes, bee pollinators tend to be less diverse and less abundant than in non-urban areas (Winfree et al. 2011). Disentangling which factors are most important is challenging but land use change and subsequent lack of native forage and nesting resources are among the main reasons for this trend. Many plant-pollinator interactions co-evolved regionally (Kearns et al. 1998) and insect pollinators are often more effective at using pollen and nectar from native plants than from exotic species (Corbet et al. 2001). When plant resources are not abundant enough or are not available at the right times as a result of land use (Kearns et al. 1998) or even climate change (Hegland et al. 2009), bees that depend on these floral resources must adjust their behavior, adapt or perish.

Human intervention can also have a significant role not only in pollination levels on a green roof but with the contribution of the roof to the necessary pollination services in the surrounding landscape. Honeybees (*Apis mellifera*), commonly kept in hives on both green and conventional roofs, can forage in many conditions and for long distances. Thus, a vertical flight of several stories poses only a minor challenge for honeybees that forage both on green roofs and in the surrounding urban landscape. While the service of pollination provided by honeybees can increase pollen deposition and seed set in some species, many plants are adapted to visit flowers. For example, beardtongue *Penstemon hirsutus*, found on green roofs in the Midwest U.S. has small flowers and depends on megachilid bees (e.g. *Hoplitis*) and species in the genus *Ceratina* for pollination and seed production (Crosswhite and Crosswhite 1966). Design for pollinators can benefit green roof plant communities, but could also improve the important presence of pollinators for flowering urban vegetation in the surrounding landscape. In this way, green roofs contribute to regional resources for many pollinating insects.

Green roofs are often promoted as habitat for pollinators, particularly, bees and butterflies (Williams et al. 2014). This is partly justified by reports of many pollinating bees, moths, butterflies, wasps and flies, presumably foraging, on green roofs (Fig. 14.3c, d, Mann 1998; Colla et al. 2009; Coffman and Waite 2011; Schindler et al. 2011; Tonietto et al. 2011; Ksiazek et al. 2012; Ksiazek et al. 2014; Braaker et al. 2013; MacIvor 2013; Madre et al. 2013). MacIvor et al. (2014) found no difference in the number of exotic and native bee visitors to a green roof in downtown Toronto, but there were significantly more large and medium sized bees than small bees, similar to studies in Chicago (Tonietto et al. 2011; Ksiazek et al. 2012; Ksiazek et al. 2014). Despite the fact that many bees visit green roofs for floral resources, few species have been observed nesting on green roofs (MacIvor 2013).

Insect pollinators play a vital role in horticulture and urban agriculture, as they are essential to the yield of food and seed crops (Klein et al. 2007). Pollination services are increasingly required and urban agriculture practices, including those on green roofs, are becoming more widespread in cities around the world (Whittinghill and Rowe 2012). Numerous successful green roof agriculture initiatives are currently underway (Gorgolewski et al. 2011) and many of the planted crops depend on insect pollination for fruit yield (peppers, tomatoes, squash, etc.). Pollinating insects are also essential to ensure adequate seed set of many flowering non-food species on green roofs (Ksiazek et al. 2012), particularly those that are self-incompatible and rely on pollen from another floral donor. In the absence of insect pollinators, these plants may produce too few seeds or seeds that are of insufficient quality to germinate and survive to reproductive maturity. For green roof owners and maintainers, successful pollination and plant reproduction ensures the future regeneration or seeding and germination of plants, including native species that may be expensive to replace.

14.3.4 Food Web Enhancement

In all ecosystems, the flow of energy between organisms can be observed in highly complex relationships known as food webs. Invertebrates are essential components in all terrestrial food webs as primary convertors of plant material to protein, providing food to organisms at higher trophic levels including mammals, reptiles and birds. Many urban bird species, being highly mobile, frequent green roofs while either passing over or foraging in urban landscapes (Fernandez-Canero and Gonzalez-Redondo 2010). As the number of green roofs increases, they may contribute to boosting regional invertebrate populations and, subsequently, the insectivorous species that depend on invertebrates as a food source.

While the food webs that are supported within the confines of a typical green roof are likely reduced compared to a diverse natural environment, the presence of thousands of invertebrate species on green roofs suggests that food webs could be quite large and complex. Nevertheless, as mobility restricts some species from gaining access roofs, green roof food webs are altered from those at ground level. For example, homogenized urban bird populations may be able to use the green roofs for feeding in areas where their mammalian predators are absent, thereby being released from this regulating pressure and allowing the population to grow rapidly (Chap. 15). Additionally, although many insectivorous bird species are generalists in their selection of prey, encouraging all insects equally as food for other species could result in dominant or nuisance species.

14.3.5 *Other Services: Providing Opportunities for Education*

Humans have an innate desire to experience and connect with nature; a phenomenon first described by E.O. Wilson (1984) as "biophilia". Green roofs can provide valuable opportunities for the general public to do so and educate them about the ecological processes. Those processes while occurring both in their neighborhoods and in large ecosystems often become only accessible and one-step removed through television, newspapers or the internet (Miller and Hobbs 2002). Essentially, local gardens foster the urban public's interest in biodiversity conservation (Andersson et al. 2007; Goddard et al. 2010). Green roofs supporting invertebrates allow for environmental educational where people might not otherwise engage in experiential, positive, and informative interactions with nature.

Broadly, conservation concern in non-urban areas is linked to individuals' experiences with urban flora and fauna (Faeth et al. 2011). However, in urban areas, where natural space is increasingly rare, disconnect between people and nature termed 'nature deficit disorder' (Louv 2008), is coupled with disinterest in nature among young people and less interest in conservation efforts by the general public (McKinney 2006; Miller 2005). There is also a widening gap in the understanding of biodiversity perpetuated by the stories presented in the media. For example, most people are familiar with honey bees due to the recent coverage of colony collapse disorder but many are unaware of the over 20,000 other species of bees worldwide. Green roofs that target invertebrate habitat may be one effective means of complementing efforts to connect citizens to nature and biodiversity in urban green spaces at ground level (Fuller et al. 2007; Lindemann-Matthies et al. 2010). Making these roofs accessible to the public for education, leisure, or, at the least, for viewing, could promote increased connectedness to nature and natural processes.

14.4 Monitoring and Managing for Success

Regardless of the design of a green roof, some invertebrates will arrive and thrive. However, many questions still remain as to which and how many species will colonize in addition to the type and duration of services they will support (Box 2). Answering these questions will provide insight into the conservation value of green roofs, as well as the measureable benefits of initiatives to enhance roofs for invertebrates. The fundamental challenges are first to distinguish between "desirable" and "undesirable" species (such as native solitary bees from aggressive paper wasps), and then to understand the ecology of these invertebrate species within the context of green roofs.

Monitoring invertebrate populations on green roofs could be achieved through both scientific and citizen-based data collection. Both methods aid in improving scientific understanding of green roof ecology. Typical monitoring incorporates complex as well as simple experimental designs. Examples include timed sweeping with an insect net on flowers, using liquid-filled pan traps set out on sunny days, or simply heading out with a camera and systematically photographing the active fauna. All methods have the potential to provide very informative data. The full utility of these data can then be realized when the sampled organisms are identified to the finest taxonomic resolution by local invertebrate experts and, perhaps most importantly, also shared with interested parties through international scientific publications, conference presentations and freely accessible sources.

Citizen science monitoring, in particular efforts that include partnering with experts in invertebrate identification at scientific institutions, could elucidate the presence of important or rare species. Good examples of this exist today that incorporate citizen science programming. Bumble bee watch (www.bumblebeewatch.org) and BeeSpotter (www.beespotter.mste.illinois.edu) are two such programs that encourages citizens to upload images of bees that are then identified by experts. This helps create digital geographic species ranges and local abundance data for climate change research in conjunction with thousands of museum specimen records without harming new specimens. Implemented on green roofs this program and other similar projects could monitor invertebrates.

Public awareness of urban biodiversity through monitoring efforts provides another way of increasing acceptance and appreciation of invertebrates by humans (Isaacs et al. 2009, Guiney and Oberhauser 2009, Dearborn and Kark 2010). Artificial nesting sites for some solitary bees and butterflies are even patented, marketed and sold with the intention of enhancing pollinator populations. Additionally, the number of urban honeybee colonies managed by hobbyists is increasing, especially those kept on green roofs. Aside from physical implementations of infrastructure to directly support the nesting requirement of some invertebrates on green roofs, signage is one useful way to make citizens aware of such projects and encourage them to become involved with monitoring and management efforts.

Box 2. Future Questions for Research

Research on urban invertebrate ecology and specifically that on green roofs is becoming more widespread but many questions remain unanswered. While not an exhaustive list, addressing these general research questions and examples of related, more-specific sub-questions below would permit future green roofs to be designed to more effectively in support of invertebrates, encouraging the ecosystem services that they provide.

1. How do interactions between invertebrates, plants and substrates contribute to green roof performance?

To what degree can herbivory impact roof cooling functions that are augmented by cover and density of vegetation?

Could substrate-burrowing invertebrates enhance horizontal water movement on green roofs, thereby enhancing plant water capture and survival?

2. To what extent do green roofs contribute to invertebrate conservation?

At what scales can green roofs influence invertebrate diversity, both on roofs and at ground level?

Can green roofs act as stepping-stone habitat throughout urban space for urban-intolerant species?

Do green roofs increase dispersal, recruitment and colonization by invasive species?

What is the conservation value of invertebrate communities on green roofs?

3. How do site-specific properties of a green roof influence its potential to serve as invertebrate habitat?

Does geographic and/or vertical isolation from natural habitat cause some green roofs designed for invertebrate diversity to become sink habitats?

In a diverse natural landscape, does designing a green roof for biodiversity have measureable impacts?

4. How do design considerations for green roof invertebrate diversity hold up empirically?

Does varying plant diversity, substrate topography, aggregate size, and other abiotic factors (e.g. Brenneisen 2006) have a significant and measureable impact on invertebrate community assembly, reproduction and persistence?

Do relationships between green roof design and invertebrate communities apply globally?

Both experimental design and citizen science monitoring of invertebrate populations can also improve techniques for design and maintenance, and help elucidate how "green" our green roofs are through the monitoring of taxa that serve as environmental indicators. Invertebrates are often used in urban landscapes as indicators of environmental quality. For example, Fountain and Hopkin (2004) used springtail diversity and the presence of a 'standard' springtail species to determine the level of metal contamination in urban soils, and Ishitani et al. (2003) suggested ground beetles to be useful indicators of environmental quality along an urban gradient in Japan. Sheffield et al. (2013) used cleptoparasites of bees as indicators, suggesting that their presence reflects both the food web's complexity and stability. Green roofs may also harbor similar indicator species but extensive research in this area has not yet been conducted.

Reporting on limitations for invertebrate success on green roofs can also help to inform future design and maintenance practices. This can be achieved by embedding designed experiments into green roof planning that permit observational and manipulated studies of the plant community and substrate (Felson and Pickett 2005). Few examples of empirical evidence currently demonstrate that green roofs designed to be 'biodiverse' provide habitat for invertebrates (Brenneisen 2006). More evidence through replicated research is required to elucidate the role of intent in green roof biodiversity design. Failure to do so runs the risk of 'biodiverse' green roofs becoming a weakly informed template for designers (e.g. "brownfield roof", "*Sedum* mat") without scientific evidence that diversity of local and desirable invertebrates and other organisms of concern are in fact supported.

Ideally, monitoring would evaluate the support that green roofs provide at all stages of invertebrate life cycles. Given the difficult environmental conditions present in green roof habitat for various species (e.g. vertical isolation and high levels of exposure to wind and sun), local factors could contribute to green roofs acting as habitat sinks for some species. In the absence of research, attempts to increase habitat value on vertically isolated green roofs may merely create a habitat façade where for particular species, crucial life cycle stages are interrupted by the extreme nature of green roof environments. This can be especially problematic on roofs with shallow substrate, and those that have low vegetative and microhabitat diversity. For example, beetles that colonize and lay eggs in the roof substrate in spring might not sense that the eggs will overheat on the roof, as a result of conditions being much hotter and drier in the middle of summer than ground level habitat (Chap. 15). Monitoring of species- and community-scale changes over time will determine how best to make adjustments to substrate topography, irrigation and planting in order to support invertebrates and other wildlife.

Green roofs intentionally designed to support diverse ecosystems may increase habitat available to invertebrates. Many new design strategies have been implemented in North America, the United Kingdom and Germany where meadows, brown fields, and even wetlands have been mimicked on extensive green roofs (Gedge and Kadas 2005; MacIvor et al. 2011b; Ksiazek 2014, Chap. 10). In Toronto, the city has published guidelines for enhancing biodiversity in green roof design (Torrance et al. 2013) to illustrate opportunities that exist to create microhabitats for invertebrates and other fauna. These strategies reference the habitat template approach described by Lundholm (2006), which suggests that green roof designers consider nearby habitats with similar microclimates to the xeric and exposure conditions on green roofs, for example, rock alvars in Southern Ontario or the coastal barrens in Atlantic Canada (Lundholm et al. 2009). Building upon the habitat template approach, green roofs have been described as opportunities for reconciliation ecology in urban landscapes; that is, they can provide habitat that supports natural floral and fauna and human infrastructure in tandem (Francis and Lorimer 2011). Recognizing the potential for green roofs as wildlife habitat early on in the design development phases can increase reconciliation of habitat space in cities (Eversham et al. 1996; Lundholm and Richardson 2010). Designing novel green roof habitat (Chap. 15) using abiotic and vegetation characteristics of regionally rare or climatically similar environments may benefit fauna more than conventional green roof installations, but more research is needed to fully understand the relationship between analogous habitats and invertebrate communities.

14.5 Conclusion

At this time, the relationship between green roof habitat and invertebrate communities is only beginning to be understood. Clearly, multi-regional, coordinated, and deliberate research efforts are needed to fully comprehend the extent to which green roofs provide habitat for invertebrates and, in turn, the extent to which these fauna contribute to ecosystem functions that positively impact green roof performance. While ecological theory predicts that a subset of species found within the larger urban environment may dominate green roof habitats, the many local and landscape factors that contribute to community structuring will result in each green roof providing a unique set of niches available for invertebrate colonization and usage.

On green roofs, as in other urban and natural habitats, invertebrates are inextricably linked with important ecosystem services. They build and stabilize soil in the substrate, control unwanted pests, pollinate crops and native plants, and act as food for other organisms in food webs. As the amount of natural habitat available for this critical group of organisms decreases, urban green roofs have a potentially larger role to play in supporting diverse fauna.

References

Altieri M, Nicholls C (2004) Biodiversity and pest management in agroecosystems. CRC Press, Boca Raton

Andersson E, Barthel S, Ahrné K (2007) Measuring social-ecological dynamics behind the generation of ecosystem services. Ecol Appl 17:1267–1278

Banaszak-Cibicka W (2014) Are urban areas suitable for thermophilic and xerothermic bee species (Hymenoptera: Apoidea: Apiformes)? Apidologie. doi:10.1007/s13592–013-0232-7

Banaszak-Cibicka W, Żmihorski M (2012) Wild bees along an urban gradient: winners and losers. J Insect Conserv 16:331–343

Blair RB, Launer AE (1997) Butterfly diversity and human land use: species assemblages along an urban gradient. Biol Conserv 80:113–125

Bolger DT, Suarez AV, Crooks KR et al (2000) Arthropods in urban habitat fragments in southern California: area, age and edge effects. Ecol Appl 10:1230–1248

Braaker S, Ghazoul J, Obrist MK, Moretti M (2013) Habitat connectivity shapes urban arthropod communities—the key role of green roofs. Ecology. doi:10.1890/13-0705.1

Brenneisen S (2006) Space for urban wildlife: designing green roofs as habitats in Switzerland. Urban Habitats 4:27–36

Brenneisen S, Hänggi A (2006) Begrünte Dächerökologische Charakterisierung eines neuen Habitattyps in Siedlungsgebieten anhand eines Vergleichs der Spinnenfauna von Dachbegrünungen mit naturschutzrelevanten Bahnarealen in Basel (Schweiz) [Green roofs—ecological characterization of a new habitat type in urban areas based on a comparison of the spider fauna on green roofs with conservation railway areas in Basel (Switzerland)]. Mitt Naturforschenden Ges beider Basel 9:99–122

Burghardt KT, Tallamy DW, Shriver WG (2009) Impact of native plants on bird and butterfly biodiversity in suburban landscapes. Conserv Biol 23:219–224

Cane JH, Minckley RL, Kervin LJ et al (2006) Complex responses within a desert bee guild (*Hymenoptera: Apiformes*) to urban habitat fragmentation. Ecol Appl 16:632–644

Chapin FS, Matson PA, Mooney HA (2002) Principles of terrestrial ecosystem ecology. Springer, New York

Clausen RT (1975) *Sedum* of North America north of the Mexican plateau. Cornell University, New York

Coffman RR, Waite T (2011) Vegetated roofs as reconciled habitats: rapid assays beyond mere species counts. Urban Habitats 6:1–10

Colding J, Barthel S (2013) The potential of 'Urban Green Commons' in the resilience building of cities. Ecol Econ 86:156–166

Colla SR, Willis E, Packer L (2009) Can green roofs provide habitat for urban bees (*Hymenoptera: Apidae*)? CATE 2:1–12

Corbet SA, Bee J, Dasmahapatra K et al (2001) Native or exotic? Double or single? Evaluating plants for pollinator-friendly gardens. Ann Bot 87:219–232

Crisp PN, Dickinson KJM, Gibbs GW (1998) Does native invertebrate diversity reflect native plant diversity? A case study from New Zealand and implications for conservation. Biol Conserv 83:209–220

Crosswhite FS, Crosswhite CD (1966) Insect pollinators of *Penstemon* series *Graciles* (Scrophulariaceae) with notes on *Osmia* and other *Megachilidae*. Am Mid Nat 76:450–467

Davies R, Simcock R, Toft R (2010) Islands in the sky urban biodiversity enhancement in NZ on indigenous living roof landscapes. In: 44th annual conference of the architectural science association, ANZAScA, Auckland, New Zealand, 24–26 November 2010

Dearborn DC, Kark S (2010) Motivations for conserving urban biodiversity. Conserv Biol 24:432–440

Deichsel R (2006) Species change in an urban setting—ground and rove beetles (*Coleoptera: Carabidae* and *Staphylinidae*) in Berlin. Urban Ecosyst 9:161–178

Duncan RP, Clemants SE, Corlett RT et al (2011) Plant traits and extinction in urban areas: a meta-analysis of 11 cities. Glob Ecol Biogeogr 20:509–519

Dunnett N, Kingsbury N (2004) Planting green roofs and living walls. Timber Press, Portland

Elton CS (1958) The ecology of invasions by animals and plants. Methuen, London

Ettema CH, Wardle DA (2002) Spatial soil ecology. Trends Ecol Evol 17:177–183

Eversham BC, Roy DB, Telfer MG (1996) Urban, industrial and other manmade sites as analogues of natural habitats for Carabidae. Ann Zoo Fennici 33:148–156

Faeth SH, Bang C, Saari S (2011) Urban biodiversity: patterns and mechanisms. Ann N Y Acad Sci 1223:69–81

Felson AJ, Pickett STA (2005) Designed experiments: new approaches to studying urban ecosystems. Front Ecol Environ 3(10):549–556

Fernandez-Canero R, Gonzalez-Redondo P (2010) Green roofs as a habitat for birds: a review. J Anim Vet Adv 9:2041–2052

Forman RTT (2014) Urban ecology: science of cities. Cambridge University Press, Cambridge

Forschungsgesellschaft Landschaftsentwicklung Landschaftsbau e.V. (FLL) (2008) Guidelines for the planning, construction and maintenance of green roofing. FLL, Bonn

Fountain MT, Hopkin SP (2004) Biodiversity of Collembola in urban soils and the use of *Folsomia candida* to assess soil 'quality'. Ecotoxicology 13:555–572

Francis RA, Lorimer J (2011) Urban reconciliation ecology: the potential of living roofs and walls. J Environ Manag 92:1429–1437

Frankie GW, Ehler LE (1978) Ecology of insects in urban environments. Ann Rev Entomol 23:367–387

Fuller RA, Irvine KN, Devine-Wright P et al (2007) Psychological benefits of greenspace increase with biodiversity. Biol Lett 3:390–394

Gaston KJ (2010) Urban ecology. Cambridge University Press, Cambridge

Gedge D, Kadas G (2005) Green roofs and biodiversity. Biologist (London) 52:161–169

Goddard MA, Dougill AJ, Benton TG (2010) Scaling up from gardens: biodiversity conservation in urban environments. Trends Ecol Evol 25:90–98
Gorgolewski M, Komisar J, Nasr J (2011) Carrot city. Monacelli, New York
Greenleaf SS, Williams NM, Winfree R et al (2007) Bee foraging ranges and their relationship to body size. Oecologia 153:589–596
Grimm NB, Faeth SH, Golubiewski NE et al (2008) Global change and the ecology of cities. Science 319:756–760
Guiney MS, Oberhauser KS (2009) Insects as flagship conservation species. Terr Arthropod Rev 1:111–123
Hahs AK, McDonnell MJ, McCarthy MA et al (2009) A global synthesis of plant extinction rates in urban areas. Ecol Lett 12:1165–1173
Hanski I, Gilpin M (1991) Metapopulation dynamics: brief history and conceptual domain. Biol J Linn Soc 42:3–16
Hartley W, Uffindell L, Plumb A et al (2008) Assessing biological indicators for remediated anthropogenic urban soils. Sci Total Environ 405:358–369
Hegland SJ, Nielsen A, Lázaro A et al (2009) How does climate warming affect plant- pollinator interactions? Ecol Lett 12:184–195
Hennig EI, Ghazoul J (2011) Plant-pollinator interactions within the urban environment. Perspect Plant Ecol Evol Syst 13:137–150
Herbold B, Moyle PB (1986) Introduced species and vacant niches. Am Nat 128:751–760
Isaacs R, Tuell J, Fiedler A et al (2009) Maximizing arthropod-mediated ecosystem services in agricultural landscapes: the role of native plants. Front Ecol Environ 7:196–203
Ishitani M, Kotze DJ, Niemelä J (2003) Changes in carabid beetle assemblages across an urban-rural gradient in Japan. Ecography 26:481–489
Jones RA (2002) Tecticolous invertebrates: a preliminary investigation of the invertebrate fauna on green roofs in urban (pp 1–36). English Nature, London
Kadas G (2006) Rare invertebrates colonizing green roofs in London. Urban Habitats 4:66–86
Kaupp A, Brenneisen S, Klausnitzer B, Nagel P (2004) Eco-faunistic characteristics of the beetle fauna of vegetated roofs (Insecta: Coleoptera). Entomol Blätter 100:47–83
Kearns CA, Inouye DW, Waser NM (1998) Endangered mutualisms: the conservation of plant-pollinator interactions. Ann Rev Ecol Syst 29(1):83–112
Kennedy CM, Lonsdorf E, Neel MC et al (2013) A global quantitative synthesis of local and landscape effects on wild bee pollinators in agroecosystems. Ecol Lett 16:584–599
Kim KC (1993) Biodiversity, conservation and inventory: why insects matter. Biodivers Conserv 2:191–214
Kim KG (2004) The application of the biosphere reserve concept to urban areas: the case of green rooftops for habitat network in Seoul. Ann N Y Acad Sci 1023:187–214
Klein AM, Vaissiere BE, Cane JH et al (2007) Importance of pollinators in changing landscapes for world crops. Proc R Soc B Biol Sci 274:303–313
Köhler M (2006) Long-term vegetation research on two extensive green roofs in Berlin. Urban Habitats 4:3–25
Ksiazek K (2014) The potential of green roofs to support urban biodiversity. In: Briz J, Köhler M, de Felipe I (eds) Green cities in the world. (Chap 6, pp 101–124). Pronatur, Madrid
Ksiazek K, Fant J, Skogen K (2012) An assessment of pollen limitation on Chicago green roofs. Landsc Urban Plan 107:401–408
Ksiazek K, Tonietto R, Ascher J (2014) Ten bee species new to green roofs in the Chicago area. Gt Lakes Entomol 47(1–2):47–52
Landis DA, Wratten SD, Gurr GM (2000) Habitat management to conserve natural enemies of arthropod pests in agriculture. Ann Rev Ent 45:175–201
Lee KE (1985) Earthworms: their ecology and relationships with soils and land use. Academic Inc., Florida
Lindemann-Matthies P, Junge X, Matthies D (2010) The influence of plant diversity on people's perception and aesthetic appreciation of grassland vegetation. Biol Conserv 143:195–202

Liu H, Stiling P (2006) Testing the enemy release hypothesis: a review and meta-analysis. Biol Invasions 8:1535–1545

Louv R (2008) Last child in the woods: saving our children from nature-deficit disorder. Algonquin Books, New York

Lundholm JT (2006) Green roofs and facades: a habitat template approach. Urban Habitats 4:87–101

Lundholm JT, Richardson PJ (2010) Habitat analogues for reconciliation ecology in urban and industrial environments. J Appl Ecol 47:966–975

Lundholm JT, MacIvor JS, Ranalli MA (2009) Benefits of green roofs on Canada's east coast. In: Proceedings of the cities alive annual green roof and wall conference, Atlanta, GA

MacArthur RH, Wilson EO (1967) The theory of island biogeography. Princeton University, New Jersey

MacIvor JS (2013) Constraints to bee and wasp nesting habitat on green roofs. In: Proceedings of the cities alive annual green roof and wall conference, San Francisco, CA

MacIvor JS, Lundholm J (2011a) Insect species composition and diversity on intensive green roofs and adjacent level-ground habitats. Urban Ecosyst 14:225–241

MacIvor JS, Lundholm J (2011b) Performance of dryland and wetland plant species on extensive green roofs. Ann Bot 107:671–679

MacIvor JS, Ruttan A, Salehi B (2014) Exotics on exotics: pollen analysis of urban bees visitingSedum on a green roof. Urban Ecosyst. doi:10.1007/s11252-014-0408-6

Madre F, Vergnes A, Machon N, Clergeau P (2013) A comparison of 3 types of green roof as habitats for arthropods. Ecol Eng 57:109–117

Majka CG, MacIvor JS (2009) *Otiorhynchus porcatus* (Coleoptera: Curculionidae): a European root weevil newly discovered in the Canadian Maritime Provinces. J Acad Entomol Soc 5:27–31

Mann G (1998) Vorkommen und Bedeutung von Bodentieren (Makrofauna) auf begrünten Dächern in Anhängigkeit von der Vegetationsform (Occurence and significance of soil organisms (macrofauna) on green roofs in response to vegetation type). Doctoral Dissertation. Eberhard, Karls Universität Tübingen

Matteson KC, Langellotto GA (2010) Determinates of inner city butterfly and bee species richness. Urban Ecosyst 13:333–347

McGuire KL, Payne SG, Palmer MI et al (2013) Digging the New York City skyline: soil fungal communities in green roofs and city parks. PloS ONE 8:e58020

McIntyre NE (2000) Ecology of urban arthropods: a review and a call to action. Ann Entomol Soc Am 93:825–835

McKinney ML (2006) Urbanization as a major cause of biotic homogenization. Biol Conserv 127:247–260

Mecke R, Grimm R (1997) Fauistichökologische Untersuchung begrüter Dachflächen im Hamburger Stadgebiet (Faunal ecology investigation of greened roofs in the Hamburg urban area). Nat schutz Landsch plan 29:297–302

Meierholfer D (2013) Ground beetles on San Francisco green roofs. In: Proceedings of the cities alive annual green roof and wall conference, San Francisco, CA

Menke SB, Guénard B, Sexton JO et al (2011) Urban areas may serve as habitat and corridors for dry-adapted, heat tolerant species: an example from ants. Urban Ecosyst 14:135–163

Michener CD (2007) The bees of the world. John Hopkins University, Baltimore

Miller JR (2005) Biodiversity conservation and the extinction of experience. Trends Ecol Evol 20:430–434

Miller JR, Hobbs RJ (2002) Conservation where people live and work. Conserv Biol 16:330–337

Myers JH, Simberloff D, Kuris AM, Carey JR (2000) Eradication revisited: dealing with exotic species. Trends Ecol Evol 15:316–320

Nagase A, Dunnett N, Choi MS (2013) Investigation of weed phenology in an establishing semi-extensive green roof. Ecol Eng 58:156–164

Niemelä J, Kotze DJ, Venn S et al (2002) Carabid beetle assemblages (Coleoptera, Carabidae) across urban-rural gradients: an international comparison. Land Ecol 17:387–401

Odum EP (1971) Fundamentals of ecology. W. B. Saunders Co., Philadelphia
Oke TR (1973) City size and the urban heat island. Atmos Environ 7:769–779
Pawelek J, Frankie GW, Thorp RW, Przybylski M (2009) Modification of a community garden to attract native bee pollinators in urban San Luis Obispo, California. CATE 2:1–7
Potts SG, Biesmeijer JC, Kremen C et al (2010) Global pollinator declines: trends, impacts and drivers. Trends Ecol Evol 25:345–353
Root RB (1973) Organization of a plant-arthropod association in simple and diverse habitats: the fauna of collards (*Brassica oleracea*). Ecol Mono 43:95–124
Rumble H, Gange AC (2013) Soil micro-arthropod community dynamics in extensive green roofs. Ecol Eng 57:197–204
Samways MJ, Steytler NS (1996) Dragonfly (*Odonata*) distribution patterns in urban and forest landscapes, and recommendations for riparian management. Biol Conserv 78:279–288
Santorufo L, Van Gestel CAM, Rocco A, Maisto G (2012) Soil invertebrates as bioindicators of urban soil quality. Environ Pollut 161:57–63
Savard JPL, Clergeau P, Mennechez G (2000) Biodiversity concepts and urban ecosystems. Landsc Urban Plan 48:131–142
Schindler BY, Griffith AB, Jones KN (2011) Factors influencing arthropod diversity on green roofs. CATE 4:1–20
Schmidt MH, Tscharntke T (2005) Landscape context of sheetweb spider (Araneae: Linyphiidae) abundance in cereal fields. J Biogeogr 32:467–473
Schrader S, Böning M (2006) Soil formation on green roofs and its contribution to urban biodiversity with emphasis on Collembolans. Pedobiologia 50:347–356
Schüepp C, Herzog F, Entling MH (2014) Disentangling multiple drivers of pollination in a landscape-scale experiment. Proc R Soc B: Biol Sci 281:20132667
Sheffield CS, Pindar A, Packer L, Kevan PG (2013) The potential of cleptoparasitic bees as indicator taxa for assessing bee communities. Apidologie 44:501–510
Shochat E, Stefanov WL, Whitehouse MEA, Faeth SH (2004) Urbanization and spider diversity: influences of human modification of habitat structure and productivity. Ecol Appl 14:268–280
Shochat E, Warren PS, Faeth SH et al (2006) From patterns to emerging processes in mechanistic urban ecology. Trends Ecol Evol 21:186–191
Siemann E (1998) Experimental tests of effects of plant productivity and diversity on grassland arthropod diversity. Ecology 79:2057–2070
Siemann E, Tilman D, Haarstad J, Ritchie M (1998) Experimental tests of the dependence of arthropod diversity on plant diversity. Am Nat 152:738–750
Simberloff D (1981) Community effects of introduced species. In: Nitecki MH (ed) Biotic crises in ecological and evolutionary time. Academic, New York, pp 53–81
Singh JS, Gupta SR (1977) Plant decomposition and soil respiration in terrestrial ecosystems. Bot Rev 43:449–528
Smith RM, Gaston KJ, Warren PH, Thompson K (2006) Urban domestic gardens (VIII): environmental correlates of invertebrate abundance. (Human Exploitation and Biodiversity Conservation) (pp 175–205). Springer, Netherlands
Sutton RK, Harrington JA, Skabelund L, MacDonagh P, Coffman RR, Koch G (2012) Prairie-based green roofs: literature, templates and analogs. J Green Build 7:143–172
Tonietto R, Fant J, Ascher J et al (2011) A comparison of bee communities of Chicago green roofs, parks and prairies. Land Urb Plan 103:102–108
Torrance S, Bass B, MacIvor JS, McGlade T (2013) City of Toronto guidelines for biodiverse green roofs. Toronto City Planning, Toronto
Whittinghill LJ, Rowe DB (2012) The role of green roof technology in urban agriculture. Renew Agric Food Syst 27:314–322
Williams NM, Winfree R (2013) Local habitat characteristics but not landscape urbanization drive pollinator visitation and native plant pollination in forest remnants. Biol Conserv 160:10–18
Williams NSG, Lundholm JT, MacIvor JS (2014) Can green roofs support biodiversity conservation goals? J Appl Ecol 51(6):1643–1649
Wilson EO (1984) Biophilia. Harvard University, Cambridge

Wilson EO (1987) The little things that run the world (the importance and conservation of invertebrates). Conserv Biol 1:344–346

Winfree R, Griswold T, Kremen C (2007) Effect of human disturbance on bee communities in a forested ecosystem. Conserv Biol 21:213–223

Winfree R, Bartomeus I, Cariveau DP (2011) Native pollinators in anthropogenic habitats. Ann Rev Ecol Evol Syst 42:1–22

Zmyslony J, Gagnon D (1998) Residential management of urban front-yard landscape: a random process? Landsc Urban Plan 40:295–307

J. Scott MacIvor is an ecologist interested in the role and relevance of biodiversity in landscape design and city planning. He is completing a PhD in the Biology Department at York University and is an instructor in the faculty of Architecture, Landscape, and Design at the University of Toronto.

Kelly Ksiazek is an urban ecologist with Plant Biology and Conservation Department, Northwestern University and the Chicago Botanic Garden, Chicago, IL, USA. Her research interests focus on native plant and pollinator conservation in cities with a focus on green roofs habitats.

Chapter 15
Placing Green Roofs in Time and Space: Scale, Recruitment, Establishment, and Regeneration

Katherine Dunster and Reid R. Coffman

Abstract The preceding chapters have followed an interconnected path through the fields of green roof research to converge in this chapter on some emerging principles for the design of green roof ecosystems that extend beyond the garden aesthetic. Understanding green roof ecosystems in time and space becomes critical to good ecological design and the desire to protect and improve biodiversity in all its forms.

Scale plays a central role in ecology, providing context to understanding patterns across the space-time continuum within the local, regional, and global landscape. Review of green roof projects and research literature indicates that beyond local concern to create authentic habitat, green roof design and research has paid scarce attention to scalar relationships. As we explain in this chapter, paying attention to scale has implications for the ecological relevance of a green roof project at both socio-political and biological levels. Contextualizing how a roof will fit into time and space establishes its place on the planet and its ecological role in the landscape. Scale also plays a role in the recruitment, establishment, and regeneration of species on a roof, that in turn sets the management path towards design success and long-term survival of the roof ecosystems created.

Keywords Scale · Biodiversity · Recruitment · Source-Sink · Disturbance · Regeneration · Management

K. Dunster (✉)
School of Horticulture, Kwantlen Polytechnic University, 12666-72 Avenue, V3W 2M8 Surrey, BC, Canada
e-mail: unfoldinglandscapes@gmail.com

R. R. Coffman
1309 Eluclid Ave., Suite 200, 44115 Cleveland, OH, USA

R. Sutton (ed.), *Green Roof Ecosystems,* Ecological Studies 223,
DOI 10.1007/978-3-319-14983-7_15

15.1 Introduction

Ecosystems interconnect dynamically through complex processes and functions including everything from pollination, migration, and dispersal to energy flow, nutrient and biogeochemical cycling, and the water cycle (Millennium Ecosystem Assessment (MEA 2005). As Harper (1977) noted, from the perspective of any species an ecosystem can be very small in scale, or as large as the entire planet depending on your size and mobility. Globally, major ecosystems, called biomes (Fig. 15.1) cover the planet with a diversity of overlapping and interdependent terrestrial and aquatic ecosystems. Nested within biomes are the landscapes and ecosystems familiar to humans at local and regional scales.

The modification of ecosystems for human purposes often results in isolated landscape site design or management decisions with unpredicted and disruptive ecological impacts elsewhere, over time and across many landscape scales, whether on the ground or above on rooftops. Whatever the context used to determine scale,

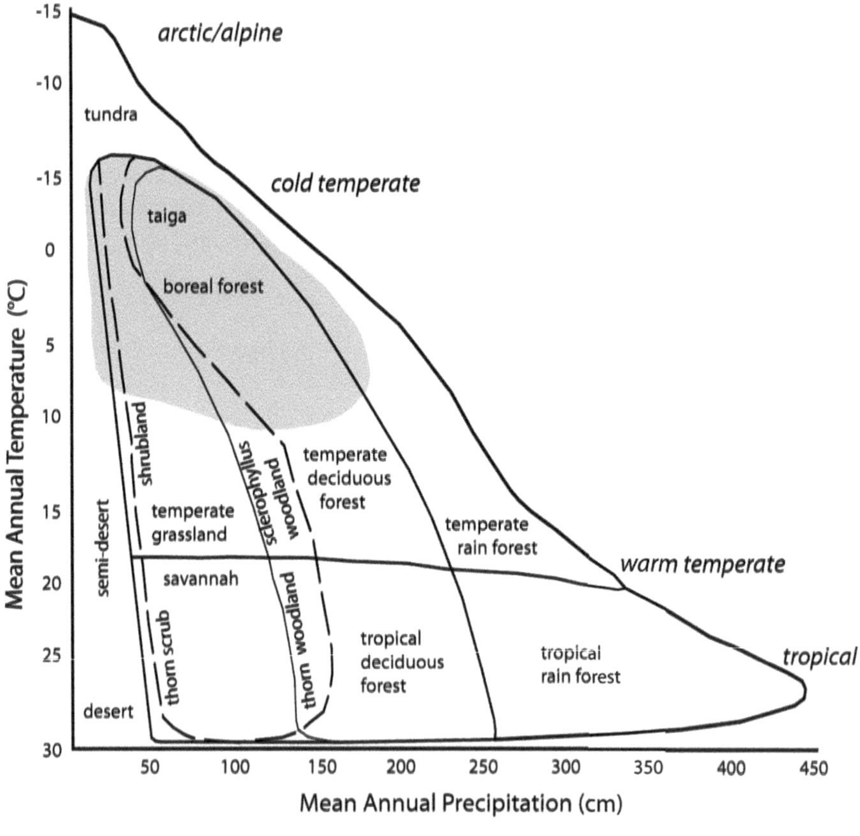

Fig. 15.1 World biomes; shaded area indicates those currently most influential in origin of plants selected for Northern Hemisphere green roofs. (Modified from www.thewildclassroom.com)

green roof designers must be challenged to see a project through the lens of multiple scales rather than a single one, in order to successfully match the scale of ecological processes to the policy or decision-making scale (ME 2003). For example, a local policy requiring biodiverse green roofs on all new structures may disregard an over-arching national or international policy to protect a regionally significant landscape or the habitat of an endangered species by avoiding altogether construction of new structures which may rise after site preparation that destroys existing habitat.

Critical to the broader success of a green roof project then is contextualization of its place in time and space. Green roof designers and urban planners following the principles of island biogeography (MacArthur and Wilson 1963, 1967) can use contextualizing to avoid creation of small isolated patches in an already fragmented landscape. Isolation addresses the spatial and temporal qualities of roof ecosystems rather than the spatial character of the roof itself, which for most roofs consists of a mosaic of small micro-habitats or ecotopes (*sensu* Whittaker et al. 1973), located within a typically well-defined and manufactured square or rectangular shape. A long-term goal in creating a rooftop ecosystem should be to avoid subsequent problems such as genetic bottlenecks caused by isolation in the absence of connectivity and inter-patch dynamics.

Rooftops are human constructs and until recently were an underutilized component of urban anthromes (anthropogenic biomes) created over the past 8000 years as humans transformed 75 % of the terrestrial wildlands on the planet (Vitousek et al. 1997; Ellis and Ramankutty 2008). Despite this change from wild to cultivated, a review of several biological conservation journals finds significant research bias towards vertebrates, forests, relatively pristine landscapes in reserves or national parks, and towards studies of single species and assemblages rather than communities or ecosystems (Fazey et al. 2005), and noted that few (12.6 %) of studies actively field tested, reviewed, or applied conservation actions.

While the imperative to protect remaining biodiversity in wildlands is a global concern articulated in the World Conservation Strategy (IUCN-UNEP-WWF 1980) and the Convention on Biological Diversity (CBD 1992), a need emerges to pay attention to conservation within anthromes, places where biodiversity flows, re-mixes, and forms new communities and novel ecosystems (Ellis 2013). Sometimes, but not always, this happens by human intervention through land abandonment, but a signature characteristic finds the novel ecosystem can function without further intervention. This provides context and the rationale for designing and placing roof ecosystems within the urban anthrome and determining whether they might be new or novel ecosystems (see Sect. 15.7).

Development of an ecosystem and recruitment of species onto a rooftop can occur intentionally by design or by happenstance. When created by design a roof top ecosystem is established entirely under the guidance of the human hand. The choice and depth of substrate materials are limited by the capacity of the structure to carry the weight of precipitation-saturated substrate and influences the initial choice of plants and the extent to which subsequent plant, insect, animal, and microbial assemblages are recruited or introduced, become established, are replaced, and thrive. When a roof is spontaneously colonised by plants arriving as seed dispersed by wind (anemochory), animals (zoochory), soil, or other means, a similar process happens unguided by humans (Chap. 10)

Size of roof area and its capacity to carry live and environmental loads on top of the roof deck such as the weight of maintenance workers, equipment, and materials including plants, substrate, drainage materials, precipitation in all forms, and movable habitat elements such as rock and branch piles becomes critical Scale comes into play here as well, as it influences design options. Plants are selected for their ability to thrive in the depth of soil the roof can support which generally removes trees and shrubs (unless dwarf) from most roof planting design palettes; the rock or branch piles must be sized to fit the design space or appear to visually overwhelm the site.

Our understanding of scale and desire to understand the world in scale is a human visual phenomenon (Sutton 2011). Placing a green roof project ecosystem within a landscape and relating it to the existing scale provides a spatial context that has both socio-political implications such as the political boundaries and policy jurisdiction that locate and govern the project, and biological context when related to a biome, bioregion, or watershed (MEA 2005). This chapter provides the reader with a grounding in the concepts of landscape scale and how it influences the design and management of green roof ecosystems.

15.2 Scale

15.2.1 Definitions

> …the problem of pattern and scale is the central problem in ecology, unifying population biology and ecosystems science, and marrying basic and applied ecology. Applied challenges… require the interfacing of phenomena that occur on very different scales of space, time, and ecological organization. (Levin 1992, p. 1943)

Space and time are measurable phenomena that allow landscapes to be defined and to examine changes in structure, function, and processes documented using a variety of spatial tools such as remote sensing, geographic information systems (GIS), and ground monitoring. This approach delineates landscapes by first identifying distinct, homogeneous boundaries clearly visible when looking at continents or biomes.

We each subjectively see and perceive landscape as a part of the larger environment (Kaplan and Kaplan 1989) is a qualitative component influencing the delineation of landscape boundaries regardless of the technology employed. Distinguishing boundaries between ecosystems and plant communities becomes more nebulous as the landscape localizes. Ecotones form where two different ecosystems meet and species distinct to either community integrate to form a different ecosystem within the ecotone boundaries (Cadenasso et al. 2003).

In landscape ecology, the paleo-ecological research of Delcourt et al. (1983) has influenced contemporary thinking about landscape as a system of ecological patterns and processes integrated within a nested series of temporal and spatial scales ranging from the micro-scale of a grassland patch to the mega-scale of the biomes that define and pattern global terrestrial vegetation. From the perspective of geological time, Delcourt and Delcourt (1988) envisioned an operational scale paradigm in which most landscape management and plant ecology work would occur within the micro-scale, spanning time from 1 to 500 hundred years across spaces that range from 1 m^2 to 100 ha in size. In their paradigm, humans can see the landscape and track it through the fossil record.

To make this paradigm meaningful to green roof design, we introduce two additional scales in Table 15.1. The nano-scale describes landscapes less than 1 m^2, typically habitat to non-vagile species such as plants and other macro-flora and fauna whereas pico-scale describes landscapes above and below the ground occupied by plant roots, seeds, and soil microorganisms (Chaps. 5, 6, 7).

At the micro-scale, dynamic functions such as seasonal weather patterns, disturbances (Sect. 15.7), and abiotic perturbations triggered by global climate change shape plant and animal populations and community structure. These changes re-mix ecosystem composition (Brown et al. 1997). Events such as a windstorm uprooting a tree, a grass fire, a stream flooding its banks, or a landslide represent brief and localized disturbances. All these processes affect and shape local landscape vegetation patterns operating on an individual in a community, the community in a small watershed, and in between.

At the meso-scale, time covers 500–10,000 years and space ranging from 100 ha (or a corridor about 1 km in width), to an area or corridor about 111 km^2 or one degree of latitude. The meso-scale captures events that have occurred during the Holocene, the most recent interglacial interval in which regional changes in abiotic and climatic processes have shaped the familiar heterogeneous landscape: an eco-regional mosaic of large and small patches and corridors interconnected within a discrete and dominant ecological matrix such as forest, grassland, savannah, desert, taiga, and tundra occurring on the most extensive landform within a region of relatively uniform climate. Disturbances affect large areas, are infrequent, and dramatically change landscape patterns and structure for significant periods of time (Foster et al. 1998; Turner and Dale 1998).

At the meso-scale landscape is important for wide-ranging species such as migratory birds, and for those that are large-area dependent such as grizzly bear (*Ursus arctos horribilis*). Changing terrestrial patch dynamics resulting from the transformation of landscape to human cultural utility dramatically impacts such species. Vagile species can migrate and adapt to new spaces, while others less so are displaced or extirpated.

Green roof ecologists designing surrogate nano-scale habitats on micro-scale rooftops raise conservation concerns about some species and ecosystems (Dunster 2007, 2010). For example, experimental green roofs in New Zealand have been designed to fit both place and space in a temporally new landscape while providing a safe-site for several ground-dwelling native skink species (*Oligosoma* spp.)

Table 15.1 Landscape types and characteristics within a continuum of spatial and temporal scales

Type	Spatial scale	Time scale	Processes	Grain
Mega	Biosphere	4.54 billion years (geologic time)	Plate tectonics	Very Coarse
			Atmospheric exchange, insulation	
			Controls primary energy inputs to climate & weather patterns	
			Climate: sun, wind & water patterns	
			Seasonality: Earth-Sun distance, annual orbit & axis tilt	
			Global patterns of temperature and rainfall	
Macro	Continental	Quaternary period (1.8 million years ago to the present)	Continental weather patterns resulting in biomes, species distributions	
Meso	Regional: 100 ha–111 km² (1 degree latitude or large watershed)	Holocene inter-glacial period (10,000 years ago to present)	Microclimate: local weather and vegetation patterns influenced by topography & geography	Coarse
			Geo-climatic changes transform landscape mosaic	
Micro	Local: 1 m²–100 ha (from small patches up to 1st order stream watersheds)	1–500 years	Geomorphic processes: soil creep, movement of sand dunes, debris avalanches, slumps, fluvial transport & deposition, and cryoturbation	Fine
			Disturbances: local to widespread of relatively short duration, such as wildfire, windthrow, & clear-cutting	
			Biotic processes: cyclic changes in populations, gap-phase replacement of plants, plant succession affects community composition from individual plants to large forest patches	
Nano	<1 m²	1–500 years	Individual environment of macroflora and fauna	
Pico	<1 cm²	<1–? years	Individual environment of microflora and fauna	Extra-fine

threatened by non-native mammalian predators that include house mouse (*Mus musculus*), rats (*Rattus* spp.), European hedgehog (*Erinaceus europaeus*), and domestic cat (*Felis cattus*) (Davies et al. 2010; Waitakere City Council 2010). This design explores the relationships between the nano-scale, where habitat is created for wildlife using indigenous plants best adapted to local conditions, and the micro-scale where regional biodiversity is protected, thus meeting national meso-scale obligations to the CBD.

Rather than visualizing landscape as a nested hierarchy of scales, landscapes can also be defined by their grain and extent (Turner et al. 1989). Grain is the size of the individual units of observation that can be distinguished through measurement. Extent is simply the spatial boundary encompassing the area of interest. The boundaries and patterns that the eye can see define grain and extent (Sutton 2011). Pattern occurs at all spatial scales, and scale affects the observed pattern (Turner et al. 1989).

In the case of roof ecosystems the finest grain could be an individual roof, or it might consist of even smaller entities such as different plant communities or habitats within the roof ecosystem. At the micro-scale, a coarser grain would see green roofs mapped across a neighbourhood. For example, in roof ecosystem management terms, the finest unit of management such as a grouping of plants defines grain while extent refers to the total area of the roof under management.

15.2.2 Placing Species and Spaces in Scale

In ecology, scale concerns both space and time, however there is a caveat regarding over-reliance on the convenience of placing a space or the life stage of a species on the scale continuum without a thorough understanding of all the variables leading to the placement decision (Addicott et al. 1987; Kareiva and Andersen 1988). Researchers, as observers, have perceptual bias conditioned by their academic training. Ecosystems show characteristic variability (heterogeneity) on a range of spatial, temporal, and organizational scales. There is no accepted standard scale at which ecosystems should be studied (Levin 1992); the scale chosen must be relative and contextual to the research question posed.

The various concepts of niche theory suggest that an organism occupies a unique ecological space (niche) that encompasses the specific environmental conditions it has adapted to that are necessary for its survival (Chase and Leibold 2003; Peterson et al. 2011). The niche then, includes both the conditions defining the habitat where the organism lives, and the functional role the organism plays in the ecosystem.

For many species it is difficult to describe niche with certainty because they do not occupy a specific landscape scale or habitat for their entire lives but rather, use different landscape scales at different times, for different processes and life stages (Harper 1977). These changes in time (see Sect. 15.4.2), space, scale, and process can be conceptually envisioned as the ecological neighbourhood (Addicott et al. 1987; Antolin and Addicott 1991).

Three factors contribute to the delineation of ecological neighbourhoods (1) the particular ecological process (e.g. foraging, migrating, reproducing, mutualistic interaction (2) the time scale appropriate to the process (e.g. day, season, year), and (3) the organism's influence or activity during this time period (Addicott et al. 1987). By tracking the organism through space, over the time period for each process or life stage event, the different landscape scales used by a species during its life determine its ecological neighbourhoods.

The ecological neighbourhood of a species for any given function or process is the space within which the species is active or has some influence during a period of time relative to the particular life-stage. The size and shape of the neighbourhood for each species may be quite different for other processes and as such is fluid rather than static. The type of ecological process such as the periodicity of disturbances such as fire (Daubenmire 1968; Suffling and Perrera 2004), or grazing (Joern 2005), defines the appropriate time and spatial scale over which to measure neighbourhood size, which may be described by physical size (area) or number of individuals (population).

The ecological neighbourhood of a bird or insect in migration may be meso- in scale, while its breeding neighbourhood may be nano or pico. For example, a butterfly egg on a larval host-specific plant is singularly local and pico in scale. When the egg hatches and the larva moves and consumes the plant, the scale (or space occupied) expands from a fixed point to the whole plant, and changes again when the adult butterfly forages across a landscape for nectar plants. These life stage events occur at different times with different behaviours and influences. In the case of larvae, the host plant is consumed to varying degrees of harm which it may or may not survive, while the adults assist pollination and regeneration of the nectar plant.

Likewise, the spatial and temporal units may not be relevant to the ecosystem structure, functions, or processes under study or design. For example, the Delcourts historical research on temporal landscape change in the Quaternary required a course filter approach which identified the meso- and mega- as the appropriate scales to study landscape. Designers placing a roof ecosystem project in the landscape generally focus on nano- and pico- scales but to maintain relevance, should relate their projects to the meso-scale as well (Table 15.1). It should also be noted that while most data is collected at small spatial scales, conservation management issues are often addressed at large spatial scales.

Instead of fixating on the niche, Addicott et al. (1987, p. 344) posit that scaling is more critical to understanding patterns, processes, and the place of species in time and space. They propose using the concept of ecological neighbourhoods to "*scale environmental patterning, thereby allowing assessment of relative patch size, relative patch isolation, relative patch duration, and relative utilization of heterogeneous patches (grain response)*." As a design tool the ecological neighbourhoods occupied by a species can be quantified across the landscape, and those spaces then used to relatively scale the pattern of the habitat and rooftop environment to accommodate the species on the roof.

The meso- and macro-scales are important from the perspectives of biodiversity conservation, landscape conservation, and planetary survival. The precedent was set by the Council of Europe (2000), who recognized the importance of landscape in the formation of local cultures and cultural diversity, as a matrix for natural heritage (biodiversity), and the essential role of landscape in promoting human well-being and sense of place. The over-arching goal is to develop an International Landscape Convention protecting both wildlands and anthromes that shape the planet, and to raise awareness of the role created places such as green roof ecosystems play in its conservation.

The Euro–North American origins of contemporary green roof design has skewed project distribution, research, and publications for several decades to work in the temperate biomes of the Northern Hemisphere (Blank et al. 2013). Many practitioners and researchers are now noting that what works in temperate regions is not particularly relevant to the biomes and bioregions (Fig. 15.1) where climate is characterised by extremes of temperature and precipitation (e.g. Waitakere City Council 2007; Williams et al. 2010; Liu et al. 2012; Caneva et al. 2013; Dvorak and Volder 2013).

As the popularity and utility of green roofs spreads around the globe, designers are tasked with placing and relating a green roof design to its specific biome, vegetated roofs are now being installed around the planet from hot and dry savannahs to hot and wet tropical rain forests—and from the cold dry tundra to cold dry deserts. As Krishnan and Hamidah (2012) note, green roofs in temperate regions are designed around the concept of a hundred year rain event which is equivalent to the annual rainfall of around 3000 mm in tropical or temperate rainforest biomes. In these wet biomes an extensive green roof built with a shallow substrate and Euro-American sedums is simply not up to the task of dealing with that much water, when urban infrastructure is unable to cope with the annual monsoon season and increasingly volatile rainfall and typhoon events.

A growing body of research describes testing and selection of indigenous species for green roofs in drier biomes, bioregions, and ecosystems. This includes the Kadas Green Roofs Ecology Centre that opened in 2012 at the University of Haifa in Israel, North American grasslands and prairies investigations (Sutton et al. 2012), in Australia (Godfree et al. 2010), and throughout the Mediterranean (e.g., Benvenuti and Bacci 2010; Caneva et al. 2013; Papafotiou et al. 2013a; Papafotiou et al. 2013b; van Mechelen et al. 2014). In Durban, South Africa the main impetus for designing green roof ecosystems is to make the city more resilient to climate change by contributing to green infrastructure, while protecting biodiversity by using species native to the region (van Niekerk et al. 2011). As the knowledge base and interest in establishing green roof ecosystems grows in all the vegetated biomes, researchers, designers, and policy-makers have opportunities to bring deeper ecological relevance to their home places, spaces, and projects by aligning goals and objectives with the meso- and macro-scale biodiversity and landscape conservation initiatives.

15.3 Diversity

The development of local government policy and guidelines on green roofs and walls articulates opportunities to protect biodiversity in local green roof projects while meeting higher level conservation objectives (English Nature 2003) and supporting efforts to protect human health and well-being (Town & Country Planning Association 2012). Several strong examples have emerged that can inform new policy development elsewhere on the planet (see for example, van Lennep and Finn 2008; Design for London 2008; van Niekerk et al. 2011; Ansel and Appl 2012; City of Melbourne 2013; City of Sydney 2014a, b; STLAi 2013; City of Toronto 2013a,

2013b). However, a general policy statement on the urgency to protect biodiversity within a local political jurisdiction is not detailed enough to make design decisions about how biodiversity and the concepts of diversity relate to the specific scale of the project.

Whittaker (1972) familiarised the concepts of scale within the study of biological diversity. The term alpha diversity (α-diversity) occurs within a small, homogeneous habitat or community. Local conservation efforts regarding it commonly focus on counting the number of species present. Species richness (S), is the term used as a simple index to express the number of species sampled in the study habitat.

Beta diversity (β-diversity) describes diversity between habitats within a landscape. It represents the number of species unique to each community, seen as the turnover in species between a pair of communities. Turnover can also be measured as similarity or dissimilarity in species composition or the fraction of species in common. The greater the β-diversity is between the pair of communities compared with the alpha diversity of each community, the greater the distinctiveness of the two communities.

Gamma diversity (γ-diversity) is the total species diversity observed in all habitats within a region, landscape, or study area. These three measures of diversity are scale dependent, so that the roof patch size a plant ecologist would consider to be one habitat (measuring α-diversity), while the same patch would comprise a mosaic of micro-habitats which ecologists studying soil micro-organisms on green roofs (McGuire et al. 2013) might count as gamma diversity.

As a simplistic demonstration of the three levels of diversity, a grassland and an open woodland are identified as two distinct habitats in a local study area. The number of species present (S) in just the grassland (10) or the woodland (15), counted independently of each other is their α-diversity. The inventory identifies five species common to both habitats and 15 species that differ between them; this value is β-diversity. The total diversity (γ-diversity) in the study landscape is therefore, two habitats and 20 species.

Fisher et al. (1943) noticed that no community exists in which all species are equally common. Instead, only a few species tend to be abundant while less so. Differences in species abundance are of concern to conservation biologists tasked with determining relative rarity of a species. Simple species counts (S) do not take into account rarity, abundance, or domination of species in a community and when abundance matters there are various indices (Magurran 2004, pp. 100–134) that can be used to calculate abundance in relation to diversity, isolation, and size of the study area (e.g., Shannon, Simpson, Gini-Simpson, Fisher's α-diversity), or they can be plotted in species relative abundance curves or species abundance distributions.

While community and species ecologists are usually familiar with three levels of diversity described above, there are three additional measures of diversity related to landscape scale at the macro- and mega- levels (Magurran 1988; Bianchelli et al. 2013). Delta (δ-diversity) is the change in large landscape diversity along major climatic or other physical or abiotic gradients. Epsilon (ε-diversity) is the species richness within biomes or very large bio-regional landscapes. Omega (Ω-diversity) is the measure of global diversity (Pielou 1979).

Researchers infer from species counts that the greater number of species sampled, the more diverse the community. Communities (*sensu stricto*) are composed of assemblages of species occupying distinct trophic groups and taxa that collectively contribute to energy flow, nutrient cycling, and species turnover. Community structure is described by species composition and the abundance or rarity of each species, resulting in richness (S) and diversity (α-diversity).

However, the science of ecology is composed of many sub-disciplines, as demonstrated in the contributions to this volume, with varying levels of interaction and collaboration. Plants are frequently used as surrogates representing all biota when describing ecosystems at many scales—from biomes to habitats. As Kitching (2013) aptly points out though, even when we think we have enumerated all the taxa in our purview, whether plants, insects, animals, or micro-organisms, we are not really studying the whole community or the ecosystem, unless we include all the taxa in a holistic enumeration that includes abiotic and cultural elements. Turning this around and looking at it purely as designers, is it even possible to determine how many species, and what keystone species are essential in the initial roof design to create a functional ecosystem?

Woodward (1994) suggests that the only places where complete measurement of richness and diversity can occur is in the extremely cold and dry biomes; places where there is little social incentive to create green roofs. Green roof policy may explicitly list acceptable plant species and the amount of uniform vegetated roof coverage required for construction permit approval (City of Toronto 2013b). However, this approach actually inhibits both biomimicry and an ecologically authentic design that might consist of a heterogeneous mosaic of patches and gaps in a substrate of varying depths and composition, the result being more of a garden than an ecosystem (Janzen 1998; Del Tredici 2004).

At all landscape scales, ecosystems are highly complex entities that can be studied in terms of structure, function, and processes. Science accepts that most community ecologists specialize and work with limited groups of organisms, focusing for example, on communities of plants, animals, or insects. Some community ecologists restrict their focus even more by studying guilds of species that cycle through life in similar ways such as insectivorous birds, or succulent plant life-forms.

Community though, refers to the collection of all species interacting in a particular space. A single species study may yield a lot of information about the behaviour of that species in an ecosystem, but not a lot about the ecosystem in its entirety. What green roof designers need to create a rooftop ecosystem requires synthesized information about individual species (autecology), their communities, neighbourhoods, and the ecosystem processes that make them function. In the absence of adequate scientific information to guide roof ecosystem design and to defend it during the design process, we recommend that the precautionary principle be applied. Tinkering with the selection of species that will populate the rooftop ecosystem without adequate knowledge could lead to sinks or traps (Sect. 15.4.1), prevent recruitment, or enable incorrect management decisions (Sect. 15.8).

15.4 Recruitment

Recruitment occurs when new individuals form a population or are added to an existing population (Harper 1977). The term is generally used to refer to a life stage where an observer can visibly detect that individuals have become established and the number of individuals in a population has increased.

The recruitment, establishment, and distribution of individuals in a population are influenced by spatially and temporally heterogeneous environmental conditions (MacArthur and Wilson 1967; Tilman and Kareiva 1997) while exhibiting dynamic traits (Royma 1992). Because communities are composed of populations, it is essential to understand the rise and decline of populations, and the factors contributing to these dynamics. The heterogeneity of the world's mosaic of landscapes challenges generalization, so it is critical that any studies of populations and population processes are conducted at the appropriate spatial scale, and any inferences made should relate to similar scales (Levin 1992; Pickett and Cadenasso 1995).

With regards to green roofs, we currently understand very little about population dynamics, particularly at the trophic level. Planting design as a gardening exercise may place plants out-of-context with other plant species in the design, resulting in a hybrid assemblage that does not for example, contribute to a biodiversity objective of increasing habitat for pollinators (see Sect. 15.7). Good roof ecosystem design considers the interactions between species, and their effect on the ecosystem. The introduction of predators can prevent or reduce competitive exclusion, and the introduction of parasites will influence interspecific interactions. For example, a keystone predator increases community diversity by foraging on the most abundant species, which benefits less abundant prey species (Paine 1966).

In a healthy-functioning ecosystem all species rely on each other and work together symbiotically, commensally, or mutually. Keystone species (Mills et al. 1993) are considered those essential to the vitality of the ecosystem and to the way all the species interrelate within the ecosystem. When a keystone species is removed from a habitat the habitat is changed, which affects all other species. The concept of keystone species is one approach to roof design that is centred on creating roof habitat for priority species for conservation.

Designing roof habitat for a keystone species will have an effect on the long term structure, functioning, and management of the ecosystem that evolves on the roof. Accomplishing the task of identifying keystone species for roof ecosystems though, is not a simple process. Determining keystone species can be difficult as many have noted (Power et al. 1996), particularly if little research has been published on an ecosystem. Ecosystems are in general complex entities that are changeable in time and variable in space (Power et al. 1996).

A species can be keystone under certain environmental conditions such as a hot dry year, but insignificant in a wet season. A species such as Garry oak (*Quercus garryana*) is the keystone defining Garry oak ecosystems at the meso-scale, but can be completely absent from the ecosystem at the nano- or pico- scales (Dunster 2007), which ultimately influences how a rooftop ecosystem is designed when located within the bioregion. The role and influence of species in community, ecosystem, and landscape must be considered when introducing species onto a roof, either as individuals or as assemblages within a habitat.

15.4.1 Sources, Sinks, and Traps

While a green roof should never be used to justify habitat removal on the ground, the potential to better design green roofs as substitute or surrogate habit to improve or increase biodiversity and fulfil local, regional, and national obligations under the CBD, presents opportunities for designers to think and act beyond aesthetics, site engineering, and green infrastructure (Berardi et al. 2014). The goal of creating high quality habitat in which species are successful in reproduction provides the greatest opportunities for a population to become a source of seeds or individuals that can disperse to other habitats and increase metapopulation size (Keagy et al. 2005; Akçakaya 2007). Forecasting scenarios of how the roof ecosystem will evolve from the initial design can guide long-term management to avoid the roof becoming a trap leading to unsuccessful breeding, or becoming a sink (Dias 1996) that causes local extinctions of species.

The identification of sources and sinks is complicated by temporal and spatial variability, density dependence in population demographics, and propagule dispersal (Dias 1996). Geographical adjacency of metapopulations of the same species avoids geographical and functional isolation and possible genetic bottlenecks (Akçakaya 2007). This requires the designer to move away from a single roof approach to design and instead, consider the location and role of the roof within the broader urban landscape.

If for some reason isolation is critical to protect a species then the roof must be designed as an island, which requires integration of green roof goals and concepts between roof projects to ensure that adjacent roofs do not act as sources for competing species that can disperse onto the isolated roof. However, if avoiding genetic bottlenecks is critical to survival of a species, then connectivity and linkages must be considered not only in terms of the metapopulation of roof habitat patches across the landscape, but also in terms of the 3-dimensional or vertical structuring of the entire urban landscape.

Urban planning can address this by changing policy from approving nano-scale green roof projects on a piecemeal basis to planning an urban rooftop landscape that maintains both proximity and connectivity to other roofs with similar habitats. Holistic urban planning for biodiversity requires integration of the rooftop landscape with the landscape on the ground. The population on one roof can then become a source, created by the outflow of individuals immigrating to other populations (Dias 1996), whether on another roof or the ground.

For vagile species this enables unimpeded movement among roof source patches by providing green corridors connecting the metapopulation of roof ecosystems across the landscape. It also provides access to different habitat elements not provided on the roof such as water bodies, large perching or nesting branches in the street tree canopy, or canopy cover to protect a species from predators or the weather.

Some species, in particular plants, lack the ability to disperse distances unassisted and the gaps between roofs present significant barriers to reaching suitable new habitat. If deaths in a population exceed births it becomes a sink and leads to extirpation. Preventing extirpation requires that the rate of successful immigrants into the population exceeds those dispersing away (emigrating) from the population (Schreiber and Kelton 2005). Immigration and emigration will be greatly facilitated by urban planning that integrates green roof designs for biodiversity across the landscape.

Ecological traps are situations where rapid environmental change, such as habitat loss due to land development, leads organisms to establish in poor quality habitats. The thinking stems from the idea that organisms actively selecting habitat must rely on environmental cues to help them identify high quality habitat (Robertson and Hutto 2006). When the environmental cues leading them to high quality habitat are lost, they are attracted and then trapped in habitat that is less suitable for reproduction and survival. Robertson and Hutto note the consequence is a change in preference to poor quality but falsely-attractive habitat, while avoiding high quality but less attractive habitat.

The implications of ecological traps for roof ecosystem designers are two-fold. Firstly, if suitable high-quality habitat is degraded, altered, or removed on the ground, organisms will be searching for alternative sites. Knowledge of the habitat requirements and what constitutes high quality habitat for these species is paramount to creating suitable new high-quality habitat whether on the ground or on a rooftop (Gilroy and Sutherland 2007). To avoid creating an ecological trap, designers must avoid altering the environmental cue set (falsely increasing its attractiveness), increase the suitability of a habitat, or do both concurrently.

15.4.2 Time and Habitat

Temporal aspects of habitat quality are also important to habitat selection and use by both reproducing adults and the individual life stages of an organism (Orians and Wittenberger 1991). High quality habitat must provide the right resources and structural elements at the right time in a life-cycle to initially attract a species, as well as continue to provide the same quality resources and elements in the future.

Green roof ecosystems have great potential to become high quality replacement habitat for some species for either specific life stages or the entire life-cycle (Kadas 2006; Olive and Minichiello 2013), and current research includes invertebrates (Jones 2002; Brenneisen 2005; Coffman and Davis 2005; Cantor 2008); birds (Duncan et al. 2001; Gedge and Kadas 2005; Baumann 2014; Fernandez-Canero and Gonzalez-Redondo 2010); bats (Pearce and Walters 2012); and reptiles (Davies 2010), as well as plants. Extensive green roofs designed as high quality habitat for a specific species are more likely to hold their attractiveness and habitat qualities. Management regimes and the roof microclimate ensure unwanted species such as volunteer trees and shrubs do not colonize, establish, and replace the desired species.

15.5 Establishment

Green roofs are built to foster, and sometimes exclude, certain types of life forms. As roof ecosystem designers we intentionally respond to the building programme requirements of architects and engineers, and consider the relationship of the roof to surrounding environmental site conditions such as microclimate (wind, sun exposure, precipitation).

We are conditioned to visualize, design, and construct roof ecosystems in layers, from 'parent material' up through the vegetation and faunal communities. For example, the FLL guidelines are premised on the idea that 'growing course' depth (soil and substrate) dictates the plant community (FLL 2008). It states that < 12 cm of growing course will allow sedums, mosses, and herbaceous plant growth, while excluding grasses, shrubs, and trees which require deeper substrate. Guidelines become problematic when they are embedded as non-negotiable requirements in local policy. Guidelines simply set out the minimum tolerances for successfully establishing plants on roofs, which has very little to do with creating a roof-top ecosystem.

For example, the City of Toronto (2013a) adopted a green-roof bylaw requiring adherence to ASTM standards, which leaves little flexibility for experimenting with ecological design techniques using for example, lower technology natural materials (Dunnett et al. 2011). The City of Melbourne (2013) however, takes a different position and sees its role as an enabler of green roofs, supporting the creation of space for experimentation by ensuring local bylaws and the building permit process do not become regulatory barriers.

For all species, but particularly for species at-risk that are declining in population size due to on-the-ground habitat loss, the imperative and primary challenge to green roof designers is to create habitat refugia that will contribute to the conservation and long-term survival of the species on the roof. This is what Davies (2010) for example, has successfully accomplished with at-risk endemic reptiles in New Zealand. To do this effectively, we need to know the environmental factors that allow the species to establish successfully whether on the ground or on a roof. We need to know how species adapt to a roof environment with such resilience and fecundity that 'next generations' can disperse off-roof as sources for population expansion to other safe sites. And, we need to know why some species are never successful at establishing viable populations on rooftops.

15.5.1 Facilitation and Competition

Key to any decision to create successful habitat for any species on a green roof is understanding the biotic and abiotic factors that influence both population dynamics in general and survival of individuals in specific habitats. This is something Oro (2013) points out is a difficult challenge for all population dynamics researchers, and there is insufficient data for many species. Oro also notes that the hidden components (such as dormant seeds) of populations and population dynamics have received less attention in the research, yet may be central to understanding why populations persist or fail.

Facilitation occurs when interactions between plants produces greater plant fitness or abundance than when plants grow alone (Bruno et al. 2003). Within time and space, facilitation and interactions between plants may vary seasonally with resource availability, life cycle stage, and whether the plant is dormant or active (Connolly et al. 1990).

Facilitation can be interspecific or intraspecific and generally involves competition for the same resources, such as light, water, or nutrients, in the ecosystem. Interactions can be above and below ground (Harper 1977). Intraspecific interactions involve individuals of the same species. Interspecific interactions are generally competitive, in which different species compete for the same resources. If the resource cannot support both species, one will thrive to the detriment of the other.

Competition between individuals of the same or different species can be direct or indirect, and positive or negative, and may vary in time and space. For example, the establishment and survival of a plant can be facilitated by the presence of other plants via nutrient sharing through common mycorrhizal networks (Hartnett and Wilson 2002). Taller plants may out-compete shorter species for sunlight. Between species this may be positive and facilitative if the shorter species requires more shade and the taller plant acts as a nurse, and negative if the shorter species is out-shaded and unable to photosynthesize (Harper 1977). Interspecific resource competition may alter the time it takes for a plant to reach reproductive maturity, impede its growth rate, and decrease seed production (Connolly et al. 1990). These factors contribute to the amount of seed fall, the density of seedlings, and seedling survival or mortality rates (Metz et al. 2010).

Many factors are considered when selecting individual plant species and species assemblages for a green roof ecosystem design. A basic understanding of a species' ecology and the environmental resources it requires during its different life stages is paramount to good design using compatible species. Species that are in direct competition should be avoided. Species that can partition site environmental resources are less likely to competitively exclude others, which would result in costly extirpations from the roof. Basic species knowledge should also be used to determine which species positively or negatively modify the roof environment to enable succession and the establishment of new species.

15.5.2 Safe Sites for Seeds and Seedlings

More often than not with conventional green roof design, recruitment refers to population increase by clonal offspring, which limits genetic diversity. In most ecosystems though, and in ecological design, the most common means of plant recruitment is by the addition of seed and seedlings to a population. Seedling recruitment includes three fundamental stages: seed germination, seedling survivorship, and seedling growth. Each of these stages depends entirely on the deposition of a seed in a safe site (*sensu* Harper 1977), a pico-scale site where it can germinate and survive threats such as predation and hostile environmental conditions such as excessive heat, light, or moisture.

The characteristics of a safe site vary from species to species. Germination may be controlled by a combination of a few or many interacting factors including light levels, presence and depth of leaf litter, proximity to allelopaths, soil qualities (organic matter, chemistry, moisture, pH, texture), and microtopography (Ahlgren and Ahlgren 1981; Beatty 1984; Fowler 1988). Microtopographic variation) Chap. 10) can be formed by abiotic events such as rain-impacted depressions, and wind and rain moving substrate to create hummocks, hollows, ridges, and rough or smooth surfaces. When seed supplies are limited or costly, knowledge of safe site qualities to optimize germination is an important consideration for roof ecosystem designers. For example, the flat-shaped seeds of common yarrow (*Achillea millefolium*), a species often used in biodiverse roof plantings, have high germination rates on moist soils when sown on an exposed smooth surface, while on dry soils are most successful when sown into soils with 3–20 mm (0.1–0.75 inch) striations (Oomes and Elberse 1976).

Biotic events also influence the formation of microtopography. For example, the weight of an ungulate foot creating a depression, a broken branch forming a windbreak or barrier to predation, an animal digging hole, the death of a plant creating a small light patch, or an upturned root ball exposing bare mineral soil. Studies have demonstrated that livestock grazing and small animal herbivory can enhance seedling recruitment (Oesterheld and Sala 1990; Milton et al. 1997; Rapp and Rabinowitz 1985; Edwards and Crawley 1999), though it may be unlikely that hoofed grazers will be employed on green roofs. Eriksson and Eriksson (1997) found that small-seeded species germinating in the autumn responded to disturbance by displaying enhanced seedling recruitment

For some species, seed production and viability is not lowered or disrupted by a poorer quality safe-site environment; the problem is seedling survival following emergence. For example, in England the habitat conditions required for seedling establishment of grey clubawn grass (*Corynephorus canescens*) are not synchronised with the actual period of seedling emergence and growth (Marshall 1968). A lack of safe sites disrupts the seedling life stage and results in a decline of new recruits to the population.

The seedling life stage forms a threshold between the dormant seed and the developing plant. Seedling survival is essential for recruitment of new individuals into a viable population consisting of juveniles, vegetative adults, and reproducing adults. The length of time a site must retain its safe characteristics is also variable depending on how much time a seed requires to break dormancy, germinate, and develop into a seedling. Seedling density can induce competition for limited resources and attract herbivores which then become factors in determining survivorship, as self-thinning removes cohorts from the seedling population which influences recruitment and regeneration.

Designers of green roof ecosystems that are planted by seeding should pay close attention to creating a roof environment that optimizes germination for the various species used (Sutton 2013). This may mean creating customized safe sites for each species and seeding individually rather than as a mix of blend suitable for a generic soil environment. Roofs left to self-organize without active management should be monitored annually if the ecosystem is lacking the disturbance processes required by some species to initiate seed germination. Without disturbance, it should be no surprise if some species disappear and are replaced by others.

15.6 Disturbance

Not all seeds germinate at the same time; some remain dormant and gravitate to the seed bank that forms below the soil surface. Attrition from the seed bank may be a result of fungal attacks, predation, decay initiated by excessive heat, cold, or moisture, and loss of viability due to age (Harper 1977; Fenner and Thompson 2005). Viable seeds remain in the seed bank until environmental conditions trigger germination. Conditions for breaking dormancy vary from species to species (Baskin and Baskin 1998). Exposure to light though, is a common requirement that requires stochastic soil disturbances to coincide with the optimum seasonality for germination and seedling survival.

Disturbance events are important to the dynamics and regeneration of populations, ecosystems, and landscapes. If the roof planting is to move from a garden to a self-organizing ecosystem the designer must know and understand what agents of disturbance influence the reference ecosystem. Watt and Gibson (1988) and Schläpfer et al. (1998) suggested that germination and seedling establishment in grassland ecosystems is higher in gaps created by trampling than in the surrounding vegetation. Simkin et al. (2004) found that the tunnelling and mound-building activities of pocket gophers in a savannah created small-scale disturbances that changed the composition and availability of soil nutrients and altered the micro-climate. He notes that other researchers have found that species in different grassland ecosystems vary in their ability to capitalize on animal disturbances, and concurs with Rapp and Rabinowitz (1985) that sometimes there are no measurable ecosystem benefits such as enhanced seedling establishment.

Collins (1987) experimented with combinations of fire and grazing disturbances in tall-grass prairie and noted a variety of effects including a positive correlation between disturbance intensity and species richness. He also found that grazing increased the number and cover of annual species in general, and diversity was significantly increased when grazed land was burned. Burning significantly reduced species diversity in un-grazed prairie, though burning prairie-based green roofs remains uncommon (Sutton 2014).

If the goal for a roof is to create a species-rich, biodiverse ecosystem then disturbances (mechanical, manual, or natural) should be planned for. Scale comes into play as the size of the roof (and insurance policy) may exclude a catastrophic disturbance event such as a fire (Keane et al. 2004), which would also risk extirpating species intended to be protected in the roof environment. The intermediate disturbance hypothesis (IDH) predicts that local species diversity is maximized when ecological disturbance is neither too rare nor too frequent (Roxburgh et al. 2004). Communities experiencing moderate levels of disturbance will have higher species richness than communities experiencing either smaller or greater amounts of disturbance because both competitive (K-selected) and opportunistic (r-selected) species have better opportunities to co-exist.

The green roof literature (research and trade) generally refers to disturbance in terms of selecting species that are drought-tolerant or wind-firm (e.g. Oberndorfer et al. 2007). Determining how much, how often, and what type of disturbance is best for a particular roof ecosystem requires familiarity with the reference ecosystem and the species selected for the initial plantings that will form associations, communities, and in time a fully functional ecosystem. The roof designer should also be aware of whether one reference ecosystem disturbance type can be substituted with another (e.g. mowing replacing grazing).

If roof disturbances are anticipated at the pico-scale, for example, attention must be paid to whether the vector of change will be abiotic (e.g. wind) or biotic (e.g. earthworms). Wind-firmness would then be less of a concern than placing plants that will uproot, in the path of prevailing winds to create more open patches for regeneration. If worm action is essential to the functioning of the roof ecosystem, then the design must accommodate deeper well-drained soils that create the best habitat for earthworm survival.

If protection and cover from wind disturbances is important for any species inhabiting the roof, placing plants that are wind-firm directly in the path of prevailing winds may be the prescription. Finally, if a species-rich, biodiverse roof ecosystem is not the overarching goal, then a lack of mechanical or manual disturbance may not matter, as a self-organizing roof ecosystem will adapt by regenerating from any natural disturbance.

15.6.1 Regeneration

Following a disturbance an ecosystem undergoes a process of reorganization and renewal in which the biota adjust to changes in their environment. This dynamic process has been termed "self-design" Odum (1983) or "self-organizing" (Odum and Odum 2003). Depending on the nature of the disturbance some species may disappear, or become more or less dominant as they adapt to the new conditions. Self-design encapsulates the change in species composition that affects structure, function, pattern, and processes within the ecosystem (Mitsch 2004).

Regeneration within an ecosystem is reached when it completes self-organization and ecosystem processes and functions have resumed sufficiently to support and sustain the biotic components. The time-scale for regeneration depends on the magnitude of the disturbance and whether the ecosystem must go through primary or secondary succession. When an ecosystem has regenerated it should be more resilient to non-catastrophic environmental stresses and disturbances. As abiotic and biotic functions and processes resume, a regenerated ecosystem will reconnect to the larger contiguous landscape and resume the dispersal and exchange of biota.

Almost all ecosystems have been disturbed to various degrees over various time-scales by anthropogenic activities. In some circumstances natural disturbances (such as fire and flooding) that trigger regeneration have been removed or suppressed to protect human populations and activities. In these instances the historical trajectory the ecosystem might naturally have followed has altered course. Understanding how humans have altered an ecosystem before attempting to introduce a replicate onto a roof is an important step in determining whether management will be required to assist regeneration. To accomplish this, we use reference ecosystems.

A reference ecosystem is an existing ecosystem that provides guidance to establish design and management goals when planning a green roof ecosystem project. The ecosystems found in parks and nature reserves often serve as tangible single-source references. The loss of many ecosystems and habitats often means the designer must turn to written and visual (photographic) descriptions to replicate a historical reference. In highly disturbed and fragmented landscapes the reference ecosystem description may be drawn from multiple sites and from many other historical sources. Given the dynamic nature of ecosystems, the multiple site reference description maybe more useful to designers than using the baseline inventory from a single site that captures a brief and static moment in the life of the ecosystem.

The designer must anticipate the moment to release control and allow self-design within the roof ecosystem to commence with occasional interventions to improve ecosystem health and integrity. Maintaining control is an acceptance that our efforts are a form of gardening as defined by Del Tredici (2004). However, Janzen (1998) points out this is not such a bad thing because gardening nurtures the ecosystem we have had a hand in creating and acknowledges the human role in protecting biodiversity.

15.6.2 Dispersal

Dispersal is the ecological process that causes gene flow through space and time (Ronce 2007). Individual living organisms disperse away from their birthplace to another favourable habitat, leaving behind their parent population. The ecological benefits of dispersal include a broadening of the species distribution, and avoidance of in-breeding and overcrowding which improves both individual and population fitness (Hanski 1999).

For plants, the dispersal of propagules (fruit, seeds, spores, and vegetative parts) happens by chance, without any control by the parent plant or the propagule. Plants have five main types of dispersal mechanisms: gravity, wind, ballistic (self-propelling), water, and animals. The type of mechanism controls the distance the individual moves through time and space, and the pattern of its dispersal. For plants, dispersal can be considered successful when a propagule lands in a safe site, germinates, and produces a seedling that matures into a reproducing adult.

Migration is a form of intentional dispersal when large numbers of an animal species move together from an unfavourable habitat to a more favourable one. The distance travelled by a species is known as the dispersal range. Migration can be one way, two way, or multiple return journeys in the case of long-lived birds moving between breeding and over-wintering grounds. Migrations can be daily, seasonal, routine, or once-in-a-lifetime (Van Dyck and Baguette 2005). Depending on the motility of the species, migration can cover short or long distances in directions that lead to or from suitable habitat for feeding, breeding, or resting. Ecological neighbourhoods can be very small or very large.

The act of dispersal involves three phases: departure, transfer, settlement, and at each phase there are different time, energy, risk, and opportunity costs and benefits (Bonte et al. 2012). Dispersal is essential for both the survival of individual species, and for maintaining ecosystem processes. Dispersal is also critical to the conservation of species in fragmented and altered landscapes (Kareiva and Wennergren 1995; Hanski and Ovaskainen 2000). For plants, attention must be given to the needs of both the plant and the vector, particularly if the vector is an animal.

For example, a roof design might include fleshy-fruited plants that are dispersed by frugivores. The design must provide habitat elements and an abundance of fleshy-fruited plants that will attract frugivores to the roof (Rey and Alcántara 2014). If frugivore dispersal is limited by landscape and meta-population fragmentation, then the designer should either reconsider the utility of including the plant if it cannot be dispersed, or promote metapopulation design that fills the gaps in the landscape with additional roof ecosystems offering similar habitat elements.

When landscapes become fragmented and altered, the distances between the sub-units of a metapopulation may deter the movement of individuals. The distribution of species and exchange of genetic material across the landscape is disrupted, potentially leading to a population decline locally in the sub-unit, and regionally within the meta-population (Hanski 1999). The designers of rooftop ecosystems must understand how an individual roof will improve and sustain the capacity of a metapopulation by reducing dispersal distances between sub-units, or by adding new favourable habitat that will serve as a source rather than a trap or sink.

15.7 Novelty and Human Agency

> The cultural landscape is fashioned from a natural landscape by a culture group. Culture is the agent, the natural area is the medium, the cultural landscape is the result. Under the influence of a given culture, itself changing through time, the landscape undergoes development, passing through phases, and probably reaching ultimately the end of its cycle of development. With the introduction of a different—that is, alien—culture, a rejuvenation of the cultural landscape sets in, or a new landscape is superimposed on the remnants of an older one. (Sauer 1925, p. 5)

In earlier sections of this chapter we introduced the urban anthrome as the environment in which most green roof projects are situated. We also explained why landscape scale is important, and why awareness of the ecological role a roof occupies in time and space will help designers place it in a scale appropriate to its place in the landscape. The urban anthrome has evolved from human (cultural) activities shaping natural elements in the landscape.

Ecological restoration is the attempt to repair and assist in the recovery of ecosystem structure, function, and processes that have been degraded, damaged, or destroyed by cultural activities (SER 2004). Formalizing the role green roof ecosystems can play within ecological restoration initiatives will give designers focus on how to improve biodiversity and connect the services provided by the roof ecosystem to the larger landscape.

While some designers are inclined to categorize green roof ecosystems as novel ecosystems *sensu* Hobbs et al. (2006), they do not fit neatly into the definition. Because humans create new species assemblages within green roof systems, green roofs are novel in the sense of being new, innovative, and different. However, the novel ecosystems Hobbs et al. (2006, 2013) envisioned are created by cultural activities and exhibit (1) novelty: new species combinations that change ecosystem functioning, (2) human agency: ecosystems as a result of deliberate or inadvertent action, and (3) self-organization: having the capacity to function and evolve without further human intervention.

Consider this scenario: a forest first converted to agriculture and actively cultivated for decades is abandoned due to improved technology allowing the same food crop to be cultivated more intensively. The initial rural cultural landscape evolves into an urban anthrome. Secondary succession commences in the abandoned area and includes a rich assemblage of both indigenous and introduced species that combine and stabilize into a functioning woodland that has adapted to local biogeoclimatic conditions. Trees grow larger while facilitating the arrival of new species that inhabit the shady understorey layers. Animals and insects find food, cover, and habitat that supports reproduction. An old tree dies and the gap is colonized by early successional plants. All this happens without human intervention.

It is unlikely that the scenario described above can occur on a green roof in temperate biomes dominated by trees and shrubs because of the need to protect the building from potential damage caused by plants and animals that are left to self-organize (Chap. 1). Biomes lacking trees and shrubs such as tundra, deserts, and grasslands have greater potential to become self-organizing ecosystems that meet the novelty criteria. The underlying question though, is whether it really matters.

Humans have been altering and adapting ecosystems for their personal use and are not the only species to do so. The spatial scale of alteration ranges from gardening for food, cultural, or ceremonial purposes at the nano-scale, to complete alteration and destruction of topographies at the hand of war or resource extraction. Rather than fixating on the semantics of novelty and novel ecosystems and whether the terms can apply to green roof ecosystems, it is more productive to focus on their fit within the anthropogenic landscape and their contributions to biodiversity, ecosystem services, and the fulfilment of socio-cultural needs.

Higgs (2012) fears technological patterning, the presumption that we can fix the original ecosystem, and thus move vigorously away from preservation and conservation of remaining intact ecosystems. This concern for the morality of restoration rationale echoes throughout the restoration ecology literature (e.g. Katz 1992; Throop 2000; Hilderbrand et al. 2005), and serves as a reminder to green roof ecosystem designers that one manufactured for a roof is no substitute for the one destroyed on the ground to create the building footprint that enables installation of the green roof ecosystem.

Hallet et al. (2013) explain that novel ecosystems offer an unrecognized way to situate humans within ecosystems and also allow for ecological functioning and evolution. Vegetated roofs are not so much a technological pattern moving away from conservation, as they are a new way to approach conservation. Construction and subsequent conservation of a new ecosystem on a roof is one more tool in the conservation spectrum that ranges from outright preservation (prevention) to restoration (treatment) of ecosystems.

Extensive green roofs intentionally exclude humans from direct physical contact, which may be a good thing when trying to protect certain species within an urban environment. However, green roof ecosystems offer new opportunities to meaningfully reconnect people with nature both passively (visual access) and actively (physical access). The cost to engineer roof loads to allow access ultimately may be a small cost to the developer, and a major benefit to society.

As Louv (2011) points out, with the global shift in human population from rural to urban and an increasing reliance on technology to meet our needs, conservation 'out there' is no longer enough. He argues that for the sake of human health and well-being, we must also restore or create natural habitats wherever humans live and work. This Louv explains, includes cities, rural fringes and farmlands, commercial and industrial areas, neighbourhoods, front and back yards, and rooftops. Protecting the biodiversity that all living creatures need, including humans, requires a bold effort to incorporate the total landscape in conservation plans. The reconnection of humans with nature and protection of biodiversity for the sake of the plant and all its inhabitants contextualizes why we really need green roof ecosystems.

15.8 Management for Biodiversity and Conservation

Many of the roofs studied in Europe over the last half century are roofs that have been left to evolve with minimal or no maintenance, and this hands-off approach helped form the definition of a classical "extensive style" of roof greening (FLL 2008); Köhler (2006) and more recently Thuring and Dunnett (2014) studied the vegetation composition and ecosystem service performance of nine extensive green roofs created in Germany between 1977 and 1991. The roofs were selected because initial maintenance contracts lapsed, leaving a ten-year period of self-organization. Human neglect allowed dynamic ecosystem processes to develop whereby spontaneous colonization was facilitated in plant communities, resulting in species turnover.

The major obstacle to allowing green roof ecosystems to self-organize without further human intervention is the fear that a laissez-faire approach will transfer to neglect of the building fabric, and potentially compromise the integrity of the building. Avoiding risk necessitates the need for monitoring the drainage and protection layers on a roof the roof ecosystem. The next step in designing green roofs as ecosystems is to develop language that permits plantings to develop into self-organizing ecosystems without maintenance, provided no harm is done to the structure,

Rooftop maintenance typically consists of prescribed and often third party maintenance regimes. Usually several times a year tasks such as inspecting for damage and roots penetrating the membrane, removal of blown-in debris, mowing and cutting, and weeding or removal of undesirable, dead or dying plants are carried out. Maintenance also includes inspecting and prescribing treatments for disease and pests, compacted or eroded growing medium, and determining any fertilizing needs. A schedule of these horticulture-based activities essentially keeps the roof plantings performing efficiently and according to a static design. Such schedules are frequently adopted in policy and guidelines as the industry performance standard for maintenance of all green roofs.

There is no mention in the literature that anything different than the maintenance norm currently occurs for roof designs with biodiversity goals. This is a significant short-coming of those researchers pioneering the design of green roof ecosystems. For example, in the case of biodiverse roofs typical maintenance activities such as weeding or soil replenishment may be counter-productive to the goals of naturalizing the rooftop for insects requiring bare patches or variable microtopography such as hummocks and hollows. Maintenance as usual may have unintended or devastating results.

As green roof designs become more driven by biodiversity goals, maintenance must be driven by long-term management plans. The moral obligation to do so is particularly crucial when at-risk species find refuge on the roof, or are provided surrogate habitat that becomes a sink or trap (Dunster 2007). Despite the age of some green roofs little attention has been given to ensuring the roof ecosystem will be managed and protected beyond the end of life for the building. Language needs to be developed that addresses how the roof ecosystem will be removed, salvaged, or transplanted if the building is demolished.

In the public realm, legal designation in England to protect rooftop ecosystems as local nature reserves (LNR) does have precedent and much promise for other jurisdictions to replicate (Natural England 2014a). LNR are part of a national framework for legal protection of landscapes and biodiversity at all scales including national parks, national nature reserves, conservation areas, sites of special scientific interest (SSSI), geological monuments, local nature reserves, and local educational nature reserves. The LNR precedent is the Sharrow School Green Roof (Chap. 10) designated as a LNR in 2009 by Sheffield City Council (Natural England 2014b) after meeting habitat action plan (HAP) targets for biodiversity values on green roofs identified by the Sheffield Local Biodiversity Action Partnership (SLBAP 2010).

The 2000 m^2 Sharrow roof was designed on three levels and provides space for locally and regionally identified priority habitats (SLBAP 2010). Habitats include local herb-rich limestone grassland, urban brown field meadows using construction rubble, unmanaged areas encouraging natural colonisation, and a small wetland area with birch woodland surrounding a wildlife pond. A management plan provides guidance for conservation activities on the Sharrow roof to meet three general objectives of (1) maintaining distinctive ecological qualities for each habitat area, (2) preventing aggressive species from dominating and displacing others, and (3) undertaking regular surveys botanical and faunal surveys.

Management plans serve as operating manuals for roof ecosystems. They articulate the significance of the roof ecosystem and the ecological values (species, habitats, communities, functions) it was designed for. The plan details why the roof ecosystem matters, what is happening and could happen to it, and the guiding principles that frame management activities over the short- and long-term. If management is prescribed to allow a self-organizing ecosystem to develop without intervention, this is clearly articulated to avoid confusion if building ownership or management changes. It includes a more detailed work program describing how and when ecological significance will be retained and adapted through access, monitoring, maintenance, repair, restoration, and research activities.

15.9 Concluding Remarks

In this chapter we have described scale as a spatial and temporal phenomenon central to ecology. Understanding how scale relates to landscape pattern, ecosystem structure and dynamics, and species' behaviour is fundamental to embarking on a roof ecosystem design project. Contextualizing how the roof fits, or is placed into the time-space continuum will guide habitat design, initial species selection, and plant assemblages, as well as determine the role the roof can play in the conservation of biodiversity. Scale also plays a key role in the recruitment, establishment, regeneration, and dispersal of species on a roof, which provides guidance for good management practices.

Putting together all the principles, concepts, and techniques described in this chapter to create a roof ecosystem requires skill born of experimentation and

Fig. 15.2 Neighbourhood location of study roof (A1) and a closer view (A2) of the study roof. (Image: Google Earth)

experience, a hallmark of many European green roof projects. The biodiverse roof featured in Fig. 15.2 was designed and installed on a commercial building near Tower Bridge by the Green Roof Consultancy in London, England to increase invertebrate and other species habitat lost at the ground level. Key ecological features are indicated in Fig. 15.3 and described in Table 15.2. The closer view (A2) in Fig. 15.2 depicts many of these features and patterns from above. While the roof is very near a major source of fresh water (River Thames), no water is provided as a habitat element on the roof. If monitoring of the roof indicates water should be provided, a hose bib on the roof is available to add and maintain this element.

Fig. 15.3 View of the study site at Tooley Street, London SE1. See Table 15.2 for description. (Image: K Dunster)

Table 15.2 Ecological features and elements of the Tooley Street green roof (Figs. 15.2 and 15.3)

Area	Feature	Elements
A	Microtopography	
A1		Ridge
A2		Rock pile/hummock
A3		Hollow
B	Microclimate	
B1		Windbreak
B2		Shade
C	Substrate	
C1		Bare mineral soil—substrate inoculated with native soil
C2		Crusted soil with mosses & lichens
D	Micro-habitat Elements	
D1		Rock piles—cover, breeding
D2		Logs and brush piles—cover, food, breeding
D3		Artificial structure: bug wall
D4		Patches and patterns
D5		Water (not provided)
E	Recruitment/Regeneration	
E1		Spontaneous colonization (*Cytisus scoparius*)
E2		Planted plugs (*Sedum* spp.)
E3		Seed heads allowed to scatter & disperse, or be collected and hand broadcast on this roof or others
F	Dispersal	
F1		Dispersal direction off roof to adjacent green roofs below

References

Addicott JF, Aho JM, Antolin MF, Padilla DK, Richardson JS, Soluk DA (1987) Ecological neighbourhoods: scaling environmental patterns. Oikos 49(3):340–346
Ahlgren CE, Ahlgren IF (1981) Some effects of different forest litters on seed germination and growth. Can J For Res 11(3):710–714
Akçakaya HR, Mills G, Doncaster CP (2007) The role of metapopulations in conservation. In: Macdonald D, Service K (eds) Key topics in conservation biology. Blackwell, Oxford, pp 64–84
Ansel W, Appl R (2012) Green roof policies—an international review of current practices and future trends. International Green Roof Association (IGRA), Nürtingen, Germany. http://www.igra-world.com/images/news_and_events/IGRA-Green-Roof-Policies.pdf. Accessed 30 April 2014
Antolin MF, Addicott JF (1991) Colonization, among shoot movement, and local population neighbourhoods of two aphid species. Oikos 61:45–53
Baskin CC, Baskin JM (1998) Seeds: ecology, biogeography, and evolution of dormancy and germination. Academic Press, San Diego
Baumann N (2014) Ground-nesting birds on green roofs in Switzerland: preliminary observations. Urban Habitats 4(1):37–50. http://www.urbanhabitats.org. Accessed 26 July 2014
Beatty SW (1984) Influence of microtopography and canopy species on spatial patterns of forest understory plants. Ecology 65(5):1406–1419
Benvenuti S, Bacci D (2010) Initial agronomic performances of Mediterranean xerophytes in simulated dry green roofs. Urban Ecosyst 13(3):349–363. http://dx.doi.org/10.1007/s11252-010-0124-129. Accessed 13 March 2014
Berardi U, GhaffarianHoseini AH, GhaffarianHoseini A (2014) State-of-the-art analysis of the environmental benefits of green roofs. Appl Energy 115:411–428. http://www.sciencedirect.com/science/journal/03062619/11. Accessed 15 April 2014
Bianchelli S, Gambi C, Mea M, Pusceddu A, Danovaro, R (2013) Nematode diversity patterns at different spatial scales in bathyal sediments of the Mediterranean Sea. Biogeosciences 10(8):5465–5479
Blank L, Vasl A, Levy S, Grant G, Kadas G, Dafni A, Blaustein L (2013) Directions in green roof research: a bibliometric study. Build Environ 66:23–28
Bonte D, Van Dyck H, Bullock JM, Coulon A, Delgado M et al (2012) Costs of dispersal. Biol Rev Camb Philos Soc 87(2):290–312
Brenneisen S (ed) (2005) The natural roof (NADA)—Research project report of the use of extensive green roofs by wild bees. University of Wädenswil, Wädenswil, Switzerland, 21 p. (English Translation by H. Waldbaum, D. Gedge (ed))
Brown JH, Valone TJ, Curtin CG (1997) Reorganization of an arid ecosystem in response to recent climate change. Proc Natl Acad Sci U S A 94(18):9729–9733
Bruno JF, Stachowicz JJ, Bertness MD (2003) Inclusion of facilitation into ecological theory. Trends Ecol Evol 18(3):119–125
Cadenasso ML, Pickett ST, Weathers KC, Jones CG (2003) A framework for a theory of ecological boundaries. BioScience 53(8):750–758
Caneva G, Kumbaric A, Savo V, Casalini R (2013) Ecological approach in selecting extensive green roof plants: a data-set of mediterranean plants. Plant Biosyst 2013. http://dx.doi.org/10.1080/11263504.2013.819819. Accessed 15 May 2014
Cantor SL (2008) Green roofs in sustainable landscape design. WW Norton, New York
Chase JM, Leibold MA (2003) Ecological niches: linking classical and contemporary approaches. University of Chicago Press, Chicago
City of Melbourne (2013) Green roofs, walls & facades policy options background paper. Draft for consultation, June 2013. Growing Green Guide for Melbourne Project, Melbourne, Australia
City of Sydney (2014a) Green roofs & walls policy, 7 April 2014. Sydney, Australia

City of Sydney (2014b) Green roofs & walls policy implementation plan, March 2014. Sydney, Australia
City of Toronto (2013a) Toronto municipal code, Chap. 492, green roofs. Toronto, Ontario
City of Toronto (2013b) Green roof construction standard supplementary guidelines. Office of the Chief Building Official, Toronto Building, Toronto, Ontario
Coffman RR, Davis G (2005) Insect and avian fauna presence on the Ford assembly plant ecoroof. In: Proceedings of 3rd North American Green Roof Conference: greening rooftops for sustainable communities, 4–6 May 2005, Washington DC, pp 457–468. The Cardinal Group, Toronto
Collins SL (1987) Interaction of disturbances in tallgrass prairie: a field experiment. Ecology 68(5):1243–1250
Connolly J, Wayne P, Murray R (1990) Time course of plant-plant interactions in experimental mixtures of annuals: density, frequency and nutrient effects. Oecologia 82(4):513–526
Convention on Biological Diversity (1992) United Nations, New York. http://www.cbd.int/. Accessed 13 March 2014
Council of Europe (2000) European landscape convention and reference documents. Council of Europe Cultural Heritage, Landscape and Spatial Planning Division, Strasbourg
Daubenmire R (1968) Ecology of fire in grasslands. Adv Ecol Res 5:209–266
Davies R (2010) Elevated enclaves: biodiversity enhancement for living roofs—a collaborative research project. Waitakere, NZ: Waitakere City Council and Unitec, New Zealand
Davies RB, Simcock R, Ussher G, Boult M (2010) Elevated enclaves living roof biodiversity enhancement through prosthetic habitats. In: Proceedings of cities alive, 8th Annual Green Roof & Wall Conference, Vancouver, 30 November–December 3, 2010. The Cardinal Group, Toronto
Del Tredici P (2004) Neocreationism and the illusion of ecological restoration. Harv Des Mag 20:1–3
Delcourt HR, Delcourt PA (1988) Quaternary landscape ecology: relevant scales in space and time. Landsc Ecol 2(1):23–44
Delcourt HR, Delcourt PA, Webb T III (1983) Dynamic plant ecology: the spectrum of vegetational change in space and time. Quat Sci Rev 1(2):153–175
Design for London (2008) Living roofs and walls technical report: supporting London plan policy. Greater London Authority, London
Dias PC (1996) Sources and sinks in population biology. Trends Ecol Evol 11(8):326–330
Duncan A, Duncan R, Rae R, Rebecca G, Stewart B (2001) Roof and ground nesting Eurasian oystercatchers in Aberdeen. J Scott Ornithol Club 22:1–8
Dunnett N, Gedge D, Little J, Snodgrass EC (2011) Small green roofs: low-tech options for greener living. Timber Press, Portland
Dunster K (2007) The Garry oak gardener's handbook. Garry Oak Ecosystems Recovery Team and Parks Canada Agency, Victoria, BC
Dunster K (2010) Improving biodiversity values on green roofs in BC: some best practices. In: Proceedings of Cities Alive, 8th Annual Green Roof & Wall Conference, Vancouver, 30 November–December 3, 2010. The Cardinal Group, Toronto
Dvorak BD, Volder A (2013) Plant establishment on unirrigated green roof modules in a subtropical climate. AoB Plants 5: pls049 first published online December 20, 2012. http://aobpla.oxfordjournals.org/content/5/pls049.full. Accessed 17 May 2014
Edwards GR, Crawley MJ (1999) Herbivores, seed bank and seedling recruitment in mesic grassland. J Ecol 87(3):423–435
Ellis EC (2013) Sustaining biodiversity and people in the world's anthropogenic biomes. Current Opinion in Environmental Sustainability 5:368–372
Ellis EC, Ramankutty N (2008) Putting people in the map: anthropogenic biomes of the world. Front Ecol Environ 6(8):439–447
English Nature (2003) Green roofs: their existing status and potential for conserving biodiversity in urban areas. English Nature Report no. 498. English Nature, Peterborough, UK
Eriksson O, Eriksson A (1997) Seedling recruitment in semi-natural pastures: the effects of disturbance, seed size, phenology and seed bank. Nordic J Bot 17(5):469–482

Fazey I, Fischer J, Lindenmayer JB (2005) What do conservation biologists publish? Biol Conserv 124(1):63–73

Fenner M, Thompson K (2005) The ecology of seeds. Cambridge University Press, Cambridge

Fernandez-Canero R, Gonzalez-Redondo P (2010) Green roofs as a babitat for birds: a review. J Anim Vet Adv 9(15):2041–2052

Fisher RA, Corbet AS, Williams CB (1943) The relation between the number of species and the number of individuals in a random sample of an animal population. J Anim Ecol 12(1):42–58

FLL (2008) Green Roofing Guidelines: 2008 Guidelines for the planning, construction and maintenance of green roofing. FLL (Forschungsgesellschaft Landschaftsentwicklung Landschaftsbau e. V.)

Foster DR, Knight DH, Franklin JF (1998) Landscape patterns and legacies resulting from large, infrequent forest disturbances. Ecosystems 1:497–510

Fowler NL (1988) What is a safe site?: neighbor, litter, germination date, and patch effects. Ecology 69(4):947–961

Gedge D, Kadas G (2005) Green roofs and biodiversity. Biologist 52(3):161–169

Gilroy JJ, Sutherland WJ (2007) Beyond ecological traps: perceptual errors and undervalued resources. Trends Ecol Evol 22(7):351–356

Godfree RC, Lepschi BJ, Carnegie MD (2010) The impact of extreme drought and climate change on the demography of plains grass populations in central New South Wales. In: Proceedings of the 25th Annual Conference of The Grassland Society of NSW, pp 75–79. http://grasslandnsw.com.au/news/publications/conference-proceedings/conferences-2000-onwards/conference-2010/. Accessed 22 May 2014

Grant G (2006) Extensive green roofs in London. Urban Habitats 4(1):51–65

Hallett LM, Standish RJ, Hulvey KB, Gardener MA, Suding KN, Starzomski BM, Murphy SM, Harris JA (2013) Toward a conceptual framework for novel ecosystems. In: Hobbs RJ, Higgs E, Hall CA (eds) Novel ecosystems: when and how do we intervene in the new ecological world order? Wiley-Blackwell, Hoboken, pp 16–28

Hanski I (1999) Metapopulation ecology. Oxford University Press, Oxford

Hanski I, Ovaskainen O (2000) The metapopulation capacity of a fragmented landscape. Nature 404(6779):755–758

Harper JL (1977) Population biology of plants. Academic Press, London

Hartnett DC, Wilson GWT (2002) The role of mycorrhizas in plant community structure and dynamics: lessons from grasslands. Plant Soil 244(1–2):319–331

Higgs E (2012) Changing nature: novel ecosystems, intervention, and knowing when to step back. In: Weinstein MP, Turner RE (eds) Sustainability science: the emerging paradigm and the urban environment. Springer, New York, pp 383–398

Hilderbrand RH, Watts AC, April M, Randle AM (2005) The myths of restoration ecology. Ecol Soc 10(1):19. http://www.ecologyandsociety.org/vol10/iss1/art19. Accessed 14 May 2014

Hobbs RJ, Arico S, Aronson J, Baron JS et al (2006) Novel ecosystems: theoretical and management aspects of the new ecological world order. Global Ecol Biogeogr 15(1):1–7

Hobbs RJ, Higgs ES, Hall CM (eds) (2013) Novel ecosystems: intervening in the new ecological world order. Wiley, Chichester

IUCN-UNEP-WWF (1980) World Conservation Strategy. Gland, Switzerland: International Union for Conservation of Nature and Natural Resources (IUCN), United Nations Environment Programme (UNEP) and the World Wildlife Fund (WWF)

Janzen D (1998) Gardenification of wildland nature and the human footprint. Science 279(5355):1312–1313

Joern A (2005) Disturbance by fire frequency and bison grazing modulate grasshopper assemblages in tallgrass prairie. Ecology 86(4) 861–873

Jones RA (2002) Tecticolous invertebrates: a preliminary investigation of the invertebrate fauna on green roofs in urban London. English Nature, Peterborough

Kadas G (2006) Rare invertebrates colonising green roofs in London. Urban Habitats 4(1):66–8

Kaplan R, Kaplan S (1989) The experience of nature: a psychological perspective. Cambridge University Press, Cambridge

Kareiva P, Andersen M (1988) Spatial aspects of species interactions. In: Hastings A (ed) Community ecology. Springer-Verlag, New York, pp 35–50

Kareiva P, Wennergren U (1995) Connecting landscape patterns to ecosystem and population processes. Nature 373(6512):299–302

Katz E (1992) The big lie: human restoration of nature. Res Philos Tech 12:231–241

Keagy JC, Schreiber SJ, Cristol DA (2005) Replacing sources with sinks: when do populations go down the drain? Restor Ecol 13(3):529–535

Keane RE, Parsons RE, Rollins MG (2004) Predicting fire regimes at multiple scales. In: Perera AJ, Buse LJ, Weber MG (eds) Emulating natural forest landscape disturbances: concepts and applications. Columbia University Press, New York, pp 55–68

Kitching RL (2013) Niches and neutrality: community ecology for entomologists. Aust J Entomol 52:1–7

Köhler M (2006) Long-term vegetation research on two extensive green roofs in Berlin. Urban Habitats 4(1):3–26

Krishnan R, Hamidah A (2012) Stormwater runoff mitigation on extensive green roof: a review on trends and factors. In: Proceedings, OS 03, 6th South East Asian Technical Universities Consortium (SEATUC) Symposium, 6–7 March 2012, Malaysia. http://teknologimalaysia.academia.edu/raymondkrishnan/Papers. Accessed 9 May 2014

Levin SA (1992) The problem of pattern and scale in ecology. Ecology 73(6):1943–1967

Liu TC, Shyu GS, Fang WT, Liu SY, Cheng BY (2012) Drought tolerance and thermal effect measurements for plants suitable for extensive green roof planting in humid subtropical climates. Energy Build 47:180–188

Louv R (2011) The nature principle: human restoration and the end of nature-deficit disorder. Algonquin Books, New York

MacArthur RH, Wilson EO (1963) An equilibrium theory of insular zoogeography. Evolution 17:373–387

MacArthur RH, Wilson EO (1967) The theory of island biogeography. Princeton University Press, Princeton

Magurran AE (1988) Ecological diversity and its measurement. Princeton University Press, Princeton

Magurran AE (2004) Measuring biological diversity. Blackwell Scientific, Oxford

Marshall JK (1968) Factors limiting the survival of *Corynephorus canescens* (L.) Beauv. in Great Britain at the northern edge of its distribution. Oikos 19(2):206–216

McGuire KL, Payne SG, Palmer MI, Gillikin CM, Keefe D et al (2013) Digging the New York City skyline: soil fungal communities in green roofs and city parks. PLoS One 8(3): e58020. http://www.plosone.org/article/info%3Adoi%2F10.1371%2Fjournal.pone.0058020

Metz MR, Sousa WP, Valencia R (2010) Widespread density-dependent seedling mortality promotes species coexistence in a highly diverse Amazonian rain forest. Ecology 91(12):3675–3685

Millennium Ecosystem Assessment (MEA) (2005) Ecosystems and human well-being, vol 5. Island Press, Washington DC

Mills LS, Soule ME, Doak DF (1993) The keystone-species concept in ecology and conservation. Bioscience 43(4):219–224

Milton SJ, Dean WRJ, Klotz S (1997) Effects of small-scale animal disturbances on plant assemblages of set-aside land in Central Germany. J Veg Sci 8(1):45–54

Mitsch W (2004) Ecological engineering and ecosystem restoration. Wiley, New York

Natural England (2014a) Local nature reserves. http://www.naturalengland.org.uk/ourwork/conservation/designations/lnr/default.aspx. Accessed 12 May 2014

Natural England (2014b) Local nature reserves: Sharrow School green roof, Sheffield. http://www.lnr.naturalengland.org.uk/Special/lnr/lnr_details.asp?C=0&N=Sharrow&ID=1627. Accessed 12 May 2014

Oberndorfer E, Lundholm J, Bass B, Coffman RR, Doshi H, Dunnett N, Gaffin S, Köhler M, Liu KKY, Rowe B (2007) Green roofs as urban ecosystems: ecological structures, functions, and services. BioScience 57(10):823–833

Odum HT (1983) Systems ecology. Wiley, New York (reprinted in 1994 by University Press of Colorado, Niwot, Colorado)
Odum HT, Odum B (2003) Concepts and methods of ecological engineering. Ecol Eng 20(5):339–361
Oesterheld M, Sala OE (1990) Effects of grazing on seedlings establishment: the role of seed and safe-site availability. Veg Sci 1:353–358
Olive A, Minichiello A (2013) Wild things in urban places: America's largest cities and multi-scales of governance for endangered species conservation. Appl Geogr 43:56–66
Oomes MJM, Elberse WT (1976) Germination of six grassland herbs in microsites with different water contents. J Ecol 64(2):745–755
Orians GH, Wittenberger JF (1991) Spatial and temporal scales in habitat selection. Am Nat 137:S29–S49
Oro D (2013) Grand challenges in population dynamics. Front Ecol Evol 1(2):1–2
Paine RT (1966) Food web complexity and species diversity. Am Nat 100(910):65–75
Papafotiou M, Pergialioti N, Tassoula L, Massas I, Kargas G (2013a) Growth of native aromatic xerophytes in an extensive Mediterranean green roof as affected by substrate type and depth and irrigation frequency. HortScience 48(10):1327–1333
Papafotiou M, Pergialioti N, Papanastassatos EA, Tassoula L, Massas I, Kargas G (2013b) Effect of substrate type and depth and the irrigation frequency on growth of semi-woody Mediterranean species in green roofs. ISHS Acta Horticulturae, II International Symposium on Woody Ornamentals of the Temperate Zone 990:481–486. http://www.actahort.org/books/990/990_62.htm
Pearce H, Walters CL (2012) Do green roofs provide habitat for bats in urban areas? Acta Chiropterologica 14(2):469–478. http://dx.doi.org/10.3161/150811012×661774
Peterson AT, Soberôn J, Pearson RG, Anderson RP, Martínez-Meyer E, Nakamura M, Araújo MB (2011) Ecological niches and geographic distributions (MPB-49). Princeton University Press, Princeton
Petrişor A-I (2008) Levels of biological diversity: a spatial approach to assessment methods. Romanian Rev Regional Stud 4(1):41–62
Pickett ST, Cadenasso ML (1995) Landscape ecology: spatial heterogeneity in ecological systems. Science 269(5222):331–334
Pielou EC (1979) Biogeography. Wiley Interscience, New York
Power ME, Tilman D, Estes JA, Menge BA, Bond WJ, Mills LS, Daily G, Castilla JC, Lubchenco J, Paine RT (1996) Challenges in the quest for keystones. Bioscience 46(8):609–620
Rapp JK, Rabinowitz D (1985) Colonization and establishment of Missouri Prairie plants on artificial soil disturbances. I. Dynamics of forb and graminoid seedlings and shoots. Am J Bot 72(10):1618–1628
Rey PJ, Alcántara JM (2014) Effects of habitat alteration on the effectiveness of plant-avian seed dispersal mutualisms: consequences for plant regeneration. Perspect Plant Ecol Evol Syst 16(1):21–31
Robertson BA, Hutto RL (2006) A framework for understanding ecological traps and an evaluation of existing evidence. Ecology 87(5):1075–1085
Ronce O (2007) How does it feel to be like a rolling stone? Ten questions about dispersal evolution. Annu Rev Ecol Evol Syst 38:231–253
Rosenzweig ML (2003) Win-win ecology: How the Earth's species can survive in the midst of human enterprise. Oxford University Press, New York
Roxburgh SH, Shea K, Wilson JB (2004) The intermediate disturbance hypothesis: patch dynamics and species coexistence. Ecology 85(2):359–371
Royma T (1992) Analytical population dynamics. Chapman and Hall, London
Sauer CO (1925) The morphology of landscape. University of California Publications (in Geography 22:19–53. Reprinted (1963) In: Leighly J). Land and life, a selection from the writings of Carl O. Sauer. University of California Press, Berkeley, pp 315–350
Schläpfer M, Zoller H, Körner C (1998) Influence of mowing and grazing on plant species composition in calcareous grassland. Bot Helv 108(1):57–67

Schreiber SJ, Kelton M (2005) Sink habitats can alter ecological outcomes for competing species. J Anim Ecol 74(6):995–1004

SER—Society for Ecological Restoration International Science & Policy Working Group (2004) The SER international primer on ecological restoration. www.ser.org & Society for Ecological Restoration International, Tucson

Sheffield Local Biodiversity Action Partnership (2010) Habitat action plan, green roofs. Sheffield City Council, UK. https://www.sheffield.gov.uk/out-about/parks-woodlands-countryside/ecology-service/biodiversity-action-plans/sheffield-local-biodiversity-action-plan.html. Accessed 12 May 2014

Simkin SM, Michener WK, Wyatt R (2004) Mound microclimate, nutrients and seedling survival. Am Midl Natl 152(1):12–24

STLAi (Scott Torrance Landscape Architect Inc.) (2013) City of Toronto guidelines for biodiverse green roofs. Toronto: Toronto City Planning Division. http://www1.toronto.ca/wps/portal/contentonly?vgnextoid=3a7a036318061410VgnVCM10000071d60f89RCRD. Accessed 20 Jan 2014

Suffling R, Perera A (2004) Characterizing natural forest disturbance regimes. In: Perera A, Buse LJ, Weber MG (eds) Emulating natural forest landscape disturbances: Concepts and applications (pp. 43–54). New York: Columbia University Press

Sutton RK (2011) A model of human scale tested on rural landscape scenes. Great Plains Res 21:215–230

Sutton RK (November 2013) Seeding green roofs with native grasses. J Living Archit 1(1) 26 pages. http://greenroofs.org/resources/JOLA[2]Sutton.pdf

Sutton R (2014) Aesthetics for green roofs and green walls. J Living Archit (March 2014), p 22

Sutton R, Harrington J, Skabelund L, MacDonagh P, Coffman R, Koch G (2012) Prairie based green roofs: literature, templates and analogs. J Green Build 7(1):143–172. http://www.journalofgreenbuilding.com/toc/jgrb/7/. Accessed 21 April 2014

Throop W (ed) (2000) Environmental restoration: ethics, theory and practice. Humanity Books, Amherst

Thuring C, Dunnett N (2014) Vegetation composition of old extensive green roofs (from 1980s Germany). Ecol Process 3:4. http://www.ecologicalprocesses.com/content/3/1/4. Accessed 10 April 2014

Tilman D, Kareiva PM (1997) Spatial ecology: the role of space in population dynamics and interspecific interactions. Princeton University Press, Princeton

Town and Country Planning Association (2012) Planning for a healthy environment—good practice guidance for green infrastructure and biodiversity. TCPA and The Wildlife Trusts, London. http://www.tcpa.org.uk/data/files/TCPA_TWT_GI-Biodiversity-Guide.pdf. Accessed 13 June 2014

Turner MG, Dale VH (1998) Comparing large, infrequent disturbances: what have we learned? Ecosystems 1:493–496

Turner MG, O'Neill RV, Gardner RH, Milne BT (1989) Effects of changing spatial scale on the analysis of landscape pattern. Landsc Ecol 3(3/4):153–162

Van Dyck H, Baguette M (2005) Dispersal behaviour in fragmented landscapes: routine or special movements? Basic Appl Ecol 6(6):535–545

van Lennep E, Finn S (2008) Green roofs over Dublin: a green roof policy guidance paper for Dublin August 5, 2008. Tepui, Dublin

Van Mechelen C, Dutoit T, Hermya M (2014) Mediterranean open habitat vegetation offers great potential for extensive green roof design. Landsc Urban Plan 121:81–91. http://www.sciencedirect.com/science/article/pii/S0169204613001898. Accessed 16 May 2014

van Niekerk M, Greenstone C, Hickman M (2011) Creating space for biodiversity in Durban: guideline for designing green roof habitats. Environmental Planning and Climate Protection Department, Thekwini Municipality Durban, SA. http://www.durban.gov.za/City_Services/development_planning_management/environmental_planning_climate_protection/Publications/Documents/Guideline%20for%20Designing%20Green%20Roof%20Habitats1.pdf. Accessed 12 May 2014

Vitousek PM, Mooney HA, Lubchenco J, Melillo JM (1997) Human domination of earth's ecosystems. Science 277(5325):494–499

Waitakere City Council (2007) Potential NZ native plants for extensive greenroofs. Waitakere City Council, NZ. http://www.waitakere.govt.nz/abtcit/ec/ecoinit/pdf/greenroof-plants.pdf. Accessed 23 April 2014

Waitakere City Council (2010) Elevated enclaves—biodiversity enhancement for living roofs. A collaborative research project with Unitec Department of Landscape Architecture, Auckland, NZ. http://www.waitakere.govt.nz/abtcit/ec/ecoinit/greenroof.asp. Accessed 10 May 2014

Watt TA, Gibson CWD (1988) The effects of sheep grazing on seedling establishment and survival in grassland. Vegetatio 78(1–2):91–98

Whittaker RH (1972) Evolution and measurement of species diversity. Taxon 21:213–251

Whittaker RH, Levin SA, Root RB (1973) Niche, habitat, and ecotope. Am Nat 107:321–338

Williams NSG, Rayner JP, Raynor KJ (2010) Green roofs for a wide brown land: opportunities and barriers for rooftop greening in Australia. Urban For Urban Green 9(3):245–251

Woodward FI (1994) How many species are required in a functional ecosystem? In: Schulze E-D, Mooney HA (eds) Biodiversity and ecosystem function. Springer-Verlag, New York, pp 271–291 (2nd printing)

Dr. Katherine Dunster RLA, RPB School of Horticulture Kwantlen Polytechnic University, B.C. leads its new Bachelor of Horticulture Science degree major in Urban Ecosystems. She is also an Adjunct Professor in the School of Architecture & Landscape Architecture at the University of British Columbia.

Dr. Reid R. Coffman heads the Novel Urban Ecological Systems (NUES) Cleveland Urban Design Collaborative at Kent State University and is a research associate professor and urban ecological designer. He specializes in the conceptualization and application of living architecture that optimizes ecological services.

Chapter 16
Eco-regional Green Roof Case Studies

Bruce Dvorak

Abstract Biological diversity and ecological functions have been observed and maintained on green roofs worldwide. Prairies, meadows, grasslands, wetlands, talus slopes, barrens, and successional plant communities are a few examples of habitat types that have been synthesized on green roofs. Nutrient cycling, plant dynamics, insect communities, evapotranspiration, and stormwater retention are some of the ecological functions studied from these communities. A group of twelve green roofs were selected as case studies to reiterate and highlight the key concepts covered in the previous chapters. This chapter provides examples of how green roofs have been designed and maintained as living, dynamic ecosystems.

Keywords Biodiversity · Ecological demonstration · Maintenance · Construction · Case study

16.1 Introduction

Previous chapters of this book introduce a variety of ecological characteristics of green roofs. This variation in part depends upon a green roof's materials, arrangements and complexity. Simple green roofs such as some low diversity sedum-based green roofs might be designed for a specific purpose such as stormwater management or cooling of rooftops. Regardless of intent, such green roofs would likely exhibit multiple ecosystem functions (Lundholm et al. 2010; Oberndorfer et al. 2007). Green roofs can also exhibit more complex vegetative structures and plant communities that attract local or mobile organisms such as migrating birds or insects (Chap. 9, 10, 14, 15). They may also require simple or complex processes for care and maintenance (Dvorak and Carroll 2008; Dvorak 2011; Snodgrass and McIntyre 2010). Although the constructed habitats of some green roofs (e.g. prairies, meadows, wetlands) can be useful in managing urban ecosystem functions and

---

B. Dvorak (✉)
Department of Landscape Architecture and Urban Planning, Texas A & M University, A305 Langford Architecture Center, College Station, TX 77843-3137, USA
e-mail: bdvorak@tamu.edu

R. Sutton (ed.), *Green Roof Ecosystems*, Ecological Studies 223,
DOI 10.1007/978-3-319-14983-7_16

services, they will likely never replace the richness and complexity of their land-based counterparts as soil horizon formation (i.e. varying depths and soil types) and connections with water tables are lacking, amongst many other variables (Sutton et al. 2012, Lundholm 2005, Aitkenhead-Petersen and Volder 2010). Therefore, preservation of ground level habitat is to be of a highest priority. Regardless, green roofs can be of immense value when they respond to and respect the ecology of a place or region and endure over time.

For a green roof to endure, it must have resilience. This means that the plant community must be robust or diverse enough to be able to respond to year-to-year climate variations in moisture and temperature and can limit or expand plant and animal species populations. For some green roofs, substrate depth, slope, drainage, and plant communities have been consciously designed to mimic a habitat. For example, the Peggy Notebaert Nature Museum rooftop garden (Fig. 16.1) consists of an engineered substrate applied at varied depths, and was originally planted with over 100 species native to Chicago area prairies and wetlands (Cantor 2008). Other green roofs such as the Laban Center in London were constructed of a rubble substrate, with no initial planting. Over time, the Laban substrate became populated with plant species from local meadows (Sect. 16.9). An eco-regional green roof, therefore, is a green roof that expands or profits the immediate urban or regional landscape ecology or ecosystem (Kowarik 2011; Forman 1995). An eco-regional green roof sustains local or regionally important plant species, plant communities, organisms, functions or services (Lundholm 2005; Sutton et al. 2012; Dvorak and

Fig. 16.1 Wetland, mesic and upland prairie green roof at the Peggy Notebaert Nature Museum in Chicago, Illinois. Over 100 native plant genera were installed in 2001. Substrate depth increases from 15 cm for the wetland (*foreground*) to 45 cm for the prairie (*background*). (Conservation Design Forum Elmhurst, IL)

Volder 2010). Any green roof likely lies within an ecological value gradient from nominal to vital based upon its elements, services and maintenance requirements.

Green roofs can differ also in their type of construction as some are made from modular plastic or biodegradable trays, monolithic engineered or natural substrates, or from pre-grown vegetated mats. They can be planted in place, or established through hydro-seeding, or left to self-populate via ambient environmental influences. There are vendor provided and warranted green roofs; custom designs with no specified warrantee; green roofs bid, constructed and maintained under public or private domain; and green roofs that remain constant over time or those that evolve and change over time. These different approaches present the concept of "green roof" not as a single type or method but as a dynamic array of possibilities. This chapter covers many of these types of green roof constructions and approaches in part to demonstrate how a widespread concept such as "green roof" can be expressed in many different ways as climate, microclimate, eco-region and many other factors shape and form green roof ecosystems and aesthetics.

This chapter will succinctly highlight some conceptions of and strategies for rooftop greening and their ecological characteristics. Twelve (12) eco-regional case studies explore key components presented in the chapters and depict important features or characteristics including: age, notable features, exemplary species maintained, depth of substrate, dominant form of vegetation, size, and other important qualities or conditions. Each case study includes an introduction and overview of the subject, descriptive overview of the project and its ecological features, a table of ecological characteristics and at least two photographs including an overview and detail photo. Many other green roofs could have been discussed—nevertheless, these twelve case studies are consistent with the previous chapter materials and provide a useful cross-section of small to large, young to old, and simple to complex green roofs from Australia, Europe and North America.

16.2 Monitoring Inputs and Outputs at the Seaton Hall Green Roofs, Manhattan, KS, U.S.A. (Skabelund et al. 2014)

Knowledge evolves from research. Green roof research and knowledge has expanded greatly since 2000 (Blank et al. 2013), and sometimes emerges from the monitoring of small-scale installations at major universities. For example, Kansas State University's first green roof (Upper Green Roof) was a 28-square-meter semi-intensive green roof implemented in 2009 on Seaton Hall, a historic campus building. Entering its sixth growing season (June, 2014) the project supports 11 species of native grasses and forbs, two succulents species, and several volunteer species. A second green roof (Lower Green Roof) was installed in 2012 to demonstrate modular and monotypic substrate types on a semi-shaded roof. The Lower Green Roof (Fig. 16.2) is irrigated with harvested rooftop rainwater in an attempt to provide full coverage of green roof substrates over the long term. Surface and sub-surface temperatures are monitored on both the upper and lower green roofs.

Fig. 16.2 Lower Seaton Hall Green Roof at Kansas State University assembled from LiveRoof™ modules. The modules are populated with 11 species of prairie plants installed in 2012 and some replanting in 2014. (Lee R. Skabelund)

Manhattan, Kansas, experiences frequent high temperatures during the summer with prolonged periods (one to three weeks) without precipitation. When summer precipitation occurs, it is often with heavy downpours and high winds. Faculty and students monitor the viability and growth of native grasses and forbs on the Upper Green Roof (Fig. 16.3), in concert with variables such as precipitation, stormwater runoff, and temperature. Grasses were evaluated for height, basal diameter, and number of flowering stalks at the end of each growing season between 2009 and 2013. Stormwater runoff was measured for most storms during six growing seasons (and for some winter storms) using a 200-gallon cistern connected to a flow meter. The green roof's vegetation, in part through evapotranspiration, retains and recycles moisture back into the atmosphere. Additional monitoring devices include a solar-charged data logger, a tipping bucket, manual rain gauge, air temperature/relative humidity sensor/wind monitor, temperature probes, and soil moisture and temperature probes. The green roof captures 30–100 % of rainfall, depending on substrate moisture levels. On the Upper Green Roof (Table 16.1) 11 of 19 native grasses and forbs planted in 2009 remain (summer 2014), despite no supplemental irrigation since mid-August 2012. Most prairie plants were planted in the 12.7-cm-deep substrate (American Hydrotech), but some plants were added to 10.8-cm-deep LiveRoof™ trays salvaged from a project in central Kansas. *Sedum kamtschaticum*, (syn. *Phedimus kamtschaticus*) native grasses and forbs are performing well. The entire roof receives supplemental irrigation, from a 300-gallon cistern (installed in May 2014) to harvest and re-use adjacent asphalt shingle rooftop runoff.

Fig. 16.3 Monitoring equipment set up on the Upper Seaton Hall Green Roof in June 2009 to capture temperature profiles, stormwater runoff, precipitation, wind speed and direction, relative humidity, and air temperature. (Lee R. Skabelund)

Table 16.1 Ecosystem characteristics for the Upper Seaton Hall Green Roof

System criteria	Characteristic
Ecoregion	Flint Hills Tallgrass prairie
Year completed	2009
System type	Extensive to semi-intensive
System provider/growing media	American Hydrotech, Inc. (Litetop™)
Size of green roof (square meters)	28.5
Approx. height above grade (meters)	9
Substrate depths (cm)	Varies from. 7.5 (edges) to 17.5 (center)
Drainage system (i.e. granular, plastic)	American Hydrotech drainage retention mat (Gardendrain GR30), with filter fabric and mineral aggregate in retention cups
Dominant form of vegetative	North American prairie with exotic succulents
Species planted; species maintained	19 (including 4 native grasses and 4 exotic sedums); 11 native herbaceous species (plus several common weeds and two sedums)
Irrigation (i.e. temporary, permanent, periodic, none)	Periodic, during dry periods for the first two growing seasons (on the west side) and the first 3 − 1/2 growing seasons (on the east side)
Notable ecosystem features	Upper Green Roof attracts local butterfly, bee, dragonfly, insect, and bird populations—with robins observed hauling off prairie grass thatch and pigeons resting in sedum
Characteristic or notable plant species sustained on the green roof	*Bouteloua curtipendula, Bouteloua gracilis* (*dominant*), *Schizachyrium scoparium, and Sporobolus heterolepis* (not irrigated since 2013) *Liatris punctata and Ratibida columnifera* are most abundant of planted forbs

16.3 Green Roof Climates and Microclimates, Burnley Green Roofs, Melbourne, Victoria, Australia (Farrell et al. 2013)

Green roof microclimates can vary from intended design characteristics because of substrate depth, composition, drainage or moisture levels. Variation occurs across designs (Simmons et al. 2008) and much can be learned about plant viability and their associative ecosystem services and climates if that variation is monitored. Achieving such variation requires either multiple green roofs, each with different designs, or multiple designs on one roof. The University of Melbourne's Burnley Green Roofs comprise a demonstration (Fig. 16.4), a research and a bio-diversity green roof (Fig. 16.5). They were designed as an outreach, teaching and research facility to demonstrate different green roof types and plant species based upon microclimates and habitat types for Melbourne. The demonstration roof includes fourteen planting zones with over 203 species of plants growing in five different substrate depths (10, 15, 20, 25, 30 cm) with varied watering treatments. Water is harvested from other roofs on campus for irrigation. The custom designed substrates include scoria, bottom ash (power station waste/Enviroagg®), and crushed roof tile based substrates.

The bio-diversity roof aims to create microhabitats to attract birds, butterflies, bees and other insect species while the fully instrumented research roof will support plant selection, hydrology and energy experiments. On the demonstration and bio-diversity roofs (Table 16.2) as of June 2014 most of the 200+species of succulents, grasses and forbs planted in 2012 remain.

Fig. 16.4 Burnley demonstration green roof demonstrates fourteen (14) different planting zones, each with different microclimates and habitats made from ascending substrate depths (front to back), watering treatments, moisture gradients and plant communities. (Les O'Rourke)

Fig. 16.5 Bio-diversity roof includes a small pond, water channel, wood and rock debris to attract local fauna. (Les O'Rourke)

Table 16.2 Ecosystem characteristics for the Burnley Demonstration and Bio-diversity green roofs

System criteria	Characteristic
Ecoregion	Grassland and Grassy Woodland
Year completed	All roofs completed in 2012
System type	Varies from extensive to semi-intensive
System provider/growing media	Zinco Drainage layers Custom substrates: Demo roof: scoria, roof tile, EnviroAgg® Biodiversity roof: scoria
Size of green roof (square meters)	166 (demonstration roof) 80 (research roof) 52 (bio-diversity roof)
Approx. height above grade (meters)	6.6
Substrate depths (cm)	10, 15, 20, 25, 30
Drainage system (i.e. granular, plastic)	Plastic
Dominant form of vegetative	Succulents, grasses and forbs
Species planted; species maintained	203; 200+
Irrigation (i.e. temporary, permanent, periodic, none)	Demo = Irrigated, deficit irrigation and unirrigated zones. Bio-diversity = unirrigated
Notable ecosystem features	Demo-plants for multiple microclimates, depths, substrate type and moisture zones Bio-diversity-Plants that attract local butterfly, bee, and bird populations
Characteristic or notable plant species sustained on the green roof	*Sedum, Mesembryanthemum, Aloe, Crassula spp. Dianella admixta, Stypandra glauca, Arthropodium milleflorum*

16.4 Water for Green Roofs, Vancouver Convention Center, Vancouver, BC, Canada (Nightingale 2010; Sutton et al. 2012)

Water is one of the limiting elements of all forms of life. Some climates experience even distributions of precipitation and some climates are defined by strong seasonal differences. British Columbia typically receives ample precipitation during the fall, winter and spring; however, summers remain dry. Coastal meadows indigenous to the Vancouver area survive droughts, but vegetated rooftops lack groundwater connections and can become drought stressed. To mitigate for the seasonal variation in precipitation, the green roof on the Vancouver Convention Center (Fig. 16.6) has an irrigation system designed to use treated wastewater from inside the convention center, and uses that water to irrigate the vegetated roof during periods of drought. About 43 km (26.7 miles) of irrigation tubing lay in the substrate and are controlled by moisture sensors. The substrate conserves moisture with materials such as lava rock, sand and composted organic matter. The roof supports a spectacular array of local coastal grassland inspired flora (Fig. 16.7) consisting of twenty-four species of local grasses and forbs including onions, sedges, fescues, poppies, strawberries, aster and stonecrop. The vegetation survives primarily without supplemental watering, limited to periods of drought (Table 16.3).

Fig. 16.6 Green roof coastal meadow ecosystem covers the sloped rooftops of the Vancouver Convention Center. The nearly 2.5 ha installation over multiple levels of green roofs and public decks manifest an elegant and seamless integration of ground level vegetation rising to the rooftops. The Vancouver skyline is in the background and overlooks the green roof meadows, Waterfront Park and Vancouver Harbor. (PWL Partnership Landscape Architects Inc.)

Fig. 16.7 View of coastal meadow ecosystem established on the sloped rooftop of the Vancouver Convention Center. As a rooftop, the meadow lacks a natural groundwater connection, however, the roof is periodically irrigated from water harvested from inside the Convention Center. (PWL Partnership Landscape Architects Inc.)

Table 16.3 Ecosystem characteristics for the Vancouver Convention Center Green Roof

System criteria	Characteristic
Ecoregion	Coastal grasslands and forests
Year completed	2009
System type	Semi-intensive
System provider/growing media	Sand from Fraser River, lava rock, garden waste; Plant supplier Holland Lands./Nats Nursery
Size of green roof (square meters)	24,821
Approx. height above grade (meters)	Varies—5–10
Substrate depths (cm)	15
Drainage system (i.e. granular, plastic)	Drainage mat
Dominant form of vegetative	Native coastal grassland
Species planted; species maintained	24
Irrigation (i.e. temporary, permanent, periodic, none)	Treated convention center blackwater supplies drip tubing irrigation buried within the substrate layer activated by moisture sensors (Nightingale 2010)
Notable ecosystem features	System is inspired from local coastland grasslands and meadows
Characteristic or notable plant species sustained on the green roof	Plants: Native fescues, grasses, Aster, Potentilla and Beach strawberry, spring bulbs. Insects: four honey bee hives approximately 60,000 bees each

16.5 Growing Substrate and Nutrient Relationships Civic Garden, Cincinnati, OH, U.S.A. (Mooney-Bullock et al. 2012)

Vegetated roofs can substantially change the water quality of rooftop runoff. Some leaching of nutrients (nitrogen, phosphorous) can be expected, depending upon the composition of growing media, organic sources used in the growing substrate (e.g., compost), and fertilizer present in some commercial substrates (Berndtsson 2010). Leaching can partly be attributed to natural processes like nutrient release during annual dieback or fertilizer applications (Aitkenhead-Peterson et al. 2010). The Cottage House green roof at the Civic Garden Center of Greater Cincinnati is one of several demonstration green roofs at the Civic Center, and is a site for research on nutrient cycling and runoff water quality (Figs. 16.8 and 16.9, Chap. 5). This roof has been monitored since 2011, including runoff amount and water quality measurements from > 100 rain events, as well as continuous monitoring (beginning 2012) of substrate moisture and temperature and local climate. During this period (roof age 1–4 years): (1) the green roof was a source of high phosphate in runoff, while nitrogen was closer to being in balance; (2) concentrations of phosphate in runoff have decreased over time; (3) strong seasonal patterns occurred, with the highest concentrations of phosphate, nitrate, calcium and other solutes leaching from the green roof in the summer (Buffam 2014). This suggests the initial nutrient content of the substrate, and the development of plant and microbial communities over time, are important in controlling nutrient cycling and leaching (Table 16.4).

Fig. 16.8 *Sedum rupestre* 'Angelina' hugging the sloped roof of the Civic Garden cottage. (Ryan Mooney-Bullock, Civic Center of Greater Cincinnati)

Fig. 16.9 Stormwater runoff from the Cottage House Green Roof (shown here) is sampled for flow rate and chemical analysis of runoff. (Ishi Buffam)

Table 16.4 Ecosystem characteristics for the Civic Garden green roofs, Cincinnati, Ohio

System criteria	Characteristic
Ecoregion	Deciduous forest
Year completed	2010
System type	Extensive
System provider/growing media	Tremco
Size of green roof (square meters)	46
Approx. height above grade (meters)	2–4
Substrate depths (cm)	10
Drainage system (i.e. granular, plastic)	Capillary geotextile layer
Dominant form of vegetative	Succulents as Sedum mat
Species planted; species maintained	9:8
Irrigation (i.e. temporary, permanent, periodic)	By hand as needed during the growing season
Notable ecosystem features	Sloped roof (21°) with soil-stabilization system
Characteristic or notable plant species sustained on the green roof	*Sedum album, Sedum sexangulare, Sedum acre, Sedum spurium, Sedum rupestre, Sedum floriferum, Sedum kamtschaticum, Sedum immergrunchen*

16.6 Soil-Based Green Roofs, BRIT Green Roof Fort Worth, TX, U.S.A.

Many green roof substrates today are comprised of an engineered mineral-based substrate with low organic levels. However, substrates of some of the first and now oldest green roofs in Germany and Europe were constructed with little more than sand and gravel. Sand and gravel substrates and some engineered substrates have the advantage of structural stability and good drainage, but are poor in nutrients. Some natural soils can be high in nutrients, but can be susceptible to compaction, excess weight, and have poor drainage characteristics. The soil-based green roof at the Botanical Research Institute of Texas (BRIT) combines the best of natural soil and engineered media. The BRIT substrate is composed of both soil from the nearby Fort Worth Prairie barrens in biodegradable trays and an engineered substrate as a sub-layer. The vegetation was grown directly in the trays, lifted to the roof, and placed on the engineered media. The green roof sustains 38 species planted in plug form, but an unknown number of species from the seed bank of the natural soil (Figs. 16.10 and 16.11). By 2012, with some irrigation during dry periods more than fifteen (15) species of volunteers and ruderal species had established including some from the seed bank, others brought in via wind and birds. The plant form diversity includes at least seven (7) forbs, four (4) grasses, and four (4) succulent species (Dvorak et al. 2013). By 2014, over one hundred (100) species had been observed (Byerley 2014).

Prior to the installation of the green roof, research and planning took place to investigate potential plant species and soil composition. Researchers, including Dr. Tony Burgess, vendors and students worked together to test arrangements of prairie plants in biodegradable trays (BioTray) and planting design alternatives for the green roof. Fifteen trays were tested over several months to identify top species to be grown from plugs or seed from the nearby prairie. For the BRIT roof installation, 6000 trays were prepared in a greenhouse prior to their installation on the roof (Fig. 16.11) (Kelly 2013) (Table 16.5).

Fig. 16.10 The sloped prairie inspired green roof at BRIT established at least fifteen native species during a historically hot and dry 2011 growing season. By 2014 (*above*) the green roof had over 100 species established. (Botanical Research Institute of Texas)

Fig. 16.11 Biodegradable coconut fiber trays were filled with a substrate mix of 20% natural soil (from nearby Fort Worth Prairie) and 80% engineered substrate in the trays and then 100% engineered substrate below the trays Trays (*above*) are being planted with pre-grown plants in plug form including the *Opuntia phaeacantha* (Prickly Pear cactus) indigenous to the Grand Prairie. (Botanical Research Institute of Texas)

Table 16.5 Ecosystem characteristics for the BRIT Prairie Green Roof

System criteria	Characteristic
Ecoregion	Grand Prairie
Year completed	2010
System type	Extensive
System provider/growing media	Substrate within the trays consisted of a lower 3.8-cm layer of 1:1 CSL and Hadite; an upper 2.5-cm layer of 1:1:2 CSL, Hadite, and biologically active Goodland Limestone topsoil harvested from a local prairie; and a 1.2-cm gravel mulch layer on top. (Dvorak et al. 2013)
Size of green roof (square meters)	1083
Approx. height above grade (meters)	7.5+
Substrate depths (cm)	12.5
Drainage system (i.e. granular, plastic)	Aggregate filled Gardendrain GR30 (American Hydrotech)
Dominant form of vegetative	Fort Worth/Grand/Goodland prairie
Species planted; species maintained	38 (plug form); 15 (plug form) plus > 50 ruderals/volunteers and from seedbank
Irrigation (i.e. temporary, permanent, periodic, none)	Drip irrigation during establishment, as needed thereafter (during dry periods)
Notable ecosystem features	Slopped barrel type roof with prairie-based green roof modeled after the local Fort Worth prairie barrens at the Fort Worth Prairie Park
Characteristic or notable plant species sustained on the green roof	*Bouteloua curtipendula* var. *curtipendula, Bouteloua dactyloides, Muhlenbergia reverchonii, Yucca pallida* and *Opuntia phaeacantha*

16.7 Plant Biodiversity on Green Roofs, Alice H. Cook and the Carl L. Becker Houses, Cornell University, Ithaca, NY, U.S.A. (Carlisle et al. 2013)

There are at least three approaches regarding planting for green roofs including: monocultures, simple plant combinations and mixtures, and plant communities (Dunnett and Kingsbury 2004). Each approach has different maintenance and aesthetic expectations and should be considered integral to the approach. Monocultures and simple plant combinations intend to maintain low plant diversity. Plant communities, however, typically include high plant diversity and the distribution and population of species may fluctuate annually or seasonally due to climate variations (Dunnett and Nagase 2007). Variation of plant communities on green roofs can be monitored and measured over time through spatial performance measures (Relevé Method) used in field ecology to measure as species richness. Designed for low maintenance, the green roofs at the Alice H. Cook and the Carl L. Becker Houses at Cornell University were the subjects of vegetative surveys over seven years to capture changes in plant community structure and learn about their spatial performance indicators including species richness, cover, and biodiversity. Compared to the initial planting, plant species richness has more than doubled. The species biodiversity on the green roof was less diverse in 2012 (True Diversity (TD) = 6.74) than at the time of planting in 2005 (TD = 8.67). Seven years after plant establishment, the Alice H. Cook House transitioned from a diverse meadow to a green roof dominated by *Schizachyrium scoparium* (57% cover); however, the roof also had 39 species, including 14 of the original 16 species. Of the other originally planted species, none were found to contribute more than 5% to total roof vegetation. Emergent species represented 31.25% of all vegetative cover (Carlisle et al. 2013) (Table 16.6) (Figs. 16.12 and 16.13).

Table 16.6 Ecosystem characteristics for House 1 and 2 green roofs at Cornell University

System criteria	Characteristic
Ecoregion	Deciduous forest
Year completed	(1) Alice H. Cook House, 2004
	(2) Carl L. Becker House, 2005
System type	(1) Intensive
	(2) Extensive
System provider/growing media	Unknown
Size of green roof (square meters)	(1) 329 (2) 418
Approx. height above grade (meters)	(1 & 2) 4.57
Substrate depths (cm)	(1) 24.13 (2) 12.70
Drainage system (i.e. granular, plastic)	(1 & 2) Water retention sheet
Dominant form of vegetative	(1) Meadow (10 forb; 5 grass; 1 shrub)
	(2) Sedum/mix (3 succulent; 1 grass; 1 forb)

Table 16.6 (continued)

System criteria	Characteristic
Species planted; species maintained	(1) 16–39 (2) 5–65
Irrigation (i.e. temp., perm., periodic, none)	Irrigated during the first year after planting only. Per LEED (USGBC) requirements, the irrigation systems were removed from both roofs after plant establishment
Notable ecosystem features	(1) Species richness has more than doubled. Roof maintains (93 %) coverage
	(2) Originally planted with three succulent; one grass; one forb species, the roof now has 65 species, and 30 plant families
Characteristic or notable plant species sustained on the green roof	(1) Domination by *Schizachyrium scoparium*, with tall emergent annual and biannual forbs, particularly along the skylights and at roof edge
	(2) Dense sedum understory punctuated by medium and tall forbs and woody plants that serve to shade the succulent understory

Fig. 16.12 Meadow and early pioneer vegetation established on the Alice H. Cook House at Cornell University. The green roof is left to natural succession, as it has not had any human interference regarding removing of species or active maintenance, with the exception of irrigation during the first year of plant establishment. These conditions have allowed for observation of species interactions, dynamics and dominance on the roof. (Halkin Mason Photography)

Fig. 16.13 Detailed plant surveys of the green roofs relied on dozens of Relevé (concise summaries of distinct vegetative units) over 7 years. Vegetation on the Alice H. Cook House during 2012, a growing season with drought conditions. (Stephanie Carlisle)

16.8 Structural and Functional Effects of Green Roof Plant Assemblages, University of St. Mary's, Halifax, Nova Scotia, Canada (Lundholm et al. 2010; MacIvor and Lundhom 2011b)

The arrangement of vegetation on green roofs can influence ecological functions (Lundholm et al. 2010). The green roof research at Saint Mary's University in Halifax, Nova Scotia (Fig. 16.14) has captured important observations about relationships between different forms of plants and their ecosystem functions. It begins to investigate questions such as, "Do green roofs with only succulents or only herbaceous or woody plants or mixed designs perform equally?" The research group investigated indigenous vegetation along with industry standard succulents for survival. Disturbance-tolerant species that can grow quickly such as *Solidago bicolor* and *Danthonia spicata* were planted and performed well, even with neighboring plants such as mosses and lichens (MacIvor and Lundholm 2011b; Heim et al. 2014). Life-forms investigated include: succulents, grasses, tall forbs, creeping forbs, mosses, lichens and creeping shrubs (Fig. 16.15) (Heim and Lundholm 2014b). It appears that the number of life-forms makes a difference. After the first growing season, most monoculture designs did not make the roof temperature cooler than the substrate-alone, however four mixed-life-form plant treatments were cooler than

Fig. 16.14 The modular green roof plots where plant, temperature, evaporation and plant combination research has taken place on the St. Mary's University in Halifax. (Jeremy Lundholm)

Fig. 16.15 Biodiverse atrium green roof research site at St. Mary's University. Vegetation such as lichens and mosses and over 200 species of insects have been identified on the roof and in Halifax. (Jeremy Lundholm)

the control, and some preserved substrate moisture (Heim and Lundholm 2014b). Time has influenced performance as well with the third growing season showing stronger effects. It appears that planting a mixture of species, especially including multiple life-forms such as grasses, succulents and forbs, mosses and lichens can be beneficial compared to many of the monoculture designs tested, but it depends upon the species canopy characteristics, cover and perhaps eco-region (Lundholm et al. 2010; Heim and Lundholm 2014a). The research group has also investigated other ecological benefits such as insect migration onto roofs. In one study, over 200 species of insects were found populating five roofs in the downtown Halifax area, including the research facility, thus expanding upon their findings of ecosystem functions and services (Table 16.7).

Table 16.7 Ecosystem characteristics for the Halifax green roofs

System criteria	Characteristic
Ecoregion	Coastal barrens and meadows
Year completed	2008; 2010 (green roofs on two buildings)
System type	Extensive and semi-intensive
System provider/growing media	Custom modular, Soprema growing substrate
Size of green roof (square meters)	65, 200
Approx. height above grade (meters)	5, 10
Substrate depths (cm)	7.5, 15, 20
Drainage system (i.e. granular, plastic)	Enkamat™
Dominant form of vegetative	Coastal barrens
Species planted; species maintained	30-varies with research. At the end of the growing season, 12 of the 15 species planted had close to 100 % survival, two species (*Arctostaphylos uva-ursi and Aster novaebelgii*) had greater than 80 %survival and one species died altogether (*V. angustifolium*)
Irrigation (i.e. temporary, permanent, periodic, none)	During establishment and during drought
Notable ecosystem features	System functions (temperature, moisture), plant combination interactions, plant viability
Characteristic or notable plant species sustained on the green roofs	Early pioneer species, lichens and mosses. Over 200 species of insects have been identified including the endangered firefly *Phospaenus hemipterus* (Coleoptera: Lampyridae)

16.9 Bio-diverse Ruderal Communities, Laban Center and Creekside Center, London, UK; and Switzerland (Gedge et al. 2012; Gedge and Kadas 2005; Gedge 2003; Brenneisen 2006)

The concept of what comprises a green roof also embraces rooftops called bio-diverse roofs. These roofs contain various materials and are reliant upon ruderal or successional vegetation. Some of these roof sites are designed as habitat to attract insects and birds such as the black redstart. Bio-diverse roofs are created from irregular layering of substrate, various size stones, branches and other materials to attract various forms of life (Figs. 16.16, 16.17) (Gedge et al. 2012; Brenneisen 2006). The rare and protected black redstart for example, travels across the UK during winter migrations, and new ground level construction on brownfield sites threaten preferred redstart habitats. Bio-diverse roofs such as the Laban Center and

Fig. 16.16 Dr. Stephan Brenneisen determined that various size stones, gravel, bare spots and a variety of vegetation forms are ideal habitat characteristics on green roofs for attracting wildlife. Monocultures and even substrate depths on green roofs can be beneficial to some wildlife, but the diversity in materials, structure and arrangement offers variety, cover and choice. (Stephan Brenneisen)

Fig. 16.17 Bio-diverse roofs give rise to a range of pioneering meadow plants, mosses, lichens, and other life-forms adapted to disturbed urban conditions without irrigation or maintenance. This photo of the Laban Center roof was taken eight years after completion, and is good habitat for arthropods such as spiders. (Dusty Gedge)

Creekside Center were set up to attract biodiversity that frequents brownfields sites including black redstarts. The mounded substrate includes gravel, stone, brick, concrete pieces and other debris. These habitats may prove important for black redstart survival, as a 2008 report by Holling et al. found only 54 breeding pairs of black redstarts (*Phoenicurus ochruros*) in the UK. Auspiciously, black redstarts have been found to frequent such roofs for resting, foraging and other activities in London, across the UK and Switzerland (Brenneisen 2006; Brenneisen 2003; Gedge 2003; Gedge and Kadas 2005). Laban was the subject of a comparative study between green roofs, brown roofs and a nearby brownfield site. The green and brown roofs at the Laban Center were much higher with invertebrate species numbers than the brownfield site. Some faunistically interesting species local to the site and some rare and scarce species were found, composing 10% of all species recorded. Some of the rare and scarce species include spiders; *Bianor aurocintus, Ostearius melanopygius, Pardosa agrestis, Zilla diodia*; beetles; *Amara curta, Microlestes minutulus, Hippodamia variegata, Athous campyloides*; and aculeate Hymenoptera (e.g. bees, wasps, ants) such as *Apis mellifera, Bombus lapidaries, Lasius flavus* and many others (Kadas 2006) (Table 16.8).

Table 16.8 Ecosystem characteristics for Laban Center Bio-diverse Roof

System criteria	Characteristic
Ecoregion	Deciduous forest and meadows
Year completed	2002
System type	Varied depth semi-intensive
System provider/growing media	Stone, gravel, sand and rubble from site
Size of green roof (square meters)	200
Approx. height above grade (meters)	25
Substrate depths (cm)	12.95–15.24
Drainage system (i.e. granular, plastic)	Stone
Dominant form of vegetative	Brownfield barrens. Ruderal
Species planted; species maintained	Not planted; varies
Irrigation (i.e. temporary, permanent, none)	None
Notable ecosystem features	Attraction of black redstarts and other animals for habitat and preservation
Characteristic or notable plant species sustained on the green roof	Populations of volunteer species from wind-blown seed

16.10 Prairie Biome on Green Roofs, Edgeland House, Austin, Texas, U.S.A. (Parker 2013)

A prairie, in south-central Texas historically meant a grassland-based biome with an incredible variety of endemic perennial and annual grasses and forbs. The structure of the prairie typically relies upon the grasses, and the forbs typically bring and array of colors throughout much of the year (Fig. 16.18, Table 16.9). The roof of a private residence in Austin, Texas occupies a former pipeline property which was previously covered with cement, rebar, gravel and other debris. Today, the property contains a prairie surrounding a new residential structure, with earth and prairie ascending to the sky. The concept is based upon the idea of a pit house, a sunken Native American earth-sheltered dwelling. The green roof was planted with over 200 species native to the Tallgrass prairie, with species local to the Edwards Aquifer prairie communities as well as species found transitioning to the drier desert communities to the west of Edwards Aquifer ecosystems. Wildlife is visible from windows throughout the house provides some evidence that the created habitat is suitable since birds, butterflies, bees, dragonflies, hawks, snakes, lizards, and frogs frequently visit the rooftop prairie. Construction took special measures as the roof required slope revetment with cables in the geotextile to prevent erosion, and a custom blended soil media with 50:50 organic to inorganic mix. Drip irrigation was installed to establish the green roof, and is used periodically throughout the summer during extended dry or hot periods (Fig. 16.19).

Although the Edgeland House is not a conventional residential structure, it is a model for how to integrate a structure with its landscape and ecological heritage. A

Fig. 16.18 Blackland Prairie plant species extend from the ground up to the vegetated and sloped rooftop for a seamless integration of land, structure and prairie. (Copyright Paul Bardagjy)

Fig. 16.19 Custom soil with cabled slope revetment geotextile, drip tube, and plant installation on the roof during construction. (Mark Simmons, Ecosystem Design Group)

Table 16.9 Ecosystem characteristics for the Edgeland House Green Roof

System criteria	Characteristic
Ecoregion	Tallgrass Prairie
Year completed	2013
System type	Semi-intensive
System provider/growing media	Ecosystem Design Group's propriety substrate mix
Size of green roof (square meters)	213.67
Approx. height above grade (meters)	0–3
Substrate depths (cm)	15.24
Drainage system (i.e. granular, plastic)	None; integral to the Geoweb slope revetment mesh
Dominant form of vegetative	Grasses and forbs
Species planted; species maintained	75; 75+/–
Irrigation (i.e. temporary, permanent, periodic, none)	Drip for establishment and supplemental in summer
Notable ecosystem features	Species for butterfly/humming bird host and food sources
Characteristic or notable plant species sustained on the green roof	*Aristida purpurea, Bothriochloa barbinodis, Bouteloua curtipendula Buchloe dactyloides, Hilaria belangeri, Leptochloa dubia, Sporobolus cryptandrus, Triden albescens*

conventional succulent or sedum-based green roof on the Edgeland House would perhaps have some nominal ecological benefits, but clearly the aesthetics and ecological services of such a roof would have been limited. Here, the prairie plants set a precedent for earth sheltered structures in central Texas and places with similar climates (Table 16.9).

16.11 Plant Dynamics and Plant Assemblages, Chicago City Hall, Chicago, IL, U.S.A.

Mature Midwestern US prairies often exhibit diverse plant and animal assemblages and communities that change over time (Packard and Mutel 1997). Restoration of prairies in the Chicago region is well known, but prior to the Chicago City Hall green roof, no investigation of prairie species occurred for green roofs. Grouped arrays (by bloom color) of primarily native prairie species were deployed across extensive and semi-intensive systems. The city wanted to learn how different species might persist on the green roof, and learn about maintenance requirements (Dvorak and Carroll 2008). Grouped arrays (by bloom color) of primarily native prairie species were deployed across extensive and semi-intensive systems. Crews maintain the roof several times each year dead-heading biomass, pruning sprawling vegetation, adding new trial plants, and removing aggressive species. Undulating substrates approximately 9 to 46 cm deep have

Fig. 16.20 Initially planted with bands of plants extending across semi-intensive (background) and extensive (foreground) substrates, plant species have now sorted to their preferred microclimates on the roof. (Bruce Dvorak)

diverse moisture gradients and solar aspects (Fig. 16.20). Persisting plants include Big Bluestem (*Andropogon gerardii*), growing on the mechanical room rooftops without irrigation, and Cardinal flower (*Lobelia cardinalis*) on the main roof with supplemental irrigation during prolonged dry periods (Fig. 16.21). On the mechanical rooms, the gravel-based drainage and limited number of drains maintain a slow-draining substrate that retains moisture. Over time, species com-

Fig. 16.21 Initially planted with 120 species of plants, the green roof now maintains up to 160 species, including Labelia cardinalis, and many species native to the Chicago area prairies. (Kevin Carroll)

Table 16.10 Ecosystem characteristics for the Chicago City Hall Pilot Project Green Roof. (Dvorak 2009; Dvorak and Carroll 2008; Yocca 2002)

System criteria	Characteristic
Ecoregion	Tallgrass Prairie
Year completed	2001
System type	Extensive and semi-intensive
System provider/growing media	Roofscapes (Optigrüen system)
Size of green roof (square meters)	3,530 (all roofs)
Approx. height above grade (meters)	33
Substrate depths (cm)	8.9, 15.24, 45.72
Drainage system (i.e. granular, plastic)	Granular-based and synthetic
Dominant form of vegetative	Gravel hill prairie, succulents
Species planted; species maintained	120; 160 (Dvorak and Carroll 2008)
Irrigation (i.e. temporary, permanent, periodic, none)	Main: drip during dry periods or drought Mechanical rooms: not irrigated
Notable ecosystem features	Main roof has an undulating substrate with diverse moisture gradients and solar aspects for self-sorting of species
Characteristic or notable plant species sustained on the green roof	*Allium cernuum, Andropogon gerardii, Amorpha canescens, Aquilegia canadensis, Asclepias incarnata, Aster laevis, Bouteloua cutipendula, Carex spp, Cassia fasiculata, Physostegia virginica, Sporobolus heterolepis, Ruellia humilis, Verbena hastate*

position and location varies as the initial radial bands have transitioned into a complex arrangement of over 160 species (see Table 16.10). Competition and adaption across varied substrate depths shows moisture loving species adapted to the flat and moisture holding semi-intensive microclimates; succulent species prefer the sloped and well-drained extensive microclimates (Fig. 16.20).

16.12 Roof Top Insect Communities, Berlin Ökowerk Nature Conservation Center, Berlin, Germany (Brown 2010; Ksiazek 2014)

Bird populations, insects, and other pollinators are being recognized on vegetated roofs across the world (Ksiazek et al. 2012; MacIvor and Lundholm 2011a; Brenneisen 2006). Insects have been observed on many green roofs, young and old. This case study examines the insect and plant communities observed on one of the world's oldest green roofs. The Ökowerk Nature Conservation Center in Berlin was completed in 1870 as a water treatment plant. The roof membrane was of bituminous material, and covered with 6 cm of sand. Over the years, ruderal species such as lichens, moss and a few species of sedum colonized and populated the roof for over one hundred years (Fig. 16.22). By 2005, the waterproofing, however, was in need of repair, and a plan was devised to replace the waterproofing while attempting to preserve the green roof's flora and fauna. The restoration included the complete removal, stockpiling and replacement of the sand substrate. Preserved clusters of plants, and the seed banks that remained in the soil were the only source of replanting and re-establishing the prior vegetative communities. As of the summer 2013, the roof is a habitat for one (1) species of fern, nine (9) grasses, fifty (50) herbs, four (4) species of mosses and lichens, and seven (7) tree species (only as seedlings)—all which provide habitat for eleven (11) species of spiders, fifteen (15) beetles, twenty-five (25) species of bees, five (5) wasps, and at least one (1) ant species (Table 16.11) (Ksiazek 2014). Thus the Ökowerk Nature center roof remains biologically diverse and serves as an example that even a very old green roof can be restored and maintained after resurfacing and repairs to the waterproofing.

Green roofs exhibiting healthy insect populations such as Ökowerk, are important to understand baseline conditions for the formation of insect communities on green roofs in large metropolitan cities such as Berlin (Fig. 16.23). As urban development continues to expand globally, habitat for insects for their own preservation or as a food source for other urban wildlife may prove critical. Bees are currently known to be in population decline across the world, and there is good evidence that green roofs can help maintain habitat for these important pollinators (Ksiazek et al. 2012; MacIvor et al. 2014; Colla et al. 2009).

Fig. 16.22 The 130-year-old ruderal green roof in Berlin, Germany has been colonized over the years by moss, lichen, grasses and over 50 species of herbs. After the green roof was removed, stockpiled, and restored in 2005, the same plant communities re-colonized the roof. (Kelly Ksiazek)

Table 16.11 Ecosystem characteristics for the Ökowerk Nature Center Green Roof

System criteria	Characteristic
Ecoregion	Mixed forest
Year completed	Around 1870; 2005 (restored waterproofing)
System type	Extensive
System provider/growing media	Sand
Size of green roof (square meters)	576
Approx. height above grade (meters)	3.65
Substrate depths (cm)	6
Drainage system (i.e. granular, plastic)	Sand. The roof deck is sloped from the center of the roof towards the edges. The roof drains to external gutters and downspouts
Dominant form of vegetative	Ruderal, meadow, and annual species
Species planted; species maintained	Unknown number from the ambient seed bank. 71 species observed on the roof (2013)
Irrigation (i.e. temporary, permanent, periodic, none)	None
Notable ecosystem features	There are species of flies, aphids, grasshoppers, snails, moths
Characteristic or notable plant species sustained on the green roof	Moss, rare lichens, and a few species of *Sedum*

Fig. 16.23 Ants such as the Formica species, here on lichen, inhabit the green roof along with many other pollinators and birds. (Kelly Ksiazek)

16.13 Conclusions

While the previous chapters in the book discuss discrete ecological aspects of green roofs at length, this chapter summarized these concepts through their built forms, parts, aesthetics and functions. For example, the restoration of the green roof at the Ökowerk Nature Conservation Center in Berlin, demonstrates ruderal plant and insect communities, and resilience. Green roof research on bird species in London and Switzerland demonstrates ecological benefits not only for the immediate roof, but for mobile and migrating species as well. The case studies vary greatly in size; however, the size of a green roof (Fig. 16.24) seems not exclusive of ecological or aesthetic values. The Vancouver Convention Center roof meadow is a large installation and may prove invaluable to the preservation of coastal meadows; but small green roofs were also vital to their environments as well such as the Edgeland House. In summary, this chapter demonstrates that there are many outstanding examples of green roofs with clear ecological and aesthetic qualities and, therefore; green roofs can be dynamic and beautiful elements of urban environments (Sutton 2014).

When land becomes urbanized, its ecological functions typically become stressed (Aitkenhead-Peterson and Volder 2010). However, the thoughtful integra-

Fig. 16.24 A houseboat green roof in the Netherlands exemplifies that no roof is too small to be greened. As urban populations continue to swell, personal green space will become more limited. The thoughtful design and placement of vegetated rooftops can trigger manifold expressions of ecological vigor and beauty. (Bruce Dvorak)

tion of green infrastructure in cities may prove useful not only for human health, but healthy urban ecology and species preservation (Kowarik 2011; Tzoulas et al. 2007; Francis and Lorimer 2011). The concepts presented in this chapter are also useful in understanding a wider integration of green roofs in cities. Across the world, there are prospects for large multi-structured developments such as the Vancouver Convention Center, with elegantly intertwined green roofs, decks, bridges and facades (Velazquez 2014). If green roofs can be conceived of and designed for improved human and ecological health, then perhaps they can begin to meaningfully transform cities. In his discussion of the Land Pyramid, Aldo Leopold asserted that if we think of the land (or a green roof in this case) as solely an economic commodity independent of its ecological context, it may be a hopelessly lopsided venture (Leopold 1966). This chapter begins to demonstrate how green roofs need not be only a nominal commodity, but can be a vital or priceless ecological gain that is richly expressive of its local ecology or region.

Acknowledgements I would like to thank the following for their review and contributors to this chapter including Lee Skabelund, Nicholas Williams, Bruce Hemstock, Ishi Buffam, Brooke Byerley Best, Stephanie Carlisle, Jeremy Lundholm, Dusty Gedge, Stephan Brenneisen, Mark Simmons, Kelly Ksiazek and Richard K. Sutton Without your invaluable comments and feedback, this chapter would not have been possible.

References

Aitkenhead-Peterson J, Volder A (2010) Urban ecosystem ecology. American Society of Agronomy. Madison

Aitkenhead-Peterson J, Dvorak B, Volder A, Stanley N (2010) Chemistry of growth medium and leachate from green roof systems in south-central Texas. Urban Ecosyst 14(1):1–17. doi:10.1007/s11252-010-0137-4

Blank L, Vasl A, Levy S, Grant G, Kadas G, Dafni A, Blaustein L (2013) Directions in green roof research: A bibliometric study. Build Environ 66(0):23–28. doi:10.1016/j.buildenv.2013.04.017

Brenneisen S (2003) The benefits of biodiversity from green roofs: Key design consequences. Paper presented at the in Proceeding of the 1st North American green roof conference: greening rooftops for sustainable communities, Chicago, Illinois,

Brenneisen S (2006) Space for urban wildlife: designing green roofs as habitats in Switzerland. Urban Habitats 4(1):27–36

Brown D (2010) Ökowerk Water Works, Berlin. Living roofs.org. Accessed March 13 2014

Buffam I (July 31, 2014) Personal communication on the substrate nutrient behavior at the Cincinnati Civic Center Cottage Garden Green Roof

Byerley B (August 7, 2014) Personal communication on the plant diversity on the BRIT green roof

Cantor S (2008) Green roofs in sustainable landscape design. Norton, New York

Carlisle S, Piana M, Timberlake K (2013) Growing resilience: long-term plant dynamics and green roof performance. Paper presented at the Cities Alive, 11th annual green roof and wall conference, San Francisco, October 23–26

Colla SR, Willis E, Packer L (2009) Can green roofs provide habitat for urban bees (Hymenoptera: Apidae)? Cities Environ 2(1):12

Dunnett N, Kingsbury N (2004) Planting green roofs and living walls. Timber Press, Portland

Dunnett N, Nagase A (2007) The dynamics and visual impact of planted and colonizing species on a green roof over 6 growing seasons 2001–2006: influence of substrate depth. Paper presented at the fifth greening rooftops for sustainable communities conference, Minneapolis, MN, April 29–30, May 1, 2007

Dvorak B (2009) The Chicago city hall green roof pilot project: a case study. Paper presented at the third international conference on smart and sustainable built environments. From building to promise building smartly in a changing climate, Delft, Netherlands, June 15–19, 2009

Dvorak B (2011) Comparative analysis of green roof guidelines and standards in Europe and North America. J Green Build 6(2):170–191. doi:10.3992/jgb.6.2.170

Dvorak B, Carroll K (2008) Chicago city hall green roof: its evolving form and care. Paper presented at the sixth annual international rooftops for sustainable communities conference, Baltimore, MD, April 30-May 2

Dvorak B, Volder A (2010) Green roof vegetation for North American ecoregions: a literature review. Landsc Urban Plan 96(4):197–213

Dvorak B, Byerley B, Volder A (2013) Plant species survival on three water conserving green roofs in a hot humid subtropical climate. J Living Archit 1(1):10

Farrell C, Szota C, Williams NG, Arndt S (2013) High water users can be drought tolerant: using physiological traits for green roof plant selection. Plant Soil:1–17. doi:10.1007/s11104-013-1725-x

Forman RT (1995) Some general principles of landscape and regional ecology. Landsc Ecol 10(3):133–142

Francis RA, Lorimer J (2011) Urban reconciliation ecology: the potential of living roofs and walls. J Environ Manag 92(6):1429–1437. doi:10.1016/j.jenvman.2011.01.012

Gedge D (2003) From rubble to redstarts. Paper presented at the greening rooftops for sustainable communities

Gedge D, Kadas G (2005) Green roofs and biodiversity. Biologist 52(3):161–169

Gedge D, Grant G, Kadas G, Dinham C (2012) Creating green roofs for invertebrates–a best practice guide. Buglife, Peterborough

Heim A, Lundholm J (2014a) The effects of substrate depth heterogeneity on plant species coexistence on an extensive green roof. Ecol Eng 68(0):184–188. doi:10.1016/j.ecoleng.2014.03.023

Heim A, Lundholm J (2014b) Species interactions in green roof vegetation suggest complementary planting mixtures. Landsc Urban Plan 130(0):125–133. doi:10.1016/j.landurbplan.2014.07.007

Heim A, Lundholm J, Philip L (2014) The impact of mosses on the growth of neighbouring vascular plants, substrate temperature and evapotranspiration on an extensive green roof. Urban Ecosyst:1–15. doi:10.1007/s11252-014-0367-y

Kadas G (2006) Rare invertebrates colonizing green roofs in London. Urban Habitats 4(1):66–86

Kelly S (2013) The Botanical Research Institute of Texas LandscapeOnlinecom (LASN–October)

Kowarik I (2011) Novel urban ecosystems, biodiversity, and conservation. Environ Pollut 159(8–9):1974–1983. doi:10.1016/j.envpol.2011.02.022

Ksiazek K (March 24, 2014) Personal communication on fanua and flora observed summer 2013 at the Ökowerk nature conservation center. Berlin

Ksiazek K, Fant J, Skogen K (2012) An assessment of pollen limitation on Chicago green roofs. Landsc Urban Plan 107(4):401–408. doi:10.1016/j.landurbplan.2012.07.008

Leopold A (1966) A sand county almanac: with essays on conservation from round river. Oxford University Press, Inc., USA

Lundholm J (2005) A habitat template approach to green building surfaces. Paper presented at the third annual greening rooftops for sustainable communities conference, Washington, D.C., May 4–6, 2005

Lundholm J, MacIvor JS, MacDougall Z, Ranalli M (2010) Plant species and functional group combinations affect green roof ecosystem functions. PLoS ONE 5(3):11

MacIvor JS, Lundholm J (2011a) Insect species composition and diversity on intensive green roofs and adjacent level-ground habitats. Urban Ecosyst 14 (2):225–241. doi:10.1007/s11252-010-0149-0

MacIvor JS, Lundholm J (2011b) Performance evaluation of native plants suited to extensive green roof conditions in a maritime climate. Ecol Eng 37 (3):407–417. doi:10.1016/j.ecoleng.2010.10.004

MacIvor Js, Lundhom J (2011c) Reconciliation ecology opportunities reach new heights: insect species composition and diversity on green roofs and adjacent ground-level habitat patches in an urban area. In: America ESo (ed) 96the ESA annual meeting, Austin, Texas, August 7–12 2011. ESA, p 1

MacIvor JS, Ruttan A, Salehi B (2014) Exotics on exotics: pollen analysis of urban bees visiting Sedum on a green roof. Urban Ecosyst:1–12. doi:10.1007/s11252-014-0408-6

Mooney-Bullock R, Buffam I, Bolan M (2012) Urban learning laboratory features four educational monitored green roofs. Paper presented at the 10th annual green roof and wall conference, Chicago, Illinois

Nightingale L (2010) West coast integrated design: how an integrative process creted innovative green roof in Vancouver. Living Architecture Monitor, vol 12, Summer 2010 edn. Green roof for healthy cities, Toronto, Onterio

Oberndorfer E, Lundholm J, Bass B, Coffman R, Doshi H, Dunnett N, Gaffin S, Köhler M, Lui K, Rowe B (2007) Green roofs as urban ecosystems: ecological structures, functions, and services. Bioscience 57(10):823–833

Packard S, Mutel CF (1997) The tallgrass restoration handbook: for prairies, savannas and woodlands. Island Press, Washington, D.C.

Parker MA (2013) Country Bunker. Dwell Vol 13. Dwell media, San Francisco

Simmons M, Gardiner B, Windhager S, Tinsley J (2008) Green roofs are not created equal: the hydrologic and thermal performance of six different extensive green roofs and reflective and non-reflective roofs in a sub-tropical climate. Urban Ecosyst 11(4):339–348

Skabelund LR, Blocksome C, Hamehkasi M, Jin Kim H, Knapp M, Brokesh D (2014) Semi-arid green roof research 2009–2014: resilience of native species. Paper presented at the cities alive 2014 12th annual green roof & wall conference, Nashville, TN, November 12–15, 2014

Snodgrass EC, McIntyre L (2010) The green roof manual: a professional guide to design, installation, and maintenance. Timber Press, London

Sutton RK (2014) Aesthetics for green roofs and green walls. J Living Archit 1(2):20

Sutton RK, Harrington JA, Skabelund L, MacDonagh P, Coffman RR, Koch G (2012) Prairie-based green roofs: literature, templates, and analogs. J Green Build 7(1):143–172. doi:10.3992/jgb.7.1.143

Tzoulas K, Korpela K, Venn S, Yli-Pelkonen V, Kaźmierczak A, Niemela J, James P (2007) Promoting ecosystem and human health in urban areas using green infrastructure: a literature review. Landsc Urban Plan 81(3):167–178. doi:10.1016/j.landurbplan.2007.02.001

Velazquez L (2014) Top 10 hot trends in greenroof & greenwall design. greenroofs.com. Accessed July 30 2014

Yocca D (2002) Chicago city hall green roof ASLA design award web site text. ASLA Web Site.

Bruce Dvorak ASLA, RLA is an associate professor in the Department of Landscape Architecture and Urban Planning at Texas A&M University where in 2009, he established the Interdisciplinary Green Roof Research Group.

Chapter 17
Green Roof Ecosystems: Summary and Synthesis

Richard K. Sutton and John Lambrinos

Abstract This book demonstrates that green roofs can and must be conceived of as ecosystems if we want to maintain their utility in providing ecosystem services and benefits while improving their effectiveness. In this chapter we summarize and synthesize previous chapters and we emphasize several overarching ecological concepts and those associated with community ecology. We then list and discuss potential future research areas and questions to improve green roof ecosystem understanding, design, management and policy.

Keywords Summary · Community ecology · Ecological applications · Green roof design · Adaptive management · Research questions

17.1 Overarching Ecological Concepts Applied to Green Roofs

Using an ecosystem perspective enhances our understanding of green roofs. They can be seen as novel, hybrid, anthropogenic structures which provide ecological services that can change the relationship between humans and our urban niche. Structure, function, and change embody ecological concepts of interest to researchers, designers, and those charged with managing green roofs. We highlight concepts with strong connections and applications to the design of green roof ecosystems, note where knowledge is limited and describe how ecosystem conscious designers might investigate questions as green roofs are created.

R. K. Sutton (✉)
Department of Agronomy & Horticulture, UN-Lincoln, 279 Plant Sciences, Lincoln, NE 68583-0915, USA
e-mail: rsutton1@unl.edu

J. Lambrinos
Oregon State University, 4017 Agriculture and Life Sciences Building, Corvallis, OR 97331-7304, USA

R. Sutton (ed.), *Green Roof Ecosystems*, Ecological Studies 223,
DOI 10.1007/978-3-319-14983-7_17

The preceding chapters highlighted a number of areas where findings and concepts from the broader ecological literature have been applied to understand the function of green roofs (Table 17.1) and as with most research many more questions remain to be explored.

Humans are Part of the Green Roof Ecosystem Historically, the study and applied management of natural biological systems such as forests developed largely independently from the study and applied management of highly human regulated systems such as urban landscapes. However, over the past two decades the distinction between natural and human dominated environments has become increasingly blurred, not the least reason being that humans now exert a strong influence on nearly all of the earth's ecosystems (Vitousek et al. 1997; Sanderson et al. 2002).

Many ecosystems are now largely novel constructions that are continuing to change at an increasingly rapid rate (Steffen et al. 2007; Blandin 2011; Haber 2011; Hobbs et al. 2013; Oldfield et al. 2014).

Despite this shift in focus we still have a relatively poor ecological understanding of intensively managed systems, and as a result, their design and management is still largely carried out oblivious to (and in some cases in direct conflict with) underlying ecological processes. This is beginning to change. The chapters in this book highlight the immense progress that has been made in understanding the ecology of green roofs. Despite being designed, constructed, and managed by humans, green roofs are nevertheless ecological systems whose characteristics and functions reflect an interaction between design, management and natural processes (Table 17.2). Green roof research design and management are beginning to examine, reflect, and even exploit these relationships.

Ecosystem processes display the interaction of abiotic and biotic features, but the simplicity of green roof ecosystems when compared to natural ones truncates the suite of ecosystem processes. For example, shallow substrates do not allow hydraulic distribution (Leffler et al. 2005; Richards and Caldwell 1987) of water during dry periods. Closely related to hydraulic distribution, evapotranspiration strongly affects the growth of plants and the very function of green roofs for water storage and dispersal between storms (Chaps. 3 and 4). Plant water use can vary with the species and its seasonal growth. For example, each plant guild filters water differently and synchronizes its phenology differently thus affecting pollinators (Chap. 4).

Interaction and biodiversity dynamics are the focus of a significant amount of ecological research devoted to understanding the relationship between biodiversity and the functioning of ecosystems. While a comprehensive model of biodiversity and ecosystem function has so far proven elusive, there is strong evidence that the functional characteristics of species strongly influence ecosystem properties, some particular combinations of species complementarily enhance ecosystem function, and having a range of species that respond differently to perturbations can stabilize ecosystems (Hooper et al. 2005). Several chapters in this book describe the evidence for these relationships in green roof systems, and explore the potential for

Table 17.1 Distribution of ecological themes and concepts in chapter papers

	Chapter													
	2	3	4	5	6	7	8	9	10	11	12	13	14	15
Overarching themes														
Humans are in the system	√	√	√	√	√	√	√	√	√	√	√	√	√	√
Ecosystem processes	√	√	√	√	√	√	√	√	√	√	√	√	√	√
Interaction & biodiversity dynamics	√			√	√	√	√	√		√	√	√	√	√
Cycling of energy and nutrients				√	√	√	√	√			√		√	
Ecosystem benefits/services		√	√			√		√		√			√	√
Wholes	√	√	√	√	√						√			
Self-organization					√	√	√			√	√			√
Emergent complexity					√			√				√		
Microbial community manipulation			√	√	√	√							√	
Wider view of ecology							√						√	√
Community ecology concepts														√
Abiotic influences	√	√	√	√	√	√	√	√	√	√	√	√	√	√
Biotic influences		√	√	√	√	√	√	√	√		√			√
Disturbance regimes		√	√				√		√	√	√	√	√	√
Functional groups						√	√	√	√	√	√	√	√	
Filters			√		√				√	√	√			√
Competition			√			√	√		√	√	√			√
Invasive species/Ruderals							√		√		√	√		
Symbioses and mutalisms			√			√	√	√						√
Pathogens						√	√							

Table 17.2 Typical degree of human and ecological regulation of green roof components

Green roof component	Degree of regulation through design or management	Degree of ecological or physical regulation	Associated physical & ecological processes
Plant community	High	Low	Abiotic sorting, dispersal and colonization, interspecific competition, facilitation, abiotic sorting
Soil/substrate	High at installation; low afterwards	Low at installation, increasing with time	Erosion, decomposition, atmospheric deposition
Soil biological community	Low	High	Abiotic sorting, dispersal and colonization, interspecific competition, facilitation, abiotic sorting
Flows of water, nutrients and energy	Intermediate	High	Climatic conditions, ecosystem engineering from vegetation (e.g. evapotranspiration, shading)

manipulating green roof biodiversity to attain specific goals such as resilience to drought stress.

Chapters 6–14 all discuss an overarching theme of biodiversity arising from the numbers and species of microbes, plants and animals that are either intentionally placed on or that autonomously colonize green roofs. Arbuscular mycorrhizal fungi (AM) and bacteria associations also have a role to play in the plant diversity that might be possible in a newly constructed environment. It is unknown what the successional microbial community might be and how it might affect the biodiversity on a green roof. Rooftop microbes may also play a role in detoxifying heavy metals. Many of the processes that drive community assembly after the initial installation of a roof (e.g. the airborne arrival of inoculum or seeds) involve a high degree of stochasticity. This randomness can complicate management, but also provides an invitation for humans to influence and direct the transition probabilities.

Ecosystems are Dynamic All ecosystems are fundamentally dynamic, patchy, and strongly influenced by periodic disturbances from frequent small impacts such as herbivory and extreme climactic events such as drought or fire that manifest themselves across a range of spatial and temporal scales (Picket and White 1985). In the past, many resource management decisions were typically based on relatively static ecosystem and community models. However, over the last several decades, natural resource managers have increasingly embraced more dynamic models that explicitly incorporate the influence of disturbance, stochasticity, and non-linear dynamics (Folke et al. 2004; Twidwell et al. 2013). In contrast, green roof designs and management specifications typically emphasize permanence and little variation from the initial design intent. As several of the chapters in this book describe, this approach can leave green roofs susceptible to perturbations such as climactic variation, and it can necessitate the use of resources to maintain initial conditions. Moreover, despite design intentions, most green roofs nevertheless change over time (Chap. 12). Studies have documented long-term changes in soil structure, fertility, species composition, and vegetation structure. Several chapters explored more resilient design approaches including using more functionally diverse initial plantings to hedge bets in the face of environmental stochasticity and accepting a degree of autonomous species turnover. However, the effects of disturbances on community structure often involve time lags and only manifest themselves over longer time frames. Few green roofs have been functioning long enough to provide a basis for evaluating the costs and benefits of these approaches (Chap. 13).

Cycling of Nutrients and Energy Cycles Green roofs largely occupy sunny locations atop buildings, thus experience high rates of insolation. While green roof vegetation must be able to survive and thrive in such conditions and it has proven difficult to find plants that require or tolerate both shade and dryness. A building's energy budget must also account for and ideally ameliorate high rates of insolation that is easily converted in to sensible heat (Chap. 2). The intervening vegetated layer provides such a service by capturing light energy for photosynthesis as well as shading roof surfaces and potentially cooling the building via evapotranspiration.

As both sources and sinks for nutrients, green roof substrate must be carefully specified to meet projected physical, chemical, and biological conditions. Initially simple, homogeneous assemblages of physical substrate materials quickly become more complex heterogeneous admixtures of nutrients, water, and biota. And while simpler than natural soils, the flow, recycling, constitution, and persistence of macro and micronutrients becomes complex (Chap. 5).

While the growth substrates typically used in green roofs are less complex than soil, over time they still support a complex community of below-ground organisms that mediate the flux of water and nutrients and influence plant productivity and community structure (Chap. 6). Chapters 3–7 all examined aspects of these complex interactions. The interaction between above ground and below ground processes is strongly influenced by environmental conditions. Unforeseen impacts occur with the use of substrate standards, materials and procedures designed and practiced on temperate climate green roofs when used in hot climates (Chap. 3). As organic matter breakdown accelerates, and freed nutrients may not be utilized if washed from the system. Macronutrients might be best understood initially in a ratio of availability since their relative abundance affects both plant and microbial performance. Rapid organic matter breakdown also reduces the substrate volume and ability to store water.

Ecosystems Provide Services and Benefits The very rationale for creating green roofs comes from ecosystems performing needed services and providing desired human benefits (Chap. 1, Table 1.2). Selection of media-plant assemblages relies on understanding their functional groupings (Chap. 9) therefore, research has attempted to isolate and measure individual plant's and plant groupings' effects on holding and cleaning stormwater, sequestrating carbon, cooling buildings and cities, capturing and denaturing air pollutants, increasing biodiversity and creating more pleasant urban views. The economic value of these services is beginning to be estimated and incorporated into cost-benefit analyses at both local and regional scales (Niu et al. 2010; Mullen et al. 2013).

Wholes, Self-organization and Emergent Complexity Green roofs are most usefully studied as wholes—i.e., entire, connected systems and we need to create models of them. For example, the impact of inter-relating factors on nutrients and their plant uptake remains poorly measured and understood both seasonally and long-term.

Wider View of Ecology While traditional ecological study has been grounded in community ecology and population ecology, it has expanded to understanding flows of matter and energy. Furthermore, as the changes and impacts on the global environment have become more obvious and in some cases threatening to humans, the field of ecology has become more interdisciplinary and has developed several sub-disciplines that seek to inform improved ecological knowledge and apply it to management of human dominated systems (Chap. 1). Green roofs as small, discrete entities can easily be seen and examined as communities with limited populations of organisms, but they also inform an applied ecological knowledge by way of their design and management.

Microbes: The Importance of Below Ground Biology Soil communities, particularly microbes, play a fundamental role in mediating the structure and function of terrestrial ecosystems (Wall et al. 2013). They are especially important in nutrient poor ecosystems where they regulate plant productivity by helping plants acquire limiting nutrients and water (Van der Heijden et al 2008). Our ability to manipulate these communities to attain specific functional goals has been hampered by a fundamental lack of understanding of soil ecology. However, this knowledge gap is beginning to narrow (Van der Putten et al. 2013). Several chapters in this book touch on the below ground ecology of green roofs. Substrate biology and microbial communities play a central role in regulating green roof function.

Using some fraction of native soil that has been vetted against abiotic requirements presents exciting possibilities for basic knowledge, design, and management. The extended case study in Chap. 6 relates such an approach, but also the frustrations of specifying any clay-sized particles in substrates as a part of the implementation process. Management inputs may actually increase when many normal soil functions such as seedbanks and soil microorganisms become traded-off between additional weight and drainage costs. Chapter 7 provides a baseline of knowledge about the presence and use of bacteria and fungi on green roofs. Unfortunately little is known about such important biological actors on green roofs, though some interest occurs regarding the use of arbuscular mycorrhizal (AM) fungal associations, to extend the root capacity of some species. No definitive assessment exists of whether AM fungi can or do readily form a symbiotic relationship with all plant species now widely planted on green roofs. AM fungal and bacteria associations also potentially influence and have a role to play in the plant diversity that might be possible in a newly constructed environment. It is unknown what the successional microbial community might be and how it might affect the biodiversity on a green roof.

17.2 Community Ecology Basics Applied to Green Roofs

Zeroing in on the microscopic world of minute organismal communities as an overarching theme provides an excellent segue to describing basic community ecology on green roofs. Traditional community ecology and its suite of key ecology concepts broaden the study of green roofs and those concepts contain important knowledge useful for green roof researchers, designers and managers. Community-focused aspects also underpin the new direction of ecological study in the twenty first Century (Schwarz and Jax 2011). Designing green roofs as experiments may yield results that further understanding of that concept.

Abiotic Influences For earlier green roof research and from literature based mostly in Europe and North America, Obderndorfer et al. (2007) reviewed interactions between green roof ecosystems and their elements: growing substrate, soil biota, and vegetation. They concentrated on relationships between structure, function, and change specifically, the abiotic environment and its moisture stress, extreme temperatures, high insolation; they also examined species dynamics and biodiversity in relation to the stresses found in green roof ecosystems. Based on what was known they made suggestions about tying those ecological areas to biodiversity policies and ecosystem services as controlled by people.

Most of the variable states for abiotic aspects of green roofs can be measured more or less precisely and being able to predict what to measure, its interactions, and then store that information as readily accessible data can be an expensive task. Using such information in tandem with biotic changes helps us understand the basics of their interrelationships in a very simple system.

What is known of carbon, nitrogen, and phosphorus in green roof substrate comes second-hand, from studying leachates. Yet, while the physical features of green roof substrate (material, depth, porosity, vertical and horizontal homogeneity etc.) are well known, the biogeochemistry, microbial biology and their interaction with non-sedum plant assemblies is not.

Water use is dynamic and complex and it interacts with regional and local microclimates, substrate composition, and plant assemblages. The hydraulics and hydrology of green roofs, in turn, importantly impact urban runoff and potentially perform environmental service or disservice. What may be a water-conserving plant cover in one season may limit rapid evapotranspiration in another season. Knowledge about plant's water use strategies can lead to more careful assembly of plants to match climate and substrate. Like substrate formulations, plant systems do not readily translate from cool climates to warm or dry ones and spawn a wider search for suitable species.

Wind and temperature regimes on green roofs can be easily measured and strongly affect both biotic and other abiotic factors.

Biotic influences represent one filter for selecting plants, for example a plant's morphology and anatomy (Chap. 11), affects flows of transpiration, air and suspended particulates, that, in turn, affect the ecosystem services a roof performs. Plant guild and functional groups impact ecosystem services differently. Identifying plant communities in naturally stressed areas that exist locally in most countries is one place to seek plants that may be adapted to roof tops and one that may already have formed interactions with other flora or fauna.

Relatively rapid plant responses such as the arrival of ruderals affect the stability and resilience of the plant assemblage and associated insects. Plant phenology, particularly flowering, impacts foraging by pollinators (Chap. 14). Roof microclimate, substrate, and plant structure are very important, because, low roof flora diversity often results in supporting only generalist fauna and rudimentary food chains. Invertebrate diversity is seasonally and yearly dynamic and strongly tied to plant biodiversity. Flows of invertebrates occur on and off roofs and affect functioning of nutrient cycling, food webs and pollination.

Disturbance regimes affect most designed green roof plantings intended to display static patterns; green roof ecosystems change over time. Many of these changes are undesirable in green roof systems that have specific design patterns and specifications. Some amount of active management is usually required to maintain important design characteristics. In addition, the degree and nature of post-installation changes reflect interactions between environmental conditions and the specific design of the green roof. Any design must respond to the initial site conditions as well as cyclical or random disturbance regimes. Over time disturbance influences the flow of material and energy as well as vegetative structure. Successional changes in assemblage composition over the long term are dynamic, therefore green roofs need to link disturbance and maintenance. Because trends on older green roofs show carbon accumulates, pH changes (moderates) fertility drops and then stabilizes, initial plantings should be functionally diverse to hedge bets regarding abiotic and biotic changes. However, many disturbance regimes have time-lag effects and become visible impacts only over longer time frames. Few green roofs have been functioning long enough to provide a basis for understanding what constitutes a reasonable (or predictable) stable state.

Functional traits of organisms in many communities reflect adaptive responses to their prevailing environment (Chap. 9). These relationships can be used to identify suites of traits that identify plant species that are suitable for green roof environments or that have desirable functional properties. Functional groups have been organized around anatomical characteristics, physiological processes, trophic levels, or behaviors to name a few. For example the importance of plants with C_3, C_4 or CAM photosynthetic processes affect a plant's need for and use of water and their ultimate suitability for green roof use. Other desirable functional traits might be found in plants with sequential bloom periods that not only support insects, but also garner human attention.

Filtered by a strongly selective of the green roof environment, plant diversity should be pre-adapted. Beyond simple genotypic and species diversity, different ways exist to measure diversity. For example, using functional group diversity, functional trait diversity and phylogenetic diversity may better stabilize green roof ecosystems, be related to the local species pool and provide wider benefits. Plant competition and facilitation occur on green roofs and help shape its diversity. Organisms, in turn, modify the physical conditions on a roof, which feedback on diversity. For example, biomass accumulation above and belowground helps create greater substrate heterogeneity that fosters plant diversity. Initially designing a more heterogeneous substrate structure may be a strategy for initiating green roof diversity.

Symbioses and mutualisms are not widely described or reported for green roofs. This may be a result from lack of robust populations of the interacting species, but researchers must look for them first. The major exception to that is the use of AM fungi to enhance the uptake of water and nutrients by symbiotic connection with green plant roots (Chap. 7, Chap. 11). AM have been added to some commercial green roof media mixes, but little is known of the most appropriate species to use.

Facilitation has been reported where both Sedums and native plantings have enhanced each other's germination and growth. Pollinators can help insure viable seeds for plant infill. Observations and experiments on green roofs are just beginning to tease apart positive biotic interactions.

Pathogens represent a small portion of the microbial community on green roofs, but it is unclear whether or not outbreaks may occur over the lifespan of a green roof, especially on roofs planted with only a few species of plants. In natural situations, pathogens are often kept in-check by other soil and endophytic microbes (Chap. 7). The bio-diversity of the substrate's microbial community is critical to pathogen suppression, but has had little study. Thus inoculation with a wide variety of microbial species may become common for future green roofs. A fear of large-scale impacts on monocultures of *Sedum* spp. on green roofs has not yet happened but easily could.

Invasive species and ruderals can both arrive on green roofs, though the invasives on a green roof would likely make it a source for wider, undesirable dispersal. Because of vegetation gaps on the green roof surface it is prone to colonization by wind-borne seeds, but these colonists typically reflect a narrow subset of the entire regional species pool. The colonizers are often ruderals whose diversity and persistence ebb and flow directly in relationship to availability of abiotic resources (Chap. 10). Ruderals, especially those present and easily dispersed in urban environments colonize newly created rubble-based substrate. Because of the uniqueness of such a site, urban ruderals do not reflect natural plant communities and characteristics that are just beginning to be understood (Chap. 10). They offer a way to diversify green roof plantings with ones adapted to repeated disturbance. Green roofs supply sites where the successional characteristics can be more closely studied.

Competition strongly drives community diversity and structure. On green roofs competition for open space, water, and nutrients favors some species over others, but the nature of that competition can vary over time. Diversity organizes largely from intra versus inter-species competition (Chap. 8, Chap. 10) and may not be apparent until the passing of several season or stress cycles. Work on biodiverse roofs (Chap. 10) has begun to tease apart the role of competition in structuring green roof pant assemblages suggesting that species labeled as competitors lose out over time because their reproduction becomes stymied by poor site resources.

17.3 Design Sets the Stage

This book was not meant to be a step-by-step design guide for green roof implementation. Other publications give a designer more background and process for doing so. However, not unsurprisingly since designers make decisions about substrate, plant assemblages, and projected management regimes, humans are seen as tightly enmeshed with green roofs. Ideas to further understand, design, and manage green roofs weave throughout this book's chapters and overarching themes identified in Chap. 1. Community ecology concepts have also been well represented (Table 17.2).

What the green roof designer should take away from this book are the following:

- Local policies, ordinances and codes also limit and shape green roofs
- Owners, developers, and users are stakeholders and learning will be required
- Every green roof has a unique abiotic setting that must be understood in detail
- Know, record, and use baseline conditions
- Selection of substrate depth and makeup precedes plant selection and diversity
- Plant selection and arrangement affects insect and avian visitors and diversity
- Every green roof is an unique experiment
- Seasonal timing for implementation is critical
- Grow-in management takes at least 3 to 5 years or longer
- Management of green roof should not negatively impact its surroundings
- Failed plantings are feedbacks that require study and adaptation.

17.4 Management Should Be Adaptive

A number of factors, not the least being the inherent variability of ecosystems, conspire to create significant uncertainty for natural resource managers. One of the most established tools for dealing with this uncertainty is the concept of adaptive management, which emphasizes flexible management based on learning, experience, and data (Westgate et al. 2013). Management in general, but especially adaptive management, requires a knowledgeable human to be closely monitoring the system and only impacting it after thoughtful and careful consideration.

Adaptive management is particularly useful for implementing new approaches or when novel circumstances are encountered. Several of the chapters advocate for an adaptive approach to both green roof design and their continued long-term management. This will require long term monitoring of green roof performance. Some green roofs have been extensively monitored in detail at least over short periods, but the potential for longer monitoring periods exists. In addition, research designs could be incorporated into more green roof management plans specifically to help develop more site-specific and regional best practices. Controlled disturbances in the form of maintenance can yield information about how certain ecological features and processes operate. Depending on the spatial size and temporal period examined, a mix of both planned anthropogenic and natural stochastic changes can be examined as experiments, observations or both (Fig. 17.1).

Ecological research has followed along similar lines. Traditional ecological research has been both observational and experimental. Observation has occurred in the Lab, Garden/Greenhouse and Field, whereas, the experimental approach has

	LABORATORY	GARDEN/ GREENHOUSE/ GREEN ROOF	FIELD
Spatial & Temporal Extent of Control	Smallest	Small	Large
Type & Amount of Control	Simplest, Maximum Control	Simple, Some Control	Complex Few Controls
Locus of Control	Strongly Anthropogenic	Anthropogenic	Natural
Approach	Largely Experimental	Experimental & Observational	Largely Observational

Fig. 17.1 Conceptual approaches to green roof research

relied largely on Lab and Garden/Greenhouse, because of difficulties in controlling field variation and complexity (e.g., Grime et al 1987; Kohler 2002)

17.5 Implications and Impact of Green Roof Policy

Initially focused on health, safety and welfare of citizens or users, individual building performance and energy conservation have worked their way into local development policies, guidelines and codes. When aggregated across many buildings across the urban milieu, benefits accrue and attract more policy implementation. This is especially true for heat island and stormwater mitigations. Led by Basel Switzerland and Toronto, Canada many cities worldwide are tackling policy for conserving and improving biodiversity and green roofs play a role (Chap. 15).

17.6 Identified Research Needs and Questions

Identified research needs and questions have not diminished since the publication by Oberndorfer et al. (2007) but they have become more sharply focused. Based on the preceding chapters it is possible to organize needs as: basic ecological understanding transferable beyond or applied to green roofs, green roof design, and green roof management. Using that framework we can then ask more detailed questions.

17.6.1 Understanding Ecology Applied to Green Roof: Future Research Questions

Our understanding of how green roofs function and change has been based on both documented experience and application of general knowledge about their biology, ecology, and environment setting. A possibility exists that green roofs could be places where even more basic information about ecosystems might be explored and discovered. This requires asking both basic and applied questions such as posed below.

Most of what is known about green roof services such as stormwater, microclimate, and noise attenuation has come from increasingly sophisticated experimental design and sensitive instrumentation.

- Are real-time monitoring systems possible?
- Can collected information be placed into researcher accessible data sets?
- Can further quantification be made of green roof benefits such as air quality?

Nutrient cycling supports the notion of self-organization and sustainability on green roofs. Since substrate and plantings can be controlled, perhaps more details of nutrient use and cycling could occur.

- How can we model nutrient cycling?
- What are the impacts of C:N:P ratios on nutrient cycling using mass-balance techniques?
- How does the complement of C3 and C4 plants cycle and enhance nutrient availability?
- What are the cumulative impacts of green roofs on climate change?

Little is known of microbial activity on green roofs, yet in established natural systems microbes provide important ecological services. Techniques to identify microbes are expensive and can only broadly be applied.

- Can we describe microbial community assembly, mutualism, parasitism, and species dynamics?
- How do abiotic versus biotic impacts shape rooftop microbial communities?
- How do we integrate monitoring of abiotic and biotic materials?

Inoculation of plants or substrates provides one method of insuring a microbial community.

- What are impacts of soil microbes on C:N:P and nutrient cycling?

With the introduction of both legumes and the properly matched species of *Rhizobium* bacteria, green roofs may produce some of their own needed nitrogen.

- What are the existence and importance of nitrogen fixation on green roofs?

Many plants associate symbiotically with specific suites of arbuscular mycorrhizal (AM) fungi. These too can be inoculated on plants or intro substrate.

- What are the specifics of mycorrhizal fungal roles on plant survival, plant performance, and plant nutrient uptake?
- Which *Sedum* spp., if any, form AM liaisons and do mycorrhizal versus non-mycorrhizal plants perform differently on green roofs?

An important ecosystem service provided by green roofs comes from diversifying the living things in an urban setting.

- Does plant diversity tend to increase over time?
- Does initial species establishment channel composition or will species composition change over time?
- What is the long-term impact of multiple microclimates on roof biodiversity and performance?
- What is the impact of diverse phylogenetic species versus simple species richness?
- How can species diversity be used to design green roofs that overcome tradeoffs in plant performance?

- Are roofs simply islands of diversity unlike their context? Or do they extend the continuity of their surroundings?
- What degree of functional diversity (structural and phylogenetic) do regional roofs possess/deliver? Or do they seem to extend the continuity of their surroundings?
- Does green roof biodiversity tend, towards or away from beta-diversity? For example are roofs simply islands of diversity unlike their context?
- To what degree can green roofs extend and disrupt local natural heritage and regional species flow paths?
- How do the effects of plant traits on green roof function vary across climates?

Most information about green roofs comes from snapshots. Longer-term information would provide a more realistic view of green roofs, but depend on rigorous collection of baseline data.

- What is the minimum type and detail of baseline data?
- How do we obtain longer term plant dynamics and resilience information such as plant function in relationship to plant species and their dynamics and its accumulated impact on roof performance?
- Is it possible to set off a predicable, long-term succession from the outset of green roof creation?
- How does inclusion of a compatible mix of shorter-term ruderals combined with longer term and more persistent species alter the later composition?
- Are regeneration characteristics of component species more important, or as important, as those that encourage persistence under stress?
- What can be learned from comparing over time geographically distinct, replications of green roofs with the same or similar functional species that are scored for their roof performance?
- How do substrates change during long-term studies in hot climates?
- Do *Sedum* spp. monocultures change the substrate compared with prairie grass and forb assemblages?

17.6.2 Design

An important aim of this book is to provide green roof designers a more grounded overview of ecology useful when creating novel, hybrid ecosystems atop built structures and to think of all green roofs as experiments. It also hopefully exposes designers to the understanding of this ecosystem's dynamics and informs more realistic selection and arrangement of substrate, plants, and outcomes of management regimes.

Arbuscular mycorrhizal (AM) symbionts offer potential plant sustenance, but little is known about them as a part of the green roof ecosystem.

- Catalog and understand AM associated with green roof plant species.
- How best to inoculate green roof plantings?

The dominant plantings on green roofs are *Sedum* spp. These have done well in northern Europe, but their narrow diversity base may not be useful in other climates.

- How can side-by-side studies of *Sedum* monocultures versus native versus mixed native and *sedum* plantings affect green roof functions?
- What species besides *Sedum* spp. are broadly adapted to various substrate, water, and temperature regimes?
- What are ET rates for various prairie forbs and grasses?
- What are the protocols and levels of performance-based assessment for overall plant assemblages?
- What are and how do we measure tradeoffs between plant traits and designs that are good for initial establishment vs. those good for longer term functioning?
- Which plant traits optimize multiple ecosystem services?

Green roof designers must not only understand the regional context, and the potential assemblage of plants, but also the growing substrate.

- What revisions need be made to the FLL substrate guideline for its use in hot, moist and hot, dry climates?
- How do substrate amendments affect outflow of water and nutrients?

Since each green roof that is designed and implemented is unique, the designer should at least collect basic, as-built information to help gage substrate and vegetation changes.

- How does designer/researcher collect baseline information and where is it stored?
- How can a designer integrate dynamics theory to observation and experimentation of green roof plant assemblages?

Since many green roofs depict strongly geometric or human created forms and patterns, it is important to understand their implication for user preference and long-term management.

- What are the causes, timing, and reactions to extinguished designed plant patterns?
- Describe and interpret human preferences for simple versus green roof patterns.
- How does ecosystem service provisioning differ between highly managed green roofs versus those where self-organization and spontaneous dynamics dominate community assembly?
- How can the encouragement of a dynamic and spontaneous approach to green roof vegetation be reconciled with human aesthetic aspirations?

17.6.3 Green Roof Maintenance and Management

Over the potential 40–60 year (or longer) life of a green roof more resources may be expended for maintenance than on the initial green roof's design and installation. Additionally, outputs from the roof may impact its nearby surroundings. Since maintenance or management extends the design intent of the green roof, such care must be understood as both static design and dynamic disturbance.

- What planting and management tactics improve plant establishment?
- Describe and explain impacts and inter-relationships between pests and green roof assemblages.
- Will weed species and pressure change over time?
- How can plant dynamics be used to inform green roof management?
- Are periodic management interventions necessary or desirable to maintain the ruderal element in a green roof?
- Does the type and timing of the disturbance/intervention affect later species composition?
- What are best nutrient management practices for green roofs? Should green roofs be fertilized or irrigated? If unfertilized, will deficiencies appear?
- How to deal with excessive biomass accumulation over time?
- What are the performance effects over time of removal of invasives or biomass?
- What can be learned from turfgrass science, range management, and prairie restoration applicable to green roofs?
- To what degree do management authorities' and citizens' actions contribute to or limit green roof services?
- Is burning on green roofs for assemblages of prairie plantings feasible and desirable?

17.6.4 Green Roof Policy

Green roofs could contribute to helping solve a number of issues facing cities worldwide such as conserving biodiversity and mitigating pollution. Cities are just beginning to craft policy that integrates green roofs into broader management strategies for the complex urban landscape.

- Do biodiverse green roof guidelines work?
- How are green roofs as best management practices integrated into urban sustainability?
- What are the private and public economic incentives and disincentives of implementing green roofs?
- Should green roofs be discouraged or their implementation regulated in some circumstances?
- Is it possible to determine how many species, and what keystone species are essential in the initial roof design to create a functional ecosystem?

References

Blandin P (2011) Ecology and biodiversity at the beginning of the twenty-first century: towards a new paradigm? In: Schwarz A, Jax K (eds) Ecology revisited: reflecting on concepts, advancing science. Springer, New York. doi:10.1007/978-90-481-9744-617

Folke C, Carpenter S, Walker B, Scheffer M, Elmqvist T, Gunderson L, Holling CS (2004) Regime shifts, resilience, and biodiversity in ecosystem management. Annu Rev Ecol Evol Syst 35:557–581

Grime J, Mackey J, Hillier S, Read D (1987) Floristic diversity in a model system using experimental microcosms. Lett Nat 328:420–422

Haber W (2011) An ecosystem view into the twenty-first century. In: Schwarz A, Jax K (eds) Ecology revisited: reflecting on concepts, advancing science. Springer, New York. doi:10.1007/978-90-481-9744-618

Hobbs RJ, Higgs ES, Hall C (2013) Novel ecosystems: intervening in the new ecological world order. Wiley, New York

Hooper DU, Chapin FS et al (2005) Effects of biodiversity on ecosystem functioning: a consensus of current knowledge. Ecol Monogr 75(1):3–35

Kohler RE (2002) Landscapes and labscapes: exploring the lab-field border in biology. University of Chicago Press, Chicago

Leffler AJ, Peek MS, Ryel RJ, Ivans CY, Caldwell MM, (2005) Hydraulic redistribution through the root systems of senesced plants. Ecology 86:633–642

Mullen JD, Lamsal M, Colson G (2013) Green roof adoption in Atlanta, Georgia: the effects of building characteristics and subsidies on net private, public, and social benefits. Environ Sci Technol 47(19):10824–10831

Niu, H, Clark C, Zhou J, Adriaens P (2010) Scaling of economic benefits from green roof implementation in Washington, DC. Environ Sci Technol 44(11):4302–4308

Oberndorfer E, Lundholm J, Bass B, Coffman R, Doshi, Dunnett N, Gaffin S, Köhler M, Liu K, Rowe DB (2007) Green roofs as urban ecosystems: ecological structures, functions, and services. BioScience 57(10):823–833

Oldfield, F, Barnosky AD, Dearing J, Fischer-Kowalski M, McNeill J, Steffen W, Zalasiewicz J (2014) The anthropocene review: its significance, implications and the rationale for a new transdisciplinary journal. Anthr Rev 1(1):3–7

Pickett STA, White SA (1985) The ecology of natural disturbance and patch dynamics. Academic, New York

Richards JH, Caldwell MM (1987) Hydraulic lift: substantial nocturnal water transport between soil layers by *Artemisia tridentata* roots. Oecologia 73:486–489

Sanderson EW, Jaiteh M, Levy MA, Redford KH, Wannebo AV, Woolmer G (2002) The human footprint and the last of the wild: the human footprint is a global map of human influence on the land surface, which suggests that human beings are stewards of nature, whether we like it or not. BioScience 52(10):891–904

Schwarz A, Jax K (eds) (2011) Ecology revisited: reflecting on concepts, advancing science. Springer, New York

Steffen W, Crutzen PJ, McNeill JR (2007) The Anthropocene: are humans now overwhelming the great forces of nature. Ambio 36(8):614–621

Twidwell D, Allred B, Fuhlendorf S (2013) National-scale assessment of ecological content in the world's largest land management framework. Ecoshpere 4(8):art9

Van der Heijden MGA, Bardgett RD, Van Straalen NM (2008) The unseen majority: soil microbes as drivers of plant diversity and productivity in terrestrial ecosystems. Ecol Lett 11:296–310

Van der Putten WH, Bardgett RD, Bever JD, Bezemer TM, Casper BB, Fukami T, Kardol P, Klironomos JN, Kulmatiski A, Schweitzer JA, Suding KN, Van de Voorde TFJ, Wardle DA (2013) Plant–soil feedbacks: the past, the present and future challenges. J Ecol 101.2(2013):265–276

Vitousek PM, Mooney HA, Lubchenco J, Melillo JM (1997) Human domination of Earth's ecosystems. Science 277(5325):494–499

Wall DH, Bardgett RD, Behan-Pelletier V, Herrick JE, Jones TH, Six J, Strong DR (eds) (2013) Soil ecology and ecosystem services. Oxford University Press, New York
Westgate MJ, Likens GE, Lindenmayer DB (2013) Adaptive management of biological systems: a review. Biol Conserv 158:128–139

Dr. Richard K. Sutton PLA, FASLA, GRP teaches landscape architecture and landscape design at the University of Nebraska-Lincoln and conducts research on green roof plants for low input, low cost green roofs.

Dr. John Lambrinos evaluates the design and function of green roofs in the Pacific Northwest. He has published regionally focused studies on appropriate green roof plant species and irrigation regimes.

Index

R. Sutton (ed.), *Green Roof Ecosystems,* Ecological Studies 223,
DOI 10.1007/978-3-319-14983-7

MIX
Papier aus verantwortungsvollen Quellen
Paper from responsible sources
FSC® C105338

If you have any concerns about our products,
you can contact us on
ProductSafety@springernature.com

In case Publisher is established outside the EU,
the EU authorized representative is:
Springer Nature Customer Service Center GmbH
Europaplatz 3, 69115 Heidelberg, Germany

Printed by Libri Plureos GmbH
in Hamburg, Germany